Lecture Notes in Computer Science 6695

Commenced Publication in 1973
Founding and Former Series Editors:
Gerhard Goos, Juris Hartmanis, and Jan van Leeuwen

Karem A. Sakallah   Laurent Simon (Eds.)

Theory and Application of Satisfiability Testing – SAT 2011

14th International Conference, SAT 2011
Ann Arbor, MI, USA, June 19-22, 2011
Proceedings

 Springer

Volume Editors

Karem A. Sakallah
University of Michigan, Department of EECS, CSE Division,
2260 Hayward Ave., Ann Arbor, MI 48109-2121, USA
E-mail: karem@umich.edu

Laurent Simon
Université Orsay Paris-Sud 11, LRI / INRIA
Parc Club Université, Bâtiment G, 4, rue Jacques Monod
91893 Orsay cedex
E-mail: simon@lri.fr

ISSN 0302-9743   e-ISSN 1611-3349
ISBN 978-3-642-21580-3   e-ISBN 978-3-642-21581-0
DOI 10.1007/978-3-642-21581-0
Springer Heidelberg Dordrecht London New York

Library of Congress Control Number: 2011928785

CR Subject Classification (1998): F.3, F.1, F.2, F.4, C.2.4, I.2, B.7

LNCS Sublibrary: SL 1 – Theoretical Computer Science and General Issues

Typesetting: Camera-ready by author, data conversion by Scientific Publishing Services, Chennai, India

Printed on acid-free paper

Springer is part of Springer Science+Business Media (www.springer.com)

Preface

This volume contains the papers presented at SAT 2011, the 14th International Conference on Theory and Applications of Satisfiability Testing (SAT). The conference was held during June 19–22, 2011 and was hosted by the Computer Science and Engineering Division of the Department of Electrical Engineering and Computer Science at the University of Michigan in Ann Arbor. Affiliated with the main conference were the workshops POS (Pragmatics of SAT), SPA (SAT for Practical Applications), CSPSAT (Workshop on the Cross-Fertilization Between CSP and SAT) and INCSAT (Workshop on Incomplete Techniques for Proving UNSAT). In addition, SAT 2011 featured three solver competitions: SAT Competition 2011, Pseudo-Boolean Competition 2011, and Max-SAT Evaluation 2011.

The International Conferences on Theory and Applications of Satisfiability Testing (SAT) originated in 1996 as a series of workshops on satisfiability. By the third meeting in 2000, the workshop had attracted a mix of theorists and experimentalists whose common interest was the enhancement of our basic understanding of the theoretical underpinnings of the satisfiability problem as well as the development of scalable algorithms for its solution in a wide range of application domains. In 2002 a competition of SAT solvers was inaugurated to spur further algorithmic and implementation developments, and to create an eclectic collection of benchmarks. The competition expanded in subsequent years to include pseudo-Boolean, QBF, and MAX-SAT solvers has become an integral part of these meetings, adding an element of excitement and anticipation. The interplay between theory and application, as well as the increased interest in satisfiability from a wider community of researchers, led to the natural evolution of these initial workshops into the current conference format. The annual SAT conference is now universally recognized as the venue for publishing the latest advances in SAT research.

This year marked the 14th SAT meeting. SAT is now interpreted in a broad sense to include not just propositional satisfiability, but also pseudo-Boolean constraint solving and optimization (PB), quantified Boolean formulae (QBF), constraint programming techniques (CP) for word-level problems and their propositional encoding, and satisfiability modulo theories (SMT). Submissions were solicited for original research on proof systems, proof complexity, search algorithms, heuristics, analysis of algorithms, hard instances, randomized formulae, problem encodings, industrial applications, solvers, simplifiers, tools, case studies and empirical results. A total of 57 submissions were received and rigorously reviewed by a 39-member international Technical Program Committee (TPC), with each paper receiving at least four independent reviews. Of these submissions, the TPC decided to accept 25 as regular papers (14 pages, 30-minute presentation) and 10 as extended abstracts (2 pages) to be presented as posters. The accepted papers were organized into eight sessions and their full text is included in these proceedings.

The conference program also featured two invited presentations. The first, by Ryan Williams, described prior and current work on connecting the art of finding good satisfiability algorithms with the art of proving complexity lower bounds. The second, by Koushik Sen, described Concolic Testing, a software verification approach that combines concrete and symbolic testing and utilizes the power of modern constraint solvers.

We would like to acknowledge several people for their help in organizing the conference and associated events. For the myriad logistical arrangements we are grateful for the superb help we received from Lauri Johnson-Rafalski. Steve Crang did an excellent job designing the SAT poster and banners as well as the packet of materials provided to the conference attendees. We appreciate the tireless efforts of the workshop organizers: Daniel Le Berre and Allen Van Gelder (POS), Carsten Sinz and Olga Tveretina (SPA), Yael Ben-Haim and Yehuda Naveh (CSPSAT), and Gilles Audemard, Gilles Dequen, and Djamal Habet (INCSAT). We also thank the competition organizers: Matti Jarvisalo, Daniel Le Berre, and Olivier Roussel (SAT Competition), Vasco Manquinho and Olivier Roussel (Pseudo-Boolean Competition), and Josep Argelich, Chu Min Li, Felip Many, and Jordi Planes (Max-SAT Evaluation). Last, but not least, we thank the members of the TPC and the additional external reviewers for their careful and thorough work, without which it would not have been possible for us to put together such an outstanding conference program.

Finally, we would like to thank Microsoft Research and Microsoft Research INRIA Joint Centre for their generous support of SAT 2011, the CSE division of the department of EECS at the University of Michigan for providing excellent facilities for hosting the conference and workshops, and the LRI of the University of Orsay Paris-Sud 11 for hosting the conference website.

April 2011   Karem A. Sakallah
                                                                      Laurent Simon

Organization

Conference and Program Chairs

Karem A. Sakallah   University of Michigan, USA
Laurent Simon   University of Orsay Paris-Sud 11, France

Program Committee

Paul Beame   University of Washington, USA
Armin Biere   Johannes Kepler University, Austria
Randal Bryant   Carnegie Mellon University, USA
Alessandro Cimatti   Istituto per la Ricerca Scientifica e Tecnologica, Italy
Nadia Creignou   Université d'Aix-Marseille, France
Leonardo De Moura   Microsoft Research, USA
John Franco   University of Cincinnati, USA
Enrico Giunchiglia   Università di Genova, Italy
Youssef Hamadi   Microsoft Research, UK
Marijn Heule   Delft University of Technology, The Netherlands
Holger Hoos   University of British Columbia, Canada
Katsumi Inoue   National Institute of Informatics, Japan
George Katsirelos   University of Paris-Sud 11, France
Hans Kleine Büning   University of Paderborn, Germany
Oliver Kullmann   University of Wales Swansea, UK
Daniel Le Berre   Université dArtois, France
Chu-Min Li   University of Picardie Jules Verne, France
Mark Liffiton   Illinois Wesleyan University, USA
Ines Lynce   Technical University of Lisbon, Portugal
Panagiotis Manolios   Northeastern University, USA
Vasco Manquinho   Technical University of Lisbon, Portugal
Felip Manyà   Spanish Council for Scientific Research, Spain
Igor Markov   University of Michigan, USA
Joao Marques-Silva   University College Dublin, Ireland
Cristopher Moore   University of New Mexico, USA
Albert Oliveras   Technical University of Catalonia, Spain
Ramamohan Paturi   University of California, San Diego, USA
Steve Prestwich   University College Cork, Ireland
Lakhdar Sais   Université dArtois, France
Roberto Sebastiani   Università di Trento, Italy
Carsten Sinz   Karlsruhe Institute of Technology, Germany
Stefan Szeider   Vienna University of Technology, Austria
Armando Tacchella   Università di Genova, Italy
Allen Van Gelder   University of California, Santa Cruz, USA

Hans Van Maaren	Delft University of Technology, The Netherlands
Toby Walsh	University of New South Wales, Australia
Xishun Zhao	Sun Yat-Sen University, China

External Reviewers

Belov, Anton	Martins, Ruben
Bordeaux, Lucas	Matthews, William
Bubeck, Uwe	Mazure, Bertrand
Chen, Huan	Micheli, Andrea
Dantchev, Stefan	Mover, Sergio
Davies, Jessica	Narizzano, Massimo
Dequen, Gilles	Narodytska, Nina
Di Rosa, Emanuele	Ordyniak, Sebastian
Egly, Uwe	Papavasileiou, Vasilis
Franzen, Anders	Piette, Cédric
Gaspers, Serge	Planes, Jordi
Goldberg, Eugene	Rajaratnam, David
Griggio, Alberto	Ridgeway, Jeremy
Gwynne, Matthew	Roussel, Olivier
Habet, Djamal	Roveri, Marco
Heras, Federico	Schaafsma, Bas
Hutter, Frank	Schuppan, Viktor
Jabbour, Said	Seidl, Martina
Jain, Mitesh	Shen, Yuping
Janota, Mikolas	Soh, Takehide
Järvisalo, Matti	Styles, James
Kim, Eun Jung	Tchaltsev, Andrei
Koshimura, Miyuki	Tichit, Laurent
Lagniez, Jean-Marie	Tomasi, Silvia
Lallouet, Arnaud	Tompkins, Dave
Lettmann	Tveretina, Olga
Lettmann, Theo	Ueda, Kazunori
Liedloff, Mathieu	Wahlström, Magnus
Lokshtanov, Daniel	Xu, Lin
Lonsing, Florian	Zdeborova, Lenka
Martin, Barnaby	Zengler, Christoph

Sponsoring Institutions

Microsoft Research, UK
Microsoft Research INRIA Joint Centre, France
University of Michigan, USA
University of Orsay Paris-Sud 11, France

Table of Contents

---

* Best Paper Candidate.

Session 3: Theoretical Analysis

Session 4: Extraction of Minimal Unsatisfiable Subsets

Session 5: SAT Algorithms

Session 6: Quantified Boolean Formulae

---

   * Best Paper Candidate.

Session 7: Model Enumeration, Local Search

Session 8: Empirical Evaluation

Extended Abstracts

Connecting SAT Algorithms
and Complexity Lower Bounds

Ryan Williams

IBM Almaden Research Center

Abstract. I will describe prior and current work on connecting the art of finding good satisfiability algorithms with the art of proving complexity lower bounds: proofs of limitations on what problems can be solved by good algorithms. Surprisingly, even minor algorithmic progress on solving the circuit satisfiability problem faster than exhaustive search can be applied to prove strong circuit complexity lower bounds. These connections have made it possible to prove new complexity lower bounds that had long been conjectured, and they suggest concrete directions for further progress.

Recent work has uncovered interesting connections between the satisfiability problem and complexity theory. Let $\mathcal{C}$ be a generic class of Boolean circuits that obey basic properties. Examples of possible $\mathcal{C}$ are the following, listed in increasing order of computational power:

- $AC^0[m]$ is the class of constant-depth, polynomial-size circuits with unbounded fan-in MODm, AND, and OR gates. (A MODm gate outputs 1 iff the sum of its inputs is divisible by m.)
- ACC is the union over all m of the classes $AC^0[m]$.
- TC^0 is the class of constant-depth, polynomial-size circuits with unbounded fan-in MAJORITY gates and NOT gates.
- NC^1 is the class of polynomial-size Boolean formulas over the connectives AND, OR, and NOT.
- P/poly consists of arbitrary polynomial-size Boolean circuits with bounded fan-in AND and OR gates, and NOT gates.

For each such class $\mathcal{C}$, we may define a corresponding $\mathcal{C}$-SAT problem: *given a generic circuit from the class $\mathcal{C}$, is it satisfiable?* Very little is known about the worst-case time complexity of this problem, even when $\mathcal{C}$ is the class of formulas in conjunctive normal form. For example, it is open whether the CNF-SAT problem can be solved in $O(1.9^n)$ time on CNF formulas with n variables and $n^{O(1)}$ clauses.

It turns out that understanding the time complexity of $\mathcal{C}$-SAT is closely related to the problem of simulating computations within the class $\mathcal{C}$. If the worst case $O(2^n)$ time bound for $\mathcal{C}$-SAT can be only slightly improved in some situations, then obstructions *against* $\mathcal{C}$-circuits can be proved. That is, somewhat weak algorithms for solving SAT on interesting circuits can be turned into strong

K.A. Sakallah and L. Simon (Eds.): SAT 2011, LNCS 6695, pp. 1–2, 2011.

complexity lower bounds for solving other problems with these interesting circuits. More formally:

Theorem. [1,2] There is a $c > 0$ such that, if C-SAT can be solved on circuits with $n + c \log n$ inputs and n^k size in $O(2^n / n^c)$ time for every k, then the class NEXP (Nondeterministic Exponential Time) contains languages that cannot be recognized with non-uniform C circuits of polynomial size.

Intuitively, the theorem states that the difficulty faced by researchers who design fast algorithms for verification of certain kinds of circuits is related to the difficulty of proving that certain problems *can't* be efficiently solved with these kinds of circuits.

Why might such a theorem be true? One intuition is that the existence of a faster C-SAT algorithm shows us a *weakness* in representing computations with circuits from C. The class C is *not* like a set of black boxes: these circuits cannot hide a satisfying input so easily. Instead, there exists a way to analyze the circuit more efficiently, finding a satisfying input or concluding there is none. Another equally valid intuition is that the existence of a faster SAT algorithm for C highlights a *strength* of algorithms that run in less-than-2^n time: they can solve nontrivial satisfiability problems.

Summing up, intuition says that a faster SAT algorithm for C simultaneously shows "less-than-2^n algorithms are strong" and "C-circuits are weak". This gives some hint as to how we might separate the two notions, and prove that some function in nondeterministic exponential time cannot be solved with small (polynomial size) C circuits. (Warning: the actual proof is much more complicated than this, and proceeds by contradiction. We assume the C-SAT algorithm exists, and that every language in NEXP has small C circuits. These two assumptions are woven together in nontrivial and unexpected ways, and eventually we derive a contradiction to a previously known complexity lower bound. Hence if the desired SAT algorithm exists, then the small C circuits cannot exist.)

New circuit complexity lower bounds have recently been proved, building on these connections. In particular, circuit size lower bounds for the class ACC have been established from the design of new algorithms for satisfiability of ACC circuits [2]. It is anticipated that further progress in complexity lower bounds will be made by studying the complexity of satisfiability.

References

1. Williams, R.: Improving exhaustive search implies superpolynomial lower bounds. In: ACM Symposium on Theory of Computing, pp. 231–240 (2010)
2. Williams, R.: Non-uniform ACC circuit lower bounds. To appear in IEEE Conference on Computational Complexity (2011)

Concolic Testing and Constraint Satisfaction

Koushik Sen

EECS Department, University of California, Berkeley, CA, USA
ksen@cs.berkeley.edu

Software testing is the most common technique used in industry to improve reliability and quality of software. Unfortunately, testing is mostly a manual process that reportedly accounts for over half of the typical cost of software development and maintenance. Symbolic execution [6,2,3,9,5] was proposed in the 70s to automate software testing by generating test inputs. During symbolic execution, the program is run with symbolic (rather than concrete) inputs and generates a *path constraint*. This path contraint is updated whenever a conditional statement is executed and encodes the constraints on the input necessary to reach a given program point. Test generation is performed by solving the collected constraints using a constraint solver.

Although symbolic execution was proposed almost 35 years ago, we have hardly seen any practical test generation tool based on this technique. There are two key reasons behind this: 1) until recently, constraint solving techniques were not powerful enough to solve constraints that arise during symbolic execution of most real-world programs, and 2) constraints generated during symbolic execution of real-world programs often fall under theories that are not decidable. The first issue has been addressed by the recent advances in SAT and SMT solving techniques.

In this talk, I will describe *concolic testing* [4,12,10,11,7] (also known as *directed automated random testing* or *dynamic symbolic execution*), a technique that addressed the second challenge associated with symbolic execution and thus paved the way for development of practical automated test generation tools. Concolic testing improves classical symbolic execution by performing symbolic execution of a program along a concrete execution path. Specifically, concolic testing executes a program starting with some given or random concrete input. It then gathers symbolic constraints on inputs at conditional statements during the execution induced by the concrete input. Finally, a constraint solver is used to infer variants of the concrete input to steer the next execution of the program towards an alternative feasible execution path. This process is repeated systematically or heuristically until all feasible execution paths are explored or a user-defined coverage criteria is met.

A key observation in concolic testing is that *intractability in symbolic execution can be alleviated using concrete values*: whenever symbolic execution generates a constraint that is beyond a decidable theory, one can simplify this constraint by replacing some of the symbolic values with concrete values. In these cases, the concolic execution degrades gracefully by leveraging concrete values to keep the path constraint decidable.

K.A. Sakallah and L. Simon (Eds.): SAT 2011, LNCS 6695, pp. 3–4, 2011.

Concolic testing and its variants are now the underlying technique of several popular testing tools: UIUC's CUTE and jCUTE[1], Stanford's KLEE[2] tool uses an approach similar to concolic testing, UC Berkeley's CREST[3] and BitBlaze[4], UCLA's SPLAT [8]. Concolic testing technology is now used in industrial practice at Microsoft (Pex[5], YOGI[6]) and IBM (Apollo [1]).

Acknowledgements

This research supported in part by Microsoft (Award #024263) and Intel (Award #024894) funding and by matching funding by U.C. Discovery (Award #DIG07-10227), by NSF Grants CNS-0720906, CCF-0747390, CCF-1018729, and CCF-1018730, and by a Sloan Foundation Fellowship.

References

1. Artzi, S., Kiezun, A., Dolby, J., Tip, F., Dig, D., Paradkar, A., Ernst, M.D.: Finding bugs in dynamic web applications. In: ISSTA 2008 (July 2008)
2. Boyer, R.S., Elspas, B., Levitt, K.N.: SELECT – a formal system for testing and debugging programs by symbolic execution. SIGPLAN Not. 10, 234–245 (1975)
3. Clarke, L.A.: A program testing system. In: Proc. of the 1976 Annual Conference, pp. 488–491 (1976)
4. Godefroid, P., Klarlund, N., Sen, K.: DART: Directed Automated Random Testing. In: PLDI 2005 (June 2005)
5. Howden, W.: Symbolic testing and the DISSECT symbolic evaluation system. IEEE Transactions on Software Engineering 3(4), 266–278 (1977)
6. King, J.C.: Symbolic execution and program testing. Commun. ACM 19, 385–394 (1976)
7. Majumdar, R., Sen, K.: Hybrid concolic testing. In: ICSE 2007 (May 2007)
8. Majumdar, R., Xu, R.-G.: Reducing test inputs using information partitions. In: Bouajjani, A., Maler, O. (eds.) CAV 2009. LNCS, vol. 5643, pp. 555–569. Springer, Heidelberg (2009)
9. Ramamoorthy, C., Ho, S.-B., Chen, W.: On the automated generation of program test data. IEEE Trans. on Software Engineering 2(4), 293–300 (1976)
10. Sen, K.: Scalable Automated Methods for Dynamic Program Analysis. PhD thesis, University of Illinois at Urbana-Champaign (June 2006)
11. Sen, K., Agha, G.: CUTE and jCUTE: Concolic unit testing and explicit path model-checking tools. In: Ball, T., Jones, R.B. (eds.) CAV 2006. LNCS, vol. 4144, pp. 419–423. Springer, Heidelberg (2006)
12. Sen, K., Marinov, D., Agha, G.: CUTE: A concolic unit testing engine for C. In: ESEC/FSE 2005 (September 2005)

---

[1] http://osl.cs.uiuc.edu/~ksen/cute/

[2] http://klee.llvm.org/

[3] http://code.google.com/p/crest/

[4] http://bitblaze.cs.berkeley.edu/

[5] http://research.microsoft.com/en-us/projects/pex/

[6] http://research.microsoft.com/en-us/projects/yogi/

Parameterized Complexity of DPLL Search Procedures*

Olaf Beyersdorff[1],[**], Nicola Galesi[2],[***], and Massimo Lauria[2]

[1] Institut für Theoretische Informatik, Leibniz Universität Hannover, Germany
[2] Dipartimento di Informatica, Sapienza Università di Roma, Italy

Abstract. We study the performance of DPLL algorithms on parameterized problems. In particular, we investigate how difficult it is to decide whether small solutions exist for satisfiability and other combinatorial problems. For this purpose we develop a Prover-Delayer game which models the running time of DPLL procedures and we establish an information-theoretic method to obtain lower bounds to the running time of parameterized DPLL procedures. We illustrate this technique by showing lower bounds to the parameterized pigeonhole principle and to the ordering principle. As our main application we study the DPLL procedure for the problem of deciding whether a graph has a small clique. We show that proving the absence of a k-clique requires $n^{\Omega(k)}$ steps for a non-trivial distribution of graphs close to the critical threshold. For the restricted case of tree-like Parameterized Resolution, this result answers a question asked in [11] of understanding the Resolution complexity of this family of formulas.

1 Introduction

Resolution was introduced by Blake [12] and since the work of Robinson [25] and Davis, Putnam, Logemann, and Loveland [20,19] has been highly employed in proof search and automated theorem proving. In the last years, the study of Resolution has gained great significance in at least two important fields of computer science. (1) *Proof complexity*, where Resolution is one of the most intensively investigated proof systems [22,30,16,6,8,13,1]. The study of lower bounds for proof length in this system has opened the way to lower bounds in much stronger proof systems [28,7]. (2) *Algorithms for the satisfiability problem* of CNF formulas, where the DPLL algorithm [19,4] is the core of the most important and modern algorithms employed for the satisfiability problem [4,5].

Parameterized Resolution was recently introduced by Dantchev, Martin, and Szeider [18] in the context of *parameterized proof complexity*, an extension of the

---

* Nominated as Best Paper candidate.
** Part of this work was done while the first author was visiting Sapienza University of Rome. This Research is part of the project "Limits of Theorem Proving" supported by grant N. 20517 by the John Templeton Foundation.
*** Partly supported by Sapienza Research Project: Complessità e Rappresentabilità Compatta di Strutture Discrete.

K.A. Sakallah and L. Simon (Eds.): SAT 2011, LNCS 6695, pp. 5–18, 2011.

proof complexity approach of Cook and Reckhow [17] to parameterized complexity. Analogously to the case of Fixed Parameter Tractable (FPT) algorithms for optimization problems, the study of Parameterized Resolution provides new approaches and insights to proof search and to proof complexity. Loosely speaking, to refute a parameterized contradiction (F, k) in Parameterized Resolution we have built-in access to new *axioms*, which encode some property on assignments. In the most common case the new axioms are the clauses forbidding assignments of hamming weight greater than k. We underline that only those axioms appearing in the proof account for the proof length. Hence Parameterized DPLL refutations can be viewed as traces of executions of a (standard) DPLL algorithm in which some branches are cut because they falsify one of the new axioms.

In spite of its recent introduction, research in this direction is already active. Gao [21] analyzes the effect of the standard DPLL algorithm on the problem of weighted satisfiability for random d-CNFs. Beyersdorff et al. [11], using an idea also developed in [15], proved that there are FPT efficient Parameterized Resolution proofs for *all* bounded-width unsatisfiable CNF formulae. The discovery of new implications for SAT-solving algorithms in Parameterized Resolution appears to be a promising research field at a very early stage of investigation.

As our first contribution, we look inside the structure of Parameterized DPLL giving a new information-theoretical characterization of proofs in terms of a two-player game, the *Asymmetric Prover-Delayer (APD) game*. The APD-game was also used in [10] to prove simplified optimal lower bounds for the pigeonhole principle in tree-like classical Resolution. Compared to [10] we present here a completely different analysis of APD-games based on an information-theoretical argument which is new and interesting by itself.

Parameterized Resolution is also a refutational proof system for parameterized contradictions. Hence proving proof length lower bounds for parameterized contradictions is important in order to understand the strength of such a proof system. Dantchev et al. [18] proved significant lower bounds for Parameterized DPLL proofs of PHP and of the ordering principle (OP). Moreover, recently the work [11] extended the PHP lower bounds to the case of parameterized dag-like bounded-depth Frege[1].

As our second contribution we provide a unified approach to reach significative lower bounds in Parameterized DPLL using the APD-game. As a simple application of our characterization, we obtain the optimal lower bounds given in [18] for PHP and OP.

It is a natural question what happens when we equip a proof system with a more efficient way of encoding the exclusion of assignments with hamming weight $\geq k$, than just adding all possible clauses with $k + 1$ negated variables. Dantchev et al. [18] proved that this is a significant point. They presented a different and more efficient encoding, and showed that under this encoding *PHP* admits efficient FPT Parameterized Resolution proofs.

---

[1] The APD-game appeared also in the technical report [9], together with a lower bound for dag-like Parameterized Resolution, but all results in [9] are subsumed and improved by [11] and the present paper.

In the previous work [11] we investigated this question further and noticed that for propositional encodings of prominent combinatorial problems like k-independent set or k-clique, the separation between the two encodings vanishes. Hence we proposed (see Question 5 in [11]) to study the performance of Parameterized Resolution on CNF encodings of such combinatorial problems and in particular to prove lower bounds. This will capture the real proof-theoretic strength of Parameterized Resolution, since it is independent of the encodings. The k-clique principle (see also [11,3] for similar principles) simply says that a given graph contains a clique of size k. When applied on a graph not containing a k-clique it is a contradiction. On the $(k-1)$-partite complete graph the k-clique principle admits efficient refutations in Parameterized Resolution.

As a third contribution, we prove significant lower bounds for the k-clique principle in the case of Parameterized DPLL. Our k-clique formula is based on random graphs distributed according to a simple variation of the Erdős-Rényi model $G(n, p)$. It is well known [23, Chapter 3] that when G is drawn according to $G(n, p)$ and $p \ll n^{-\frac{2}{k-1}}$, with high probability G has no k-clique.

The paper is organized as follows. Section 2 contains all preliminary notions and definitions concerning fixed-parameter tractability, parameterized proof systems, and Parameterized Resolution. In Section 3 we define our asymmetric Prover-Delayer game and establish its precise relation to the proof size in tree-like Parameterized Resolution. In Section 4, as an example of the application of the APD-game, we give a simplified lower bound for the pigeonhole principle in tree-like Parameterized Resolution. In Section 5 we introduce the formula $Clique(G, k)$ which is satisfiable if and only if there is a k-clique in the graph G and we show that on a certain distribution of random graphs the following holds with high probability: G has no k-clique and the size of the shortest refutation of $Clique(G, k)$ is $n^{\Omega(k)}$. From an algorithmic perspective, this result can be formulated as: any algorithm for k-clique which (i) cleverly selects a vertex and branches in whether it is in the clique or not, (ii) deletes all its non-neighbors and (iii) stops branching when there are no vertices left, must use at least $n^{\Omega(k)}$ steps for most random graphs with a certain edge probability.

2 Preliminaries

Parameterized complexity is a branch of complexity theory where problems are analyzed in a finer way than in the classical approach: we say that a problem is *fixed-parameter tractable* (FPT) with parameter k if it can be solved in time $f(k)n^{O(1)}$ for some computable function f of arbitrary growth. In this setting classically intractable problems may have efficient solutions, assuming the parameter is small, even if the total size of the input is large. Parameterized complexity also has a completeness theory: many parameterized problems that appear not to be fixed-parameter tractable have been classified as being complete under fpt-reductions for complexity classes in the so-called weft hierarchy $W[1] \subseteq W[2] \subseteq W[3] \subseteq \dots$.

Consider the problem WEIGHTED CNF SAT of finding a satisfying assignment of Hamming weight at most k for a formula in conjunctive normal form. Many combinatorial problems can be naturally encoded in WEIGHTED CNF SAT: finding a vertex cover of size at most k; finding a clique of size at least k; or finding a dominating set of size at most k. In the theory of parameterized complexity, the hardness of the WEIGHTED CNF SAT problem is reflected by the fact that it is W[2]-complete (see [18, 11]).

Dantchev, Martin, and Szeider [18] initiated the study of *parameterized proof complexity*. After considering the notions of propositional *parameterized tautologies* and *fpt-bounded* proof systems, they laid the foundations for the study of complexity of proofs in a parameterized setting. The problem WEIGHTED CNF SAT leads to parameterized contradictions:

Definition 1 (Dantchev et al. [18]). *A* parameterized contradiction *is a pair* (F, k) *consisting of a propositional formula* F *and* $k \in \mathbb{N}$ *such that* F *has no satisfying assignment of weight* $\leq k$.

The notions of a parameterized proof system and of fpt-bounded proof systems were also developed in [18]:

Definition 2 (Dantchev et al. [18]). *A* parameterized proof system *for a parameterized language* $L \subseteq \Sigma^* \times \mathbb{N}$ *is a function* $P : \Sigma^* \times \mathbb{N} \to \Sigma^* \times \mathbb{N}$ *such that* $rng(P) = L$ *and* $P(x, k)$ *can be computed in time* $O(f(k)|x|^{O(1)})$ *for some computable function* f. *The system* P *is* fpt-bounded *if there exist computable functions* s *and* t *such that every* $(x, k) \in L$ *has a* P-proof (y, k') *with* $|y| \leq s(k)|x|^{O(1)}$ *and* $k' \leq t(k)$.

The main motivation behind the work of [18] was that of generalizing the classical approach of Cook and Reckhow [17] to the parameterized case and that of working towards a separation of complexity classes as FPT and W[2] by techniques developed in proof complexity.

2.1   Parameterized Resolution and Parameterized DPLL

A *literal* is a positive or negated propositional variable and a *clause* is a set of literals. The *width* of a clause is the number of its literals. A clause is interpreted as the disjunction of its literals and a set of clauses as the conjunction of the clauses. Hence clause sets correspond to formulas in CNF. The *Resolution system* is a refutation system for the set of all unsatisfiable CNF. Resolution gets its name from its only rule, the *Resolution rule* $\frac{\{x\} \cup C \quad \{\neg x\} \cup D}{C \cup D}$ for clauses C, D and a variable x. The aim in Resolution is to demonstrate unsatisfiability of a clause set by deriving the empty clause. If in a derivation every derived clause is used at most once as a prerequisite of the Resolution rule, then the derivation is called *tree-like*, otherwise it is called *dag-like*. The *size* of a Resolution proof is the number of its clauses where multiple occurrences of the same clause are counted separately.

For the remaining part of this paper we will concentrate on *Parameterized Resolution* as introduced by Dantchev, Martin, and Szeider [18]. Parameterized Resolution is a refutation system for the set of parameterized contradictions (cf.

Definition 1). Given a set of clauses F in variables $x_1, \ldots, x_n$, a *Parameterized Resolution refutation* of (F, k) is a Resolution refutation of the set of clauses $F \cup \{\neg x_{i_1} \vee \cdots \vee \neg x_{i_{k+1}} \mid 1 \leq i_1 < \cdots < i_{k+1} \leq n\}$. Thus, in Parameterized Resolution we have built-in access to all parameterized clauses of the form $\neg x_{i_1} \vee \cdots \vee \neg x_{i_{k+1}}$. All these clauses are available in the system, but when measuring the size of a refutation we only count those which occur in the refutation.

If refutations are tree-like we speak of *tree-like Parameterized Resolution*. Running parameterized DPLL procedures on parameterized contradictions produces tree-like Parameterized Resolution refutations, thus tree-like Resolution proof lengths are connected with the running time of DPLL procedures. Exactly as in usual tree-like Resolution, a tree-like Parameterized refutation of (F, k) can equivalently be described as a *boolean decision tree* where inner nodes are labeled with variables from F and leaves are labeled either with clauses from F or with parameterized clauses $\neg x_{i_1} \vee \cdots \vee \neg x_{i_{k+1}}$.

3 Asymmetric Prover-Delayer Games for DPLL

The original Prover-Delayer game for tree-like Resolution has been developed by Pudlák and Impagliazzo [24], and arises from the well-known fact that a tree-like Resolution refutation for a CNF F can be viewed as a decision tree which solves the search problem of finding a clause of F falsified by a given assignment. In the game, Prover queries a variable and Delayer either gives it a value or leaves the decision to Prover and receives *one* point. The number of Delayer's points at the end of the game bounds from below the height of the proof tree. Our new game, in contrast, assigns points to the Delayer asymmetrically ($\log c_0$ and $\log c_1$) according to two functions c_0 and c_1 (s.t. $c_0^{-1} + c_1^{-1} = 1$) which depend on the principle, the variable queried, and the current partial assignment. In fact, the original Prover-Delayer game of [24] is the case where $c_0 = c_1 = 2$.

Loosely speaking, we interpret the inverse of the score functions as a way to define a distribution on the choices made by the DPLL algorithm. Under this view the Delayer's score at each step is just the entropy of the bit encoding the corresponding choice. Since root-to-leaf paths are in bijection with leaves, this process induces a distribution on the leaves. Hence the entropy collected on the path is the entropy of the corresponding leaf choice. In this interpretation, the asymmetric Prover-Delayer game becomes a challenge between Prover, who wants to end the game giving up little entropy, and Delayer, who wants to get a lot of it. This means that the average score of the Delayer is a measure (actually a lower bound) of the number of leaves. In our setup the DPLL algorithm decides the Prover queries, and the score function defines the distribution on paths. The Delayer role corresponds to a conditioning on such distribution.

Let (F, k) be a parameterized contradiction where F is a set of clauses in n variables $x_1, \ldots, x_n$. We define a Prover-Delayer game: Prover and Delayer build a (partial) assignment to $x_1, \ldots, x_n$. The game is over as soon as the partial assignment falsifies either a clause from F or a parameterized clause $\neg x_{i_1} \vee \cdots \vee \neg x_{i_{k+1}}$ where $1 \leq i_1 < \cdots < i_{k+1} \leq n$. The game proceeds in rounds.

In each round, Prover suggests a variable x_i, and Delayer either chooses a value 0 or 1 for x_i or leaves the choice to the Prover. In this last case the Prover sets the value and the Delayer gets some points. The number of points Delayer earns depends on the variable x_i, the assignment α constructed so far in the game, and two functions c_0 and c_1. More precisely, the number of points that Delayer will get is

$$\begin{array}{ll} 0 & \text{if Delayer chooses the value,} \\ \log c_0(x_i, \alpha) & \text{if Prover sets } x_i \text{ to 0, and} \\ \log c_1(x_i, \alpha) & \text{if Prover sets } x_i \text{ to 1.} \end{array}$$

Moreover, the functions c_0 and c_1 are non negative and are chosen in such a way that for each variable x and assignment α

$$\frac{1}{c_0(x, \alpha)} + \frac{1}{c_1(x, \alpha)} = 1 \tag{1}$$

holds. We remark that (1) is not strictly necessary for all α and x, but it must hold at least for those assignments α and choices x of the Delayer that can actually occur in any game with the Delayer strategy. We call this game the (c_0, c_1)-game on (F, k). The connection of this game to size of proofs in tree-like Parameterized Resolution is given by the next theorem:

Theorem 3. *Let (F, k) be a parameterized contradiction and let c_0 and c_1 be two functions satisfying (1) for all partial assignments α to the variables of F. If (F, k) has a tree-like Parameterized Resolution refutation of size at most S, then for each (c_0, c_1)-game played on (F, k) there is a Prover strategy (possibly dependent on the Delayer) that gives the Delayer at most $\log S$ points.*

Proof. Let (F, k) be a parameterized contradiction using variables $x_1, \ldots, x_n$. Choose any tree-like Parameterized Resolution refutation of (F, k) of size S and interpret it as a boolean decision tree T for F. The decision tree T completely specifies the query strategy for Prover: at the first step he will query the variable labeling the root of T. Whatever decision is made regarding the value of the queried variable, Prover moves to the root of the corresponding subtree and queries the variable which labels it. This process induces a root-to-leaf walk on T, and such walks are in bijection with the set of leafs.

To completely specify Prover's strategy we need to explain how Prover chooses the value of the queried variable in case Delayer asks him to. A game position is completely described by the partial assignment α computed so far, and by the variable $x \notin dom(\alpha)$ queried at that moment. If the Prover is asked to answer the query for x, the answer will be: $\begin{cases} 0 & \text{with probability } \frac{1}{c_0(x, \alpha)} \\ 1 & \text{with probability } \frac{1}{c_1(x, \alpha)} \end{cases}$. Thus we are dealing with a randomized Prover strategy. In a game played between our randomized Prover and a specific Delayer D, we denote by $p_{D,\ell}$ the probability of such a game to end at a leaf ℓ. We call π_D this distribution on the leaves. To prove the theorem the following observation is crucial:

If the game ends at leaf ℓ, then Delayer D scores exactly $\log \frac{1}{p_{D,\ell}}$ points.

Before proving this claim, we show that it implies the theorem. The expected score of a Delayer D is

$$H(\pi_D) = \sum_{\ell} p_{D,\ell} \log \frac{1}{p_{D,\ell}}$$

which is the information-theoretic entropy of π_D. Since the support of π_D has size at most S, we obtain $H(\pi_D) \leq \log S$, because the entropy is maximized by the uniform distribution. By fixing the random choices of the Prover, we can force Delayer D to score at most $\log S$ points.

To prove the claim consider a leaf ℓ and the unique path that reaches it. W.l.o.g. we assume that this path corresponds to the ordered sequence of assignments $x_1 = \epsilon_1, \ldots, x_m = \epsilon_m$. The probability of reaching the leaf is

$$p_{D,\ell} = p_1 p_2 \cdots p_m$$

where p_i is the probability of setting $x_i = \epsilon_i$ conditioned on the previous choices. If Prover chooses the value of the variable x_i, the score Delayer D gets at step i is

$$\log c_{\epsilon_i}(x_i, \{x_1 = \epsilon_1, x_2 = \epsilon_2, \ldots, x_{i-1} = \epsilon_{i-1}\})$$

which is exactly $\log \frac{1}{p_i}$. If Delayer makes the choice at step i, then $p_i = 1$ and the score is 0, which is also $\log \frac{1}{p_i}$. Thus the score of the game play is

$$\sum_{i=1}^{m} \log \frac{1}{p_i} = \log \frac{1}{\prod_{i=1}^{m} p_i} = \log \frac{1}{p_{D,\ell}} \ ,$$

and this concludes the proof of the claim and the theorem.   □

4   An Application of the Lower Bound Method

We will illustrate the use of asymmetric Prover-Delayer games with an application to the *pigeonhole principle* PHP_n^{n+1}. Variable $x_{i,j}$ for $i \in [n+1]$ and $j \in [n]$ indicates that pigeon i goes into hole j. PHP_n^{n+1} consists of the clauses $\bigvee_{j \in [n]} x_{i,j}$ for all pigeons $i \in [n+1]$ and $\neg x_{i_1,j} \vee \neg x_{i_2,j}$ for all choices of distinct pigeons $i_1, i_2 \in [n+1]$ and holes $j \in [n]$. We prove that PHP_n^{n+1} is hard for tree-like Parameterized Resolution.

Theorem 4. *Any tree-like Parameterized Resolution refutation of (PHP_n^{n+1}, k) has size $n^{\Omega(k)}$.*

Proof. Let α be a partial assignment to the variables $\{x_{i,j} \mid i \in [n+1], j \in [n]\}$. Let $z_i(\alpha) = |\{j \in [n] \mid \alpha(x_{i,j}) = 0\}|$, i.e., $z_i(\alpha)$ is the number of holes already excluded by α for pigeon i. We define

$$c_0(x_{i,j}, \alpha) = \frac{n - z_i(\alpha)}{n - z_i(\alpha) - 1} \quad \text{and} \quad c_1(x_{i,j}, \alpha) = n - z_i(\alpha)$$

which clearly satisfies (1). We now describe Delayer's strategy in a (c_0, c_1)-game played on (PHP_n^{n+1}, k). If Prover asks for a value of $x_{i,j}$, then Delayer decides as follows:

set $\alpha(x_{i,j}) = 0$ if there exists $i' \in [n+1] \setminus \{i\}$ such that $\alpha(x_{i',j}) = 1$ or if there exists $j' \in [n] \setminus \{j\}$ such that $\alpha(x_{i,j'}) = 1$

set $\alpha(x_{i,j}) = 1$ if there is no $j' \in [n]$ with $\alpha(x_{i,j'}) = 1$ and $z_i(\alpha) \geq n - k$

let Prover decide   otherwise.

Intuitively, Delayer leaves the choice to Prover as long as pigeon i does not already sit in a hole, there are at least k holes free for pigeon i, and there is no other pigeon sitting already in hole j. If Delayer uses this strategy, then clauses from PHP_n^{n+1} will not be violated in the game, i.e., a contradiction will always be reached on some parameterized clause. To verify this claim, let α be a partial assignment constructed during the game with $w(\alpha) \leq k$ (we denote the the weight of α by $w(\alpha)$). Then, for every pigeon which has not been assigned to a hole yet, there are at least k holes where it could go, and only $w(\alpha)$ of these are already occupied by other pigeons. Thus α can be extended to a one-one mapping of exactly k pigeons to holes.

Therefore, at the end of the game exactly $k + 1$ variables have been set to 1. Let us denote by p the number of variables set to 1 by Prover and let d be the number of 1's assigned by Delayer. As argued before $p + d = k + 1$. Let us check how many points Delayer earns in this game. If Delayer assigns 1 to a variable $x_{i,j}$, then pigeon i was not assigned to a hole yet and, moreover, there must be $n - k$ holes which are already excluded for pigeon i by α, i.e., for some $J \subseteq [n]$ with $|J| = n - k$ we have $\alpha(x_{i,j'}) = 0$ for all $j' \in J$. Most of these 0's have been assigned by Prover, as Delayer has only assigned a 0 to $x_{i,j'}$ when some other pigeon was already sitting in hole j', and there can be at most k such holes. Thus, before Delayer sets $\alpha(x_{i,j}) = 1$, she has already earned points for at least $n - 2k$ variables $x_{i,j'}$, $j' \in J$, yielding at least

$$\sum_{z=0}^{n-2k-1} \log \frac{n-z}{n-z-1} = \log \prod_{z=0}^{n-2k-1} \frac{n-z}{n-z-1} = \log \frac{n}{2k} = \log n - \log 2k$$

points for the Delayer. Note that because Delayer never allows a pigeon to go into more than one hole, she will earn at least the number of points calculated above for *each* of the d variables which she sets to 1.

If, conversely, Prover sets variable $x_{i,j}$ to 1, then Delayer gets $\log(n - z_i(\alpha))$ points for this, but she also receives points for most of the $z_i(\alpha)$ variables set to 0 before that. Thus, in this case Delayer earns on pigeon i at least

$$\log(n - z_i(\alpha)) + \sum_{z=0}^{z_i(\alpha)-k-1} \log \frac{n-z}{n-z-1} =$$

$$\log(n - z_i(\alpha)) + \log \frac{n}{n - z_i(\alpha) + k} = \log n - \log \frac{n - z_i(\alpha) + k}{n - z_i(\alpha)} \geq \log n - \log k$$

points. In total, Delayer gets at least

$$d(\log n - \log 2k) + p(\log n - \log k) \geq k(\log n - \log 2k)$$

points in the game. By Theorem 3, we obtain $\left(\frac{n}{2k}\right)^k$ as a lower bound to the size of each tree-like Parameterized Resolution refutation of (PHP_n^{n+1}, k).   □

As a second example we discuss the DPLL performance on the parameterized *ordering principle OP*, also called *least element principle*. The principle claims that any finite partially ordered set has a minimal element. There is a direct propositional translation of OP to a family OP_n of unsatisfiable CNFs. Each CNF OP_n expresses that there exists a partially ordered set of size n such that any element has a predecessor. The ordering principle has the following clauses:

$$\neg x_{i,j} \vee \neg x_{j,i} \qquad \text{for every } i, j \qquad \text{(Antisymmetry)}$$
$$\neg x_{i,j} \vee \neg x_{j,k} \vee x_{i,k} \qquad \text{for every } i, j, k \qquad \text{(Transitivity)}$$
$$\bigvee_{j \in [n] \setminus \{i\}} x_{j,i} \qquad \text{for every } i \qquad \text{(Predecessor)}$$

With respect to parameterization the ordering principles are interesting. Both OP and the *linear ordering principle (LOP)*, which additionally assumes the order to be total, do not admit short tree-like Resolution refutations [14] and have general Resolution refutations of polynomial size [29]. In the parameterized setting things are different: LOP has short tree-like refutations (see [11]) while OP does not and provides a separation between tree-like and dag-like Parameterized Resolution. The following theorem has been first proved in [18]. Their proof is based on a model-theoretic criterion, while ours is based on the Prover-Delayer game. The proof will appear in the full version of this paper (see also [9]).

Theorem 5. *Any tree-like Parameterized Resolution refutation of (OP_n, k) has size $n^{\Omega(k)}$.*

5   DPLL and the Decision Tree Complexity of k-Clique

Instead of adding parameterized clauses of the form $\neg x_{i_1} \vee \cdots \vee \neg x_{i_{k+1}}$, there are also more succinct ways to enforce only satisfying assignments of weight $\leq k$. One such method was considered in [18] where for a formula F in n variables $x_1, \ldots, x_n$ and a parameter k, a new formula $M = M(F, k)$ is computed such that $F \wedge M$ is satisfiable if and only if F has a satisfying assignment of weight at most k. The formula M uses new variables $s_{i,j}$, where $i \in [k]$ and $j \in [n]$, and consists of the clauses

$$\neg x_j \vee \bigvee_{i=1}^{k} s_{i,j} \quad \text{and} \quad \neg s_{i,j} \vee x_j \qquad \text{for } i \in [k] \text{ and } j \in [n] \qquad (2)$$

$$\neg s_{i,j} \vee \neg s_{i,j'} \qquad \text{for } i \in [k] \text{ and } j \neq j' \in [n] \qquad (3)$$

$$\neg s_{i,j} \vee \neg s_{i',j} \qquad \text{for } i \neq i' \in [k] \text{ and } j \in [n]. \qquad (4)$$

The clauses (2) express the fact that an index i is associated to a variable x_j if and only if such variable is set to true. The fact that the association is an injective function is expressed by the clauses (3) and (4).

In [11] we argue that the clique formulas are "invariant" with respect to this transformation, thus its classical proof complexity is equivalent to its parameterized proof complexity (in both the formulation with explicit parameterized axioms and the succinct encoding). Therefore in [11] we posed the question of determining the complexity of the clique formulas in Resolution. Theorem 7 below provides an answer to this question for the tree-like case.

Our study focuses on the average-case complexity of proving the absence of a k-clique in random graphs distributed according to a variation of the Erdős-Rényi model $G(n, p)$. It is known that k-cliques appear at the threshold probability $p^* = n^{-\frac{2}{k-1}}$. If $p < p^*$, then with high probability there is no k-clique; while for $p > p^*$ with high probability there are many. For $p = p^*$ there is a k-clique with constant probability.

The complexity of k-clique has been already studied in restricted computational models by Rossman [26, 27]. He shows that in these models any circuit which succeeds with good probability on graph distributions close to the critical threshold requires size $\Omega(n^{\frac{k}{4}})$, and even matching upper bounds exist in these models [2, 27]. Since we want to study negative instances of the clique problem, we focus on probability distributions with $p < p^*$. To ease the proof presentation we will prove a lower bound for a slightly sparser distribution. We now give the CNF formulation of a statement claiming that a k-clique exists in a graph.

Definition 6. *Given a graph $G = (V, E)$ and a parameter k, $Clique(G, k)$ is a formula in conjunctive normal form containing the following clauses*

$$\bigvee_{v \in V} x_{i,v} \qquad \text{for every } i \in [k] \tag{5}$$

$$\neg x_{i,u} \vee \neg x_{j,v} \qquad \text{for every } i, j \in [k],\ i \neq j \text{ and every } \{u, v\} \notin E \tag{6}$$

$$\neg x_{i,u} \vee \neg x_{i,v} \qquad \text{for every } u \neq v \in V. \tag{7}$$

Clearly, the formula $Clique(G, k)$ is satisfiable if and only if the graph G has a clique of size k.

We now describe a family of hard graph instances for k-clique: such graphs have a simplified structure to make the proof more understandable. We also restrict the formula, which makes it easier. This only strengthens eventual lower bounds. We consider a random graph G on kn vertices. The set of vertices V is divided into k blocks of n vertices each, named $V_1, V_2, \ldots, V_k$. Edges may be present only between vertices of different blocks. The random variable in the graph is the set of edges. For any constant ϵ and any pair of vertices (u, v) with $u \in V_i$, $v \in V_{i'}$ and $i < i'$, the edge $\{u, v\}$ is present with probability

$$p = n^{-(1+\epsilon)\frac{2}{k-1}}.$$

We call this distribution of graphs $\mathcal{G}_\epsilon$. Notice that all graphs in $\mathcal{G}_\epsilon$ are properly colorable with k colors. Later we will focus on a specific range for ϵ.

In a k-colorable graph, each clique contains at most one vertex per color class. Because of this observation we can simplify the k-clique formula in the following way, which we call $h(G)$

$$\bigvee_{v \in V_i} x_v \qquad \text{for every } i \in [k] \qquad (8)$$

$$\neg x_u \vee \neg x_v \qquad \text{for every } \{u, v\} \notin E(G). \qquad (9)$$

We omit the parameter k in the notation of h to keep notation as simple as possible. We now see that a lower bound to the size of a (tree-like) Resolution refutation of $h(G)$ transfers to the same lower bound for $Clique(G, k)$.

Fact 1. *Let G be a k-colorable graph. Then each (tree-like) Resolution refutation of $Clique(G, k)$ can be transformed into a (tree-like) Resolution refutation of $h(G)$ of the same size (with the partition in $h(G)$ induced by the coloring).*

A comment regarding the encoding is required. In [3] formulas similar to *Clique* (G, k) and $h(G)$ have been studied for the dual problem of independent sets. They study the case of $k = \Omega(n)$, so the former encoding has a lower bound because it contains clauses of a non-trivial pigeonhole principle. In the parameterized framework this is not necessarily true, since k is small and PHP_{k-1}^k is feasible here.

We will now show that for a random graph $G \in \mathcal{G}_\epsilon$ any decision tree which proves unsatisfiability of k-clique has size $n^{\Omega(k(1-\epsilon))}$ with high probability. To show that k-clique requires refutations of size $n^{\Omega(k(1-\epsilon))}$ it suffices to exhibit two score functions c_0 and c_1 and a Delayer strategy such that the Delayer is guaranteed to score $\Omega(k(1 - \epsilon) \log n)$ points in any game played against any Prover.

Theorem 7. *For any $0 < \epsilon < 1$. For a random $G \in \mathcal{G}_\epsilon$ the k-clique CNF requires tree-like Parameterized Resolution refutations of size $n^{\Omega(k(1-\epsilon))}$ with high probability.*

Proof. Let G be a random graph distributed according to $\mathcal{G}_\epsilon$. For a set S of vertices, let $\Gamma^c(S)$ be the set of common neighbors of S. We first show that with high probability the following properties hold:

1. G has no clique of size k;
2. For any set S of less than $\frac{k}{4}$ vertices in distinct blocks, $|\Gamma^c(S) \cap V_b| \geq n^{\Omega(1-\epsilon)}$ for any block V_b disjoint from S.

For item 1: the expected number of k-cliques in G is $n^k p^{\binom{k}{2}} = n^{-k\epsilon}$. By Markov inequality, the probability of the existence of a single k-clique is at most the expected value.

For item 2: it is sufficient to show the statement for sets of size exactly $\frac{k}{4} - 1$. Fix any such set S, and fix any block V_b which does not contain vertices in this set. We denote by X_i the random variable which is 1 when $i \in \Gamma^c(S)$, and 0 otherwise. Thus the size of $V_b \cap \Gamma^c(S)$ is the sum of n independent variables. Notice that X_i is 1 with probability $p^{\frac{k}{4}-1} \geq n^{-\frac{1+\epsilon}{2}}$. Thus the expected value is at least $n^{\frac{1-\epsilon}{2}}$. We define $T = \frac{n^{\frac{1-\epsilon}{2}}}{2}$. Since $T = n^{\Omega(1-\epsilon)}$ and T is a constant fraction of the expected value, by the Chernoff bound we obtain that $V_b \cap \Gamma(S)$ has size less than T with probability at most $e^{-n^{\Omega(1-\epsilon)}}$. By the union bound on the choices of block V_b and of set S of size $\frac{k}{4} - 1$ we get item 2.

We now define functions c_0 and c_1 which are legal cost functions for an asymmetric Prover-Delayer game played on the k-clique formula of the graph G. We also show a Delayer strategy which is guaranteed to score $\Omega(k \log T)$ points. This, together with Theorem 3, implies the main statement.

For any partial assignment α we consider the set of vertices "chosen by α", which is $\{u \mid \alpha(x_u) = 1\}$; any vertex which is the common neighbor of the chosen set is called "good for α". Notice that a good vertex for α can be set to 1 without causing an immediate contradiction. Notice also that α may set to 0 some good vertices. In particular we denote by $R_b(\alpha)$ the vertices of the block V_b which are good for α, but are nevertheless set to 0 in α.

When asked for a variable x_v, for some $v \in V_b$, the Delayer behaves according to the following strategy:

- If α contains at least $\frac{k}{4}$ variables set to 1, the Delayer surrenders;
- if there is u such that $\alpha(x_u) = 1$ and $\{u, v\} \notin E(G)$, the Delayer answers 0;
- if $R_b(\alpha)$ has size at least $T - 1$, then the Delayer answers 1;
- otherwise the Delayer leaves the answer to the Prover.

During the game the invariant $|R_b(\alpha)| < T$ holds for every $b \in [k]$: the only way such a set can increase in size is when Prover sets a good vertex in V_b to 0. Thus the size of $R_b(\alpha)$ can only increase one by one. When it reaches $T - 1$ and the Delayer is asked for a variable in that block, she will reply 1, so the size of $R_b(\alpha)$ won't increase any more.

Another important property of the Delayer strategy is that her decision to answer 1 never falsifies a clause, since all blocks contain at least T good vertices at any moment during the game. This follows from item 2 and from the fact that the Delayer surrenders after $\frac{k}{4}$ vertices are set in α. This proves that no clause in (8) can be falsified during the game.

Neither clauses in (9) can be falsified during the game: the Delayer imposes answer 0 whenever a vertex is not good for α, which means that, if chosen, it would not form a clique with the ones chosen before. It is also not possible that the game ends by violating a parameterized clause as these are just weakenings of the clauses (9). Therefore, the game only ends when the Delayer gives up.

For an assignment α and a vertex $v \in V_b$, let

$$c_0 = \frac{T - |R_b(\alpha)|}{T - |R_b(\alpha)| - 1} \qquad \text{and} \qquad c_1 = T - |R_b(\alpha)|.$$

Because of the previous observations the values of c_0 and c_1 are always nonnegative. Furthermore notice that when $|R_b(\alpha)| = T - 1$ Delayer never leaves the choice to Prover, thus c_0 is always well defined when the Delayer scores.

Consider a game play and the set of $\frac{k}{4}$ vertices chosen by the final partial assignment α. We show that for any chosen vertex, the Delayer scores $\log T$ points for queries in the corresponding block.

Fix the block b of a chosen vertex u. Consider the assignment α which corresponds to the game step when x_u is set to 1. Consider $R = R_b(\alpha)$. We identify partial assignments $\alpha_0 \subset \alpha_1 \subset \ldots \subset \alpha_{|R|-1} \subset \alpha$ corresponding to the moments in the game when Prover sets to 0 one of the variables indexed by R. For such rounds the Delayer gets at least

$$\sum_{i=0}^{|R|-1} \log \frac{T - |R_b(\alpha_i)|}{T - |R_b(\alpha_i)| - 1} \geq \sum_{i=0}^{|R|-1} \log \frac{T - i}{T - i - 1} = \log(T) - \log(T - |R|)$$

points. Here the first inequality follows from the fact that any vertex which is good at some stage of the game is also good in all previous stages. Thus $|R_b(\alpha_i)| \geq i$.

Now we must consider two cases: either $x_u = 1$ is set by Prover, or it is set by Delayer. In the former case Delayer gets $\log(T - |R|)$ points for Prover setting $x_u = 1$. Together with the points for the previous zeros this yields $\log T$ points. In the latter case Delayer gets 0 points as she set $x_u = 1$ by herself, but now $|R| = T - 1$ and she got already $\log T$ points for all the zeros assigned by Prover. In both cases the total score of the Delayer is $\log T = \frac{1-\epsilon}{2} \log n$.

Since this score is obtained in at least $\frac{k}{4}$ blocks, we are done.   □

Acknowledgments

We thank the anonymous referees for their insightful suggestions which helped to improve the paper.

References

1. Alekhnovich, M., Razborov, A.A.: Resolution is not automatizable unless W[P] is tractable. SIAM Journal on Computing 38(4), 1347–1363 (2008)
2. Amano, K.: Subgraph isomorphism on AC^0 circuits. Computational Complexity 19(2), 183–210 (2010)
3. Beame, P., Impagliazzo, R., Sabharwal, A.: The resolution complexity of independent sets and vertex covers in random graphs. Comput. Complex. 16(3), 245–297 (2007)
4. Beame, P., Karp, R.M., Pitassi, T., Saks, M.E.: The efficiency of resolution and Davis-Putnam procedures. SIAM J. Comput. 31(4), 1048–1075 (2002)
5. Beame, P., Kautz, H.A., Sabharwal, A.: Towards understanding and harnessing the potential of clause learning. J. Artif. Intell. Res. 22, 319–351 (2004)
6. Beame, P., Pitassi, T.: Simplified and improved resolution lower bounds. In: Proc. 37th IEEE Symposium on the Foundations of Computer Science, pp. 274–282 (1996)
7. Beame, P.W., Impagliazzo, R., Krajíček, J., Pitassi, T., Pudlák, P.: Lower bounds on Hilbert's Nullstellensatz and propositional proofs. Proc. London Mathematical Society 73(3), 1–26 (1996)

8. Ben-Sasson, E., Wigderson, A.: Short proofs are narrow - resolution made simple. Journal of the ACM 48(2), 149–169 (2001)

9. Beyersdorff, O., Galesi, N., Lauria, M.: Hardness of parameterized resolution. Technical Report TR10-059, Electronic Colloquium on Computational Complexity (2010)

10. Beyersdorff, O., Galesi, N., Lauria, M.: A lower bound for the pigeonhole principle in tree-like resolution by asymmetric prover-delayer games. Information Processing Letters 110(23), 1074–1077 (2010)

11. Beyersdorff, O., Galesi, N., Lauria, M., Razborov, A.: Parameterized bounded-depth Frege is not optimal. Technical Report TR10-198, Electronic Colloquium on Computational Complexity (2010)

12. Blake, A.: Canonical expressions in boolean algebra. PhD thesis, University of Chicago (1937)

13. Bonet, M.L., Esteban, J.L., Galesi, N., Johannsen, J.: On the relative complexity of resolution refinements and cutting planes proof systems. SIAM Journal on Computing 30(5), 1462–1484 (2000)

14. Bonet, M.L., Galesi, N.: Optimality of size-width tradeoffs for resolution. Computational Complexity 10(4), 261–276 (2001)

15. Chen, Y., Flum, J.: The parameterized complexity of maximality and minimality problems. Annals of Pure and Applied Logic 151(1), 22–61 (2008)

16. Chvátal, V., Szemerédi, E.: Many hard examples for resolution. J. ACM 35(4), 759–768 (1988)

17. Cook, S.A., Reckhow, R.A.: The relative efficiency of propositional proof systems. The Journal of Symbolic Logic 44(1), 36–50 (1979)

18. Dantchev, S.S., Martin, B., Szeider, S.: Parameterized proof complexity. In: Proc. 48th IEEE Symposium on the Foundations of Computer Science, pp. 150–160 (2007)

19. Davis, M., Logemann, G., Loveland, D.W.: A machine program for theorem-proving. Commun. ACM 5(7), 394–397 (1962)

20. Davis, M., Putnam, H.: A computing procedure for quantification theory. Journal of the ACM 7, 210–215 (1960)

21. Gao, Y.: Data reductions, fixed parameter tractability, and random weighted d-CNF satisfiability. Artificial Intelligence 173(14), 1343–1366 (2009)

22. Haken, A.: The intractability of resolution. Theor. Comput. Sci. 39, 297–308 (1985)

23. Janson, S., Łuczak, T., Ruciński, A.: Random Graphs. Wiley, Chichester (2000)

24. Pudlák, P., Impagliazzo, R.: A lower bound for DLL algorithms for SAT. In: Proc. 11th Symposium on Discrete Algorithms, pp. 128–136 (2000)

25. Robinson, J.A.: A machine-oriented logic based on the resolution principle. Journal of the ACM 12, 23–41 (1965)

26. Rossman, B.: On the constant-depth complexity of k-clique. In: Proc. 40th ACM Symposium on Theory of Computing, pp. 721–730 (2008)

27. Rossman, B.: The monotone complexity of k-clique on random graphs. In: Proc. 51th IEEE Symposium on the Foundations of Computer Science, pp. 193–201. IEEE Computer Society, Los Alamitos (2010)

28. Segerlind, N., Buss, S.R., Impagliazzo, R.: A switching lemma for small restrictions and lower bounds for k-DNF resolution. SIAM Journal on Computing 33(5), 1171–1200 (2004)

29. Stalmark, G.: Short resolution proofs for a sequence of tricky formulas. Acta Informatica 33, 277–280 (1996)

30. Urquhart, A.: Hard examples for resolution. J. ACM 34(1), 209–219 (1987)

Satisfiability Certificates Verifiable in Subexponential Time[*]

Evgeny Dantsin[1] and Edward A. Hirsch[2,**]

[1] Department of Computer Science, Roosevelt University, USA
[2] Steklov Institute of Mathematics at St.Petersburg, Russia

Abstract. It is common to classify satisfiability problems by their time complexity. We consider another complexity measure, namely the length of certificates (witnesses). Our results show that there is a similarity between these two types of complexity if we deal with certificates verifiable in subexponential time. In particular, the well-known result by Impagliazzo and Paturi [IP01] on the dependence of the time complexity of k-SAT on k has its counterpart for the certificate complexity: we show that, assuming the exponential time hypothesis (ETH), the certificate complexity of k-SAT increases infinitely often as k grows. Another example of time-complexity results that can be translated into the certificate-complexity setting is the results of [CIP06] on the relationship between the complexity of k-SAT and the complexity of SAT restricted to formulas of constant clause density. We also consider the certificate complexity of CircuitSAT and observe that if CircuitSAT has subexponential-time verifiable certificates of length cn, where $c < 1$ is a constant and n is the number of inputs, then an unlikely collapse happens (in particular, ETH fails).

1 Introduction

If we assume $\mathbf{P} \neq \mathbf{NP}$, the question of refined complexity classification of **NP**-complete problems remains open. For example, what is the best possible running time for deciding k-SAT, SAT, or CircuitSAT? Is it possible to solve k-SAT in subexponential time? Is it possible to solve SAT or even CircuitSAT faster than using the trivial enumeration of all assignments? Although the questions like those above seem far enough from being resolved, many interesting results shedding more light on such questions have been appeared for the past two decades, see surveys in [DH09, PP10].

In this paper, we compare a time-complexity classification of problems in **NP** with a classification based on the length of certificates (witnesses). Note an asymmetry between these complexity measures. Any problem in **NP** can be

---

[*] Nominated as Best Paper candidate.

[**] Supported in part by Federal Target Programme "Scientific and scientific-pedagogical personnel of the innovative Russia" 2009-2013, by the grant NSh-5282.2010.1 from the President of RF for Leading Scientific Schools, by the Programme of Fundamental Research of RAS, and by RFBR grant 11-01-00760.

K.A. Sakallah and L. Simon (Eds.): SAT 2011, LNCS 6695, pp. 19–32, 2011.
© Springer-Verlag Berlin Heidelberg 2011

trivially solved by enumerating all possible candidates for a certificate. Therefore, if the certificate length is upper bounded by a function ℓ then the running time is upper bounded by 2^ℓ up to the time needed for verifying a candidate. On the other hand, if the running time is upper bounded by a function t then it is not necessarily true that the certificate length is upper bounded by $\lg t$ (unless $\mathbf{E} \subseteq \mathbf{NP}$, where $\mathbf{E}$ is the complexity class for exponential time with linear exponent).

We observe a similarity between the two types of complexity classifications for satisfiability problems. More specifically, we show that many known results on the time complexity of k-SAT, SAT$_\Delta$ (the restriction of SAT to formulas whose clause density is at most Δ), and CircuitSAT have their counterparts for the certificate complexity. It is important to note that this similarity holds for certificates defined as certificates *verifiable in subexponential time* (although the polynomial-time verification suffices for some cases). Precise definitions for the subexponential-time verification are given in Sect. 2. Our main results can be summarized as follows.

Certificate complexity of k-SAT. It is well known that k-SAT can be solved in time $O(2^{cn})$ where n is the number of variables and $c < 1$ is a constant depending on k. This bound was obtained using different approaches: critical clauses [PPZ97, PPSZ98], local search [Sch99], covering codes [DGH+02]. The proof based on covering codes can be adapted to show that k-SAT has certificates of length cn (we include this adapted proof for self-containedness).

Another known result on k-SAT is the result by Impagliazzo and Paturi [IP01] on increasing the time complexity of k-SAT as k grows. They defined the sequence $\{s_k\}_{k \geq 3}$ where

$$s_k = \inf\{s \mid k\text{-SAT can be solved by an } O(2^{sn})\text{-time algorithm}\}.$$

The conjecture that $s_k > 0$ for all $k \geq 3$ is called the *Exponential Time Hypothesis* (ETH). Note that ETH is stronger than the $\mathbf{P} \neq \mathbf{NP}$ conjecture. It is shown in [IP01] that if ETH is true then $\{s_k\}$ increases infinitely often. We define the sequence $\{c_k\}_{k \geq 3}$ by

$$c_k = \inf\{c \mid k\text{-SAT has certificates of length } cn\}$$

and we show that if ETH is true then $\{c_k\}$ increases infinitely often too. To index the search space appearing in the proof of [IP01] by certificates of appropriate length, we use the *combinatorial* (also called *binomial*) *number system*, see e.g. [Knu05].

It is an intriguing open question whether $s_k = c_k$.

Certificate complexity of SAT$_\Delta$. Using Schuler's reduction from SAT$_\Delta$ to k-SAT [Sch05], it was shown that SAT$_\Delta$ can be solved in time $O(2^{cn})$ with $c < 1$ [CIP06]. We translate this result into the certificate settings: SAT$_\Delta$ has certificates of length cn. The combinatorial number system is again used in our proof.

The time complexity of $\texttt{SAT}_\Delta$ is characterized by the sequence $\{d_\Delta\}$ where

$$d_\Delta = \inf\{d \mid \texttt{SAT}_\Delta \text{ can be solved by an } O(2^{dn})\text{-time algorithm}\}.$$

It was shown in [CIP06] that this sequence is interwoven with $\{s_k\}$ and thus $s_\infty = d_\infty$, where $s_\infty = \lim_{k\to\infty} s_k$ and $d_\infty = \lim_{\Delta\to\infty} d_\Delta$. We characterize the certificate complexity of $\texttt{SAT}_\Delta$ by the sequence $\{b_\Delta\}$, where

$$b_\Delta = \inf\{b \mid \texttt{SAT}_\Delta \text{ has certificates of length } bn\},$$

and we show that the relationship between the certificate complexities $\{c_k\}$ and $\{b_\Delta\}$ is similar to the relationship between the time complexities $\{s_k\}$ and $\{d_\Delta\}$. In particular, $\lim_{k\to\infty} c_k = \lim_{\Delta\to\infty} b_\Delta$.

Nondeterministic subexponential time and $\texttt{CircuitSAT}$. The class **SE** consists of all parameterized problems that can be solved in time subexponential in the parameter [IPZ01]. In Sect. 5, we define the class **NSE** to be the class of all parameterized problems that have subexponential-time verifiable certificates of length bounded by the parameter. Note that there is an analogy between the pair **P** versus **NP** and the pair **SE** versus **NSE**. We also define a subexponential-time reducibility that preserves the certificate length and we observe that

- **NSE** is closed under this reducibility;
- $\texttt{CircuitSAT}$ with the number of inputs as the parameter is complete for **NSE** under this reducibility.

It follows from the completeness of $\texttt{CircuitSAT}$ that if $\texttt{CircuitSAT}$ has certificates of length cn, where n is the number of inputs and $c < 1$ is a constant, then **NSE** collapses to **SE**. Therefore, since ETH is a stronger assumption than **SE** $\neq$ **NSE**, ETH also implies that $\texttt{CircuitSAT}$ has no certificates shorter than the number of inputs.

This observation can be viewed as a certificate offset of recent results on the time complexity of $\texttt{CircuitSAT}$. For example, it is shown by Paturi and Pudlák [PP10] that $\texttt{CircuitSAT}$ cannot be solved by a one-sided probabilistic polynomial-time algorithm with success probability better than $2^{-n+o(n)}$ unless some unlikely complexity containments hold. On the other hand, Williams [Wil10] shows that even a slight improvement in the running time over exhaustive search for $\texttt{CircuitSAT}$ implies a proof of **NEXP** $\not\subseteq$ **P/poly**.

2   Definitions

Definition 1 (parameterized problem, [FG06]). *A parameterized problem is a pair (L, p) consisting of a language $L \in \{0,1\}^*$ and a polynomial-time computable parameterization function $p : \{0,1\}^* \to \mathbb{N}$.*

Definition 2 (verifier and certificate). *A* verifier *for a parameterized problem (L, p) is an algorithm V such that*

$$x \in L \iff \exists w \in \{0,1\}^* \, (|w| \le p(x) \text{ and } V \text{ accepts the pair } (x, w))$$

where the string w is called a certificate *for x.*

Remark 1. In the definition above and throughout the paper, we use the word "algorithm" to denote a deterministic algorithm. However, all results of the paper hold if "algorithm" means a randomized algorithm.

Definition 3 (subexponential verification scheme). *A subexponential verification scheme for a parameterized problem (L, p) is a family $\{V_t\}_{t \in \mathbb{N}}$ of verifiers for (L, p) such that for each verifier V_t, the running time of V_t on (x, w) is*

$$|x|^{O(1)} \, 2^{p(x)/t}$$

where the polynomial $|x|^{O(1)}$ may depend on t. If (L, p) has a subexponential verification scheme, we also say that L has subexponential-time verifiable certificates of length p.

Remark 2. It would be more common if we defined subexponential verification schemes as a family of verifiers $V_\epsilon(x, w)$ like, for example, the definition of a family of subexponential reductions (SERF) in [IPZ01]. These two versions are equivalent, however we prefer the version with $1/t \to 0$ instead of $\epsilon \to 0$ to avoid discussions on the representation of ϵ (especially when it is given as a function of other parameters).

Remark 3. An important special case of subexponential verification schemes is the case where all verifiers V_t are the same and each of them runs in time polynomial in both p and $|x|$. If so, we say that L has *polynomial-time verifiable* certificates of length p. An obvious example of this special case is the polynomial-time verification for (SAT, n): a certificate for a satisfiable formula is an n-bit string that encodes a satisfying assignment. Less obvious examples are given in Theorems 1 and 3 below.

Remark 4. All certificates considered in this paper are verifiable in subexponential time. To simplify the terminology, we omit the words "subexponential-time verifiable". Thus, throughout the paper, when we write "L has certificates of length p", this means "L has subexponential-time verifiable certificates of length p".

3   Shortest Certificates for k-SAT

The time complexity of k-SAT for $k \geq 3$ is characterized by the sequence $\{s_k\}_{k \geq 3}$ where

$$s_k = \inf\{s \mid k\text{-SAT can be solved by an } O(2^{sn})\text{-time algorithm}\}.$$

The current knowledge and open questions about this sequence can be described as follows:

- We know that $s_k < 1$. More exactly, $s_k \leq (1 - \mu/k)$ for some constant $\mu > 0$. This bound is obtained using critical clauses [PPZ97, PPSZ98], local search [Sch99], covering codes [DGH+02, MS11].

- We do not know whether $s_k = 0$. The conjecture that $s_k > 0$ for all $k \geq 3$ is called the *Exponential Time Hypothesis* (ETH).
- If ETH holds then $\{s_k\}$ increases infinitely often [IP01].
- Let $s_\infty = \lim_{k \to \infty} s_k$. The conjecture that $s_\infty = 1$ is called the *Strong Exponential Time Hypothesis* (SETH). The relationship between s_∞ and the complexity of SAT is also unknown, where the complexity of SAT is the minimum number s such that SAT can be solved in time 2^{sn} up to a polynomial in the input size.

The certificate complexity of k-SAT is defined below through a sequence similar to $\{s_k\}$.

Definition 4 (certificate complexity for k-SAT). *For each $k \geq 3$, let*

$$c_k = \inf\{c \mid k\text{-SAT has certificates of length } cn\}.$$

The limit of $\{c_k\}$ as $k \to \infty$ is denoted c_∞.

Note that $s_k \leq c_k$ for all $k \geq 3$ and $s_\infty \leq c_\infty$.

3.1   Upper Bound on Certificate Length for k-SAT

The following theorem shows that $c_k < 1$ and, moreover, this inequality holds even for polynomial-time verifiable certificates.

Theorem 1. *For each $k \geq 3$ and for each $\epsilon > 0$, k-SAT has polynomial-time verifiable certificates of length $\left(1 - \lg \frac{k+1}{k} + \epsilon\right) n$.*

Certificates of the claimed length can be extracted from the algorithm that solves k-SAT in time $O\left(2^{\left(1 - \lg \frac{k+1}{k} + \epsilon\right)n}\right)$ using covering codes [DGH+02]. Such a certificate includes the number of the ball containing a satisfying assignment and the index of this assignment in a search tree inside the ball. Although the proof essentially repeats that of [DGH+02], we include it here for the sake of self-containedness.

Proof. Let F be a satisfiable k-CNF formula over n variables. We show that a satisfying assignment for F can be encoded using less than n bits. Each assignment for F is identified with a point in the Boolean cube $\{0,1\}^n$. The first step of the encoding is to cover the cube with Hamming balls of radius ρn, where a value for ρ will be chosen later. It is known that any such covering must contain at least $2^{(1-H(\rho))n}$ balls, where H is the binary entropy function. An "almost" optimal covering (with at most $2^{(1-H(\rho)+\epsilon)n}$ balls for any $\epsilon > 0$) is constructed in [DGH+02] as follows.

The centers of the balls are viewed as a covering code for the cube. For any $\epsilon > 0$, we need a covering code of radius ρn that contains at most $2^{(1-H(\rho)+\epsilon)n}$ codewords. Consider a partition of n bits into n/b blocks of size b, where b is a constant (without loss of generality, we can assume that n is divisible by b and n is sufficiently large). Using a brute-force enumeration, we can find an optimal covering code of radius ρb for each block. Let $\mathcal{C} = \{w_1, \ldots, w_M\}$ be such a code, where M is at most $2^{(1-H(\rho))b}$ up to a polynomial in b. The direct sum of n/b copies of $\mathcal{C}$ is a covering code of radius ρn for the cube. It is easy to see that given ρ and ϵ, a value for b can chosen such that this direct sum (denoted $\mathcal{C}^{n/b}$) has at most $2^{(1-H(\rho)+\epsilon)n}$ codewords. We encode each codeword $w_i \in \mathcal{C}$ by an integer i. Then each codeword in $\mathcal{C}^{n/b}$ can be encoded by a concatenation of n/b integers from 1 to M each. The length of this encoding is at most $(1-H(\rho)+\epsilon)n$. Moreover, given such a concatenation, the corresponding codeword (or, equivalently, the corresponding ball center) can be computed in time polynomial in n.

Assume that F has a satisfying assignment in a ball of radius ρn centered at an assignment A. Then the encoding of A (with at most $(1 - H(\rho) + \epsilon)n$ bits) is the first part of a certificate for F. To construct the second part, we again refer to [DGH⁺02] where it is shown how to search for a satisfying assignment inside a ball. This search is essentially a recursive procedure that modifies F and A using the following approach: if the current assignment α does not satisfy the current formula ϕ, take the first unsatisfied clause $l_1 \vee \ldots \vee l_h$ in ϕ and consider pairs $(\phi_1, \alpha_1), \ldots, (\phi_h, \alpha_h)$ where each α_i is obtained from α by flipping the value of the literal l_i and each ϕ_i is obtained from ϕ by substituting the new value for l_i in ϕ. This procedure starts with (F, A) and builds a recursion tree T of depth at most ρn. Since F is a k-CNF formula, the degree of each node in T is at most k. At least one leaf in T is a pair (ϕ, α) where α satisfies ϕ. Hence, α satisfies F.

Thus, a satisfying assignment α in a ball of radius ρn centered at A can be encoded by a path from the root to a leaf in T. Such a path is determined by a sequence of literals chosen in unsatisfied clauses. If we choose a literal l_i in a clause $l_1 \vee \ldots \vee l_h$, we encode this choice by the integer i. The entire path can thus be encoded by a sequence of integers $i_1, \ldots, i_{\lfloor \rho n \rfloor}$ where $1 \leq i_j \leq k$ for each j. Removing the leading 1s in binary representation of these integers, we encode the path by a concatenation of $\lfloor \rho n \rfloor$ bit strings of length $\lfloor \lg k \rfloor$ each.

Finally, a certificate for F is a pair, where the first element encodes the center of a ball containing a satisfying assignment and the second element encodes a path in T. For any ϵ, the total length of this certificate is at most $(1 - H(\rho) + \epsilon)n + \rho n \lg k$. Taking $\rho = 1/(k + 1)$, we have:

$$(1 - H(\rho) + \epsilon)n + \rho n \lg k = \left(1 - \lg \frac{k + 1}{k} + \epsilon\right) n.$$

To verify it polynomial time, just compute the center A of the ball from a given index and use a given path to modify A to a satisfying assignment.   □

3.2 The Growth of Certificate Lengths for k-SAT

It is proved in [IP01] that ETH implies the following relationship between s_k and s_∞:

$$s_k \leq s_\infty(1 - \sigma/(ek)), \tag{1}$$

where σ is the solution of $H(\sigma) = s_\infty/2$ on $(0; 1/2]$. Therefore, if ETH holds then $\{s_k\}$ increases infinitely often. We prove a similar result for $\{c_k\}$.

Theorem 2. *If ETH holds then*

$$c_k \leq c_\infty(1 - \gamma/(ek)) \qquad (2)$$

where γ is the solution of $H(\gamma) = c_\infty/2$ on $(0; 1/2]$.

This theorem is proved using the following lemma from [IP01]:

Lemma 1 ([IP01]). *Let F be a formula in k-CNF such that F is not satisfiable by any assignment of weight[1] at most δn. For any $\epsilon > 0$, there exists k' such that the following holds: The satisfiability of F is equivalent to the satisfiability of the disjunction $F_1 \vee \ldots \vee F_N$, where $N \leq 2^{\epsilon n}$ and each F_i is a formula in k'-CNF on at most $n(1 - \delta/(ek))$ variables. Moreover, this disjunction can be computed from F in time $n^{O(1)} 2^{\epsilon n}$.*

Proof (of Theorem 2). We mimic the proof of inequality (1) in [IP01]. The proof shows how to construct an $O(2^{cn})$-time algorithm for k-SAT using an $O(2^{c'n})$-time algorithm for k'-SAT for certain $k' > k$ and $c' > c$. We must make sure that the decrease in the running time is accompanied by the decrease in the length of a certificate verifiable in subexponential time.

The algorithm constructed in [IP01] tests satisfiability of a given k-CNF formula F as follows (here $\epsilon > 0$ and $w = \lfloor \sigma n \rfloor$):

1. Use exhaustive search to check all assignments of weight at most w. If at least one of them satisfies F, return "satisfiable".
2. Apply Lemma 1 (with $\delta = w/n$) to obtain k'-CNF formulas $F_1, \ldots, F_N$ on at most $n(1 - w/(ekn))$ variables each, where $N \leq 2^{\epsilon n}$.
3. Apply a k'-SAT algorithm to F_i's; if at least one of them is satisfiable, return "satisfiable"; otherwise return "unsatisfiable".

In the certificate settings, we take $w = \lfloor \gamma n \rfloor$ and we bound the length of certificates considering two cases: the case of a satisfying assignment of low weight ($\leq w$), and the case of application of Lemma 1.

1. If F is satisfied by an assignment of weight at most w then F has a certificate of length

$$\left\lceil \lg \binom{n}{w} \right\rceil + O(\lg n).$$

Such a certificate can be obtained using the *combinatorial* (also called *binomial) number system*, see e.g. [Knu05].

(a) Consider the lexicographic order of all assignments (n-bit strings) of weight exactly w and consider the numbering of assignments in this list by numbers from 0 to $\binom{n}{w} - 1$. Let A be an assignment with 1s on positions $n > a_w > \ldots > a_1 \geq 0$ and 0s on all other positions. We encode A by

---

[1] An assignment is identified with a bit string; the *weight* of an assignment is the number of 1s in the string.

its number N_A in the lexicographic order, where N_A can be computed as the following sum:

$$N_A = \binom{a_w}{w} + \ldots + \binom{a_1}{1}.$$

Obviously, the decoding can be done efficiently: first, find a_w, then proceed to lower terms.

(b) To encode an assignment of weight $w - i$, we first encode i and then append the number

$$\binom{a_{w-i}}{w-i} + \ldots + \binom{a_1}{1}.$$

The encoding of i has length $O(\lg n)$ if we encode i as follows: $1\ldots10\langle i\rangle$ where $\langle i\rangle$ is i written in binary and the number of 1s is equal to the length of the binary representation of i.

2. In the case of application of Lemma 1, we specify the index i of the first satisfiable formula F_i by $\lceil \epsilon n \rceil$ bits. The formula itself can be found by running the procedure in Lemma 1, which takes time $2^{\epsilon n} n^{O(1)}$. These $\lceil \epsilon n \rceil$ bits are appended with the the certificate for F_i. By definition of $c_{k'}$, its length is bounded by $(c_{k'} + \epsilon)$ times the number of variables in F_i. Finally, we put leading 0 on top of all that to indicate that this is the case of application of Lemma 1.

In total, we have the following upper bound on the certificate length:

$$\begin{aligned}
\max\{\lceil \lg \binom{n}{w} \rceil + O(\lg n), 1 + \lceil \epsilon n \rceil + (c_{k'} + \epsilon)\lceil n(1 - w/(ekn)) \rceil\} &= \\
n \cdot \max\{H(w/n), c_{k'}(1 - w/(ekn)) + 2\epsilon\} + O(1) &= \\
n \cdot \max\{c_\infty/2, c_\infty(1 - \gamma/(ek)) + 2\epsilon\} + O(1) &= \\
n \cdot (c_\infty(1 - \gamma/(ek)) + 2\epsilon) + O(1). &
\end{aligned}$$

$\square$

Corollary 1. *If ETH holds then the sequence $\{c_k\}$ increases infinitely often as k grows.*

Proof. Straightforwardly follows from (2). $\square$

4   Shortest Certificates for SAT_Δ

The *clause density* of a CNF formula with m clauses over n variables is the ratio m/n. For any positive constant Δ, we write SAT_Δ to denote the restriction of SAT to formulas whose clause density is at most Δ.

Lemma 2. *For each $\Delta > 0$, $k \geq 3$, and $c > 0$, if k-SAT has (polynomial-time verifiable) certificates of length cn then SAT_Δ has (polynomial-time verifiable) certificates of length*

$$\left(c + \frac{(\Delta + 1/k)\lg e}{2ck}\right)n + o(n).$$

Proof. Let F be a satisfiable formula in CNF with $m/n \leq \Delta$. We build a certificate for F using Schuler's reduction [Sch05] which transforms any CNF formula into an equivalent disjunction of an exponential number of k-CNF formulas. This reduction can be represented as a labeled binary tree in which the root is labeled by F and the leaves are labeled by k-CNF formulas [CIP06]. Any path from the root to a leaf is given by a sequence of choices:

- either choose a left branch where a clause is reduced to a k-clause;
- or choose a right branch where the number of variables is decreased by k variables.

The maximum number of branchings to the left is m; the maximum number of branchings to the right is n/k (without loss of generality we can assume that n is divisible by k).

Consider a path from the root to a leaf such that the path contains exactly r branchings to the right. Then the k-CNF formula at the leaf has $n - kr$ variables. Let P_r be the set of all such paths. Any path in P_r can be identified with a bit string of length $m + n/k$ that has exactly r ones. We encode these strings using the combinatorial number system [Knu05], like we encoded assignments of fixed weight in the proof of Theorem 2. Then any path in P_r is encoded by a bit string of length

$$\left\lfloor \lg \binom{m + n/k}{r} \right\rfloor + 1$$

and the decoding can be done in polynomial time.

Given a path from the root to a leaf, the k-CNF formula at this leaf can be computed in time polynomial in the size of F. Therefore, a certificate for F is a path to a leaf L labeled by a satisfiable k-CNF formula F_L plus a certificate for F_L. If the path to L has r branchings to the right then a certificate for F can be defined as the concatenation of the following three strings:

- the encoding of the integer r with $\lfloor \lg(n/k) \rfloor + 1$ bits;
- the encoding of the path to L with $\lfloor \lg \binom{m+n/k}{r} \rfloor + 1$ bits;
- the encoding of a certificate for F_L with $\lfloor c(n - kr) \rfloor + 1$ bits.

Now we show

$$\lg(n/k) + \lg \binom{m + n/k}{r} + c(n - kr) \leq \left(c + \frac{(\Delta + 1/k) \lg e}{2^{ck}} \right) n + o(n).$$

Since the first term in the left-hand side is sublinear, it suffices to upper bound the sum of the other two terms. We estimate it using the same way as in [CIP06]:

$$\begin{aligned}
\lg \binom{m+n/k}{r} + c(n - kr) &\leq \lg \left(\sum_{r=0}^{m+n/k} \binom{m+n/k}{r} 2^{c(n-kr)} \right) \\
&\leq \lg \left(2^{cn} \left(1 + 2^{-ck} \right)^{m+n/k} \right) \\
&\leq cn + (m + n/k) \lg \left(e^{2^{-ck}} \right) \\
&\leq cn + \frac{(m+n/k) \lg e}{2^{ck}} \\
&\leq \left(c + \frac{(\Delta + 1/k) \lg e}{2^{ck}} \right) n.
\end{aligned}$$

Given a certificate described above, the verification of satisfiability of F consists of two steps. The first step is to decode the certificate into a k-CNF formula G and a certificate of satisfiability of G. This can be done in polynomial time. The second step is to verify satisfiability of G. If a certificate for G is verifiable in polynomial time then this step can also be done in polynomial time.   □

Theorem 3. *For each $\Delta > 0$, there exists $b < 1$ such that SAT_Δ has polynomial-time verifiable certificates of length bn.*

Proof. We apply Lemma 2 choosing k and c such that

$$c + \frac{(\Delta + 1/k)\lg e}{2^{ck}} < 1.$$

Namely, if $c = 1 - \lg(1 + 1/k) + \epsilon$ for some $\epsilon > 0$ (Theorem 1) then

$$\frac{(\Delta + 1/k)\lg e}{2^{ck}} \le \frac{O(\Delta)}{2^k}.$$

Now if we take $k = r\lg(\Delta + 2)$, where r is a sufficiently large constant, we have

$$
\begin{aligned}
c + \tfrac{(\Delta + 1/k)\lg e}{2^{ck}} &\le 1 - \lg\left(1 + \tfrac{1}{r\lg(\Delta+2)}\right) + \epsilon + \tfrac{O(\Delta)}{2^{r\lg(\Delta+2)}} \\
&\le 1 - \tfrac{O(1)}{r\lg(\Delta+2)} + \epsilon + \tfrac{O(1)}{(\Delta+2)^{r-1}} < 1.
\end{aligned}
$$

□

Without loss of generality, we can assume that the clause density Δ is a positive integer. Then, similarly to the case of k-SAT, the time complexity of SAT_Δ is characterized by the sequence $\{d_\Delta\}_{\Delta \ge 1}$ where

$$d_\Delta = \inf\{d \mid \mathrm{SAT}_\Delta \text{ can be solved by an } O(2^{dn})\text{-time algorithm}\}.$$

It is known that $d_\Delta < 1$ for all Δ. More exactly, SAT can be solved in time $2^{(1-1/O(\lg \Delta))n}$ up to a factor polynomial in the size of the input formula [CIP06, DH09]. It is also known that $\{d_\Delta\}$ is interwoven with $\{s_k\}$. Namely, as shown in [CIP06],

- for any k and for any $\epsilon > 0$, there exists Δ such that $s_k \le d_\Delta + \epsilon$;
- for any Δ and for any $\epsilon > 0$, there exists k such that $d_\Delta \le s_k + \epsilon$.

Therefore, $s_\infty = d_\infty$ where $d_\infty = \lim_{\Delta \to \infty} d_\Delta$.

We define an analog of $\{d_\Delta\}$ in the certificate settings and show a similarity between the two sequences.

Definition 5 (certificate complexities for SAT_Δ). *For each $\Delta \ge 1$, let*

$$b_\Delta = \inf\{b \mid \mathrm{SAT}_\Delta \text{ has certificates of length } bn\}.$$

Similarly to d_∞, we define $b_\infty = \lim_{\Delta \to \infty} b_\Delta$.

Lemma 3. *For each $\Delta > 0$ and $\epsilon > 0$, there exists k such that $b_\Delta \leq c_k + \epsilon$.*

Proof. Consider two cases: $c_\infty > 0$ and $c_\infty = 0$. In the case of $c_\infty > 0$, we apply Lemma 2 with k such that $c_k > 0$. Then we have

$$b_\Delta \leq c_k + \frac{(\Delta + 1/k)\lg e}{2^{c_k k}} + o(1)$$

for each $\Delta > 0$. Taking k sufficiently large, we can make the fraction in the right-hand side arbitrarily small. If $c_\infty = 0$, we can apply Lemma 2 with arbitrarily small $c > 0$. In particular, if we take c as a function of k such that $ck \to \infty$ as $k \to \infty$, we can make the right-hand side arbitrarily small. Hence $b_\Delta = 0$ in this case. □

Corollary 2. $b_\infty \leq c_\infty$

Proof. Take $\Delta, k \to \infty$ and $\epsilon \to 0$. □

Lemma 4 (Sparsification Lemma, [IPZ01]). *Let F be a formula in k-CNF. There is a function $f(k, \epsilon)$ upper bounded by a polynomial in $\frac{1}{\epsilon}$ such that for any $\epsilon > 0$, the satisfiability of F is equivalent to the satisfiability of the disjunction $F_1 \vee \ldots \vee F_N$ over the same set of variables, where $N \leq 2^{\epsilon n}$ and each F_i is a k-CNF formula in which every variable occurs at most $f(k, \epsilon)$ times. Moreover, this disjunction can be computed from F in time $n^{O(1)} 2^{\epsilon n}$.*

Lemma 5. *For any $k \geq 3$ and for any $\epsilon > 0$, we have $c_k \leq b_\infty + \epsilon$.*

Proof. Similarly to [CIP06, Corollary 2], the proof proceeds by application of Lemma 4. Given $k \geq 3$ and $\epsilon > 0$, we show that k-SAT has certificates of length $(b_\infty + \epsilon)n$. Namely, we construct a subexponential verification scheme $\{V_t\}$, where each verifier V_t runs in time

$$|F|^{O(1)} 2^{(b_\infty + \epsilon)n/t} \tag{3}$$

where $|F|$ is the size of the input k-CNF formula F.

Each V_t starts with sparsifying F by Lemma 4. The parameter of the sparsification procedure is chosen so that the procedure runs in time

$$|F|^{O(1)} 2^{(b_\infty + \epsilon)n/2t}.$$

Let $\Delta = \Delta(k, \epsilon)$ be the maximum clause density of the formulas $F_1, \ldots, F_s$ returned by the sparsification procedure. The input string w for V_t is interpreted as a certificate of satisfiability for some F_j. Therefore, V_t tests each formula F_i: whether w is a certificate for F_i. This test is done using a subexponential verification scheme $\{U_t\}$ for $(\text{SAT}_\Delta, b_\Delta + \epsilon)$. More exactly, the verifier V_t uses U_{2t} and, thus, the test of F_i runs in time

$$|F|^{O(1)} 2^{(b_\Delta + \epsilon)n/2t}.$$

Since $b_\Delta \leq b_\infty$, the overall running time of V_t is (3). □

Corollary 3. $c_\infty \leq b_\infty$

Proof. Take $k \to \infty$ and $\epsilon \to 0$. $\square$

Theorem 4. $c_\infty = b_\infty$

Proof. Corollaries 2 and 3. $\square$

Theorem 5. *If ETH holds then the sequence* $\{b_\Delta\}$ *of certificate complexities for* SAT$_\Delta$ *increases infinitely often.*

Proof. Suppose that $b_{\Delta_0} = b_\infty$ for some Δ_0. Then, by Lemma 3, there exists k_0 such $c_{k_0} \geq b_\infty$. Since $b_\infty = c_\infty$ and $\{c_k\}$ is nondecreasing, we have $c_k = c_\infty$ for all $k \geq k_0$, which contradicts Theorem 2. $\square$

5   Shortest Certificates for CircuitSAT

Definition 6 (subexponential time). *We say that a parameterized problem* (L, p) *can be solved in* subexponential time *if for any* $t \in \mathbb{N}$, *there exists an algorithm that decides* L *in time* $|x|^{O(1)} 2^{p(x)/t}$, *where* x *is an instance. The class* **SE** *consists of all parameterized problems* (L, p) *that can be solved in subexponential time.*

Definition 7 (nondeterministic subexponential time). *The class* **NSE** *consists of all parameterized problems* (L, p) *that have subexponential verification schemes.*

Remark 5. Note that **NSE** is to **SE** as **NP** is to **P**: the larger class requires a verifiable certificate to accept a "yes" instance. There are two differences:

- subexponential time versus polynomial time;
- the bound $|w| \leq p(x)$ on the certificate length in the case of parameterized problems $(L, p) \in$ **NSE** versus the bound $|w| \leq |x|^{O(1)}$ in the case of problems in **NP**.

The class **SE** is closed under reducibility defined in [IPZ01] and called *subexponential reduction families* (SERFs for short). Informally, a SERF from (L, p) to (L', p') is a collection of Turing reductions R_t from L to L' such that each reduction runs in time $|x|^{O(1)} 2^{p(x)/t}$ and allows at most a linear increase of the parameter. We define a "strict" version of SERFs under which **NSE** is closed.

Definition 8 (strict SERF). *We say that* R *is a* strict subexponential reduction family *(strict SERF) from a parameterized problem* (L, p) *to a parameterized problem* (L', p') *if* R *is a sequence of algorithms* R_t *such that*

- *each algorithm* R_t *takes a string* $x \in \{0, 1\}^*$ *as input and outputs strings* $y_1, \ldots, y_m$, *where* $m \leq 2^{p(x)/t}$;
- *each* R_t *runs in time* $|x|^{O(1)} 2^{p(x)/t}$;
- $p'(y_i) \leq p(x)$ *for all* $1 \leq i \leq m$;

– *for every $x \in \{0,1\}^*$, we have*

$$x \in L \iff \bigvee_{1 \leq i \leq m} (y_i \in L').$$

Remark 6. A strict SERF is a special case of a SERF, where the word "strict" alludes to two refinements:

– a strict SERF is a disjunctive truth table reduction, while a SERF is a Turing reduction;
– a strict SERF does not increase the parameter, while a SERF allows multiplying the parameter by an arbitrary constant.

Note also that if we allowed a slight increase of the parameter

$$p'(y_i) \leq p(x) + o(p(x)),$$

we would have an equivalent definition.

Theorem 6. **NSE** *is closed under strict SERFs: if (L, p) has a strict SERF to $(L', p') \in$ **NSE**, then $(L, p) \in$ **NSE**.*

Proof. A certificate for x is a certificate for a y_i such that $y_i \in L'$. The verification of this certificate includes generating $y_1, \ldots, y_m$ with checking each of them: whether the given certificate is a certificate for y_j. □

Theorem 7. `CircuitSAT` *with the number of inputs as the parameter is complete for* **NSE** *under strict SERFs.*

Proof. Consider $(L, p) \in$ **NSE**. Let $t \in \mathbb{N}$. Consider a Turing machine that verifies certificates of length $p(x)$ in time $|x|^{O(1)} 2^{p(x)/2t}$. It is well-known that the machine can be transformed into a circuit with $p(x)$ inputs (after hardwiring a specific x) and size polynomial in the length of the machine's input and quadratic in the running time. The reduction R_t outputs this circuit. □

Corollary 4. *If* `CircuitSAT` *has certificates of length cn, where n is the number of inputs and $c < 1$ is a constant, then* **SE** $=$ **NSE**.

Proof. Suppose that `CircuitSAT` has certificates of length cn. We show that if $(L, p) \in$ **NSE** then $(L, p) \in$ **SE**. Since (L, p) has a strict SERF to `CircuitSAT` with p inputs, L has certificates of length cp. That is, $(L, cp) \in$ **NSE** and therefore (L, cp) has a strict SERF to `CircuitSAT` with cp inputs. Using the supposition again, we obtain $(L, c^2p) \in$ **NSE**. Continuing, we can conclude that L has certificates of arbitrarily small length. Hence, L can be solved in subexponential time. □

Remark 7. It follows from Corollary 4 that if ETH is true then there is no constant $c < 1$ such that `CircuitSAT` has certificates of length cn. Indeed, $(\text{3-SAT}, n) \in$ **NSE** where n is the number of variables. However, if ETH is true then $(\text{3-SAT}, n) \notin$ **SE**, i.e., ETH implies **SE** $\neq$ **NSE**.

Acknowledgements

This work was done in part when the authors attended Dagstuhl Seminar 10441, Exact Complexity of NP-Hard Problems, November 2010. The authors are grateful to its participants for numerous enlightening discussions.

References

[CIP06]   Calabro, C., Impagliazzo, R., Paturi, R.: A duality between clause width and clause density for SAT. In: Proceedings of the 21st Annual IEEE Conference on Computational Complexity, CCC 2006, pp. 252–260. IEEE Computer Society, Los Alamitos (2006)

[DGH$^+$02] Dantsin, E., Goerdt, A., Hirsch, E.A., Kannan, R., Kleinberg, J., Papadimitriou, C.H., Raghavan, P., Schöning, U.: A deterministic $(2 - 2/(k + 1))^n$ algorithm for k-SAT based on local search. Theoretical Computer Science 289(1), 69–83 (2002)

[DH09]   Dantsin, E., Hirsch, E.A.: Worst-case upper bounds. In: Handbook of Satisfiability, ch. 12, pp. 403–424. IOS Press, Amsterdam (2009)

[FG06]   Flum, J., Grohe, M.: Parameterized Complexity Theory. Texts in Theoretical Computer Science. An EATCS Series. Springer, Heidelberg (2006)

[IP01]   Impagliazzo, R., Paturi, R.: On the complexity of k-SAT. Journal of Computer and System Sciences 62(2), 367–375 (2001)

[IPZ01]   Impagliazzo, R., Paturi, R., Zane, F.: Which problems have strongly exponential complexity. Journal of Computer and System Sciences 63(4), 512–530 (2001)

[Knu05]   Knuth, D.E.: Generating All Combinations and Partitions. The Art of Computer Programming, fascicle 3, vol. 4, pp. 1–6. Addison-Wesley, Reading (2005)

[MS11]   Moser, R.A., Scheder, D.: A full derandomization of Schöning's k-SAT algorithm. In: Proceedings of the 43rd Annual ACM Symposium on Theory of Computing, STOC 2011. ACM, New York (2011) (to appear)

[PP10]   Paturi, R., Pudlák, P.: On the complexity of circuit satisfiability. In: Proceedings of the 42nd Annual ACM Symposium on Theory of Computing, STOC 2010, pp. 241–250. ACM, New York (2010)

[PPSZ98]   Paturi, R., Pudlák, P., Saks, M.E., Zane, F.: An improved exponential-time algorithm for k-SAT. In: Proceedings of the 39th Annual IEEE Symposium on Foundations of Computer Science, FOCS 1998, pp. 628–637 (1998)

[PPZ97]   Paturi, R., Pudlák, P., Zane, F.: Satisfiability coding lemma. In: Proceedings of the 38th Annual IEEE Symposium on Foundations of Computer Science, FOCS 1997, pp. 566–574 (1997)

[Sch99]   Schöning, U.: A probabilistic algorithm for k-SAT and constraint satisfaction problems. In: Proceedings of the 40th Annual IEEE Symposium on Foundations of Computer Science, FOCS 1999, pp. 410–414 (1999)

[Sch05]   Schuler, R.: An algorithm for the satisfiability problem of formulas in conjunctive normal form. Journal of Algorithms 54(1), 40–44 (2005)

[Wil10]   Williams, R.: Improving exhaustive search implies superpolynomial lower bounds. In: Proceedings of the 42nd Annual ACM Symposium on Theory of Computing, STOC 2010, pp. 231–240. ACM, New York (2010)

On Variables with Few Occurrences in Conjunctive Normal Forms

Oliver Kullmann[1] and Xishun Zhao[2,*]

[1] Computer Science Department
Swansea University
O.Kullmann@Swansea.ac.uk
http://cs.swan.ac.uk/~csoliver
[2] Institute of Logic and Cognition
Sun Yat-sen University, Guangzhou, 510275, P.R.C.

Abstract. We consider the question of the existence of variables with few occurrences in boolean conjunctive normal forms (clause-sets). Let $\mu\mathrm{vd}(F)$ for a clause-set F denote the minimal variable-degree, the minimum of the number of occurrences of variables. Our main result is an upper bound $\mu\mathrm{vd}(F) \leq \mathrm{nM}(\sigma(F)) \leq \sigma(F) + 1 + \log_2(\sigma(F))$ for *lean clause-sets* F in dependency on the *surplus* $\sigma(F)$. Lean clause-sets, defined as having no non-trivial autarkies, generalise minimally unsatisfiable clause-sets. For the surplus we have $\sigma(F) \leq \delta(F) = c(F) - n(F)$, using the deficiency $\delta(F)$ of clause-sets, the difference between the number of clauses and the number of variables. $\mathrm{nM}(k)$ is the k-th "non-Mersenne" number, skipping in the sequence of natural numbers all numbers of the form $2^n - 1$. As an application of the upper bound we obtain that clause-sets F violating $\mu\mathrm{vd}(F) \leq \mathrm{nM}(\sigma(F))$ must have a non-trivial autarky (so clauses can be removed satisfiability-equivalently by an assignment satisfying some clauses and not touching the other clauses). It is open whether such an autarky can be found in polynomial time.

1 Introduction

We study the existence of "simple" variables in boolean conjunctive normal forms, considered as clause-sets. "Simple" here means a variable occurring not very often. A major use of the existence of such variables is in inductive proofs of properties of minimally unsatisfiable clause-sets, using splitting on a variable to reduce n, the number of variables, to $n - 1$: here it is vital that we have control over the changes imposed by the substitution, and so we want to split on a variable occurring as few times as possible. The background for these considerations is the enterprise of classifying minimal unsatisfiable clause-sets F in dependency on the deficiency $\delta(F) := c(F) - n(F)$, the difference between the number $c(F) := |F|$ of clauses of F and the number $n(F) := |\mathrm{var}(F)|$ of variables of F. The most basic fact is $\delta(F) \geq 1$, as first shown in [1]. For $\delta(F) = 1$ the structure is completely known ([1,2,6], for $\delta(F) = 2$ the structure after reduction

---

* Supported by NSFC Grant 60970040.

K.A. Sakallah and L. Simon (Eds.): SAT 2011, LNCS 6695, pp. 33–46, 2011.
© Springer-Verlag Berlin Heidelberg 2011

of singular variables (occurring in one sign only once) is known ([4]), while for $\delta(F) \in \{3, 4\}$ only basic cases have been classified ([15]).

The starting point of our investigation is Lemma C.2 in [6], where it is shown that a minimally unsatisfiable clause-set F must have a variable v with at most $\delta(F)$ positive and at most $\delta(F)$ negative occurrences; we write this as $\mathrm{ld}_F(v) \leq \delta(F)$ and $\mathrm{ld}_F(\overline{v}) \leq \delta(F)$, using the notion of *literal degrees* (the number of occurrences of the literal). Thus we have $\mathrm{vd}_F(v) \leq 2\delta(F)$, using the *variable degree* $\mathrm{vd}_F(v) := \mathrm{ld}_F(v) + \mathrm{ld}_F(\overline{v})$. Using the *minimum variable degree* (min-var-degree) $\mu\mathrm{vd}(F) := \min_{v \in \mathrm{var}(F)} \mathrm{vd}_F(v)$ of F, this becomes $\mu\mathrm{vd}(F) \leq 2\delta(F)$. In this article we show a sharper bound on $\mu\mathrm{vd}(F)$ for a larger class of clause-sets F. More precisely, we show that the worst-cases $\mathrm{ld}_F(v), \mathrm{ld}_F(\overline{v}) \leq \delta(F)$ can not occur at the same time (for a suitable variable), but actually $\mathrm{ld}_F(v) + \mathrm{ld}_F(\overline{v}) - \delta(F)$ only grows logarithmically in $\delta(F)$, and this for a larger class of formulas.

The larger class of clause-sets considered is the class $\mathcal{LEAN}$ of *lean clause-sets*, which are clause-sets having no non-trivial autarky. For an overview on the theory of minimally unsatisfiable clause-sets and on the theory of autarkies see [5]. The deficiency $\delta(F) \in \mathbb{Z}$ of clause-sets is replaced by the *surplus* $\sigma(F) \in \mathbb{Z}$, which is the minimal deficiency over all clause-sets $F[V]$ for non-empty variable sets $V \subseteq \mathrm{var}(F)$, where $F[V]$ is obtained from F by removing clauses which have no variables in V, and restricting the remaining clauses to V; see [11] for more information on the surplus of (generalised) clause-sets. We need to count multiple occurrences of clauses here (which might arise during the process of removing literals with variables not in V), and thus actually multi-clause-sets F are used here. Note that by considering $V = \mathrm{var}(F)$ we have $\sigma(F) \leq \delta(F)$, and by considering $V = \{v\}$ for $v \in \mathrm{var}(F)$ we get $\sigma(F) \leq \mu\mathrm{vd}(F) - 1$. Now the main result of this article (Theorem 11) is

$$\mu\mathrm{vd}(F) \leq \mathrm{nM}(\sigma(F))$$

for lean F, where $\mathrm{nM} : \mathbb{N} \to \mathbb{N}$ (see Definition 2) is a super-linear function with $\mathrm{nM}(k) \leq k + 1 + \log_2(k)$. As an application we obtain (Corollary 12), that if a (multi-)clause-set F has no variable occurring with degree at most $\delta(F) + 1 + \log_2(\delta(F))$, then F has a non-trivial autarky. It is an open problem whether such an autarky can be found in polynomial time (for arbitrary clause-sets F); we conjecture (Conjecture 13) that this is possible.

Related work. This article appears to be the first systematic study of the problem of minimum variable occurrences in minimally unsatisfiable clause-sets and generalisations, in dependency on the deficiency, asking for the existence of a variable occurring "infrequently" in general, or for extremal examples where all variables occur not infrequently. The problem of maximum variable occurrences (asking for the existence of a variable occurring frequently in general, or for extremal examples where all variables occur not frequently) in uniform (minimally) unsatisfiable clause-sets, in dependency on the (constant) clause-length, has been studied in the literature, starting with [14]; for a recent article see [3].

Overview. In Section 2 basic notions and concepts regarding clause-sets, autarkies and minimal unsatisfiability are reviewed. Section 3 introduces the numbers nM(k) and proves exact formulas and sharp lower and upper bounds. Section 4 contains the main results. First in Subsection 4.1 the bound is shown for minimally unsatisfiable clause-sets (Theorem 15). In Subsection 4.2 the bound then is lifted to lean clause-sets, proving Theorem 11. The immediate corollary of Theorem 11 is, that if the asserted upper bound on the minimal variable degree is not fulfilled, then a non-trivial autarky must exist (Corollary 12). In Subsection 4.3 the problem of finding such autarky is discussed, with Conjecture 13 making precise our believe that one can find such autarkies efficiently. In Section 5 we discuss the sharpness of the bound, and the possibilities to generalise it further. Finally, in Section 6 open problems are stated, culminating in the central Conjecture 25 about the classification of unsatisfiable hitting clause-sets (or "disjoint/orthogonal tautologies" in the terminology of DNFs).

2 Preliminaries

We follow the general notations and definitions as outlined in [5], where also further background on autarkies and minimal unsatisfiability can be found. We use $\mathbb{N} = \{1, 2, \ldots\}$ and $\mathbb{N}_0 = \mathbb{N} \cup \{0\}$.

Complementation of literals x is denoted by $\overline{x}$, while for a set L of literals we define $\overline{L} := \{\overline{x} : x \in L\}$. A **clause** C is a finite and clash-free set of literals (i.e., $C \cap \overline{C} = \emptyset$), while a **clause-set** is a finite set of clauses. We use $\mathrm{var}(F) := \bigcup_{C \in F} \mathrm{var}(C)$ for the set of variables of F, where $\mathrm{var}(C) := \{\mathrm{var}(x) : x \in C\}$ is the set of variables of clause C, while $\mathrm{var}(x)$ is the underlying variable for a literal x. For a clause-set F we denote by $n(F) := |\mathrm{var}(F)| \in \mathbb{N}_0$ the number of variables and by $c(F) := |F| \in \mathbb{N}_0$ the number of clauses. The **deficiency** of a clause-set is denoted by $\delta(F) := c(F) - n(F) \in \mathbb{Z}$. We call a clause C **full** for a clause-set F if $\mathrm{var}(C) = \mathrm{var}(F)$, while a clause-set F is called full if every clause is full. For a finite set V of variables let $A(V)$ be the set of all $2^{|V|}$ full clauses over V. Thus full clause-sets are exactly the sub-clause-sets of some $A(V)$. A **partial assignment** is a map $\varphi : V \to \{0, 1\}$ for some (possibly empty) set V of variables. The application of a partial assignment φ to a clause-set F is denoted by $\varphi * F$, which yields the clause-set obtained from F by removing all satisfied clauses (which have at least one literal set to 1), and removing all falsified literals from the remaining clauses. A clause-set F is satisfiable iff there is a partial assignment φ with $\varphi * F = \top := \emptyset$, otherwise F is unsatisfiable. All $A(V)$ are unsatisfiable.

These notions are generalised to **multi-clause-sets**, which are pairs (F, m), where F is a clause-set and $m : F \to \mathbb{N}$ determines the multiplicity of the clauses. Now $c((F, m)) := \sum_{C \in F} m(C)$, while the application of partial assignments φ to a multi-clause-set F yields a *multi*-clause-set $\varphi * F$, where the multiplicity of a clause C in $\varphi * F$ is the sum of all multiplicities of clauses in F which are shortened to C by φ. For example if φ is a total assignment for F (assigns all variables of F) which does not satisfying F (i.e., $\varphi * F \neq \top$), then $\varphi * F$ is $(\{\bot\}, (f)_{C \in \{\bot\}})$, where $\bot := \emptyset$ is the empty clause, while $f \in \mathbb{N}$ is the number of clauses (with their multiplicities) of F falsified by φ.

For the number of occurrences of a literal x in a (multi-)clause-set (F, m) we write $\mathrm{ld}_F(x) := \sum_{C \in F, x \in C} m(C)$, called the **literal-degree**, while the **variable-degree** of a variable v is defined as $\mathrm{vd}_F(v) := \mathrm{ld}_F(v) + \mathrm{ld}_F(\overline{v})$. A **singular variable** in a (multi-)clause-set F is a variable occurring in one sign only once (i.e., $1 \in \{\mathrm{ld}_F(v), \mathrm{ld}_F(\overline{v})\}$). A (multi-)clause-set is called **non-singular** if it does not have singular variables.

For a set V of variables and a multi-clause-set F by $F[V]$ the **restriction** of F to V is denoted, which is obtained by removing clauses from F which have no variables in common with V, and removing from the remaining clauses all literals where the underlying variable is not in V (note that this can increase multiplicities of clauses).

An **autarky** for a clause-set F is a partial assignment φ which satisfies every clause $C \in F$ it touches, i.e., with $\mathrm{var}(\varphi) \cap \mathrm{var}(C) \neq \emptyset$. The empty partial assignment is always an autarky for every F, the **trivial autarky**. If φ is an autarky for F, then $\varphi * F \subseteq F$ holds, and thus $\varphi * F$ is satisfiability-equivalent to F. A clause-set F is **lean** if there is no non-trivial autarky for F. A weakening is the notion of a **matching-lean** clause-set F, which has no non-trivial **matching autarky**, which are special autarkies given by a matching condition (for every clause touched, a unique variable underlying a satisfied literal must be selectable). The process of applying autarkies as long as possible to a clause-set is confluent, yielding the **lean kernel** of a clause-set. Computation of the lean kernel is NP-hard, but the **matching-lean kernel**, obtained by applying matching autarkies as long as possible, which is also a confluent process, is computable in polynomial time. Note that a clause-set F is lean resp. matching lean iff the lean resp. matching-lean kernel is F itself. For every matching-lean multi-clause-set $F \neq \top$ we have $\delta(F) \geq 1$, while in general a multi-clause-set $F \neq \top$ is matching lean iff $\sigma(F) \geq 1$, where the **surplus** $\sigma(F) \in \mathbb{Z}$ is defined as the minimum of $\delta(F[V])$ for all $\emptyset \neq V \subseteq \mathrm{var}(F)$. Note that while w.r.t. general autarkies there is no difference between a multi-clause-set and the underlying clause-set, for matching autarkies there is a difference, due to the matching condition. For more information on autarkies see [5,11].

The set of minimally unsatisfiable clause-sets is $\mathcal{MU}$, the set of all clause-sets which are unsatisfiable, while removal of any clause makes them satisfiable. Furthermore the set of saturated minimally unsatisfiable clause-sets is $\mathcal{SMU} \subset \mathcal{MU}$, which is the set of minimally unsatisfiable clause-sets such that addition of any literal to any clause renders them satisfiable. We recall the fact that every minimally unsatisfiable clause-set $F \in \mathcal{MU}$ can be **saturated**, i.e., by adding literal occurrences to F we obtain $F' \in \mathcal{SMU}$ with $\mathrm{var}(F') = \mathrm{var}(F)$ such that there is a bijection $\alpha : F \to F'$ with $C \subseteq \alpha(C)$ for all $C \in F$. Some basic properties of $\mathcal{MU}$ and $\mathcal{SMU}$ w.r.t. the application of partial assignments are given in the following lemma.

Lemma 1. *For all clause-sets F we have:*

1. *$F \in \mathcal{SMU}$ iff for all $v \in \mathrm{var}(F)$ and $\varepsilon \in \{0, 1\}$ we have $\langle v \to \varepsilon \rangle * F \in \mathcal{MU}$.*
2. *If for some variable v holds $\langle v \to 0 \rangle * F \in \mathcal{SMU}$ and $\langle v \to 1 \rangle * F \in \mathcal{SMU}$, then $F \in \mathcal{SMU}$.*

3. *If for some variable v holds $\langle v \to 0 \rangle * F \in \mathcal{MU}$ and $\langle v \to 1 \rangle * F \in \mathcal{MU}$,*
   then $F \in \mathcal{MU}$.

For more information on minimal unsatisfiability see [5,12].

3   Non-mersenne Numbers

Splitting in minimally unsatisfiable clause-sets on variables with minimum occurrence will lead by Theorem 15 to the following recursion. The understanding of this recursion is the topic of this section. On a first reading, only Definition 2 and the main results, Lemma 9 and Corollary 10, need to be considered.

Definition 2. *For $k \in \mathbb{N}$ let $\mathrm{nM}(k) := 2$ if $k = 1$, while else*

$$\mathrm{nM}(k) := \max_{i \in \{2,\ldots,k\}} \min(2 \cdot i, \mathrm{nM}(k - i + 1) + i).$$

Remarks

1. This is sequence http://oeis.org/A062289 in the "On-Line Encyclopedia of Integer Sequences". It can be defined as the enumeration of those natural numbers containing the string "10" (at consecutive positions). The sequence leaves out exactly the number of the form $2^n - 1$ for $n \in \mathbb{N}$, and thus the name. The sequence consists of arithmetic progressions of slope 1 and length $2^m - 1$, $m = 1, 2, \ldots$, each such progression separated by an additional step of $+1$. The recursion in Definition 2 is new, and so we can not use these characterisations, but must directly prove the basic properties.
2. $\mathrm{nM}(k)$ for $k = (1), (2, 3, 4), (5, \ldots, 11), (12, \ldots, 26)$ is $(2), (4, 5, 6), (8, \ldots, 14)$, $(16, \ldots, 30)$.
3. For $k \geq 2$ we have $\mathrm{nM}(k) \geq 4$. This holds since $\mathrm{nM}(2) = 4$, while the induction step for $k \geq 3$ is $\mathrm{nM}(k) = \max_{i \in \{2,\ldots,k\}} \min(2i, \mathrm{nM}(k-i+1)+i) \geq \min(4, \min(4 + 2, 1 + 3)) = 4$.
4. By induction and by definition we have $k + 1 \leq \mathrm{nM}(k) \leq 2 \cdot k$ for $k \in \mathbb{N}$.

For a sequence $a : \mathbb{N} \to \mathbb{R}$ and $k \in \mathbb{N}$ let $\boldsymbol{\Delta a(k)} := a(k+1) - a(k)$ be the step in the value of the sequence from k to $k + 1$. The next number in the sequence of non-Mersenne numbers is obtained by adding 1 or 2 to the previous number:

Lemma 3. *For $k \in \mathbb{N}$ holds $\Delta \mathrm{nM}(k) \in \{1, 2\}$.*

Proof. For $k = 1$ we get $\Delta \mathrm{nM}(1) = 2$. Now consider $k \geq 2$. We have $\mathrm{nM}(k+1) = \max(\min(4, \mathrm{nM}(k)+2), \max_{i \in \{3,\ldots,k+1\}} \min(2i, \mathrm{nM}(k-i+2)+i)) = \max_{i \in \{3,\ldots,k+1\}} \min(2i, \mathrm{nM}(k - i + 2) + i) = \max_{i \in \{2,\ldots,k\}} \min(2(i + 1), \mathrm{nM}(k - (i + 1) + 2) + (i + 1)) = \max_{i \in \{2,\ldots,k\}} \min(2i + 2, \mathrm{nM}(k - i + 1) + i + 1) = 1 + \max_{i \in \{2,\ldots,k\}} \min(2i + 1, \mathrm{nM}(k - i + 1) + i)$.

Thus on the one hand we have $\mathrm{nM}(k+1) \geq 1 + \max_{i \in \{2,\ldots,k\}} \min(2i, \mathrm{nM}(k-i+1)+i) = 1 + \mathrm{nM}(k)$, and on the other hand $\mathrm{nM}(k+1) \leq 1 + \max_{i \in \{2,\ldots,k\}} \min(2i + 1, \mathrm{nM}(k - i + 1) + i + 1) = 2 + \mathrm{nM}(k)$. □

Corollary 4. $\mathrm{nM} : \mathbb{N} \to \mathbb{N}$ *is strictly increasing.*

Corollary 5. *We have $\mathrm{nM}(a + b) \geq \mathrm{nM}(a) + b$ for $a \in \mathbb{N}$ and $b \in \mathbb{N}_0$, and thus $\mathrm{nM}(a - b) \leq \mathrm{nM}(a) - b$ for $b \leq a$.*

Instead of considering the maximum over $k - 1$ cases $i \in \{2, \ldots, k\}$ to compute $\mathrm{nM}(k)$, we can now simplify the recursion to only one case $i(k) \in \{2, \ldots, k\}$, and for that case also consideration of the minimum is dispensable:

Lemma 6. *For $k \in \mathbb{N}$, $k \geq 2$, let $i(k) \in \mathbb{N}$ be the smallest $i \in \{2, \ldots, k\}$ with $i \geq \mathrm{nM}(k - i + 1)$ (note that $k \geq \mathrm{nM}(k - k + 1) = 2$, and thus $i(k)$ is well-defined). For example we have $i(2) = 2$, $i(3) = 3$, $i(4) = 4$ and $i(5) = 4$. Then we have:*

1. $i(k) - \mathrm{nM}(k - i(k) + 1) \leq 2$.
2. $\mathrm{nM}(k) = \mathrm{nM}(k - i(k) + 1) + i(k)$.
3. $\Delta i(k) \in \{0, 1\}$.

Proof. We have $i(k) = 2$ iff $k = 2$, while for $k = 2$ the assertions hold trivially; so assume $k \geq 3$ and $i(k) \geq 3$. Part 1 follows by Lemma 3 from the facts that the sequence $i \in \{2, \ldots, k\} \mapsto i$ moves up in steps of $+1$, while the sequence $i \in \{2, \ldots, k\} \mapsto \mathrm{nM}(k - i + 1)$ moves down in steps of -1 or -2. It remains to show Part 2. By Lemma 3 the sequence $i \in \{2, \ldots, k\} \mapsto \mathrm{nM}(k - i + 1) + i$ is monotonically decreasing, and thus by definition we obtain $\mathrm{nM}(k) = \max(2 \cdot (i(k) - 1), \mathrm{nM}(k - i(k) + 1) + i(k))$. That the maximum here is actually always attained in the second component follows by Part 1. Finally Part 3 follows again from Lemma 3. □

After these preparations we are able to characterise the "jump positions", the set $J \subset \mathbb{N}$ of $k \in \mathbb{N}$ with $\Delta \mathrm{nM}(k) = 2$. Thus $\Delta \mathrm{nM}(k) = 1$ iff $k \notin J$, and $J = \{1, 4, 11, 26, \ldots\}$. Note $\mathrm{nM}(k) = 1 + k + |\{k' \in J : k' < k\}|$.

Lemma 7. *Let $i'(k) := k - i(k) + 1$ and $h(k) := \mathrm{nM}(i'(k))$ for $k \in \mathbb{N}$, $k \geq 2$. Thus $\Delta i'(k) \in \{0, 1\}$ and $\Delta i(k) = 1 - \Delta i'(k)$. Furthermore we have $\mathrm{nM}(k) = h(k) + i(k)$, thus $\Delta \mathrm{nM}(k) = \Delta h(k) + \Delta i(k)$, and $i(k) - h(k) \in \{0, 1, 2\}$. Consider $k \geq 2$.*

1. *If $\Delta i(k) = 0$, then:*
   (a) $\Delta i(k + 1) = 1$
   (b) $i(k) \neq h(k)$.
   (c) $i(k + 1) = h(k + 1)$.
2. *If $\Delta i(k) = 1$, then:*
   (a) $\Delta h(k) = 0$, and so $k \notin J$
   (b) $i(k) \neq h(k) + 2$.
3. *The following conditions are equivalent:*
   (a) $k \in J$
   (b) $\Delta h(k) = 2$
   (c) $i(k) = h(k) + 2$
   (d) $\Delta i(k - 1) = 1$ and $i(k - 1) = h(k - 1) + 1$
   (e) $\Delta i(k - 2) = \Delta i(k - 1) = 1$
   (f) $i'(k) = i'(k - 1) = i'(k - 2)$ and $i'(k) \in J$.
4. *If $k \in J$, then $i'(k) = \max(k' \in J : k' < k)$.*

Proof. Part 1a follows by definition. For Part 1b note $i(k + 1) = i(k)$ while $h(k + 1) \geq h(k) + 1$. For Part 1c assume $i(k + 1) > h(k + 1)$. Then we have $i(k) = h(k) + 2$ and $h(k + 1) = h(k) + 1$. However then $i(k) - 1 = h(k) + 1 = h(k + 1) = \mathrm{nM}(k - (i(k) - 1) + 1)$ contradicting the definition of $i(k)$. For Part 2a assume $i(k) = i(k+1) = i(k+2)$. We have $i(k) \geq h(k+2) = \mathrm{nM}(k-i(k)+3)$, while $i(k) - 1 < \mathrm{nM}(k - (i(k)-1)+1) = \mathrm{nM}(k-i(k)+2)$, i.e., $i(k) \leq \mathrm{nM}(k-i(k)+2)$, contradicting the strict monotonicity of nM. Part 2b follows by $i(k+1) \leq h(k+1) + 2$ and $i(k+1) = i(k)+1$, $h(k+1) = h(k)$. Now consider Part 3.

Condition 3a implies condition 3b due to $\Delta i(k) = 0$ in case of $k \in J$ by Part 2a. Condition 3b implies condition 3c, since $\Delta h(k) = 2$ implies $\Delta i(k) = 0$ (otherwise we had $\Delta \mathrm{nM}(k) = 3$), and so by Part 1c we have $i(k) = i(k+1) = h(k+1)$, while the assumption says $h(k+1) = h(k) + 2$. In turn condition 3c implies condition 3a, since by Part 2b we get $\Delta i(k) = 0$, and thus $\Delta \mathrm{nM}(k) = \Delta h(k)$, while in case of $\Delta h(k) \leq 1$ we would have $i(k) - 1 \geq \mathrm{nM}(k - (i(k)-1)+1)$ contradicting the definition of $i(k)$, due to $\mathrm{nM}(k - (i(k)-1)+1) = \mathrm{nM}((k+1) - i(k+1)+1) = h(k+1) \leq h(k)+1 = i(k) - 1$. So now we can freely use the equivalence of these three conditions.

Condition 3c implies condition 3d, since we have $\Delta i(k) = 0$, and thus $\Delta i(k-1) = 1$ with Part 1a, from which we furthermore get $i(k) = i(k-1) + 1$ and $h(k-1) = h(k)$, and so $i(k-1) = i(k) - 1 = h(k)+1 = h(k-1)+1$. Condition 3d implies condition 3e, since in case of $\Delta i(k-2) = 0$ we had $i(k-1) = h(k-1)$ with Part 1c. In turn condition 3e implies condition 3c, since $i(k) = i(k-1)+1 = i(k-2)+2$, while $h(k) = h(k-1) = h(k-2)$, where by definition $i(k-2) \geq h(k-2)$ holds, whence $i(k) \geq h(k) + 2$, which implies $i(k) = h(k) + 2$. So now the first five conditions have been shown to be equivalent.

Now condition 3e implies condition 3f, since it only remains to show $i'(k) \in J$, which follows with condition 3b (using $\Delta i(k) = 0$). In turn condition 3f implies immediately condition 3e.

Finally, we prove Part 4 by induction on k (regarding the enumeration of J). We have $i'(4) = 1$, and so the induction holds for $k = 4$, the smallest jump position $k \geq 2$. Now assume that the assertion holds for all elements of $J \cap \{1, \ldots, k-1\}$, where $k > 4$, and we have to show the assertion for k. By Part 3f we know $i'(k) \in J$, where $2 \leq i'(k) < k$. Assume there is $k' \in J$ with $i'(k) < k' < k$. Now by induction hypothesis we get $i'(k) \leq i'(k') < k'$. However by Part 1 we get $\Delta i'(k') = 1$, and thus $i'(k) > i'(k')$ (since $k > k'$). $\square$

Corollary 8. *We have $J = \{2^{m+1} - m - 2 : m \in \mathbb{N}\}$.*

Proof. Let k_m for $m \in \mathbb{N}$ be the mth element of J; so the assertion is $k_m = 2^{m+1} - m - 2$. We have $k_1 = 4 - 1 - 2 = 1 = \min J$; in the remainder assume $m \geq 2$. We prove the assertion by induction, in parallel with $i(k_m) = 2^{m+1} - 2^m$. For $m = 2$ we have $k_2 = 8 - 2 - 2 = 4 = \min J \setminus \{1\}$, while $i(4)$ is the smallest $i \in \{2, 3, 4\}$ with $i \geq \mathrm{nM}(5-i)$, which yields $i(4) = 4 = 2^3 - 2^2$. Now we consider the induction step, from $m - 1$ to m. The induction hypothesis yields $k_{m-1} = 2^m - m - 1$ and $i(k_{m-1}) = 2^m - 2^{m-1}$. Lemma 7, Part 4 yields $i'(k_m) = k_{m-1}$, from which by $i'(k_m) = k_m - i(k_m) + 1$ follows $k_m = 2^m - m - 2 + i(k_m)$. By definition we get $i(k_m) = \Delta i(k_m - 1) + \cdots + \Delta i(k_{m-1}) + i(k_{m-1})$. By Lemma 7, Parts 1 - 3 the sequence of Δ-values has the form (starting with the lowest index) $0, 1, 0, 1, \ldots, 0, 1, 1$, and thus their sum has the value $\frac{1}{2}(k_m - k_{m-1} - 1) + 1$. So we get $i(k_m) = \frac{1}{2}(k_m - k_{m-1} - 1) + 1 + i(k_{m-1}) = \frac{1}{2}(2^m - m - 2 + i(k_m) - 2^m + m + 1 -$

$1) + 1 + 2^m - 2^{m-1} = \frac{1}{2}i(k_m) - 1 + 1 + 2^m - 2^{m-1}$, from which $i(k_m) = 2^{m+1} - 2^m$ follows. Finally $k_m = 2^m - m - 2 + 2^{m+1} - 2^m = 2^{m+1} - m - 2$.   □

Now the closed formula for $\mathrm{nM}(k)$ can be proven (using $\mathrm{ld}(x) := \log_2(x)$):

Lemma 9. *For $k \in \mathbb{N}$ let $\mathrm{fld}(k) := \lfloor \mathrm{ld}(k) \rfloor$ ("floor of logarithm dualis"). Then we have for $k \in \mathbb{N}$ the equality $\mathrm{nM}(k) = k + \mathrm{fld}(k + 1 + \mathrm{fld}(k + 1))$.*

Proof. Let $g(k) := \mathrm{fld}(k + 1 + \mathrm{fld}(k + 1))$ and $f(k) := k + g(k)$ (so $\mathrm{nM}(k) = f(k)$ is to be shown, for $k \geq 1$). We have $f(1) = 1 + \mathrm{fld}(2 + \mathrm{fld}(2)) = 1 + \mathrm{fld}(3) = 2 = \mathrm{nM}(1)$. We will now prove that the function $g(k)$ changes values exactly at the transitions $k \mapsto k + 1$ for $k \in J$, that is, for indices $k = k_m := 2^{m+1} - m - 2$ (using Corollary 8) with $m \in \mathbb{N}$ we have $\Delta g(k_m) = 1$, while otherwise we have $\Delta g(k_m) = 0$, from which the assertion follows (by the definition of J).

We have $g(1) = 1$ and $g(2) = 2$. Now consider $m \in \mathbb{N}$ and $k_m + 1 \leq k \leq k_{m+1}$. We show $g(k) = m + 1$, which proves the claim. Note that $g(k)$ is monotonically increasing. Now $g(k) \geq g(k_m + 1) = \lfloor \mathrm{ld}(2^{m+1} - m + \lfloor \mathrm{ld}(2^{m+1} - m) \rfloor) \rfloor = \lfloor \mathrm{ld}(2^{m+1} - m + m) \rfloor = m + 1$ and $g(k) \leq g(k_{m+1}) = \lfloor \mathrm{ld}(2^{m+2} - m - 2 + \lfloor \mathrm{ld}(2^{m+2} - m - 2) \rfloor) \rfloor \leq \lfloor \mathrm{ld}(2^{m+2} - m - 2 + m + 1) \rfloor = \lfloor \mathrm{ld}(2^{m+2} - 1) \rfloor = m + 1$.   □

As a result, we obtain very precise bounds:

Corollary 10. $k + \mathrm{fld}(k + 1) \leq \mathrm{nM}(k) \leq k + 1 + \mathrm{fld}(k)$ *holds for $k \in \mathbb{N}$.*

Proof. The lower bound follows trivially. The upper bound holds (with equality) for $k \leq 2$, so assume $k \geq 3$. We have to show $g(k) = \mathrm{fld}(k + 1 + \mathrm{fld}(k + 1)) \leq 1 + \mathrm{fld}(k)$, which follows from $\mathrm{ld}(k + 1 + \mathrm{fld}(k + 1)) \leq 1 + \mathrm{ld}(k)$. Now $\mathrm{ld}(k + 1 + \mathrm{fld}(k + 1)) \leq \mathrm{ld}(k + 1 + \mathrm{ld}(k + 1)) \leq \mathrm{ld}(k + k) = 1 + \mathrm{ld}(k)$.   □

4   Lean Clause-Sets and the Surplus

In this section we prove the main result of this paper, Theorem 11. The proof consists in first handling a special case, minimally unsatisfiable clause-sets instead of lean clause-sets, in Subsection 4.1, and then lifting the result to the general case in Subsection 4.2. In Subsection 4.3 we consider the algorithmic implications of this result.

Theorem 11. *We have $\mu\mathrm{vd}(F) \leq \mathrm{nM}(\sigma(F))$ for a lean multi-clause-set F with $n(F) > 0$. More precisely, there exists a variable $v \in \mathrm{var}(F)$ with $\mathrm{vd}_F(v) \leq \mathrm{nM}(\sigma(F))$ and $\mathrm{ld}_F(v), \mathrm{ld}_F(\overline{v}) \leq \sigma(F)$.*

We obtain a sufficient criterion for the existence of a non-trivial autarky.

Corollary 12. *Consider a multi-clause-set F with $n(F) > 0$. If $\sigma(F) \leq 0$, then F has a non-trivial matching autarky. So assume $\sigma(F) \geq 1$. If we have $\mu\mathrm{vd}(F) > \mathrm{nM}(\sigma(F))$, then for every $\emptyset \neq V \subseteq \mathrm{var}(F)$ with $\delta(F[V]) = \sigma(F)$ we have an autarky φ for F with $\mathrm{var}(\varphi) = V$ (and thus F has a non-trivial autarky).*

The quantities $\mu\mathrm{vd}(F)$ and $\mathrm{nM}(\sigma(F))$ (resp. $\mathrm{nM}(\delta(F))$) are computable in polynomial time, and so the applicability of Corollary 12 can be checked in polynomial time. We conjecture that also "constructivisation" of Corollary 12 can be done in polynomial time:

Conjecture 13. There is a poly-time algorithm for computing a non-trivial au-tarky in case of $\mu\mathrm{vd}(F) > \mathrm{nM}(\sigma(F))$ (or $\mu\mathrm{vd}(F) > \mathrm{nM}(\delta(F))$) for matching-lean clause-sets F.

See Subsection 4.3 for more discussion on Conjecture 13 (there also the remaining details of Corollary 12 are proven).

4.1   The Special Case of Minimally Unsatisfiable Clause-Sets

The main auxiliary lemma is the following statement, which receives its impor-tance from the fact that every minimally unsatisfiable clause-set can be saturated (this method was first applied in [6]).

Lemma 14. *Consider $F \in \mathcal{SMU}_{\delta=k}$ for $k \in \mathbb{N}$ and a variable $v \in \mathrm{var}(F)$ realising the minimal var-degree (i.e., $\mathrm{vd}_F(v) = \mu\mathrm{vd}(F)$). Using $m_0 := \mathrm{ld}_F(\overline{v})$ and $m_1 := \mathrm{ld}_F(v)$ we have $\langle v \to \varepsilon \rangle * F \in \mathcal{MU}_{k-m_\varepsilon+1}$ for $\varepsilon \in \{0,1\}$, where $n(\langle v \to \varepsilon \rangle * F) = n(F) - 1$. Since minimally unsatisfiable clause-sets have deficiency at least one, we get $m_\varepsilon \leq k$.*

Proof. We have $n(\langle v \to \varepsilon \rangle * F) = n(F) - 1$ since F contains no pure variable, while v realises the minimum of var-degrees. Thus $\delta(\langle v \to \varepsilon \rangle * F) = \delta(F) - m_\varepsilon + 1$, while $\langle v \to \varepsilon \rangle * F \in \mathcal{MU}$ by Lemma 1, Part 1.   □

Theorem 15. *For all $k \in \mathbb{N}$ and $F \in \mathcal{MU}_{\delta \leq k}$ we have $\mu\mathrm{vd}(F) \leq \mathrm{nM}(k)$. More precisely, for $n(F) > 0$ there exists a variable $v \in \mathrm{var}(F)$ with $\mathrm{vd}_F(v) \leq \mathrm{nM}(k)$ and $\mathrm{ld}_F(v), \mathrm{ld}_F(\overline{v}) \leq k$.*

Proof. The assertion is known for $k = 1$, so assume $k > 1$, and we apply in-duction on k. Assume $\delta(F) = k$ (due to $k > 1$ we have $n(F) > 1$). Satu-rate F and obtain F'. Consider a variable $v \in \mathrm{var}(F')$ realising the min-var-degree of F'. If $\mathrm{vd}_{F'}(v) = 2$ then we are done, so assume $\mathrm{vd}_{F'}(v) \geq 3$. Let $i := \max(\mathrm{ld}_{F'}(v), \mathrm{ld}_{F'}(\overline{v}))$; so $\mathrm{vd}_{F'}(v) \leq 2i$. W.l.o.g. assume that $i = \mathrm{ld}_{F'}(v)$. By Lemma 14 we get $2 \leq i \leq k$. Applying the induction hypothesis and Lemma 14 we obtain a variable $w \in \mathrm{var}(G)$ for $G := \langle v \to 1 \rangle * F$ with $\mathrm{vd}_G(w) \leq \mathrm{nM}(k - i + 1)$. By definition we have $\mathrm{vd}_{F'}(w) \leq \mathrm{vd}_G(w) + \mathrm{ld}_{F'}(v)$. Altogether we get $\mu\mathrm{vd}(F) \leq \min(2i, \mathrm{nM}(k - i + 1) + i) \leq \mathrm{nM}(k)$.   □

It is interesting to generalise Theorem 15 for generalised clause-sets (see [11,12] for a systematic study, and [10] for the underlying report). Generalised clause-sets have literals "$v \neq \varepsilon$" for variables v with domains D_v and values $\varepsilon \in D_v$, and the deficiency is generalised by giving every variable a weight $|D_v| - 1$ (which is 1 in the boolean case). The base case of deficiency $k = 1$ is handled in Lemma 5.4 in [12], showing that for generalised clause-sets we have here $\mu\mathrm{vd}(F) \leq \max_{v \in \mathrm{var}(F)} |D_v|$. But $k \geq 2$ requires more work:

1. The basic method of saturation is not available for generalised clause-sets, as discussed in Subsection 5.1 in [12]. Thus the proofs for the boolean case seem not to be generalisable.
2. Stipulating the effects of saturation via the "substitution stability parameter regarding irredundancy", in Corollary 5.10 in [12] one finds a first approach

towards generalising the basic bound $\mu\mathrm{vd}(F) \leq 2\delta(F)$ (for the boolean case) by $\mu\mathrm{vd}(F) \leq \max_{v \in \mathrm{var}(F)} |D_v| \cdot \delta(F)$.

3. Another approach uses translations to boolean clause-sets. The "generic translation scheme" (see [9,12]) allows (for certain instances) to preserve the deficiency and the other structures relevant here. So we get general upper bounds for the minimum number of occurrences of variables in generalised clause-sets from the boolean case. But further investigations are needed in these bounds.

4.2   Proof of the General Case

Now consider an arbitrary (multi-)clause-set F. Consider a set of variables $\emptyset \neq V \subseteq \mathrm{var}(F)$ realising the surplus of F, i.e., such that $\delta(F[V])$ is minimal. If $F[V]$ would be satisfiable, then a satisfying assignment would give a non-trivial autarky for F. Assuming that F is lean thus yields that $F[V]$ must be unsatisfiable. So there exists a minimally unsatisfiable $F' \subseteq F[V]$. If now $\mathrm{var}(F') \neq \mathrm{var}(F[V]) = V$ would be the case, then we would loose control over the deficiency of F'. Fortunately this can not happen, as the following lemma shows.

Lemma 16. *Consider a multi-clause-set F with $\sigma(F) = \delta(F)$. Then for every unsatisfiable sub-multi-clause-set $F' \leq F$ we have $\mathrm{var}(F') = \mathrm{var}(F)$.*

Proof. Assume $\mathrm{var}(F') \subset \mathrm{var}(F)$, and consider a minimally unsatisfiable sub-clause-set $F'' \subseteq F'$. By definition we have $\delta(F'') + \delta(F[\mathrm{var}(F) \backslash \mathrm{var}(F'')]) \leq \delta(F)$, where $\delta(F[\mathrm{var}(F) \backslash \mathrm{var}(F'')]) \geq \sigma(F) = \delta(F)$, from which we conclude $\delta(F'') \leq 0$, but $\delta(F'') \geq 1$ must hold since F'' is minimally unsatisfiable. □

Finally we are able to prove Theorem 11. Recall that F is a lean multi-clause-set with $n(F) > 0$, and we have to show the existence of a variable v with $\mathrm{vd}_F(v) \leq \mathrm{nM}(\sigma(F))$ and $\mathrm{ld}_F(v), \mathrm{ld}_F(\bar{v}) \leq \sigma(F)$.

Consider $\emptyset \neq V \subseteq \mathrm{var}(F)$ with $\delta(F[V]) = \sigma(F)$, and let $F' := F[V]$. F' is unsatisfiable, since F is lean. Because of $\delta(F') = \sigma(F)$ we have $\delta(F') = \sigma(F')$. Consider some minimally unsatisfiable $F'' \subseteq F'$. By Lemma 16 we have $\mathrm{var}(F'') = \mathrm{var}(F')$. So we get $\delta(F'') = \delta(F') - (c(F') - c(F''))$. By Theorem 15 there is $v \in \mathrm{var}(F'')$ with $\mathrm{vd}_{F''}(v) \leq \mathrm{nM}(\delta(F'')) = \mathrm{nM}(\delta(F') - (c(F') - c(F''))) \leq \mathrm{nM}(\delta(F')) - (c(F') - c(F''))$ and $\mathrm{ld}_{F''}(v), \mathrm{ld}_{F''}(\bar{v}) \leq \delta(F'') = \delta(F') - (c(F') - c(F''))$. Finally we have $\mathrm{vd}_F(v) \leq \mathrm{vd}_{F''}(v) + (c(F') - c(F''))$ (note that all occurrences of v in F are also in F'), and similarly for the literal degrees. QED

Corollary 17. $\mu\mathrm{vd}(F) \leq \mathrm{nM}(\delta(F))$ *for lean multi-clause-sets F with $n(F) > 0$.*

Corollary 18. *Consider a lean multi-clause-set F.*

1. *$\sigma(F) = 1$ holds if and only if $\mu\mathrm{vd}(F) = 2$ holds.*
2. *$\mu\mathrm{vd}(F) = 3$ implies $\sigma(F) = 2$.*
3. *$\sigma(F) = 2$ implies $\mu\mathrm{vd}(F) \in \{3, 4\}$.*

Proof. First consider Part 1. If $\sigma(F) = 1$ (so $n(F) > 0$), then by Theorem 11 we have $\mu\mathrm{vd}(F) \leq \mathrm{nM}(1) = 2$, while in case of $\mu\mathrm{vd}(F) = 1$ there would be a matching autarky for F. If on the other hand $\mu\mathrm{vd}(F) = 2$ holds, then by definition $\sigma(F) \leq 2 - 1 = 1$, while $\sigma(F) \geq 1$ holds since F is matching lean. For Part 2 note that due to $\sigma(F) + 1 \leq \mu\mathrm{vd}(F)$ we have $\sigma(F) \leq 2$, and then the assertion follows by Part 1. Part 3 also follows by Part 1. □

Remarks

1. An example for $\mu\mathrm{vd}(F) = 4$ in Part 3 is given by the full unsatisfiable clause-set with 2 variables.
2. Is there a minimally unsatisfiable F with $\mu\mathrm{vd}(F) = 4$ and $\sigma(F) = 3$?
3. More generally, is there for every $k \in \mathbb{N}$ a minimally unsatisfiable F with $\sigma(F) = k$ and $\mu\mathrm{vd}(F) = k+1$?

4.3   On Finding the Autarky

The following lemma together with Theorem 11 yields the proof of Corollary 12:

Lemma 19. *Consider a matching-lean multi-clause-set F with $n(F) > 0$. If we have $\mu\mathrm{vd}(F) > \mathrm{nM}(\sigma(F))$, then all $F[V]$ for $\emptyset \subset V \subseteq \mathrm{var}(F)$ with $\delta(F[V]) = \sigma(F)$ are satisfiable.*

Proof. If some $F[V]$ would be unsatisfiable, then by the proof of Theorem 11 in Subsection 4.2 there would be a variable v with $\mathrm{vd}_F(v) \leq \mathrm{nM}(\sigma(F))$. $\square$

Now consider a matching-lean multi-clause-set F with $n(F) > 0$, where Corollary 12 is applicable (recall that we have $\sigma(F) \geq 1$), that is, we have $\mu\mathrm{vd}(F) > \mathrm{nM}(\sigma(F))$. So we know that F has a non-trivial autarky. Conjecture 13 states that finding such a non-trivial autarky in this case can be done in polynomial time (recall that finding a non-trivial autarky in general is NP-complete, which was shown in [7]).

The task of actually finding the autarky can be considered as finding a satisfying assignment for the following class $\mathcal{MLCR} \subset \mathcal{SAT} \cap \mathcal{MLEAN}$ of satisfiable(!) clause-sets F, obtained by considering all $F[V]$ for minimal sets of variables V with $\delta(F[V]) = \sigma(F)$ (where "CR" stands for "critical"):

Definition 20. *Let $\mathcal{MLCR}$ be the class of clause-sets F fulfilling the following three conditions:*

1. *F is matching-lean, has at least one variable, and does not contain the empty clause.*
2. *The only $\emptyset \neq V \subseteq \mathrm{var}(F)$ with $\delta(F[V]) = \sigma(F)$ is $V = \mathrm{var}(F)$ (and thus we have $\delta(F) = \sigma(F)$).*
3. *$\mu\mathrm{vd}(F) > \mathrm{nM}(\sigma(F))$.*

It is sufficient to find a non-trivial autarky for this class of satisfiable clause-sets. Constructing elements of $\mathcal{MLCR}$ seems a non-trivial task.

Lemma 21. *Conjecture 13 is equivalent to the statement, that finding a non-trivial autarky for clause-sets in $\mathcal{MLCR}$ can be achieved in polynomial time.*

5   On Strengthening the Bound

For a class $\mathcal{C}$ of clause-sets let $\mu\mathrm{vd}(\mathcal{C})$ be the supremum of $\mu\mathrm{vd}(F)$ for $F \in \mathcal{C}$ with $n(F) > 0$. So by Theorem 15 we have $\mu\mathrm{vd}(\mathcal{MU}_{\delta=k}) \leq \mathrm{nM}(k)$ for all $k \in \mathbb{N}$.

The task of precisely determining $\mu\mathrm{vd}(\mathcal{MU}_{\delta=k})$ for all k will be pursued in the forthcoming [13]; we need more theory for minimally unsatisfiable clause-sets (especially for unsatisfiable hitting clause-sets), and so here we can only mention some results connected with this article.

- We can show for infinitely many k that $\mu\mathrm{vd}(\mathcal{MU}_{\delta=k}) = \mathrm{nM}(k)$.
- We can also show that the smallest k where we don't have equality is $k = 6$, namely $\mu\mathrm{vd}(\mathcal{MU}_{\delta=6}) = 8 = \mathrm{nM}(6) - 1$.
- Let $\mathrm{nM}_1 : \mathbb{N} \to \mathbb{N}$ be defined by the recursion as in Definition 2, however with different start values, namely $\mathrm{nM}_1(k) := \mathrm{nM}(k)$ for $1 \le k \le 5$, while $\mathrm{nM}_1(6) := \mathrm{nM}(6) - 1 = 8$. We have $\mathrm{nM}_1(k) = \mathrm{nM}(k)$ for $k \notin \{2^m - m + 1 : m \in \mathbb{N}, m \ge 3\}$, while for $k = 2^m - m + 1$ we have $\mathrm{nM}_1(k) = \mathrm{nM}(k) - 1 = 2^m$.
- With the same proof as for Theorem 15 we can show $\mu\mathrm{vd}(\mathcal{MU}_{\delta=k}) \le \mathrm{nM}_1(k)$ for all $k \in \mathbb{N}$.
- It seems that this bound can not be generalised to lean clause-sets (as in Theorem 11).

Conjecture 22. For all $k \in \mathbb{N}$ we have $\mu\mathrm{vd}(\mathcal{MU}_{\delta=k}) \ge \mathrm{nM}(k) - 1$.

Conjecture 23. For all $k \in \mathbb{N}$ we have $\mu\mathrm{vd}(\mathcal{LEAN}_{\delta=k}) = \mathrm{nM}(k)$.

Now we consider the question whether the bound holds for a larger class of clause-sets, that is, whether Theorem 11 can be generalised further, incorporating non-lean clause-sets. We consider the large class $\mathcal{MLEAN}$ of matching lean clause-sets, as introduced in [7], which is natural, since a basic property of $F \in \mathcal{MU}$ used in the proof of Theorem 11 is $\delta(F) \ge 1$ for $F \ne \top$, and this actually holds for all $F \in \mathcal{MLEAN}$. We will construct for arbitrary deficiency $k \in \mathbb{N}$ and $K \in \mathbb{N}$ clause-sets $F \in \mathcal{MLEAN}$ of deficiency k where every variable occurs positively at least K times. Thus neither the upper bound $\max(\mathrm{ld}_F(v), \mathrm{ld}_F(\overline{v})) \le f(\delta(F))$ nor $\mathrm{ld}_F(v) + \mathrm{ld}_F(\overline{v}) = \mathrm{vd}_F(v) \le f(\delta(F))$ for some chosen variable v and for any function f does hold for $\mathcal{MLEAN}$.

An example for $F \in \mathcal{MLEAN}_{\delta=1}$ with $\mu\mathrm{ld}(F) \ge 2$ (and thus $\mu\mathrm{vd}(F) \ge 4$) is given in Section 5 in [8], displaying a "star-free" (thus satisfiable) clause-set F with deficiency 1. In Subsection 9.3 in [11] it is shown that this clause-set is matching lean. "Star-freeness" in our context means, that there are no singular variables (occurring in one sign only once). Our simpler construction pushes the number of positive occurrences arbitrary high, but there are variables with only one negative occurrence (i.e., there are singular variables).

For a finite set V of variables let $M(V) \subseteq A(V)$ be the full clause-set over V containing all full clauses with at most one complementation. Obviously $\delta(F) = 1$ holds, and it is easy to see that $M(V) \in \mathcal{MLEAN}$ (for every $\emptyset \ne F' \subset F \subseteq A(V)$ we have $\delta(F') < \delta(F)$, and thus a full clause-set F is matching lean iff $\delta(F) \ge 1$). Furthermore by definition we have $\mathrm{ld}_{M(V)}(v) = |V|$ and $\mathrm{ld}_{M(V)}(\overline{v}) = 1$ for $v \in V$.

Lemma 24. *For $k \in \mathbb{N}$ and $K \in \mathbb{N}$ there are clause-sets $F \in \mathcal{MLEAN}_{\delta=k}$ such that for all variables $v \in \mathrm{var}(F)$ we have $\mathrm{ld}_F(v) \ge K$.*

Proof. For $k = 1$ we can set $F := M(\{v_1, \ldots, v_K\})$; so assume $k \geq 2$. Consider any clause-set $G \in \mathcal{MLEAN}_{\delta=k-1}$ with $n := n(G) \geq K$ (for example we could use $F \in \mathcal{MU}_{\delta=k-1}$), and let $V := \mathrm{var}(G)$. Consider a disjoint copy of V, that is a set V' of variables with $V' \cap V = \emptyset$ and $|V'| = |V|$, and consider two enumerations of the clauses $M(V) = \{C_1, \ldots, C_{n+1}\}$, $M(V') = \{C'_1, \ldots, C'_{n+1}\}$. Now

$$F := G \cup \{ C_i \cup C'_i : i \in \{1, \ldots, n+1\} \}$$

has no matching autarky: If φ is a matching autarky for F, then $\mathrm{var}(\varphi) \cap V = \emptyset$ since G is matching lean, whence $\mathrm{var}(\varphi) \cap V' = \emptyset$ since $M(V')$ is matching lean, and thus φ must be trivial. Furthermore we have $n(F) = 2n$ and $c(F) = c(G) + n + 1$, and thus $\delta(F) = c(G) + n + 1 - 2n = \delta(G) + 1 = k$. By definition for all variables $v \in \mathrm{var}(F)$ we have $\mathrm{ld}_F(v) \geq n$. □

Remarks

1. It is open whether for deficiency $k \in \mathbb{N}$ there are examples $F \in \mathcal{MLEAN}_{\delta=k}$ with $\mu\mathrm{ld}(F) \geq k + 1$ (the above mentioned star-free clause-sets shows that this is the case for $k = 1$), or stronger, $\mu\mathrm{ld}(F) \geq K$ for arbitrary $K \in \mathbb{N}$.
2. The clause-sets F constructed in Lemma 24 are not elements of $\mathcal{MLCR}_{\delta=k}$ for $k \geq 2$, since $\delta(F[V']) = n + 1 - n = 1$, thus $\sigma(F) = 1$, and so Condition 2 of Definition 20 is not fulfilled. The corresponding autarky is a satisfying assignment of $M(V')$, which is easy to find.

6   Conclusion and Open Problems

We have shown the upper bound $\mu\mathrm{vd}(F) \leq \mathrm{nM}(\sigma(F))$ for lean clause-sets (Theorem 11). The function $\mathrm{nM}(k)$ has been characterised in Lemma 9 and Corollary 10. We presented first initial results regarding the sharpness of the bound and regarding the constructive aspects of the bound (i.e., what happens if the bound is violated). There remain several open problems:

1. Prove Conjecture 13, which says that such an autarky, which must exist if a clause-set does not fulfil the upper bound on the minimum variable degree of Theorem 11, can be found in polynomial time. See Subsection 4.3 for more information on this topic.
2. Generalise Theorem 15 to clause-sets with non-boolean variables; see the discussion after Theorem 15.
3. See the remarks to Corollary 18 (an underlying question is to understand better the quantity "surplus").
4. Strengthen the bound on the minimum variable degree according to Conjectures 22, 23 (see the forthcoming [13]).
5. Strengthen the construction of Lemma 24 (perhaps completely different constructions are needed).

As mentioned in the introduction, a major motivation for us is the project of the classification of minimally unsatisfiable clause-sets for deficiencies $\delta =$

$1, 2, \ldots$ Especially the classification of unsatisfiable hitting clause-sets in dependency on the deficiency seems very interesting (recall that a hitting clause-set F is defined by the condition that every two clauses $C, C' \in F$, $C \neq C'$, clash in at least one variable, that is $|C \cap \overline{C'}| \geq 1$). The main conjecture is:

Conjecture 25. For every deficiency $k \in \mathbb{N}$ there are only finitely many isomorphism types of non-singular unsatisfiable hitting clause-sets.

For $k \leq 2$ this conjecture follows from known results, while recently we were able to prove it for $k = 3$.

References

1. Aharoni, R., Linial, N.: Minimal non-two-colorable hypergraphs and minimal unsatisfiable formulas. Journal of Combinatorial Theory, A 43, 196–204 (1986)
2. Davydov, G., Davydova, I., Büning, H.K.: An efficient algorithm for the minimal unsatisfiability problem for a subclass of CNF. Annals of Mathematics and Artificial Intelligence 23, 229–245 (1998)
3. Gebauer, H., Szabo, T., Tardos, G.: The local lemma is tight for SAT. Technical Report arXiv:1006.0744v1 [math.CO], arXiv.org (June 2010)
4. Büning, H.K.: On subclasses of minimal unsatisfiable formulas. Discrete Applied Mathematics 107, 83–98 (2000)
5. Büning, H.K., Kullmann, O.: Minimal unsatisfiability and autarkies. In: Biere, A., Heule, M.J.H., van Maaren, H., Walsh, T. (eds.) Handbook of Satisfiability. Frontiers in Artificial Intelligence and Applications, vol. 185, ch.11, pp. 339–401. IOS Press, Amsterdam (2009)
6. Kullmann, O.: An application of matroid theory to the SAT problem. In: Fifteenth Annual IEEE Conference on Computational Complexity (2000), pp. 116–124. IEEE Computer Society, Los Alamitos (July 2000)
7. Kullmann, O.: Lean clause-sets: Generalizations of minimally unsatisfiable clause-sets. Discrete Applied Mathematics 130, 209–249 (2003)
8. Kullmann, O.: On some connections between linear algebra and the combinatorics of clause-sets. In: Franco, J., Giunchiglia, E., Kautz, H., Büning, H.K., van Maaren, H., Selman, B., Speckenmeyer, E. (eds.) Sixth International Conference on Theory and Applications of Satisfiability Testing, Santa Margherita Ligure – Portofino, Italy, May 5-8, pp. 45–59 (2003)
9. Kullmann, O.: Green-tao numbers and SAT. In: Strichman, O., Szeider, S. (eds.) SAT 2010. LNCS, vol. 6175, pp. 352–362. Springer, Heidelberg (2010)
10. Kullmann, O.: Constraint satisfaction problems in clausal form. Technical Report arXiv:1103.3693v1 [cs.DM], arXiv (March 2011)
11. Kullmann, O.: Constraint satisfaction problems in clausal form I: Autarkies and deficiency. Fundamenta Informaticae 109 (to appear, 2011)
12. Kullmann, O.: Constraint satisfaction problems in clausal form II: Minimal unsatisfiability and conflict structure. Fundamenta Informaticae 109 (to appear, 2011)
13. Kullmann, O., Zhao, X.: Bounds for variables with few occurrences in conjunctive normal forms (in preparation, 2011)
14. Tovey, C.A.: A simplified NP-complete satisfiability problem. Discrete Applied Mathematics 8, 85–89 (1984)
15. Zhao, X., Decheng, D.: Two tractable subclasses of minimal unsatisfiable formulas. Science in China (Series A) 42(7), 720–731 (1999)

Satisfiability of Acyclic and almost Acyclic CNF Formulas (II)[*]

Sebastian Ordyniak[1], Daniel Paulusma[2], and Stefan Szeider[1]

[1] Institute of Information Systems, Vienna University of Technology, A-1040 Vienna, Austria
sebastian.ordyniak@kr.tuwien.ac.at, stefan@szeider.net
[2] School of Engineering and Computing Sciences, Durham University, Durham, DH1 3LE, UK
daniel.paulusma@durham.ac.uk

Abstract. In the first part of this work (FSTTCS'10) we have shown that the satisfiability of CNF formulas with β-acyclic hypergraphs can be decided in polynomial time. In this paper we continue and extend this work. The decision algorithm for β-acyclic formulas is based on a special type of Davis-Putnam resolution where each resolvent is a subset of a parent clause. We generalize the class of β-acyclic formulas to more general CNF formulas for which this type of Davis-Putnam resolution still applies. We then compare the class of β-acyclic formulas and this superclass with a number of known polynomial formula classes.

1   Introduction

We continue our study [12] of the SATISFIABILITY (SAT) problem on classes of CNF formulas (formulas in Conjunctive Normal Form) with restrictions on their associated hypergraphs, which are obtained from these formulas by ignoring negations and considering clauses as hyperedges on variables.

Because many computationally hard problems can be solved efficiently on acyclic instances, it is a natural to consider SAT for CNF formulas with acyclic hypergraphs. There are several notions of acyclicity for hypergraphs as described by Fagin [6]: α-acyclicity, β-acyclicity, γ-acyclicity, and Berge-acyclicity, which are strictly ordered with respect to their generality, i.e., we have

$$\alpha\text{-}\text{ACYC} \supsetneq \beta\text{-}\text{ACYC} \supsetneq \gamma\text{-}\text{ACYC} \supsetneq \text{Berge-}\text{ACYC}$$

where X-ACYC denotes the class of X-acyclic hypergraphs, which are in 1-to-1 correspondence to a class of CNF formulas called X-acyclic formulas. It is known that SAT is NP-complete for α-acyclic formulas [13], and that Berge-ACYC-SAT is solvable in polynomial time [7,13]. In a recent paper [12] we completed the complexity classification of these four classes by showing that SAT can be solved in polynomial time for β-acyclic formulas, and consequently, for γ-acyclic formulas as well.

New results. The first aim of our paper is to generalize our polynomial-time algorithm for β-acyclic formulas [12]. This algorithm is based on the so-called Davis-Putnam Procedure [5], which successively eliminates variables using Davis-Putnam Resolution. In

---

[*] Ordyniak and Szeider's research was funded by the ERC (COMPLEX REASON, 239962). Paulusma's research was funded by EPSRC (EP/G043434/1).

K.A. Sakallah and L. Simon (Eds.): SAT 2011, LNCS 6695, pp. 47–60, 2011.

general, this procedure is not efficient, because the number of clauses may increase after each application of Davis-Putnam Resolution. However, the special structure of β-acyclic formulas allows us to compute an elimination ordering of the variables, such that this does not happen. Hence, we can solve SAT in polynomial time for β-acyclic formulas. In fact, the elimination ordering produced this way has the special property that each obtained resolvent is a subset of a parent clause. This type of resolution is known as subsumption resolution [11]. In Section 3 we show that there are CNF formulas that are not β-acyclic but that still admit an elimination ordering of their variables based on subsumption resolution, such that the Davis-Putnam procedure takes polynomial time. We call such an elimination ordering *DP-simplicial*. This leads to a new class DPS of CNF formulas that contains the class of β-acyclic formulas. In Section 4, we show that testing membership in this class is an NP-complete problem. The reason for the NP-hardness is that a formula may have several so-called *DP-simplicial* variables, one of which must be chosen to be eliminated but we do not know which one. In Section 5, we show how to work around this obstacle to some extent, i.e., we identify a subclass of DPS that is a proper superclass of the class of β-acyclic formulas for which SAT is polynomial-time solvable.

The second aim of our paper is to make a comparison between the class of β-acyclic formulas and other known polynomial classes of CNF formulas. We do this in Section 6, and our results show that the class of β-acyclic formulas is *incomparable* with all considered classes. Hence, β-acyclic formulas form a new "island of tractability" for SAT.

2 Preliminaries

We assume an infinite supply of propositional *variables*. A *literal* is a variable x or a negated variable $\overline{x}$; if $y = \overline{x}$ is a literal, then we write $\overline{y} = x$. For a set S of literals we put $\overline{S} = \{\, \overline{x} \mid x \in S \,\}$; S is *tautological* if $S \cap \overline{S} \neq \emptyset$. A *clause* is a finite non-tautological set of literals. A finite set of clauses is a *CNF formula* (or *formula*, for short). A variable x *occurs* in a clause C if $x \in C \cup \overline{C}$; $\mathrm{var}(C)$ denotes the set of variables which occur in C. A variable x *occurs* in a formula F if it occurs in one of its clauses, and we put $\mathrm{var}(F) = \bigcup_{C \in F} \mathrm{var}(C)$.

Let F be a formula and $X \subseteq \mathrm{var}(F)$. A *truth assignment* is a mapping $\tau : X \to \{0,1\}$ defined on some set X of variables; we write $\mathrm{var}(\tau) = X$. For $x \in \mathrm{var}(\tau)$ we define $\tau(\overline{x}) = 1 - \tau(x)$. A truth assignment τ *satisfies* a clause C if C contains some literal x with $\tau(x) = 1$; τ satisfies a formula F if it satisfies all clauses of F. A formula is *satisfiable* if it is satisfied by some truth assignment; otherwise it is *unsatisfiable*. Two formulas F and F' are *equisatisfiable* if either both are satisfiable or both are unsatisfiable. The SATISFIABILITY (SAT) problem asks whether a given CNF formula is satisfiable.

Let C, D be two clauses such that $C \cap \overline{D} = \{x\}$ for a variable x. The clause $(C \cup D) \setminus \{x, \overline{x}\}$ is called the *x-resolvent* (or *resolvent*) of C and D; the clauses C and D are called *parent clauses* of the *x*-resolvent. Note that by definition any two clauses have at most one resolvent. Let F be a formula. A sequence $C_1, \dots, C_n$ is a *resolution derivation* of C_n from F if every C_i is either in F or the resolvent of two clauses C_j

and $C_{j'}$ for some $1 \leq j < j' \leq i - 1$. The derivation is *minimal* if we cannot delete a clause from it and still have a resolution derivation of C_n from F. We call a clause C_n a *resolution descendant* of a clause $C_1 \in F$ if there is a minimal resolution derivation $C_1, \ldots, C_n$ of C_n from F.

Consider a formula F and a variable x of F. Let $\mathrm{DP}_x(F)$ denote the formula obtained from F after adding all possible x-resolvents and removing all clauses in which x occurs. We say that $\mathrm{DP}_x(F)$ is obtained from F by *Davis-Putnam Resolution*, and that we *eliminated* x. It is well known (and easy to show) that F and $\mathrm{DP}_x(F)$ are equisatisfiable.

For an ordered sequence of variables $x_1, \ldots, x_k$ of F, we set $\mathrm{DP}_{x_1, \ldots, x_k}(F) = \mathrm{DP}_{x_k}(\cdots(\mathrm{DP}_{x_1}(F))\cdots)$ and $\mathrm{DP}_\emptyset(F) = F$. The Davis-Putnam Procedure [5] considers an ordering of the variables $x_1, \ldots, x_n$ of a formula F and checks whether $\mathrm{DP}_{x_1, \ldots, x_n}(F)$ is empty or contains the empty clause. In the first case F is satisfiable, and in the second case F is unsatisfiable. However, $\mathrm{DP}_x(F)$ contains in general more clauses than F. Hence, repeated application of Davis-Putnam Resolution to F may cause an exponential growth in the number of clauses. As a result, the Davis-Putnam Procedure has an exponential worst-case running time.

3   Generalizing β-Acyclic Formulas

A *hypergraph* H is a pair (V, E) where V is the set of *vertices* and E is the set of *hyperedges*, which are subsets of V. A hypergraph is α-*acyclic* if it can be reduced to the empty hypergraph $(\emptyset, \emptyset)$ by repeated application of the following reduction rules:

1. Remove hyperedges that are empty or contained in other hyperedges.
2. Remove vertices that appear in at most one hyperedge.

A hypergraph H is β-*acyclic* if it is α-acyclic and remains α-acyclic after removing an arbitrary number of hyperedges. Thus β-acyclicity is the hereditary variant of α-acyclicity. The *hypergraph $H(F)$* of a formula F has vertex set $\mathrm{var}(F)$ and hyperedge set $\{\, \mathrm{var}(C) \mid C \in F \,\}$. We say that F is α-*acyclic* or β-*acyclic* if $H(F)$ is α-acyclic or β-acyclic, respectively. It is known that SAT is NP-complete for the class of α-acyclic formulas [13]. However, β-acyclicity makes SAT polynomial.

Theorem 1 ([12]). SAT *can be solved in polynomial time for β-acyclic formulas.*

The proof of Theorem 1 is based on the following [12]. A vertex x of a hypergraph H is *weakly simplicial* if the hyperedges of H that contain x form a chain under set inclusion. A nontrivial β-acyclic hypergraph always contains a weakly simplicial vertex. After deletion of this vertex the hypergraph remains β-acyclic. Thus, by repeated deletion of weakly simplicial vertices we can eliminate all vertices of a β-acyclic hypergraph, producing a *weakly simplicial elimination ordering* of its vertices. Because we can find a weakly simplicial vertex in polynomial time, we can compute a weakly simplicial elimination ordering for a β-acyclic hypergraph in polynomial time. Once we have this ordering, we apply the Davis-Putnam procedure. This results in a sequence of formulas with a *non-increasing* number of clauses. As such, the Davis-Putnam procedure runs in polynomial time. Consequently, Theorem 1 holds.

Besides that it is possible to identify a "suitable" vertex in polynomial time, the other key observation in the proof of Theorem 1 is that the number of clauses must not increase by applying Davis-Putnam resolution. We can ensure this by requiring the following property that is more general than being weakly simplicial. Let F be a formula. We say that a variable $x \in \text{var}(F)$ is *DP-simplicial* in F if

(*) for any two clauses $C, D \in F$ that have an x-resolvent, this x-resolvent is a subset of C or a subset of D.

We observe that whenever an x-resolvent is a subset of a parent clause C then it is equal to $C \setminus \{x, \overline{x}\}$. If x is DP-simplicial in F, then $|\text{DP}_x(F)| \leq |F|$, as desired. An ordering $x_1, \ldots, x_n$ of the variables of F is a *DP-simplicial elimination ordering* if x_i is DP-simplicial in $\text{DP}_{x_1,\ldots,x_{i-1}}(F)$ for all $1 \leq i \leq n$. We let DPS denote the class of all formulas that admit a DP-simplicial elimination ordering, and we let BAC denote the class of all β-acyclic formulas. We observe that every weakly simplicial elimination ordering of $H(F)$ is a DP-simplicial elimination ordering of F. This means that BAC $\subseteq$ DPS. However, due to Example 3.1, the reverse is not true. Hence, DPS is a proper superclass of BAC.

Given an DP-simplicial ordering, the Davis-Putnam procedure runs in polynomial time. Hence we obtain the following result.

Proposition 1. *Let $F \in$ DPS. If a DP-simplicial elimination ordering of the variables in $\text{var}(F)$ is given, then* SAT *can be solved in polynomial time for F.*

In fact, if a DP-simplicial elimination ordering of the variables in $\text{var}(F)$ is given, we can even compute a certificate for the (un)satisfiability of F in polynomial time. This holds, because we can obtain a satisfying truth assignment of F from a satisfying truth assignment of $\text{DP}_x(F)$, and we can obtain a resolution refutation of F from a resolution refutation of $\text{DP}_x(F)$.

3.1 An Example

We give an example of a formula in DPS $\setminus$ BAC. Consider the formula F that has variables y, z, b, b', b^* and c and clauses $\{y, z, \overline{b}, b'\}, \{\overline{y}, \overline{z}, \overline{b}, b^*\}, \{y, \overline{b}\}, \{\overline{y}, \overline{b}\}, \{z, \overline{b}\}, \{\overline{z}, \overline{b}\}, \{y, b, b^*, c\}, \{\overline{y}, b, b', \overline{c}\}, \{\overline{b'}, b^*\}, \{\overline{b^*}, b'\}, \{c, b', b^*\}, \{\overline{c}, b', b^*\}$ and $\{\overline{b}, b'\}$.

We observe first that none of the variables of F are weakly simplicial. Consequently, there is no weakly simplicial elimination ordering of F. Hence $F \notin$ BAC. However, we will show below that y, b, b', b^*, c, z is a DP-simplicial elimination ordering of F. Then $F \in$ DPS, as desired.

We find that y is DP-simplicial in F and obtain $\text{DP}_y(F) = \{\{z, \overline{b}, b'\}, \{\overline{z}, \overline{b}, b^*\}, \{z, \overline{b}\}, \{\overline{z}, \overline{b}\}, \{\overline{b'}, b^*\}, \{\overline{b^*}, b'\}, \{c, b', b^*\}, \{\overline{c}, b', b^*\}, \{\overline{b}, b'\}\}$. We then find that b is DP-simplicial in $\text{DP}_y(F)$ and obtain $\text{DP}_{y,b}(F) = \{\{\overline{b'}, b^*\}, \{\overline{b^*}, b'\}, \{c, b', b^*\}, \{\overline{c}, b', b^*\}\}$. We then find that b' is DP-simplicial in $\text{DP}_{y,b}(F)$ and obtain $\text{DP}_{y,b,b'}(F) = \{\{c, b', b^*\}, \{\overline{c}, b', b^*\}\}$. We then find that b^* is DP-simplicial in $\text{DP}_{y,b,b'}(F)$ and obtain $\text{DP}_{y,b,b',b^*}(F) = \emptyset$. Hence, y, b, b', b^*, c, z is a DP-simplicial elimination ordering of F.

We note that z is also DP-simplicial in F. Suppose that we started with z instead of y. We first derive that $\text{DP}_z(F) = \{\{y, \overline{b}, b'\}, \{\overline{y}, \overline{b}, b^*\}, \{y, \overline{b}\}, \{\overline{y}, \overline{b}\}, \{y, b, b^*, c\},$

$\{\overline{y}, b, b', \overline{c}\}$, $\{\overline{b'}, b^*\}$, $\{\overline{b^*}, b'\}$, $\{c, b', b^*\}$, $\{\overline{c}, b', b^*\}$, $\{\overline{b}, b'\}\}$. In contrast to $DP_y(F)$, the clauses $\{y, b, b^*, c\}$ and $\{\overline{y}, b, b', \overline{c}\}$ are still contained in $DP_z(F)$. This implies that $DP_z(F)$ has no DP-simplicial variables. Consequently, F has no DP-simplicial elimination ordering that starts with z.

We conclude that in contrast to weakly simplicial elimination orderings it is important to choose the right variable when we want to obtain a DP-simplicial elimination ordering. In the next section we will extend this consideration and show that making the right choice is in fact an NP-hard problem.

4   The NP-Completeness Result

We prove that the problem of testing whether a given CNF formula belongs to the class DPS, i.e., admits a DP-simplicial elimination ordering, is NP-complete. This problem is in NP, because we can check in polynomial time whether an ordering of the variables of a CNF formula is a DP-simplicial elimination ordering. In order to show NP-hardness we reduce from SATISFIABILITY. In Section 4.1 we construct a CNF formula F' from a given CNF formula F. We also show a number of properties of F'. In Section 4.2 we use these properties to prove that F is satisfiable if and only if F' admits a DP-simplicial elimination ordering.

4.1   The Gadget and Its Properties

For a given CNF formula F with variables $x_1, \ldots, x_n$ called the x-*variables* and clauses $C_1, \ldots, C_m$, we construct a CNF formula F' as follows. For every x_i we introduce two variables y_i and z_i. We call these variables the y-variables and z-variables, respectively. For every C_j we introduce a variable c_j. We call these variables the c-variables. We also add three new variables b, b' and b^* called the b-variables. We let $var(F')$ consist of all b-variables, c-variables, y-variables, and z-variables.

Let C_j be a clause of F. We replace every x-variable in C by its associated y-variable if the occurrence of x in C is positive; otherwise we replace it by its associated z-variable. This yields a clause D_j. For instance, if $C_j = \{x_1, \overline{x_2}, x_3\}$ then $D_j = \{y_1, z_2, y_3\}$.

We let F' consist of the following $6n + 4m + 3$ clauses:

- $\{y_i, \overline{b}\}$ and $\{\overline{y_i}, \overline{b}\}$ for $i = 1, \ldots, n$ called by-clauses

- $\{z_i, \overline{b}\}$ and $\{\overline{z_i}, \overline{b}\}$ for $i = 1, \ldots, n$ called bz-clauses

- $\{y_i, z_i, \overline{b}, b'\}$ and $\{\overline{y_i}, \overline{z_i}, \overline{b}, b^*\}$ for $i = 1, \ldots, n$ called byz-clauses

- $\{c_j, b', b^*\}$ and $\{\overline{c_j}, b', b^*\}$ for $j = 1, \ldots, m$ called bc-clauses

- $D_j \cup \{b, b^*, c_j\} \cup \{\overline{c_k} \mid k \neq j\}$ and $\overline{D_j} \cup \{b, b', c_j\} \cup \{\overline{c_k} \mid k \neq j\}$ for $j = 1, \ldots, m$ called bcD-clauses

- $\{\overline{b}, b'\}$, $\{\overline{b'}, b^*\}$ and $\{b', \overline{b^*}\}$ called b-clauses.

We call a pair $D_j \cup \{b, b^*, c_j\} \cup \{\overline{c_k} \mid k \neq j\}$ and $\overline{D_j} \cup \{b, b', c_j\} \cup \{\overline{c_k} \mid k \neq j\}$ for some $1 \leq j \leq m$ a bcD-*clause pair*. We call a CNF formula M a yz-*reduction* *formula* of F' if there exists a sequence of variables $v^1, \ldots, v^k$, where every v^i is either a y-variable or a z-variable, such that $DP_{v^1,\ldots,v^k}(F') = M$, and v^i is DP-simplicial in $DP_{v^1,\ldots,v^{i-1}}(F')$ for $i = 1, \ldots, k$. We say that two clauses C and D *violate* (*) if they have a resolvent that is neither a subset of C nor a subset of D, i.e., $C \cap \overline{D} = \{v\}$ for some variable v but neither $(C \cup D) \setminus \{v, \overline{v}\} = C \setminus \{v\}$ nor $(C \cup D) \setminus \{v, \overline{v}\} = D \setminus \{\overline{v}\}$. We will now prove five useful lemmas valid for yz-reduction formulas.

Lemma 1. *Let M be a yz-reduction formula of F'. If M contains both clauses of some bcD-clause pair, then no b-variable and no c-variable is DP-simplicial in M.*

Proof. Let $E_1 = D_j \cup \{b, b^*, c_j\} \cup \{\overline{c_k} \mid k \neq j\}$ and $E_2 = \overline{D_j} \cup \{b, b', c_j\} \cup \{\overline{c_k} \mid k \neq j\}$ for some $1 \leq j \leq m$ be a bcD-clause pair in M. We observe that by definition M contains all b-clauses and bc-clauses. This enables us to prove the lemma. Let v be a b-variable or c-variable. Then we must distinguish 5 cases. If $v = b$, then $\{\overline{b}, b'\}$ and E_1 violate (*). If $v = b'$, then $\{\overline{b'}, b^*\}$ and E_2 violate (*). If $v = b^*$, then $\{b', \overline{b^*}\}$ and E_1 violate (*). If $v = c_j$, then $\{\overline{c_j}, b', b^*\}$ and E_1 violate (*). If $v = c_k$ for some $1 \leq k \leq m$ with $k \neq j$, then $\{c_k, b', b^*\}$ and E_1 violate (*). $\square$

Lemma 2. *Let M be a yz-reduction formula of F'. Then $y_i \in var(M)$ or $z_i \in var(M)$ for $i = 1, \ldots, n$.*

Proof. Suppose that M does not contain y_i or z_i for some $1 \leq i \leq m$, say $y_i \notin var(M)$. We show that $z_i \in var(M)$. Let M' be the formula obtained from F' just before the elimination of y_i. Because M is a yz-reduction formula, M' is a yz-reduction formula as well. Hence, $var(M')$ contains all b-variables. Because y_i and z_i are in $var(M')$, we then find that M' contains the clauses $\{y_i, z_i, \overline{b}, b'\}$, $\{\overline{y_i}, \overline{b}\}$, $\{\overline{y_i}, \overline{z_i}, \overline{b}, b^*\}$ and $\{y_i, \overline{b}\}$. Because the first two clauses resolve into $\{z_i, \overline{b}, b'\}$, and the last two resolve into $\{\overline{z_i}, \overline{b}, b^*\}$, we obtain that $DP_{y_i}(M')$ contains $\{z_i, \overline{b}, b'\}$ and $\{\overline{z_i}, \overline{b}, b^*\}$, which violate (*). Because M contains all b-variables by definition, z_i will never become DP-simplicial when we process $DP_{y_i}(M')$ until we obtain M. Hence, $z_i \in var(M)$, as desired. $\square$

Lemma 3. *Let M be a yz-reduction formula of F', and let $1 \leq j \leq m$. If there is a variable that occurs in D_j but not in M, then M neither contains $D_j \cup \{b, b^*, c_j\} \cup \{\overline{c_k} \mid k \neq j\}$ nor $\overline{D_j} \cup \{b, b', c_j\} \cup \{\overline{c_k} \mid k \neq j\}$ nor their resolution descendants.*

Proof. Let v be a variable that occurs in D_j but not in M. We may assume without loss of generality that v is the first variable in D_j that got eliminated and that $v = y_i$ for some $1 \leq i \leq n$. Let S be the set that consists of all clauses $D_{j'} \cup \{b, b^*, c_{j'}\} \cup \{\overline{c_k} \mid k \neq j'\}$ and $\overline{D_{j'}} \cup \{b, b', c_{j'}\} \cup \{\overline{c_k} \mid k \neq j'\}$ in which y_i occurs.

Let M' be the formula obtained from F' just before the elimination of y_i. Because M is a yz-reduction formula, M' is a yz-reduction formula as well. Hence, by definition, all b-variables and all c-variables occur in M'. Then the clauses in M', in which y_i occurs, are $\{y_i, \overline{b}\}$, $\{\overline{y_i}, \overline{b}\}$, $\{y_i, z_i, \overline{b}, b'\}$, $\{\overline{y_i}, \overline{z_i}, \overline{b}, b^*\}$, together with clauses that are either from S or a resolution descendant of a clause in S. Note that these resolution descendants still contain all their b-variables and c-variables.

When we eliminate y_i, we remove all clauses in M' in which y_i occurs. Hence, $DP_{y_i}(M')$, and consequently, M neither contains $E_1 = D_j \cup \{b, b^*, c_j\} \cup \{\overline{c_k} \mid k \neq j\}$

nor $E_2 = \overline{D_j} \cup \{b, b', c_j\} \cup \{\overline{c_k} \mid k \neq j\}$. We show that $\mathrm{DP}_{y_i}(M')$ does not contain a resolvent of one of these two clauses either. This means that M' does not contain one of their resolution descendants, as desired. We only consider E_1, because we can deal with E_2 in the same way. There is no y_i-resolvent of E_1 and a clause C from $\{\{y_i, \overline{b}\}, \{\overline{y_i}, \overline{b}\}, \{y_i, z_i, \overline{b}, b'\}, \{\overline{y_i}, \overline{z_i}, \overline{b}, b^*\}\}$, because $E_1 \cap \overline{C}$ contains b. There is no y_i-resolvent of E_1 and a (resolution descendant from a) clause C of S either, because $E_1 \cap \overline{C}$ contains c_j. □

Lemma 4. *Let M be a yz-reduction formula of F', and let $1 \leq i \leq n$. If $var(M)$ contains y_i and z_i, then both y_i and z_i are DP-simplicial in M.*

Proof. By symmetry, we only have to show that y_i is DP-simplicial in M. Let S be the set of all clauses $D_{j'} \cup \{b, b^*, c_{j'}\} \cup \{\overline{c_k} \mid k \neq j'\}$ and $\overline{D_{j'}} \cup \{b, b', c_{j'}\} \cup \{\overline{c_k} \mid k \neq j'\}$ in which y_i occurs. By definition, $var(M)$ contains all b-variables and all c-variables. This has the following two consequences. First, as $var(M)$ also contains y_i and z_i, we find that M contains the clauses $\{y_i, \overline{b}\}$, $\{\overline{y_i}, \overline{b}\}$, $\{y_i, z_i, \overline{b}, b'\}$, and $\{\overline{y_i}, \overline{z_i}, \overline{b}, b^*\}$. Second, by Lemma 3, the other clauses of M in which y_i occurs form a subset of S. This means that there are only 3 pairs of clauses C_1, C_2 in M with $C_1 \cap \overline{C_2} = \{y_i\}$, namely the pair $\{y_i, \overline{b}\}, \{\overline{y_i}, \overline{b}\}$, the pair $\{y_i, \overline{b}\}, \{\overline{y_i}, \overline{z_i}, \overline{b}, b^*\}$, and the pair $\{\overline{y_i}, \overline{b}\}, \{y_i, z_i, \overline{b}, b'\}$. Each of these pairs satisfies (*). This completes the proof of Lemma 4. □

Lemma 5. *Let M be a yz-reduction formula of F'. If M contains neither bcD-clauses nor resolution descendants of such clauses, then M has a DP-simplicial elimination ordering $b, c_1, \ldots, c_m, b', b^*, v^1, \ldots, v^\ell$, where $v^1, \ldots, v^\ell$ form an arbitrary ordering of the y-variables and z-variables in $var(M)$.*

Proof. By our assumptions, the only clauses in M in which b occurs are by-clauses, bz-clauses, byz-clauses, and the clause $\{\overline{b}, b'\}$. In all these clauses b occurs as $\overline{b}$. Hence, b is (trivially) DP-simplicial in M. We then find that $\mathrm{DP}_b(M)$ consists of $\{\overline{b'}, b^*\}$, $\{b', \overline{b^*}\}$ and all bc-clauses. For every c_j, there exists exactly one bc-clause, namely $\{c_j, b', b^*\}$, in which c_j occurs as c_j, and exactly one bc-clause, namely $\{\overline{c_j}, b', b^*\}$, in which c_j occurs as $\overline{c_j}$. Hence, c_j is DP-simplicial in $\mathrm{DP}_{b, c_1, \ldots, c_{j-1}}(M)$ for $j = 1, \ldots, m$. We deduce that $\mathrm{DP}_{b, c^1, \ldots, c^m}(M) = \{\{b', b^*\}, \{\overline{b'}, b^*\}, \{b', \overline{b^*}\}\}$. Then b' is DP-simplicial in $\mathrm{DP}_{b, c_1, \ldots, c_m}(M)$, and we find that $\mathrm{DP}_{b, c_1, \ldots, c_m, b'}(M) = \{\{b^*\}\}$. Then b^* is DP-simplicial in $\mathrm{DP}_{b, c_1, \ldots, c_m, b'}(M)$, and we find that $\mathrm{DP}_{b, c_1, \ldots, c_m, b', b^*}(M) = \emptyset$. Consequently, v^i is DP-simplicial in $\mathrm{DP}_{b, c_1, \ldots, c_m, b', b^*, v^1, \ldots, v^{i-1}}(M)$ for $i = 1, \ldots, \ell$. This concludes the proof of Lemma 5. □

4.2 The Reduction

We are now ready to prove the main result of Section 4.

Theorem 2. *The problem of deciding whether a given CNF formula admits a DP-simplicial elimination ordering is* **NP**-*complete.*

Proof. Recall that the problem is in **NP**. Given a CNF formula F that has variables $x_1, \ldots, x_n$ and clauses $C_1, \ldots, C_m$, we construct in polynomial time the CNF formula F'. We claim that F is satisfiable if and only if F' admits a DP-simplicial elimination ordering.

First suppose that F is satisfiable. Let τ be a satisfying truth assignment of F. We define functions f and g that map every x-variable to a y-variable or z-variable in the following way. If $\tau(x_i) = 1$, then $f(x_i) = y_i$ and $g(x_i) = z_i$. If $\tau(x_i) = 0$, then $f(x_i) = z_i$ and $g(x_i) = y_i$. Let $x_1, \ldots, x_n$ be the x-variables in an arbitrary ordering. Then, for every $1 \leq i \leq n$, the formula $\mathrm{DP}_{f(x_1), \ldots, f(x_i)}(F')$ is a yz-reduction formula. From Lemma 4 we deduce that $f(x_i)$ is DP-simplicial in $\mathrm{DP}_{f(x_1), \ldots, f(x_{i-1})}(F')$ for every $1 \leq i \leq n$. Because τ satisfies F, $\mathrm{var}(D_j)$ contains a variable that is not in $\mathrm{var}(\mathrm{DP}_{f(x_1), \ldots, f(x_n)}(F'))$, for every $1 \leq j \leq m$. Lemma 3 implies that M does not contain any bcD-clause or any of their resolution descendants. Then, by Lemma 5, we find that $f(x_1), \ldots, f(x_n), b, c_1, \ldots, c_m, b', b^*, g(x_1), \ldots, g(x_n)$ is a DP-simplicial elimination ordering of F'.

Now suppose that F' admits a DP-simplicial elimination ordering $v^1, \ldots, v^{|\mathrm{var}(F')|}$. Let v^k be the first variable that is neither a y-variable nor a z-variable. Then $M = \mathrm{DP}_{v^1, \ldots, v^{k-1}}(F')$ is a yz-reduction formula. Let $A = \{v^1, \ldots, v^{k-1}\}$, and let X consist of all x-variables that have an associated y-variable or z-variable in A. We define a truth assignment $\tau : X \to \{0, 1\}$ by setting $\tau(x_i) = 1$ if $y_i \in A$ and $\tau(x_i) = 0$ if $z_i \in A$, for every $x_i \in X$. By Lemma 2, we find that τ is well defined. Because v^k is a DP-simplicial b-variable or a DP-simplicial c-variable in M, we can apply Lemma 1 and find that, for every $1 \leq j \leq m$, at least one of the two clauses $D_j \cup \{b, b^*, c_j\} \cup \{\overline{c_k} \mid k \neq j\}$ and $\overline{D_j} \cup \{b, b', c_j\} \cup \{\overline{c_k} \mid k \neq j\}$ is not in M. This means that every clause C_j contains a literal x with $\tau(x) = 1$. Hence, F is satisfiable. This completes the proof of Theorem 2.   □

5   Intermediate Classes

We discuss a possibility for coping with the NP-hardness result of the previous section. The ultimate reason for this hardness is that a formula may have several DP-simplicial variables, and it is hard to choose the right one. A simple workaround is to assume a fixed ordering of the variables and always choose the DP-simplicial variable which comes first according to this ordering. In this way we loose some generality but win polynomial time tractability. This idea is made explicit in the following definitions.

Let Ω denote the set of all strict total orderings of the propositional variables. Let $\prec \in \Omega$ and F be a CNF formula. A variable $x \in \mathrm{var}(F)$ is $\prec$-*DP-simplicial* in F if x is DP-simplicial in F, and $\mathrm{var}(F)$ contains no variable $y \prec x$ that is DP-simplicial in F. A strict total ordering $x_1, \ldots, x_n$ of the variables of F is a $\prec$-*DP-simplicial elimination ordering* if x_i is $\prec$-DP-simplicial in $\mathrm{DP}_{x_1, \ldots, x_{i-1}}(F)$ for all $1 \leq i \leq n$. We let $\mathrm{DPS}_\prec$ denote the class of all CNF formulas that admit a $\prec$-DP-simplicial elimination ordering, and we set $\mathrm{DPS}_\forall = \bigcap_{\prec \in \Omega} \mathrm{DPS}_\prec$.

Proposition 2. $\mathrm{DPS}_\prec$ *can be recognized in polynomial time for every* $\prec \in \Omega$. *More precisely, it is possible to find in polynomial time a* $\prec$-*DP-simplicial elimination ordering for a given CNF formula* F, *or else to decide that* F *has no such ordering.*

Proof. Let $x_1, \ldots, x_n$ be the variables of F, ordered according to $\prec$. We check whether x_i is DP-simplicial in F, for $i = 1, \ldots, n$. Each check is clearly polynomial. When we have found the first DP-simplicial variable x_i, we replace F by $\mathrm{DP}_{x_i}(F)$. We iterate this procedure as long as possible. Let F' be the formula we end up with. If $\mathrm{var}(F') = \emptyset$ then $F \in \mathrm{DPS}_\prec$ and the sequence of variables as they have been eliminated provides a $\prec$-DP-simplicial elimination ordering. If $\mathrm{var}(F') \neq \emptyset$ then $F \notin \mathrm{DPS}_\prec$.   □

Proposition 3. $\mathrm{BAC} \subsetneq \mathrm{DPS}_\forall \subsetneq \mathrm{DPS} = \bigcup_{\prec \in \Omega} \mathrm{DPS}_\prec$.

Proof. First we show that $\mathrm{BAC} \subsetneq \mathrm{DPS}_\forall$. Let $F \in \mathrm{BAC}$ and $\prec \in \Omega$. We use induction on the number of variables of F to show that $F \in \mathrm{DPS}_\prec$. The base case $|\mathrm{var}(F)| = 0$ is trivial. Let $|\mathrm{var}(F)| \geq 1$. Because $F \in \mathrm{BAC}$ and $\mathrm{var}(F) \neq \emptyset$, we find that F has at least one weakly simplicial variable. Recall that each weakly simplicial variable is DP-simplicial. Consequently, F has at least one DP-simplicial variable. Let x be the first DP-simplicial variable in the ordering $\prec$. By definition, x is a $\prec$-DP-simplicial variable. We consider $F' = \mathrm{DP}_x(F)$. Because a β-acyclic hypergraph remains β-acyclic under vertex and hyperedge deletion, $F' \in \mathrm{BAC}$. Because F' has fewer variables than F, we use the induction hypothesis to conclude that $F' \in \mathrm{DPS}_\prec$. Hence $\mathrm{BAC} \subseteq \mathrm{DPS}_\prec$ follows. Because $\prec \in \Omega$ was chosen arbitrarily, $\mathrm{BAC} \subseteq \mathrm{DPS}_\forall$ follows.

In order to see that $\mathrm{BAC} \neq \mathrm{DPS}_\forall$, we take a hypergraph H that is not β-acyclic and consider H as a CNF formula with only positive clauses. All variables of H are DP-simplicial and can be eliminated in an arbitrary order. Thus $H \in \mathrm{DPS}_\forall \setminus \mathrm{BAC}$.

Next we show that $\mathrm{DPS}_\forall \subsetneq \mathrm{DPS}$. Inclusion holds by definition. In order to show that the inclusion is strict, we consider the formula F of the example in Section 3.1. In that section we showed that y, b, b', b^*, c, z is a DP-simplicial elimination ordering of F. Hence, $F \in \mathrm{DPS}_\prec$ for any ordering $\prec$ with $y \prec b \prec b' \prec b^* \prec c \prec z$. We also showed that z is DP-simplicial in F but that F has no DP-simplicial ordering starting with z. Hence, $F \notin \mathrm{DPS}_{\prec'}$ for any ordering $\prec'$ with $z \prec' y$. We conclude that $F \in \mathrm{DPS} \setminus \mathrm{DPS}_\forall$. Finally, the equality $\mathrm{DPS} = \bigcup_{\prec \in \Omega} \mathrm{DPS}_\prec$ holds by definition.   □

5.1 Grades of Tractability

What properties do we require from a class $\mathcal{C}$ of CNF formulas to be a "tractable class" for SAT? Clearly we want $\mathcal{C}$ to satisfy the property:

1. Given a formula $F \in \mathcal{C}$, we can decide in polynomial time whether F is satisfiable.

This alone is not enough, since even the class of all satisfiable CNF formulas has this property. Therefore a tractable class $\mathcal{C}$ should also satisfy the property:

2. Given a formula F, we can decide in polynomial time whether $F \in \mathcal{C}$.

However, if $\mathcal{C}$ is not known to satisfy property 2, then it may still satisfy the property:

3. There exists a polynomial-time algorithm that either decides where a given a formula F is satisfiable or not, or else shows that F does not belong to $\mathcal{C}$.

The algorithm mentioned in property 3 may decide the satisfiability of some formulas outside of $\mathcal{C}$, hereby avoiding the recognition problem. Such algorithms are called *robust algorithms* [16]. In addition we would also assume from a tractable class $\mathcal{C}$ to be closed under isomorphisms, i.e., to satisfy the property:

4. If two formulas differ only in the names of their variables, then either both or none belong to $\mathcal{C}$.

This leaves us with two notions of a tractable class for SAT, a *strict* one where properties 1, 2, and 4 are required, and a *permissive* one where only properties 3 and 4 are required. Every strict class is permissive, but the converse does not hold in general. For instance, the class of Horn formulas is strictly tractable, but the class of extended Horn formulas is only known to be permissively tractable [14].

Where are the classes from our paper located within this classification? As a result of Theorem 1, we find that BAC is strictly tractable. By Theorem 2, DPS is not strictly tractable (unless $P = NP$). The classes $DPS_\prec$ do not satisfy property 4. Hence they are not considered as tractable classes. However, $DPS_\forall$ is permissively tractable, because an algorithm for $DPS_\prec$ for an arbitrary ordering $\prec$ is a robust algorithm for $DPS_\forall$. It remains open whether DPS is permissively tractable.

6   Comparisons

We compare the classes of our paper with other known (strictly or permissively) tractable classes. Due to Proposition 3, we only need to consider the boundary classes BAC and DPS. We say that two classes $\mathcal{C}_1$ and $\mathcal{C}_2$ of CNF formulas are *incomparable* if for every n larger than some fixed constant there exist formulas in $\mathcal{C}_1 \setminus \mathcal{C}_2$ and in $\mathcal{C}_2 \setminus \mathcal{C}_1$ with at least n variables. We show that BAC and DPS are each incomparable with a wide range of classes of CNF formulas, in particular with all the tractable classes considered in Speckenmeyer's survey [15], and classes based on graph width parameters [9].

We first introduce some terminology. The *incidence graph* $I(H)$ of a hypergraph $H = (V, E)$ is the bipartite graph where the sets V and E form the two partitions, and where $e \in E$ is incident with $v \in V$ if and only if $v \in e$. The *incidence graph* of a formula F is the bipartite graph $I(F)$ with vertex set $\text{var}(F) \cup F$ and edge set $\{ \{C, x\} \mid C \in F \text{ and } x \in \text{var}(C) \}$. A graph is *chordal bipartite* if it is bipartite and has no induced cycle on 6 vertices or more. There exists a useful relationship between β-acyclic formulas and chordal bipartite graphs, due to Tarjan and Yannakakis [17]. They presented this relationship in terms of β-acyclic hypergraphs, whereas we use the formulation of our previous paper [12].

Proposition 4 ([17]). *For a CNF formula F, statements (i)-(iii) are equivalent:*

   (i) F is β-acyclic;
   (ii) $I(H(F))$ is chordal bipartite;
   (iii) $I(F)$ is chordal bipartite.

The following four families of formulas will be sufficient for showing most of our incomparability results. Here, $n \geq 1$ is an integer, $x_1, \dots, x_n$ and $y_1, \dots, y_{2^n}$ are variables, and $C_1, \dots, C_{2^n}$ are all possible clauses with variables $x_1, \dots, x_n$.

$$F_a(n) = \{C_1, \ldots, C_{2^n}\}$$
$$F_s(n) = \{\{x_1, \ldots, x_{\lceil \frac{n}{2} \rceil}\}, \{x_{\lceil \frac{n}{2} \rceil}, \ldots, x_n\}\}$$
$$F_c(n) = \{\{x_i, \overline{x}_{i+1}\} \mid 1 \le i \le n-1\} \cup \{\{x_n, \overline{x}_1\}\}$$
$$F_{ac}(n) = \{\{y_{j-1}, \overline{y}_j\} \cup C_j \mid 1 < j \le 2^n\} \cup \{\{y_{2^n}, \overline{y}_1\} \cup C_1\} \cup$$
$$\{\{y_j, y_{j+1}\} \cup C_j \mid 1 \le j \le 2^n\} \cup \{\{y_{2^n}, y_1\} \cup C_{2^n}\}.$$

We observe that every $I(F_a(n))$ is a complete bipartite graph with partition classes of size n and 2^n, respectively, and that every $I(F_s(n))$ is a tree. Because complete bipartite graphs and trees are chordal bipartite, we can apply Proposition 4 to obtain the following lemma.

Lemma 6. $F_a(n), F_s(n) \in$ BAC *for all $n \ge 1$.*

By the following lemma, the other two classes of formulas do not intersect with DPS.

Lemma 7. $F_c(n), F_{ac}(n) \notin$ DPS *for all $n \ge 3$.*

Proof. Throughout the proof we compute indices of modulo n for the vertices x_i, and modulo 2^{n+1} for the vertices y_j.

First we show that $F_c(n) \notin$ DPS. The clauses $C = \{x_i, \overline{x}_{i+1}\}$ and $C' = \{x_{i-1}, \overline{x}_i\} \in F_c(n)$ have the x_i-resolvent $\{x_{i-1}, \overline{x}_{i+1}\}$ which is not a subset of C or C'. Hence, C and C' violate (*). Consequently, x_i is not DP-simplicial for any $1 \le i \le n$. Because $F_c(n)$ has no other resolvents, $F_c(n)$ has no DP-simplicial variables. Because $\mathrm{var}(F_c(n)) \ne \emptyset$ either, we conclude that $F_c(n) \notin$ DPS for all $n \ge 3$.

Next we show that $F_{ac}(n) \notin$ DPS. Let $1 \le i \le n$ for some $n \ge 3$. Let $1 \le j_1, j_2 \le 2^n$ such that $C_{j_1} \cap C_{j_2} = \{x_i\}$. By definition, $F_{ac}(n)$ contains the clauses $C = \{y_{j_1}, y_{j_1+1}\} \cup C_{j_1}$ and $C' = \{y_{j_2}, y_{j_2+1}\} \cup C_{j_2}$, which have x_i-resolvent $C^* = \{y_{j_1}, y_{j_1+1}, y_{j_2}, y_{j_2+1}\} \cup (C_{j_1} \cup C_{j_2}) \setminus \{x_i, \overline{x}_i\}$. However, since $\{y_{j_1}, y_{j_1+1}\} \ne \{y_{j_2}, y_{j_2+1}\}$, we find that C^* is not a subset of C or C'. Hence, C and C' violate (*). Consequently, x_i is not DP-simplicial for any $1 \le i \le n$.

Let $1 \le j \le 2^n$ for some $n \ge 3$. Then $F_{ac}(n)$ contains the two clauses $C = \{y_j, y_{j+1}\} \cup C_j$ and $C' = \{y_{j-1}, \overline{y}_j\} \cup C_j$, which have y_j-resolvent $C^* = \{y_{j-1}, y_{j+1}\} \cup C_j$. However, $y_{j-1} \in C^* \setminus C$ and $y_{j+1} \in C^* \setminus C'$. Hence, C^* is not a subset of C or C'. Consequently y_j is not DP-simplicial for any $1 \le j \le 2^n$. Because $F_{ac}(n)$ has no other resolvents, $F_{ac}(n)$ has no DP-simplicial variables. Because $\mathrm{var}(F_{ac}(n)) \ne \emptyset$ either, we conclude that $F_{ac}(n) \notin$ DPS for all $n \ge 3$. $\square$

Suppose that we want to show that BAC and DPS are incomparable with a class $\mathcal{C}$ of CNF formulas. Then, Proposition 3 combined with Lemmas 6 and 7 implies that we only have to show the validity of the following two statements:

(i) $F_a(n) \notin \mathcal{C}$ or $F_s(n) \notin \mathcal{C}$ for every n larger than some fixed constant;
(ii) $F_c(n) \in \mathcal{C}$ or $F_{ac}(n) \in \mathcal{C}$ for every n larger than some fixed constant.

6.1   Easy Classes

We use (i) and (ii) to show that BAC and DPS are incomparable with the classes considered by Speckenmeyer [15]. For example, consider the class of *2-CNF formulas*, i.e., CNF formulas where every clause contains at most two literals. For every $n \geq 3$, $F_a(n)$ is not a 2-CNF formula. This shows (i). Furthermore, (ii) follows from the fact that $F_c(n)$ is a 2-CNF formula for every $n \geq 3$. Consequently, the class of 2-CNF formulas is incomparable with BAC and DPS.

As a second example we consider the class of *hitting formulas*, i.e., CNF formulas where $C \cap \overline{C'} \neq \emptyset$ holds for any two of their clauses [15]. Now, for every $n \geq 3$ the formula $F_s(n)$ is not a hitting formula. This shows (i). It is not difficult to see that for $n \geq 3$, $F_{ac}(n)$ is a hitting formula. This shows (ii). Consequently, the class of hitting formulas is incomparable with BAC and DPS. The proofs for other classes of formulas considered in [15] are similar. In particular, for the classes *Horn, renameable Horn, extended Horn, CC-balanced, Q-Horn, SLUR, Matched, bounded deficiency, nested, co-nested, and BRLR$_k$ formulas* we can utilize the formulas $F_a(n)$ to show (i) and the formulas $F_c(n)$ to show (ii).

6.2   Classes of Bounded Width

The SATISFIABILITY problem is tractable for various classes of formulas that are defined by bounding certain width-measures of graphs associated with formulas. Besides the incidence graph $I(F)$, the two other prominent graphs associated with a CNF formula F are the *primal graph* $P(F)$ and the *directed incidence graph* $D(F)$. The graph $P(F)$ has vertex set var(F) and edge set $\{\,\{x,y\} \mid x,y \in \mathrm{var}(C) \text{ for some } C\,\}$. The graph $D(F)$ is the directed graph with vertex set var$(F) \cup F$ and arc set $\{\,(C,x) \mid C \in F \text{ and } x \in C\,\} \cup \{\,(x,C) \mid C \in F \text{ and } \overline{x} \in C\,\}$. We restrict our scope to the graph invariants *treewidth* (tw), and *clique-width* (cw). For their definitions we refer to other sources [9], as we do not need these definitions here.

For a graph invariant π, a graph representation $G \in \{P, I, D\}$ and an integer k, we consider the class $\mathrm{CNF}_k^G(\pi)$ of CNF formulas F with $\pi(G(F)) \leq k$. It is known [9] that for every fixed $k \geq 0$, SAT can be solved in polynomial time for the classes $\mathrm{CNF}_k^P(\mathrm{tw})$, $\mathrm{CNF}_k^I(\mathrm{tw})$, and $\mathrm{CNF}_k^D(\mathrm{cw})$.

Proposition 5. *For every $k \geq 2$, $\mathrm{CNF}_k^P(tw)$ is incomparable with BAC and DPS.*

Proof. We prove that (i) and (ii) hold with respect to $\mathrm{CNF}_k^P(\mathrm{tw})$. Because $P(F_a(n))$ is the complete graph on n vertices, it has treewidth $n - 1$ [1,10]. Hence, $F_a(n) \notin \mathrm{CNF}_k^P(\mathrm{tw})$ for all $n \geq k + 2$. This proves (i). Because $P(F_c(n))$ is a cycle of length n, it has treewidth 2 [1,10]. Hence, $F_c(n) \in \mathrm{CNF}_2^P(\mathrm{tw})$. This proves (ii).   □

Proposition 6. *For every $k \geq 2$, $\mathrm{CNF}_k^I(tw)$ is incomparable with BAC and DPS.*

Proof. We prove that (i) and (ii) hold with respect to $\mathrm{CNF}_k^I(\mathrm{tw})$. Because $I(F_a(n))$ is a complete bipartite graph with partition classes of size n and 2^n, respectively, it has treewidth n [1,10]. Hence, $F_a(n) \notin \mathrm{CNF}_k^I(\mathrm{tw})$ for all $n \geq k + 1$. This proves (i). Because $I(F_c(n))$ is a cycle of length $2n$, it has treewidth 2 [1,10]. Hence, $F_c(n) \in \mathrm{CNF}_2^I(\mathrm{tw})$. This proves (ii).   □

Proposition 7. *For every* $k \geq 4$, $\mathrm{CNF}_k^D(cw)$ *is incomparable with* BAC *and* DPS.

Proof. First we show that $\mathrm{BAC} \setminus \mathrm{CNF}_k^D(cw)$ contains formulas with an arbitrary large number of variables. For all $n \geq 1$, Brandstädt and Lozin [3] showed that there is a bipartite permutation graph $G(n)$ with clique-width n. We do not need the definition of a bipartite permutation graph; it suffices to know that bipartite permutation graphs are chordal bipartite [16].

Let $G'(n) = (U_n \cup W_n, E_n)$ denote the graph obtained from $G(n)$ by deleting twin vertices as long as possible (two vertices are twins if they have exactly the same neighbors). The deletion of twins does not change the clique-width of a graph [4]. Hence, $G'(n)$ has clique-width n. It is well known and easy to see that the clique-width of a bipartite graph with partition classes of size r and s, respectively, is not greater than $\min(r, s) + 2$. Hence $|U_n| \geq n - 2$. Because we only deleted vertices, $G'(n)$ is also chordal bipartite.

Let $F(n) = \{\, N(w) \mid w \in W_n \,\}$ where $N(w)$ denotes the set of neighbors of w in $G'(n)$. Then $G'(n)$ is the incidence graph of $F(n)$, because $G'(n)$ has no twins. Hence $F(n) \in \mathrm{BAC}$ follows from Proposition 4. Recall that the clique-width of $G'(n) = I(F(n))$ is n and that $|U_n| \geq n - 2$. Since all clauses of $F(n)$ are positive, $I(F(n))$ and $D(F(n))$ have the same clique-width. We conclude that $F(n)$ is a formula on at least $n - 2$ variables that belongs to $\mathrm{BAC} \setminus \mathrm{CNF}_k^D(cw)$ for $n \geq k + 1$.

For the converse direction we observe that $D(F_c(n))$ is an oriented cycle and clearly has clique-width at most 4. This means that $D(F_c(n)) \in \mathrm{CNF}_4^D(cw)$. By Lemma 7, we have that $D(F_c(n)) \notin \mathrm{DPS}$ for all $n \geq 3$. We then conclude that $\mathrm{CNF}_4^D(cw) \setminus \mathrm{DPS}$ contains $D(F_c(n))$ for all $n \geq 3$. We are left to apply Proposition 3 to complete the proof of Proposition 7.   □

Results similar to Propositions 5-7 also hold for the graph invariants *branchwidth* and *rank-width*, since a class of graphs has bounded branchwidth if and only if it has bounded treewidth [1], and a class of directed graphs has bounded rank-width if and only if it has bounded clique-width [8].

7   Conclusion

We have studied new classes of CNF formulas: the strictly tractable class BAC, the permissively tractable class $\mathrm{DPS}_\forall$, and the hard-to-recognize class DPS. Our results show that the classes are incomparable with previously studied classes. Our results establish an interesting link between SAT and algorithmic graph theory: the formulas in BAC are exactly the formulas whose incidence graphs belong to the class of chordal bipartite graphs, a prominent and well-studied graph class. It would be interesting to consider other classes of bipartite graphs, e.g., the classes described by Brandstädt, Le and Spinrad [2], and determine the complexity of SAT restricted to CNF formulas whose incidence graphs belong to the class under consideration. Of particular interest are minimal super-classes of the class of chordal bipartite graphs.

References

1. Bodlaender, H.L.: A partial k-arboretum of graphs with bounded treewidth. Theoret. Comput. Sci. 209(1-2), 1–45 (1998)
2. Brandstädt, A., Le, V.B., Spinrad, J.P.: Graph classes: a survey. SIAM Monographs on Discrete Mathematics and Applications. SIAM, Philadelphia (1999)
3. Brandstädt, A., Lozin, V.V.: On the linear structure and clique-width of bipartite permutation graphs. Ars Combinatoria 67, 273–281 (2003)
4. Courcelle, B., Olariu, S.: Upper bounds to the clique-width of graphs. Discr. Appl. Math. 101(1-3), 77–114 (2000)
5. Davis, M., Putnam, H.: A computing procedure for quantification theory. J. ACM 7(3), 201–215 (1960)
6. Fagin, R.: Degrees of acyclicity for hypergraphs and relational database schemes. J. ACM 30(3), 514–550 (1983)
7. Fischer, E., Makowsky, J.A., Ravve, E.R.: Counting truth assignments of formulas of bounded tree-width or clique-width. Discr. Appl. Math. 156(4), 511–529 (2008)
8. Ganian, R., Hlinený, P., Obdržálek, J.: Better algorithms for satisfiability problems for formulas of bounded rank-width. In: Lodaya, K., Mahajan, M. (eds.) IARCS Annual Conference on Foundations of Software Technology and Theoretical Computer Science, FSTTCS 2010, Chennai, India, December 15-18. LIPIcs, vol. 8, pp. 73–83. Schloss Dagstuhl - Leibniz-Zentrum fuer Informatik (2010)
9. Gottlob, G., Szeider, S.: Fixed-parameter algorithms for artificial intelligence, constraint satisfaction, and database problems. The Computer Journal 51(3), 303–325 (2006), Survey paper
10. Kloks, T., Bodlaender, H.: Approximating treewidth and pathwidth of some classes of perfect graphs. In: Ibaraki, T., Iwama, K., Yamashita, M., Inagaki, Y., Nishizeki, T. (eds.) ISAAC 1992. LNCS, vol. 650, pp. 116–125. Springer, Heidelberg (1992)
11. Kullmann, O., Luckhardt, H.: Algorithms for SAT/TAUT decision based on various measures (1999) (manuscript)
12. Ordyniak, S., Paulusma, D., Szeider, S.: Satisfiability of acyclic and almost acyclic CNF formulas. In: Lodaya, K., Mahajan, M. (eds.) IARCS Annual Conference on Foundations of Software Technology and Theoretical Computer Science, FSTTCS 2010, Chennai, India, December 15-18. LIPIcs, vol. 8, pp. 84–95. Schloss Dagstuhl - Leibniz-Zentrum fuer Informatik (2010)
13. Samer, M., Szeider, S.: Algorithms for propositional model counting. J. Discrete Algorithms 8(1), 50–64 (2010)
14. Schlipf, J.S., Annexstein, F.S., Franco, J.V., Swaminathan, R.P.: On finding solutions for extended Horn formulas. Information Processing Letters 54(3), 133–137 (1995)
15. Speckenmeyer, E.: Classes of easy expressions. In: Biere, A., Heule, M., van Maaren, H., Walsh, T. (eds.) Handbook of Satisfiability, ch. 13, Section 1.19, pp. 27–31. IOS Press, Amsterdam (2009)
16. Spinrad, J.P.: Efficient Graph Representations. Fields Institute Monographs. AMS, Providence (2003)
17. Tarjan, R.E., Yannakakis, M.: Simple linear-time algorithms to test chordality of graphs, test acyclicity of hypergraphs, and selectively reduce acyclic hypergraphs. SIAM J. Comput. 13(3), 566–579 (1984)

BDDs for Pseudo-Boolean Constraints – Revisited

Ignasi Abío, Robert Nieuwenhuis, Albert Oliveras,
and Enric Rodríguez-Carbonell*

Technical University of Catalonia (UPC), Barcelona

Abstract. Pseudo-Boolean constraints are omnipresent in practical applications, and therefore a significant effort has been devoted to the development of good SAT encoding techniques for these constraints. Several of these encodings are based on building Binary Decision Diagrams (BDDs) and translating these into CNF. Indeed, BDD-based encodings have important advantages, such as *sharing* the same BDD for representing many constraints.

Here we first prove that, unless NP = Co-NP, there are Pseudo-Boolean constraints that admit no variable ordering giving a polynomial (Reduced, Ordered) BDD. As far as we know, this result is new (in spite of some misleading information in the literature). It gives several interesting insights, also relating proof complexity and BDDs.

But, more interestingly for practice, here we also show how to overcome this theoretical limitation by *coefficient decomposition*. This allows us to give the first polynomial arc-consistent BDD-based encoding for Pseudo-Boolean constraints.

1 Introduction

In this paper we study Pseudo-Boolean constraints (PB constraints for short), that is, constraints of the form $a_1 x_1 + \cdots + a_n x_n \,\#\, K$, where the a_i and K are integer coefficients, the x_i are Boolean (0/1) variables, and the relation operator $\#$ belongs to $\{<, >, \leq, \geq, =\}$. We will assume that $\#$ is $\leq$ and the a_i and K are positive since other cases can be easily reduced to this one (see [ES06]).

Such a constraint is a Boolean function $C\colon \{0,1\}^n \to \{0,1\}$ that is monotonic decreasing in the sense that any solution for C remains a solution after flipping inputs from 1 to 0. Therefore these constraints can be expressed by a set of clauses with only negative literals. For example, each clause could simply define a (minimal) subset of variables that cannot be simultaneously true. Note however that not every such a monotonic function is a PB constraint. For example, the function expressed by the two clauses $\overline{x}_1 \vee \overline{x}_2$ and $\overline{x}_3 \vee \overline{x}_4$ has no (single) equivalent PB constraint $a_1 x_1 + \cdots + a_n x_n \leq K$ (since wlog. $a_1 \geq a_2$ and $a_3 \geq a_4$, and then

---

* All the authors are partially supported by Spanish Min. of Educ. and Science through the LogicTools-2 project (TIN2007-68093-C02-01). Abío is also partially supported by FPU grant.

K.A. Sakallah and L. Simon (Eds.): SAT 2011, LNCS 6695, pp. 61–75, 2011.
© Springer-Verlag Berlin Heidelberg 2011

also $\overline{x}_1 \vee \overline{x}_3$ is needed). Hence, even among the monotonic Boolean functions, PB constraints are a rather restricted class (see also [J.S07]).

PB constraints are omnipresent in practical SAT applications, not just in typical 0-1 linear integer problems, but also as an ingredient in new SAT approaches to, e.g., cumulative scheduling [SFSW09], so it is not surprising that a significant number of SAT encodings for these constraints have been proposed in the literature. Here we are interested in encoding a PB constraint C by a clause set S (possibly with auxiliary variables) that is not only equisatisfiable, but also (generalized) *arc-consistent*: given a partial assignment A, if x_i is false in every extension of A satisfying C, then unit propagating A on S sets x_i to false.

To our knowledge, the only polynomial arc-consistent encoding so far was given by Bailleux, Boufkhad and Roussel [BBR09]. Other existing encodings are based on building (forms of) Binary Decision Diagrams (BDDs) and translating these into CNF. Although [BBR09] is not BDD-based, our motivation to revisit BDD-based encodings is twofold:

Example 1. Consider the constraint $3x_1 + 2x_2 + 4x_3 \leq 5$ and the constraint $30001x_1 + 19999x_2 + 39998x_3 \leq 50007$. Both are clearly equivalent: the Boolean function they represent can be expressed, e.g., by the clauses $\overline{x}_1 \vee \overline{x}_3$ and $\overline{x}_2 \vee \overline{x}_3$. However, encodings like the one of [BBR09] heavily depend on the concrete coefficients of each constraint, and generate a significantly larger SAT encoding for the second one. Since, given a variable ordering, (Reduced, Ordered) BDDs are a canonical representation for Boolean functions [Bry86], i.e., each Boolean function has a unique ROBDD, a ROBDD-based encoding will treat both constraints equivalently.   □

The second reason for revisiting BDDs is that in practical problems numerous PB constraints exist that share variables among each other. Representing them all as a single BDD has the potential of generating a much more compact SAT encoding that is moreover likely to have better propagation properties.

Related work. The same authors of [BBR09] proposed an encoding "very close to those using a BDD and translating it into clauses" [BBR06]. It is arc-consistent, but an example of a PB constraint family is given in [BBR06] for which their kind of *non-reduced* BDDs, with *their concrete variable ordering* is exponentially large. However, as we show here, ROBDDs for this family are polynomial. Their method works as follows. Given the PB constraint $a_1x_1 + \cdots + a_nx_n \leq K$ with coefficients ordered from small to large, the root node is labelled with variable $D_{n,K}$, expressing that the sum of the first n terms is no more than K. Its two children are $D_{n-1,K}$ and $D_{n-1,K-a_n}$, which correspond to setting x_n to false and true, respectively, etc. Two binary and two ternary clauses per node express the relationships between the variables.

Example 2. The encoding of [BBR06] on $2x_1 + \cdots + 2x_{10} + 5x_{11} + 6x_{12} \leq 10$ is illustrated in Figure 1. Node $D_{10,5}$ represents $2x_1 + 2x_2 + \cdots 2x_{10} \leq 5$, whereas node $D_{10,4}$ represents $2x_1 + 2x_2 + \cdots 2x_{10} \leq 4$. The method fails to identify that both these PB constraints are equivalent and hence subtrees B and C will not be merged, yielding a much larger representation than with ROBDDs.   □

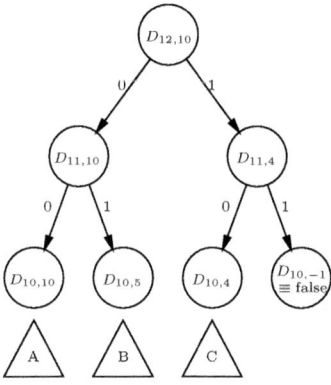

Fig. 1. Tree-like construction of [BBR06] for $2x_1 + \cdots + 2x_{10} + 5x_{11} + 6x_{12} \leq 10$

On the other hand, Eén and Sörensson use ROBDDs in MiniSAT+ [ES06]. Their encoding uses six three-literal clauses per BDD node and is arc-consistent, but the proof of arc-consistency relies on a particular variable ordering. Regarding the size of their ROBDDs, they cite [BBR06] to say *"It is proven that in general a PB-constraint can generate an exponentially sized BDD [BBR06]"* which, as we have seen, cannot be concluded from that paper since they do not use ROB-DDs. Apart from their BDD-based encoding, [ES06] also suggests two alternative methods: one based on adder networks ($\mathcal{O}(n)$ in size but not arc-consistent) and another one based on sorting networks ($\mathcal{O}(n \log n)$ in size and not arc-consistent).

Finally, as we have already mentioned, [BBR09] presents an arc-consistent and polynomial BDD-based SAT encoding (size $\mathcal{O}(n^2 \log n \log a_{max})$, i.e., it depends on the size of the coefficients) based on a network of unary adders.

Main Contributions and Organization of This Paper

- Subsection 3.2: The first, to our knowledge, PB constraint family for which ROBDDs with small-to-large variable ordering are exponential in size (and also for the large-to-small ordering).
- Subsection 3.3: A proof that, unless NP=co-NP, there are PB constraints that admit no polynomial-size ROBDD, independently of the variable order.
- Subsection 4.1: A proof that PB constraints whose coefficients are powers of two do admit polynomial-size BDDs.
- Subsections 4.2 and 4.3: An arc-consistent and polynomial (size $\mathcal{O}(n^3 \log a_{max})$) BDD-based encoding for PB constraints.
- Section 5: An arc-consistent SAT encoding of BDDS for monotonic functions, a more general class of Boolean functions than PB constraints. This encoding uses only one binary and one ternary clause per node (the standard if-then-else encoding for BDDs used in, e.g., [ES06], requires six ternary clauses per node). Moreover, this translation works for any BDD variable ordering.

2   Preliminaries

We assume the reader is familiar with the basic notions of propositional logic. Otherwise, basic definitions can be found in [BHvMW09]. Pseudo-Boolean constraints (PB constraints for short) are constraints of the form $a_1x_1 + \cdots + a_nx_n \# K$, where the a_i and K are integer coefficients, the x_i are Boolean $(0/1)$ variables, and the relation operator $\#$ belongs to $\{<, >, \leq, \geq, =\}$. We will assume that $\#$ is $\leq$ and the a_i and K are positive, since other cases can be easily reduced to this one [1]: (i) changing into $\leq$ is straightforward if coefficients can be negative; (ii) replacing $-ax$ by $a(1-x) - a$; (iii) replacing $(1-x)$ by $\bar{x}$. Negated variables like $\bar{x}$ can be handled as positive ones or, alternatively, replaced by a fresh x' and adding the clauses $x \vee x'$ and $\bar{x} \vee \bar{x}'$.

Our main goal is to find SAT encodings for PB constraints. That is, given a PB-constraint C, construct an equisatisfiable clause set (a CNF) S such that any model for S restricted to the variables of C is a model of C. Two extra properties are sought: (i) *consistency checking by unit propagation* or simply *consistency*: whenever a partial assignment A cannot be extended to a model for C, unit propagation on S and A produces a contradiction (a literal l and its negation $\bar{l}$); and (ii) (generalized) *arc-consistency* (again by unit propagation): given an assignment A that can be extended to a model of C, but such that $A \cup \{x\}$ cannot, unit propagation on S and A produces $\bar{x}$. More concretely, we will use BDDs for finding such encodings, as illustrated by the following example.

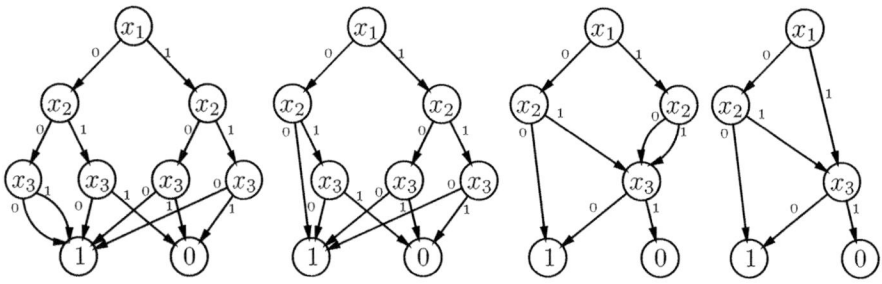

Fig. 2. Construction of a BDD for $2x_1 + 3x_2 + 5x_3 \leq 6$

Example 3. Figure 2 explains (one method for) the construction of a ROBDD for the PB constraint $2x_1 + 3x_2 + 5x_3 \leq 6$ and the ordering $x_1 < x_2 < x_3$. The root node has as *selector variable* x_1. Its *false child* represents the PB constraint assuming $x_1 = 0$ (i.e., $3x_2 + 5x_3 \leq 6$) and its *true child* represents $2 + 3x_2 + 5x_3 \leq 6$, that is, $3x_2 + 5x_3 \leq 4$. The two children have the next variable in the ordering (x_2) as selector, and the process is repeated until we reach the last variable in the sequence. Then, a constraint of the form $0 \leq K$ is the *True* node (1 in the figure) if $K \geq 0$ is positive, and the *False* node (0)

---

[1]   An $=$-constraint can be split into a $\leq$-constraint and a $\geq$-constraint. Here we consider (arc-)consistency for the latter two isolatedly, not for the original $=$-constraint.

if $K < 0$. This construction (leftmost in the figure), is known as an Ordered BDD. For obtaining a Reduced Ordered BDD (BDD for short in the rest of the paper), two reductions are applied until fixpoint: removing nodes with identical children (as done with the leftmost x_3 node in the second BDD of the figure), and merging isomorphic subtrees, as done for x_3 in the third BDD. The fourth final BDD is a fixpoint. For a given ordering, BDDs are a canonical representation of Boolean functions: each Boolean function has a unique BDD. BDDs can be encoded in CNF by introducing an auxiliary variable a for every node. If the selector variable of the node is x and the auxiliary variables for the false and true child are f and t, respectively, add the if-then-else clauses:

$$\overline{x} \wedge \overline{f} \to \overline{a} \qquad x \wedge \overline{t} \to \overline{a} \qquad \overline{f} \wedge \overline{t} \to \overline{a}$$
$$\overline{x} \wedge f \to a \qquad x \wedge t \to a \qquad f \wedge t \to a \qquad\qquad \square$$

In what follows, the *size* of a BDD is its number of nodes. We will say that a BDD *represents* a PB constraint if they represent the same Boolean function. Given an assignment A over the variables of a BDD, we define the *path induced by A* as the path that starts at the root of the BDD and at each step, moves to the false (true) child of a node iff its selector variable is false (true) in A.

3   Exponential BDDs for PB Constraints

In this section, we prove that, unless NP=co-NP, there are PB constraints whose BDDs are all exponential, regardless of the variable ordering. We start by defining the notion of the *interval* of a PB constraint. After that, we consider two families of PB constraints and we study the size of their BDDs. Finally, we prove the main result of this section.

3.1   Intervals

Example 4. Consider the constraint $2x_1 + 3x_2 + 5x_3 \leq 6$. Since no combination of its coefficients adds to 6, the constraint is equivalent to $2x_1 + 3x_2 + 5x_3 < 6$, and hence to $2x_1 + 3x_2 + 5x_3 \leq 5$. This process cannot be repeated again since 5 can be obtained with the existing coefficients.

Similarly, we could try to increase the right-hand side of the constraint. However, there is a combination of the coefficients that adds 7, which implies that the constraint is not equivalent to $2x_1 + 3x_2 + 5x_3 \leq 7$. All in all, we can state that the constraint is equivalent to $2x_1 + 3x_2 + 5x_3 \leq K$ for any $K \in [5, 6]$. It is trivial to see that the set of valid K's is always an interval. $\qquad\square$

Definition 1. *Let C be a constraint of the form $a_1x_1 + \cdots + a_nx_n \leq K$. The interval of C consists of all integers M such that $a_1x_1 + \cdots + a_nx_n \leq M$, seen as a Boolean function, is equivalent to C.*

In the following, given a BDD representing a PB constraint and a node ν, we will refer to *the interval of ν* as the interval of the constraint represented by the BDD rooted at ν. Unless stated otherwise, the ordering used in the BDD will be $x_1 < x_2 < \ldots < x_n$.

Proposition 2. *If $[\beta, \gamma]$ is the interval of a node ν with selector variable x_i then:*

1. *There is an assignment $\{x_j = v_j\}_{j=i}^{n}$ such that $a_i v_i + \cdots + a_n v_n = \beta$.*
2. *There is an assignment $\{x_j = v_j\}_{j=i}^{n}$ such that $a_i v_i + \cdots + a_n v_n = \gamma + 1$.*
3. *There is an assignment $\{x_j = v_j\}_{j=1}^{i-1}$ such that $K - a_1 v_1 - a_2 v_2 - \cdots - a_{i-1} v_{i-1} \in [\beta, \gamma]$*
4. *Take $h < \beta$. There exists an assignment $\{x_j = v_j\}_{j=i}^{n}$ such that $a_i v_i + \cdots + a_n v_n > h$ and its path goes from ν to True.*
5. *Take $h > \gamma$. There exists an assignment $\{x_j = v_j\}_{j=i}^{n}$ such that $a_i v_i + \cdots + a_n v_n \leq h$ and its path goes from ν to False.*
6. *The interval of the True node is $[0, \infty)$.*
7. *The interval of the False node is $(-\infty, -1]$. Moreover, it is the only interval with negative values.*

We now prove that, given a BDD for a PB constraint, one can easily compute the intervals for every node bottom-up. We first start with an example.

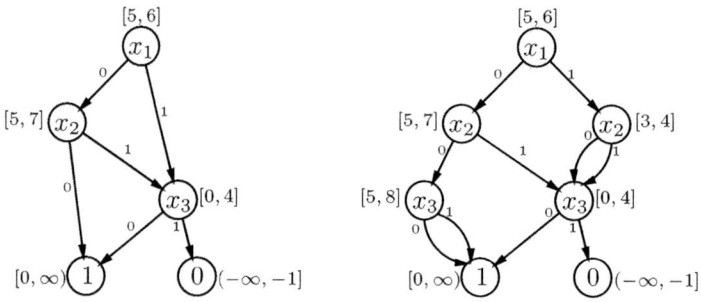

Fig. 3. Intervals of the BDD for $2x_1 + 3x_2 + 5x_3 \leq 6$

Example 5. Let us consider again the constraint $2x_1 + 3x_2 + 5x_3 \leq 6$. Assume that all variables appear in every path from the root to the leaves (otherwise, add extra nodes as in the rightmost BDD of Figure 3). Assume now that we have computed the intervals for the two children of the root (rightmost BDD in Figure 3). This means that the false child of the root is the BDD for $3x_2 + 5x_3 \leq [5, 7]$ and the true child the BDD for $3x_2 + 5x_3 \leq [3, 4]$. Assuming x_1 to be false, the false child would also represent the constraint $2x_1 + 3x_2 + 5x_3 \leq [5, 7]$, and assuming x_1 to be true, the true child would represent the constraint $2x_1 + 3x_2 + 5x_3 \leq [5, 6]$. Taking the intersection of the two intervals, we can infer that the root node represents $2x_1 + 3x_2 + 5x_3 \leq [5, 6]$. □

More formally, the interval of every node can be computed as follows:

Proposition 3. *Let $a_1 x_1 + a_2 x_2 + \cdots + a_n x_n \leq K$ be a constraint, and let $\mathcal{B}$ be its BDD with the order $x_1 < \ldots < x_n$. Consider a node ν with selector variable*

x_i, false child ν_f (with selector variable x_f and interval $[\beta_f, \gamma_f]$) and true child ν_t (with selector variable x_t and interval $[\beta_t, \gamma_t]$). The interval of ν is $[\beta, \gamma]$, with:

$$\beta = \max\{\beta_f + a_{i+1} + \cdots + a_{f-1}, \ \beta_t + a_i + a_{i+1} + \cdots + a_{t-1}\},$$
$$\gamma = \min\{\gamma_f, \ \gamma_t + a_i\}.$$

If in every path from the root to the leaves of the BDD all variables were present, the definition of β would be much simpler ($\beta = \max\{\beta_f, \beta_t + a_i\}$). The other coefficients are necessary to account for the variables that have been removed due to the BDD reduction process.

3.2 Some Families of PB Constraints and Their BDD Size

We start by revisiting the family of PB constraints given in [BBR06], where it is proved that, for their concrete variable ordering, their non-reduced BDDs grow exponentially for this family. Here we prove that ROBDDs are polynomial for this family, and that this is even independent of the variable ordering. The family is defined by considering a, b and n positive integers such that $\sum_{i=1}^{n} b^i < a$. The coefficients are $\omega_i = a + b^i$ and the right-hand side of the constraint is $K = a \cdot n/2$. We will first prove that the constraint $C : \omega_1 x_1 + \cdots + \omega_n x_n \leq K$ is equivalent to the cardinality constraint $C' : x_1 + \cdots + x_n \leq n/2 - 1$. For simplicity, we assume that n is even.

- Take an assignment satisfying C'. In this case, there are at most $n/2 - 1$ variables x_i assigned to true, and the assignment also satisfies C since: $\omega_1 x_1 + \cdots + \omega_n x_n \leq \sum_{i=n/2+2}^{n} \omega_n = (n/2 - 1)a + \sum_{i=n/2+2}^{n} b^n < K - a + \sum_{i=1}^{n} b^i < K.$

- Consider now an assignment not satisfying C'. In this case, there are at least $n/2$ true variables in the assignment and it does not satisfy C either: $\omega_1 x_1 + \cdots + \omega_n x_n \geqslant \sum_{i=1}^{n/2} \omega_i = (n/2) \cdot a + \sum_{i=1}^{n/2} b^i > (n/2) \cdot a = K.$

Since the two constraints are equivalent and BDDs are canonical, the BDD representation of C and C' are the same. But the BDD of C' is known to be of quadratic size because it is a cardinality constraint (see, for instance, [BBR06]).

Theorem 4. *There exists a family of PB constraints parameterized by n, whose ROBDDs grow exponentially in n when ordering the variables according to their coefficients from small to large. The same happens ordering from large to small.*

Proof. We consider constraints of the form $a_1 x_1 + \cdots + a_{4n} x_{4n} \leq K$. It is convenient to describe the coefficients in binary notation:

$$
\begin{array}{rlll}
 & & \overbrace{}^{2n} & \\
a_1 & = & 0\,0\,0\quad 0\,0\ \cdots\ 0\,1 & = 1 \\
a_2 & = & 0\,0\,0\quad 0\,0\ \cdots\ 1\,0 & = 2 \\
\cdots & & & \\
a_{2n-1} & = & 0\,0\,0\quad 0\,1\ \cdots\ 0\,0 & \\
a_{2n} & = & 0\,0\,0\quad 1\,0\ \cdots\ 0\,0 & = 2^{2n-1} \\
a_{2n+1} & = & 1\,0\,0\quad 0\,0\ \cdots\ 0\,1 & \\
a_{2n+2} & = & 1\,0\,0\quad 0\,0\ \cdots\ 1\,0 & \\
\cdots & & & \\
a_{4n-1} & = & 1\,0\,0\quad 0\,1\ \cdots\ 0\,0 & \\
a_{4n} & = & 1\,0\,0\quad 1\,0\ \cdots\ 0\,0 & \\
\end{array}
$$

$$
K = d_m \ \ldots \ d_0\,0\,0\quad 1\,1\ \cdots\ 1\,1
$$

where $d_m \ldots d_0$ is the binary representation of n. Note that, to sum to exactly K, one needs exactly n coefficients of the bottom half (between a_{2n+1} and a_{4n}) to obtain the digits $d_m \ldots d_0$, and that, once such a subset is chosen, a unique subset of exactly n coefficients of the top half exists that will complete the $11 \ldots 11$ suffix of K. Reversely, for each subset of size n of the top half, a unique subset of size n of the bottom half exists that complements it to sum exactly K. Now consider a BDD ordered $x_1 < \cdots < x_{4n}$, and any two distinct assignments T and T' for $x_1 \ldots x_{2n}$ that set exactly n variables to true. Then T and T' induce paths that necessarily lead to different nodes of the BDD. To see this, wlog., assume that the sum of coefficients corresponding to true variables in T is smaller than the one of T'. Consider the assignment B to $x_{2n} \ldots x_{4n}$ that sets to true the unique size-n subset of the bottom half coefficients that sums to K for T (and hence exceeds K for T'). Then the PB constraint satisfies $T \cup B$, but not $T' \cup B$; hence B *distinguishes* the nodes for T and T'. Altogether, the BDD must have at least as many nodes as distinct assignments setting exactly n variables of the top half to true, i.e., an exponential number, $\binom{2n}{n}$. For the large-to-small ordering exactly the same reasoning applies[2].   □

For the PB constraint family of the previous proof, it can be shown that the "interleaved" ordering $x_1 < x_{2n+1} < x_2 < x_{2n+2} < \ldots < x_{2n} < x_{4n}$ leads to a polynomial-sized BDD (the proof is non-trivial, but we had to omit it here due to space limitations). The next natural step would be to present a concrete family of PB constraints whose BDDs are always exponential regardless of the variable ordering. We have not been able to find such a family. But in the next section we prove that, unless NP=co-NP, such a family must exist.

3.3   Probably There Are No Small BDDs for All PB Constraints

Our goal is now to prove that, unless NP=co-NP, there are PB constraints all whose BDDs are exponential, independently of the variable ordering. The

---

[2] We thank Guillem Godoy for his help with this example.

main ingredient is an algorithm that, given a BDD $\mathcal{B}$ and a PB constraint C : $a_1x_1 + \cdots + a_nx_n \leq K$ over the same set of variables, allows one to decide, in time polynomial in the size of the BDD, whether $\mathcal{B}$ represents C. Again, w.l.o.g., we assume that the BDD ordering is $x_1 < x_2 < \ldots < x_n$.

Given the BDD, the algorithm first computes, in a bottom-up manner, an interval for every node of the BDD, as explained in Proposition 3 and points 6 and 7 of Proposition 2. Note that the cost of computing a single interval is $\mathcal{O}(n)$ and hence computing all intervals takes $\mathcal{O}(nm)$ time, where m is the BDD's size. After that, we know that $\mathcal{B}$ is a representation of C if and only if K belongs to the interval of the BDD root.

Theorem 5. $\mathcal{B}$ *is a BDD representing a PB constraint* $a_1x_1 + \cdots + a_nx_n \leq K$ *if, and only if, K belongs to the interval of the root of $\mathcal{B}$ computed by our algorithm.*

Proof. If $\mathcal{B}$ is a BDD representing C, then K belongs to the interval of the root by definition of interval (Def. 1). Moreover, Proposition 3 guarantees that our algorithm correctly computes such an interval.

Let us now assume that $\mathcal{B}$ is not a BDD representing C. Then, there exists an assignment $\{x_1 = v_1, \ldots, x_n = v_n\}$ that either satisfies C but leads to the False node in $\mathcal{B}$ or does not satisfy C but leads to the True node in $\mathcal{B}$.

Let us assume that the assignment satisfies C. The other case is analog to this one. In this case, we will prove that $\gamma_1 < K$, where $[\beta_1, \gamma_1]$ is the interval computed for the root node.

We define a sequence of nodes $\nu_1, \nu_2, \ldots, \nu_n, \nu_{n+1}$ as follows: ν_1 is the root of $\mathcal{B}$. If the selector variable of ν_1 is not x_1, $\nu_2 = \nu_1$. Otherwise, ν_2 is its false child if $v_1 = 0$ or its true child if $v_1 = 1$, and so on. By definition of the assignment, ν_{n+1} is the False node. If we let $[\beta_i, \gamma_i]$ be the computed interval for the node ν_i, we want to prove that every node ν_i satisfies $\gamma_i < a_{i+1}v_{i+1} + \cdots + a_nv_n$.

Since ν_{n+1} is the False node and its theoretical interval is $(-\infty, -1]$, it holds that $\gamma_{n+1} < 0$. Assume that it is true for every $k > i'$, and let us prove it for i'.

Let us assume that x_i is the selector variable of $\nu_{i'}$ (in this case, $i' \leq i$ by construction). There are two cases:

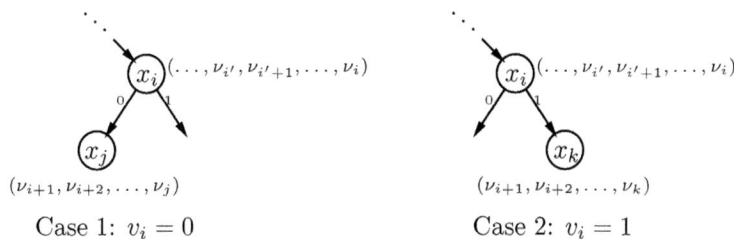

Fig. 4. Several ν's refer to the same BDD node. ν_j and ν_k are the last in the sequence.

– $v_i = 0$. Let us take j such that ν_j is the false child of $\nu_{i'}$ and the selector variable of ν_j is x_j (see Figure 4). Then, $\gamma_{i'} \leq \gamma_j$ by definition of the algorithm.

Using the induction hypothesis:

$$\gamma_{i'} \leq \gamma_j < a_{j+1}v_{j+1} + \cdots + a_n v_n \leq a_{i'+1}v_{i'+1} + \cdots + a_n v_n.$$

- $v_i = 1$. Similarly, let us take k such that ν_k is the true child of $\nu_{i'}$ and the selector variable of ν_k is x_k (see Figure 4). Then, $\gamma_{i'} \leq \gamma_k + a_i$ by definition of the algorithm. Using the induction hypothesis and that $v_i = 1$:

$$\gamma_{i'} \leq a_i + \gamma_k < a_i + a_{k+1}v_{k+1} + \cdots + a_n v_n \leq a_{i'+1}v_{i'+1} + \cdots + a_n v_n.$$

Therefore, it holds that $\gamma_i < a_{i+1}v_{i+1} + \cdots + a_n v_n$ for every i. In particular, it holds for $i = 1$. Since the assignment satisfies the PB constraint by hypothesis, we have

$$\gamma_1 < a_1 v_1 + \cdots + a_n v_n \leq K,$$

and hence K does not belong to the theoretical interval of the root node.   □

Notice that if $\mathcal{B}$ is not the BDD of C some of the computed intervals might be empty. However, the algorithm will be able to compute the remaining intervals and, since the interval of the root node will be empty, the algorithm will also be correct. We are now ready for the following result.

Theorem 6. *Unless NP=co-NP, there are PB constraints that do not admit polynomial BDDs.*

Proof. A well-known NP-complete problem is the following (variant of the) *subset sum* problem: given a set of integers $\{a_1, \ldots, a_n\}$ and an integer K, decide whether there exists a subset of $\{a_1, \ldots, a_n\}$ that sums to exactly K. Here we prove that if a polynomial-size BDD existed for every PB constraint then for every unsatisfiable subset sum problem a polynomial-size unsatisfiability certificate would exist, that could moreover be verified in polynomial time, thus collapsing NP and co-NP. Indeed, obviously, a subset sum problem $(\{a_1, \ldots, a_n\}, K)$ is unsatisfiable if, and only if, the PB constraints $a_1 x_1 + \cdots + a_n x_n \leq K$ and $a_1 x_1 + \cdots + a_n x_n \leq K - 1$ are equivalent, i.e., they are the same Boolean function. So if the subset sum problem $(\{a_1, \ldots, a_n\}, K)$ is unsatisfiable, and a polynomial-size BDD for $a_1 x_1 + \cdots + a_n x_n \leq K$ existed, this BDD would also represent $a_1 x_1 + \cdots + a_n x_n \leq K - 1$, which, as we proved in the previous theorem, can be checked in polynomial time (for both PB constraints at once).   □

We find it quite surprising that, even for the limited kind of monotonic functions that can be represented by a single PB constraint, the existence of polynomial-size BDDs would imply NP=co-NP. As said, to our knowledge it remains unknown whether there exists a family of PB constraints that admit no polynomial-size BDD. This situation is analogous to what happens with extended resolution in Cook's program for propositional proof complexity: it is unknown whether there exists a family of propositional problems that admit no polynomial-size extended resolution proof. So, finding successively more compact unsatisfiability certificates for subset sum might be an interesting alternative to Cook's program for attacking the NP vs co-NP question.

4   Avoiding Exponential BDDs

In this section we introduce our positive results. We restrict ourselves to a particular class of PB constraints, where all coefficients are powers of two. As we will show below, these constraints admit polynomial BDDs. Moreover, any PB constraint can be reduced to this class.

Example 6. Let us take the PB constraint $9x_1 + 8x_2 + 3x_3 \leq 10$. Considering the binary representation of the coefficients, this constraint can be rewritten into $(2^3 x_{3,1} + 2^0 x_{0,1}) + (2^3 x_{3,2}) + (2^1 x_{1,3} + 2^0 x_{0,3}) \leq 10$ if we add the binary clauses expressing that $x_{i,r} = x_r$ for appropriate i and r.   □

4.1   Power-of-Two PB Constraints Do Have Polynomial-Size BDDs

Let us consider a PB constraints of the form:
$$C: \begin{array}{l} 2^0 \cdot \delta_{0,1} \cdot x_{0,1} + 2^0 \cdot \delta_{0,2} \cdot x_{0,2} + \cdots + 2^0 \cdot \delta_{0,n} \cdot x_{0,n} + \\ 2^1 \cdot \delta_{1,1} \cdot x_{1,1} + 2^1 \cdot \delta_{1,2} \cdot x_{1,2} + \cdots + 2^1 \cdot \delta_{1,n} \cdot x_{1,n} + \\ \qquad\qquad\qquad \cdots \qquad\qquad\qquad\qquad\qquad\qquad + \\ 2^m \cdot \delta_{m,1} \cdot x_{m,1} + 2^m \cdot \delta_{m,2} \cdot x_{m,2} + \cdots + 2^m \cdot \delta_{m,n} \cdot x_{m,n} \leq K, \end{array}$$

where $\delta_{i,r} \in \{0,1\}$ for all i and r. Notice that every PB constraint whose coefficients are powers of 2 can be expressed in this way. Let us consider its BDD representation with the ordering $x_{0,1} < x_{0,2} < \ldots < x_{0,n} < x_{1,1} < \ldots < x_{m,n}$.

Lemma 7. *Let $[\beta, \gamma]$ be the interval of a node with selector variable $x_{i,r}$. Then 2^i divides β and $0 \leq \beta < (n + r - 1) \cdot 2^i$.*

Proof. By Proposition 2.1, β can be expressed as a sum of coefficients all of which are multiples of 2^i, and hence β itself is a multiple of 2^i. By Proposition 2.7, the only node whose interval contains negative values is the False node, and hence $\beta \geqslant 0$. Now, using Proposition 2.3, there must be an assignment to the variables $\{x_{0,1}, \ldots, x_{i,r-1}\}$ such that $2^0 \delta_{0,1} x_{0,1} + \cdots + 2^i \delta_{i,r-1} x_{i,r-1}$ belongs to the interval. Therefore:

$$\beta \leq 2^0 \delta_{0,1} x_{0,1} + \cdots + 2^i \delta_{i,r-1} x_{i,r-1} \leq 2^0 + 2^0 + \cdots + 2^i$$
$$= n2^0 + n2^1 + \cdots + n2^{i-1} + (r-1) \cdot 2^i = n(2^i - 1) + 2^i(r-1)$$
$$< 2^i(n + r - 1)$$

Corollary 8. *The number of nodes with selector variable $x_{i,r}$ is bounded by $n + r - 1$. In particular, the size of the BDD belongs to $\mathcal{O}(n^2 m)$.*

Proof. Let $\nu_1, \nu_2, \ldots, \nu_t$ be all the nodes with selector variable $x_{i,r}$. Let $[\beta_j, \gamma_j]$ the interval of ν_j. Note that such intervals are pair-wise disjoint since a non-empty intersection would imply that there exists a constraint represented by two different BDDs. Hence we can assume, w.l.o.g., that $\beta_1 < \beta_2 < \cdots < \beta_t$. Due to Lemma 7, we know that $\beta_j - \beta_{j-1} \geqslant 2^i$. Hence $2^i(n + r - 1) > \beta_t \geqslant \beta_{t-1} + 2^i \geqslant \cdots \geqslant \beta_1 + 2^i(t-1) \geqslant 2^i(t-1)$ and we can conclude that $t < n+r$.   □

4.2   A Consistent Encoding for PB Constraints

Let us now take an arbitrary PB constraint $C : a_1x_1 + \cdots a_nx_n \leq K$ and assume that a_M is the largest coefficient. For $m = \log a_M$, we can rewrite C splitting the coefficients into powers of two as shown in Example 6:

$$
\begin{aligned}
\tilde{C} : \quad & 2^0 \cdot \delta_{0,1} \cdot x_{0,1} & + \quad & 2^0 \cdot \delta_{0,2} \cdot x_{0,2} & + \cdots + \quad & 2^0 \cdot \delta_{0,n} \cdot x_{0,n} & + \\
& 2^1 \cdot \delta_{1,1} \cdot x_{1,1} & + \quad & 2^1 \cdot \delta_{1,2} \cdot x_{1,2} & + \cdots + \quad & 2^1 \cdot \delta_{1,n} \cdot x_{1,n} & + \\
& & & \cdots & & & + \\
& 2^m \cdot \delta_{m,1} \cdot x_{m,1} & + \quad & 2^m \cdot \delta_{m,2} \cdot x_{m,2} & + \cdots + \quad & 2^m \cdot \delta_{m,n} \cdot x_{m,n} & \leq K,
\end{aligned}
$$

where $\delta_{m,r} \, \delta_{m-1,r} \cdots \delta_{0,r}$ is the binary representation of a_r. Notice that C and $\tilde{C}$ represent the same constraint if we add clauses expressing that $x_{i,r} = x_i$ for appropriate i and r. This process is called *coefficient decomposition* of the PB constraint. A similar idea can be found in [BBR03].

The important remark is that, using a consistent SAT encoding of the BDD for $\tilde{C}$ (e.g. the one given in [ES06] or the one presented in the next section) and adding clauses expressing that $x_{i,r} = x_i$ for appropriate i and r, we obtain a consistent encoding for the original constraint C using $\mathcal{O}(n^2 \log a_M)$ auxiliary variables and clauses.

This is not difficult to see. Take an assignment A over the variables of C which cannot be extended to a model of C. This is because the coefficients corresponding to the variables true in A add more than K. Using the clauses for $x_{i,r} = x_i$, unit propagation will produce an assignment to the $x_{i,r}$'s that cannot be extended to a model of $\tilde{C}$. Since the encoding for $\tilde{C}$ is consistent, a false clause will be found. Conversely, if we consider an assignment A over the variables of C than can be extended to a model of C, this assignment can clearly be extended to a model for $\tilde{C}$ and the clauses expressing $x_{i,r} = x_i$. Hence, unit propagation on those clauses and the encoding of $\tilde{C}$ will not detect a false clause.

4.3   An Arc-Consistent Encoding for PB Constraints

Unfortunately, the previous approach does not produce an arc-consistency encoding. The intuitive idea can be seen in the following example:

Example 7. Let us consider the constraint $3x_1 + 4x_2 \leq 6$. After splitting the coefficients into powers of two, we obtain $C' : x_{0,1} + 2x_{1,1} + 4x_{2,2} \leq 6$. If we set $x_{2,2}$ to true, C' implies that either $x_{0,1}$ or $x_{1,1}$ have to be false, but the encoding cannot exploit the fact that both variables will receive the same truth value and hence both should be propagated. Adding clauses stating that $x_{0,1} = x_{1,1}$ does not help in this sense.   □

In order to overcome this limitation, we follow the method presented in [BKNW09, BBR09]. Let $C : a_1x_1 + \cdots + a_nx_n \leq K$ be an arbitrary PB constraint. We denote as C_i the constraint $a_1x_1 + \cdots + a_i \cdot 1 + \cdots + a_nx_n \leq K$, i.e., the constraint assuming x_i to be true. For every i with $1 \leq i \leq n$, we encode C_i as in Section 4.2 and, in addition, we add the binary clause $r_i \vee \neg x_i$, where r_i

is the root of the BDD for C_i. This clause helps us to preserve arc-consistency: given an assignment A such that $A \cup \{x_i\}$ cannot be extended to a model of C, literal $\bar{r}_i$ will be propagated using A (because the encoding for C_i is consistent). Hence the added clause will allow us to propagate $\bar{x}_i$.

All in all, the suggested encoding is arc-consistent and uses $\mathcal{O}(n^3 \log(a_M))$ clauses and auxiliary variables, where a_M is the largest coefficient.

5   SAT Encodings of BDDs for Monotonic Functions

In this section we consider a BDD representing a monotonic function F and we want to encode it into SAT. As expected, we want the encoding to be as small as possible and arc-consistent.

As usual, the encoding introduces an auxiliary variable for every node. Let ν be a node with selector variable x and auxiliary variable n. Let f be the variable of its false child and t be the auxiliary variable of its true child. Only two clauses per node are needed:

$$\neg f \rightarrow \neg n \qquad \neg t \wedge x \rightarrow \neg n.$$

Furthermore, we add a unit clause with the variable of the True node and another one with the negation of the variable of the False node.

Theorem 9. *The encoding is consistent in the following sense: a partial assignment A cannot be extended to a model of F if and only if $\neg r$ is propagated by unit propagation, where r is the root of the BDD.*

Proof. We prove the theorem by induction on the number of variables of the BDD. If the BDD has no variables, then the BDD is either the True node or the False node and the result is trivial.

Assume that the result is true for BDDs with less than k variables, and let F be a function whose BDD has k variables. Let r be the root node, x_1 its selector variable and f, t respectively its false and true children (note that we abuse the notation and identify nodes with their auxiliary variable). We denote by F_1 the function $F|_{x_1=1}$ (i.e., F after setting x_1 to true) and by F_0 the function $F|_{x_1=0}$.

- Let A be a partial assignment that cannot be extended to a model of F.
  - Assume $x_1 \in A$. Since A cannot be extended, the assignment $A \setminus \{x_1\}$ cannot be extended to a model of F_1. By definition of the BDD, the function F_1 has t as a BDD. By induction hypothesis, $\neg t$ is propagated, and since $x_1 \in A$, $\neg r$ is also propagated.
  - Assume $x_1 \notin A$. Then, the assignment $A \setminus \{\neg x_1\}$ cannot be extended to a model of F_0. Since F_0 has f as a BDD, by induction hypothesis $\neg f$ is propagated, and hence $\neg r$ also is.
- Let A be a partial assignment, and assume $\neg r$ has been propagated. Then, either $\neg f$ has also been propagated or $\neg t$ has been propagated and $x_1 \in A$ (note that x_1 has not been propagated because it only appears in one clause which is already true).

- Assume that $\neg f$ has been propagated. Since f is the BDD of F_0, by induction hypothesis the assignment $A \setminus \{x_1, \neg x_1\}$ cannot be extended to a model of F_0. Since the function is monotonic, $A \setminus \{x_1, \neg x_1\}$ neither can be extended to a model of F. Therefore, A cannot be extended to a model of F.
- Assume that $\neg t$ has been propagated and $x_1 \in A$. Since t is the BDD of F_1, by induction hypothesis $A \setminus \{x_1\}$ cannot be extended to a model of F_1, so neither can A be extended to a model of F.   □

For obtaining an arc-consistent encoding, we only have to add a unit clause.

Theorem 10. *If we add a unit clause forcing the variable of the root node to be true, the previous encoding becomes arc-consistent.*

Proof. We will prove it by induction on the variables of the BDD. The case $n = 0$ is trivial, so let us prove the induction case.

As before, let r be the root node, x_1 its selector variable and f, t its false and true children. We denote by F_1 and F_1 the functions $F_{|x_1=1}$ and $F_{|x_1=0}$.

Let A be a partial assignment that can be extended to a model of F. Assume that $A \cup \{x_i\}$ cannot be extended. We want to prove that $\overline{x}_i$ will be propagated.

- Let us assume that $x_1 \in A$. In this case, t is propagated due to the clause $\neg t \wedge x_1 \to \neg n$ and the unit clause n. Since $x_1 \in A$ and $A \cup \{x_i\}$ cannot be extended to a model of F, $A \setminus \{x_1\} \cup \{x_i\}$ neither can be extended to an assignment satisfying F_1. By induction hypothesis, since t is the BDD of the function F_1, $\neg x_i$ is propagated.
- Let us assume that $x_1 \notin A$ and $x_i \neq x_1$. Since F is monotonic, $A \cup \{x_i\}$ cannot be extended to a model of F if and only if it cannot be extended to a model of F_0. Notice that f is propagated thanks to the clause $\neg f \to \overline{n}$ and the unit clause n. By induction hypothesis, the method is arc-consistent for F_0, so $\overline{x}_i$ is propagated.
- Finally, assume that $x_1 \notin A$ and $x_i = x_1$. Since $A \cup \{x_1\}$ cannot be extended to a model of F, A cannot be extended to model of F_1. By Theorem 9, $\neg t$ is propagated and, due to $\neg t \wedge x_1 \to \neg n$ and n, also is $\neg x_1$.   □

6   Conclusions and Future Work

Both theoretical and practical contributions have been made. Regarding the theoretical part, we have proved that, unless NP=co-NP, there are PB constraints that do not admit polynomial BDDs. The existence of a concrete PB constraint family for which no polynomial BDDs exist remains an open problem, with interesting connections to the area of proof complexity. One of our aims is to continue working on this open question in the near future.

At the practical level, we have introduced a BDD-based polynomial and arc-consistent encoding of PB constraints and we have developed a BDD-based arc-consistent encoding of monotonic functions that only uses two clauses per BDD node. Indeed our initial motivation for this work has been practical, and we are currently working on implementation and experimental comparison of our encodings with other existing approaches on realistic problems.

References

[BBR06]   Bailleux, O., Boufkhad, Y., Roussel, O.: A Translation of Pseudo Boolean Constraints to SAT. JSAT 2(1-4), 191–200 (2006)

[BBR09]   Bailleux, O., Boufkhad, Y., Roussel, O.: New Encodings of Pseudo-Boolean Constraints into CNF. In: Kullmann, O. (ed.) SAT 2009. LNCS, vol. 5584, pp. 181–194. Springer, Heidelberg (2009)

[BBR03]   Bartzis, C., Bultan, T.: Construction of efficient BDDs for Bounded Arithmetic Constraints. In: Garavel, H., Hatcliff, J. (eds.) TACAS 2003. LNCS, vol. 2619, pp. 394–408. Springer, Heidelberg (2003)

[BHvMW09]   Biere, A., Heule, M.J.H., van Maaren, H., Walsh, T. (eds.): Handbook of Satisfiability. IOS Press, Amsterdam (2009)

[BKNW09]   Bessiere, C., Katsirelos, G., Narodytska, N., Walsh, T.: Circuit Complexity and Decompositions of Global Constraints. In: IJCAI 2009, pp. 412–418 (2009)

[Bry86]   Bryant, R.E.: Graph-based algorithms for boolean function manipulation. IEEE Trans. Computers 35(8), 677–691 (1986)

[ES06]   Eén, N., Sörensson, N.: Translating Pseudo-Boolean Constraints into SAT. JSAT 2, 1–26 (2006)

[J.S07]   Smaus, J.: On Boolean Functions Encodable as a Single Linear Pseudo-Boolean Constraint. In: Van Hentenryck, P., Wolsey, L.A. (eds.) CPAIOR 2007. LNCS, vol. 4510, pp. 288–302. Springer, Heidelberg (2007)

[SFSW09]   Schutt, A., Feydy, T., Stuckey, P.J., Wallace, M.: Why Cumulative Decomposition Is Not as Bad as It Sounds. In: Gent, I.P. (ed.) CP 2009. LNCS, vol. 5732, pp. 746–761. Springer, Heidelberg (2009)

DPLL+ROBDD Derivation Applied to Inversion of Some Cryptographic Functions

Alexey Ignatiev and Alexander Semenov

Institute for System Dynamics and Control Theory SB RAS, Irkutsk, Russia
alexey.ignatiev@gmail.com, biclop@rambler.ru

Abstract. The paper presents logical derivation algorithms that can be applied to inversion of polynomially computable discrete functions. The proposed approach is based on the fact that it is possible to organize DPLL derivation on a small subset of variables appeared in a CNF which encodes the algorithm computing the function. The experimental results showed that arrays of conflict clauses generated by this mode of derivation, as a rule, have efficient ROBDD representations. This fact is the departing point of development of a hybrid DPLL+ROBDD derivation strategy: derivation techniques for ROBDD representations of conflict databases are the same as those ones in common DPLL (variable assignments and unit propagation). In addition, compact ROBDD representations of the conflict databases can be shared effectively in a distributed computing environment.

1 Introduction

We consider the problem of inverting functions that form a family of type

$$f_n : \{0,1\}^n \to \{0,1\}^*,$$

where $\{0,1\}^n$ is the set of all possible binary sequences of the length n, $n \in N_1$,

$$\{0,1\}^* = \bigcup_{n \in N_1} \{0,1\}^n .$$

Assume that there exists a program M for deterministic Turing machine which computes an arbitrary function f_n of the considered family, and this program is polynomial time. The problem of inverting a function f_n at point $y \in range\ f_n$ is the problem of finding such (an arbitrary) $x \in \{0,1\}^n$ that $f_n(x) = y$.

There exists an effective procedure (polynomial time in n) reducing this problem to SAT problem. With the use of Tseitin transformations [22] this procedure constructs a CNF-encoding of a circuit $S(f_n)$ over $\{\&, \neg\}$ (any other complete basis could be here) which emulates M on all the possible inputs of $\{0,1\}^n$. By $X = \{x_1, \ldots, x_n\}$ we denote a set of Boolean variables corresponding to n

K.A. Sakallah and L. Simon (Eds.): SAT 2011, LNCS 6695, pp. 76–89, 2011.
© Springer-Verlag Berlin Heidelberg 2011

inputs of $S(f_n)$. For each logic gate G some new auxiliary variable $v(G)$ is introduced. Every AND-gate G is encoded by a CNF-representation of a Boolean function $v(G) \leftrightarrow u\&w$. Every NOT-gate G is encoded by a CNF-representation of a Boolean function $v(G) \leftrightarrow \neg u$. Here u and w are the variables corresponding to inputs of G. The CNF-encoding of $S(f_n)$ is

$$\underset{G \in S(f_n)}{\&} C(G),$$

where $C(G)$ is a CNF-encoding of G. Then

$$C_y(f_n) = \left(\underset{G \in S(f_n)}{\&} C(G) \right) \cdot y_1^{\sigma_1} \cdot \ldots \cdot y_m^{\sigma_m}$$

is a CNF encoding the invertion problem of the function f_n at point $y = (\sigma_1, \ldots, \sigma_m)$. Here

$$z^\sigma = \begin{cases} \bar{z}, \text{if } \sigma = 0 \\ z, \text{if } \sigma = 1 \end{cases}$$

and $y_1, \ldots, y_m$ are Boolean variables corresponding to outputs of $S(f_n)$. If $y \in range\ f_n$ then CNF $C_y(f_n)$ is satisfiable, and in any of its satisfying assignments one can find effectively such a vector $x \in \{0,1\}^n$ that $f_n(x) = y$.

It is well-known that while searching for a satisfying assignment for $C_y(f_n)$ it is possible to restrict DPLL derivation to a set of variables denoting an input for $S(f_n)$. We refer to this derivation strategy as "core-DPLL". Along with clause learning and restarts core-DPLL is complete for CNFs which encode the inversion of discrete functions of the class described above.

It is shown in [12] that, generally speaking, core-DPLL cannot polynomially simulate DPLL (even without clause learning and restarts). Our aim is to show that, nevertheless, the use of core-DPLL in inversion of discrete functions provides a number of additional (or rather useful) technical capabilities. In particular, a number of problems difficult for modern DPLL-solvers can though be solved without removing learnt clauses. One can also observe that arrays of conflict clauses learnt by core-DPLL generally have small ROBDD representations (even for CNFs which encode cryptographic algorithms). The size of these ROBDD representations are hundreds of times smaller than the original clauses form. Therefore, it is possible to share effectively arrays of conflict clauses (in the ROBDD form) accumulated at various nodes of distributed computing environments.

A brief outline of the paper is given below. In the first section, we describe basic logical derivation mechanisms combining core-DPLL and a derivation technique for ROBDD representations of conflict databases. The second section describes a parallel implementation of DPLL+ROBDD solver made with the use of MPI. In the third section, we present results of numerical experiments on inversion of some cryptographic functions by the described solver.

2   Basic Mechanisms of DPLL+ROBDD Derivation

Let $C_y(f_n)$ be a CNF encoding the problem of inverting a discrete function f_n (of the class described above) at an arbitrary point $y \in range\ f_n$. In this section we use binary decision diagrams (more precisely, ROBDDs) to represent arrays of conflict clauses accumulated by core-DPLL while finding a satisfying assignment for $C_y(f_n)$. General ideas to represent exhaustive DPLL derivation in the form of binary decision diagrams were considered in [11]. It should be also noted that there are examples of hybrid approaches combining DPLL with BDDs [1,8,4,9]. The methods we suggest here are based on the empirical fact that ROBDDs do compress arrays of conflict clauses learnt during the core-DPLL derivation.

Binary decision diagrams (BDDs) were introduced by C. Y. Lee in the article [14]. The importance of this fundamental data structure for discrete mathematics was realized after R. Bryant's work [3] coming out. In that paper he described a family of algorithms manipulating Boolean functions with the use of BDDs. One of the main theoretical results in [3] is the theorem about canonical representation of Boolean functions in the form of ROBDDs (a ROBDD is a reduced BDD without repeatable fragments). ROBDDs are often able to represent Boolean functions arising in applications in a very compact form.

Next, we will use two algorithms described in [3]. The first one is *Apply* which constructs a ROBDD representation of a function $f_1 * f_2$ using ROBDD representations $B(f_1)$ and $B(f_2)$ of functions f_1 and f_2, where "$*$" is an arbitrary binary logical operation. If the variable orderings in $B(f_1)$ and $B(f_2)$ are identical, then time complexity of *Apply* is $O\left(|B(f_1)| \cdot |B(f_2)|\right)$ (here and below by $|B|$ we denote a number of vertices in B). The second one is *Restrict*. Algorithm *Restrict* takes $(B(f), x, \alpha)$ as an input. Here $B(f)$ is a ROBDD representation of a function f defined by a Boolean formula $L(f)$, x is a variable appeared in $L(f)$ and α is a constant of $\{0, 1\}$. This algorithm produces a ROBDD representation of a function $f|_{x=\alpha}$ defined by a formula $L(f)|_{x=\alpha}$. Time complexity of *Restrict* is $O\left(|B(f)|\right)$.

Consider a CNF

$$C_y(f_n) \cdot D_1\left(x_1^1, \ldots, x_{r_1}^1\right) \cdot \ldots \cdot D_q\left(x_1^q, \ldots, x_{r_q}^q\right),$$

where $D_i\left(x_1^i, \ldots, x_{r_i}^i\right)$, $i \in \{1, \ldots, q\}$ are the conflict clauses learnt during q restarts of core-DPLL for $C_y(f_n)$. Thus,

$$\bigcup_{i=1}^{q} \{x_1^i, \ldots, x_{r_i}^i\} \subseteq X,$$

where $X = \{x_1, \ldots, x_n\}$ is a set of input variables for a circuit $S(f_n)$. Let's denote a ROBDD representation of a function defined by the formula

$$D_1\left(x_1^1, \ldots, x_{r_1}^1\right) \cdot \ldots \cdot D_q\left(x_1^q, \ldots, x_{r_q}^q\right) \tag{1}$$

as B^*. We have the following fact.

Theorem 1. *Let* $x \in \{0,1\}^n$ *be a solution of the inversion problem for* f_n *at some point* $y \in range\ f_n$. *Then there exists such a path* π *in* B^* *from the root to the terminal "1", that* $x \in A(\pi)$, *where* $A(\pi)$ *is a subset of* $\{0,1\}^n$ *specified by* π.

Proof sketch. Let $x \in \{0,1\}^n$ be an arbitrary solution of the inversion problem considered. Suppose that there is no such a path from the root of B^* to "1", which contains x. Therefore, if we substitute x into (1) we get 0. Note that each clause $D_i\left(x_1^i, \ldots, x_{r_i}^i\right)$, $i \in \{1, \ldots, q\}$, is a logical consequence of CNF $C_y(f_n)$. However, if we substitute x into $C_y(f_n)$ then the satisfying assignment for $C_y(f_n)$ results from unit propagation [6]. Thereby, CNF $C_y(f_n)$ is made true by some assignment and CNF (1) (which is a logical consequence of $C_y(f_n)$) is made false by the same assignment. This contradicts our assumption, so we are forced to conclude that there is a path from the root of B^* to "1", which contains x.   □

This theorem provides a basis for the general hybrid DPLL+ROBDD derivation strategy considered below. During the derivation process a ROBDD representation of conflict databases is regarded as a formula. Therefore, one can assign some variables in the ROBDD, and certain variables can be implied from a unit propagation similarity. Just as in DPLL, the result of every conflict is some conflict clause learnt. In our case, every conflict clause contains only literals over a set of input variables for a function. The resulting conflict clauses are added to the ROBDD representation of a conflict database using *Apply* procedure. Let $B(f)$ be a ROBDD representation of an arbitrary Boolean function $f(x_1, \ldots, x_n)$. Each path from the root of $B(f)$ to a terminal vertex defines a family of sets of truth values for $x_1, \ldots, x_n$.

Let's put in correspondence each variable x_i, $i \in \{1, \ldots, n\}$, and terminal vertex "0" with a set of the variable's truth values defined by all the paths in $B(f)$ from the root to "0". We denote this set by $\Delta^0(x_i)$. One can define $\Delta^1(x_i)$ in a similar manner.

Suppose, that in ROBDD $B(f)$ the following conditions for a variable $x_k \in X$, $X = \{x_1, \ldots, x_n\}$, hold:

1. Every path π in $B(f)$ from the root to "1" passes through a vertex marked by x_k.
2. $|\Delta^1(x_k)| = 1$.

Then variable x_k may take on exactly one value (the value of $\Delta^1(x_k)$) in any truth assignment over X that makes f assign true.

Definition 1. *The situation defined by conditions 1–2 is called a ROBDD-based consequence of a value of variable* x_k.

A ROBDD-based consequence of some variable in $B(f)$ presenting an array of conflict clauses is a similarity of unit propagation used in DPLL derivation. Further we make use of a modified version of *Restrict* which could assign a set of variables of X into $B(f)$ at the same time. As noted by R. Bryant in [3], time complexity of this algorithm is the same as time complexity of the original

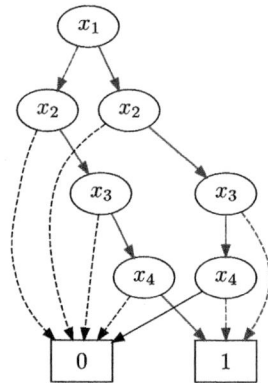

Fig. 1. ROBDD representation of a function $x_2 \cdot (x_1 \oplus x_3 \cdot x_4)$ using the variable ordering $x_1 \prec x_2 \prec x_3 \prec x_4$. We have a ROBDD-based consequence of variable x_2 ($x_2 = 1$) here because each path from the root to "1" passes through a vertex marked by x_2 and $|\Delta^1(x_2)| = 1$.

Restrict, i.e. $O(|B(f)|)$. The basic idea of the procedure described below was proposed in [5]. However, the authors of that paper did not estimate its time complexity.

Theorem 2. *For a ROBDD $B(f)$ and the values $x_{i_1} = \alpha_{i_1}, \ldots, x_{i_m} = \alpha_{i_m}$, $m \leq n$, $\alpha_{i_j} \in \{0, 1\}$, $j \in \{1, \ldots, m\}$, time complexity of the procedure which substitutes given values into $B(f)$ and checks for ROBDD-based consequences of other variables is $O(|B(f)|)$.*

Proof sketch. Let's substitute $x_{i_1} = \alpha_{i_1}, \ldots, x_{i_m} = \alpha_{i_m}$ into $B(f)$. As it was said above, this process takes time bounded by $O(|B|)$. After making the substitutions we check for ROBDD-based consequences. Note that ROBDD-based consequence of some variable x_k results in exactly one of the following:

1. each vertex marked by x_k has "0" as the high-child;
2. each vertex marked by x_k has "0" as the low-child.

Therefore, we have a ROBDD-based consequence of $x_i = 1$ if and only if each vertex marked by x_i has "0" as the low-child and every path from "1" to the root passes through a vertex marked by x_i.

Using this fact, we go from "1" towards the root of the ROBDD. Let $V(1)$ be a set containing parents of "1". We also denote a set of variables marking vertices of $V(1)$ by $X(1) = \{x_{i_1}, \ldots, x_{i_r}\}$. We can choose from $V(1)$ all the vertices marked by such a variable x_{i_*} that $x_{i_*} \prec x_j \; \forall j \in \{i_1, \ldots, i_r\} \setminus \{i_*\}$ (according to the variable ordering in the ROBDD). Variable x_{i_*} is referred to as a minimal variable in $X(1)$ with respect to the variable ordering. It is obvious that for any variable of $X(1) \setminus \{x_{i_*}\}$ a ROBDD-based consequence is not possible. By $\tilde{V}(1)$ we denote a set of all vertices marked by variables of $X(1) \setminus \{x_{i_*}\}$. Next, move up from each vertex in $\tilde{V}(1)$ toward the root of the

ROBDD until the first vertex marked by variable x_k appears, such that either $x_k = x_{i_*}$, or $x_k \prec x_{i_*}$. A set of the ROBDD vertices generated in this sense by set $V(1)$ is denoted by $V(x_{i_*})$, and a set of variables to mark vertices of $V(x_{i_*})$ is denoted by $X(x_{i_*}) = \{x_{k_1}, \ldots, x_{k_s}\}$. If $x_{k_1} = \ldots = x_{k_s} = x_{i_*}$, then we check for each vertex of $V(x_{i_*})$ whether its low-child (or high-child) is "0". If yes, then we have a ROBDD-consequence of variable x_{i_*}. If not, then we should go on the procedure. It is not difficult to understand that the described algorithm finds all the possible ROBDD-consequences in one pass through the ROBDD. Hence, time complexity of the procedure which makes substitutions into $B(f)$ and checks for every possible ROBDD-based consequence is $O\left(|B(f)|\right)$. $\square$

This theorem implies the next corollary.

Theorem 3. *If some substitution into $B(f)$ implies a ROBDD-based conse-*
quence of $x_k = \alpha_k$, $\alpha_k \in \{0,1\}$ for some $x_k \in X$, then substitution of $x_k = \alpha_k$
in $B(f)$ cannot imply another ROBDD-based consequence.

Proof sketch. Suppose, that some substitution into $B(f)$ implies a ROBDD-based consequence $x_k = \alpha_k$. Assume without loss of generality that $\alpha_k = 1$. In accordance with the above (see the first paragraph of theorem's 2 proof) this assumption means that the low-child of each vertex marked by x_k is "0". Substitution of $x_k = 1$ in $B(f)$ means that each vertex $u(x_k)$ hands over its high-child to its parents. However, the low-child of $u(x_k)$, that is, the terminal "0", is not handed over to any vertex. Thus, substitution of $x_k = 1$ into $B(f)$ cannot cause such a vertex in $B(f)$ *to appear*, that some of its children is "0" (but it does not mean that there is no such a vertex before the substitution). Similar arguments hold if $\alpha_k = 0$. $\square$

This fact shows a very useful feature of ROBDD considered as an array of Boolean constraints. It's known that substituting a variable's value into a CNF may lead to a situation where unit clause rule can be used several times. The procedure implementing iterative unit clause rule is the so-called Boolean constraint propagation (BCP). In the general case, BCP passes through the CNF many times. The obtained feature of ROBDDs means that ROBDD-based consequences implied by an arbitrary substitution cannot imply a new ROBDD-based consequence and, therefore, all the information implied by the substitution comes out as a result of a single pass through a ROBDD (see Fig. 2).

Another positive property of the hybrid derivation is the possibility to easily implement lazy computations (an analogue of well-known data structures used in BCP, i.e. "watched literals", [15]) using ROBDDs.

Let's consider the conditions determining a situation which in some sense is ambivalent to a ROBDD-based consequence.

3. For a variable $x_q \in X$, $X = \{x_1, \ldots, x_n\}$, every path π in $B(f)$ from the root to "0" passes through a vertex marked by x_q.
4. $|\Delta^0(x_q)| = 1$.

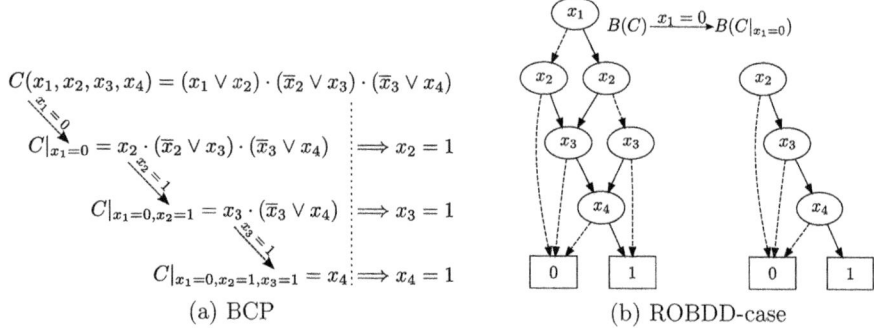

$$C(x_1, x_2, x_3, x_4) = (x_1 \vee x_2) \cdot (\overline{x}_2 \vee x_3) \cdot (\overline{x}_3 \vee x_4)$$

$$C|_{x_1=0} = x_2 \cdot (\overline{x}_2 \vee x_3) \cdot (\overline{x}_3 \vee x_4) \implies x_2 = 1$$

$$C|_{x_1=0,x_2=1} = x_3 \cdot (\overline{x}_3 \vee x_4) \implies x_3 = 1$$

$$C|_{x_1=0,x_2=1,x_3=1} = x_4 \implies x_4 = 1$$

(a) BCP   (b) ROBDD-case

Fig. 2. On the left we show the BCP process in CNF $(x_1 \vee x_2) \cdot (\overline{x}_2 \vee x_3) \cdot (\overline{x}_3 \vee x_4)$ started by assigning $x_1 = 0$; on the right we demonstrate the result of substituting $x_1 = 0$ into ROBDD representation of the considered CNF — here a single pass through the ROBDD is required.

Theorem 4. *Let $B(f)$ be an arbitrary ROBDD and there be such a variable x_q in $B(f)$ so that conditions 3–4 hold for x_q. Then there are no possible ROBDD-based consequences of any variable from $X \setminus \{x_q\}$ in $B(f)$. Time complexity of procedure which checks whether conditions 3–4 hold is $O(|B(f)|)$.*

Proof sketch. Let conditions 3–4 hold for some variable $x_q \in X$ in $B(f)$. Assume without loss of generality that $\Delta^0(x_q) = \{1\}$. By analogy with the proof of theorem 2 the assumption means that each vertex marked by x_q has "1" as the low-child.

Let $x_p \in X \setminus \{x_q\}$ be an arbitrary variable. There are two possible alternatives for its location relative to x_q with respect to the variable ordering in $B(f)$ (variable x_1 marks the root of $B(f)$):

$$1 : x_1 \prec \ldots \prec x_q \prec \ldots \prec x_p \prec \ldots$$
$$2 : x_1 \prec \ldots \prec x_p \prec \ldots \prec x_q \prec \ldots$$

Consider the first case. As it was said above, the low-child of each vertex marked by x_q is the terminal "1". This means that there is such a path from the root of $B(f)$ to "1" that does not pass through vertices marked by x_p. In other words, the ROBDD-based consequence of x_p is not possible.

Consider the second case. Assume there is a ROBDD-based consequence of x_p in $B(f)$, i.e. conditions 1–2 hold for x_p. Then one of the children of each vertex marked by x_p is "0". However, this means that there are such paths from vertices marked by x_p to "0" which do not pass through vertices marked by x_q. This contradicts the fact that conditions 3–4 hold for x_q.

It is not difficult to understand that validity of conditions 3–4 can be checked by a procedure which is similar to the procedure described in the proof of theorem 2 and has the same time complexity — $O(|B(f)|)$.   □

This theorem provides a possibility to formulate mechanisms of lazy computations while assigning variables implied during the hybrid derivation process.

If conditions 3–4 hold for some x_q in a ROBDD B^*, and a value of x_k, $k \neq q$, is derived from a CNF, then it is not necessary to substitute this value into B^* because no new ROBDD based consequences will be implied. It is sensible to store up all the variables to assign until the moment of assigning x_q and after that to substitute them all into B^* at the same time checking every possible ROBDD-based consequence (see theorem 2).

3   Parallel DPLL+ROBDD Solver Sharing Arrays of Conflict Clauses in the ROBDD Form

As already mentioned, in practice even for hard cryptographic tests core-DPLL generates conflict databases which have compact ROBDD representations (one can use the variable ordering defined by a current state of accumulated variable activities [17]). This fact leads us to an idea of a parallel solver to accumulate arrays of conflict clauses in the ROBDD form at different computing nodes and to share them effectively between the nodes. It is a small size of a ROBDD representation of conflict clauses that provides the efficiency.

In more detail, the solver consists of two components. A core-DPLL component is implemented as a modification of MiniSat-C v1.14.1 [7] named "coresat". Conflict analysis made by coresat uses information only on those function's input variables which are responsible for a conflict. It is based on the use of characteristic vectors (this technique is similar to the one proposed in [13]). As a result, there is no need to use an implication graph [16] to determine a reason for the conflict. Another solver's component encloses the process constructing ROBDD representations of conflict databases and derivation procedures for ROBDDs based on the algorithms described above.

The interaction between core-DPLL and ROBDD components of the hybrid solver is implemented in compliance with the schema shown in Fig. 3.

Under this schema, the hybrid DPLL+ROBDD solver is an iterated procedure determined by the actions listed below.

1. At the initial stage only coresat operates. The result is an array of learnt conflict clauses, each contains literals over a set of input variables for the function.
2. The solver suspends coresat and starts to construct a ROBDD representation of the array of conflict clauses learnt during the first step (for this purpose we use algorithm *Apply* by R. Bryant). It is reasonable to use the variable ordering defined by variable activities which were accumulated by this moment.
3. The result of each iteration is a new ROBDD obtained using *Apply* to a previous one and the ROBDD representation of the conflict database constructed during the current iteration (see step 2). The variable ordering can differ in the two ROBDDs. Therefore, before running *Apply* we need to re-order the old ROBDD according to the new variable ordering.
4. The process continues iteratively and is terminated if a satisfying assignment is found or it is proven that the CNF instance is unsatisfiable.

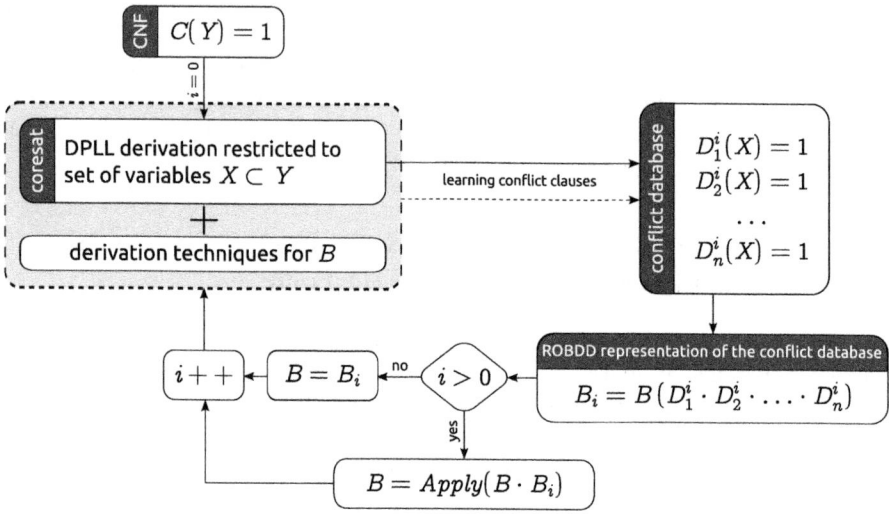

Fig. 3. Schema of the hybrid DPLL+ROBDD solver

A sequential variant of the hybrid solver is referred to as "hsat". A parallel version of the hybrid solver (we name it "mhsat") is implemented as an MPI application and is a bunch of hsat instances, which work simultaneously and periodically share their conflict databases in the ROBDD form. To ensure that hsat instances start to solve the problem differently from each other, we choose unique initial variable activities for each of them.

Operating of mhsat can be seen as a serial implementation of the following steps:

1. The stage of accumulating conflict clauses in the ROBDD form. Each node generates conflict clauses irrespective of each other and constructs its local ROBDD in accordance with its current variable activities.
2. The stage of merging accumulated conflict databases. There is a number of alternatives on how to make this step. Here we describe the simplest one:
   (a) Exchanging local variable activities to construct the common variable ordering;
   (b) Reconstructing each local ROBDD according to the common variable ordering;
   (c) Exchanging conflict databases and joining them (we use recursive doubling [21] and *Apply* for this purpose). The result of this stage is a final ROBDD which is constructed on some computing node and represents the complete array of conflict clauses with respect to the common variable ordering;
   (d) Sharing the final ROBDD to all the other nodes;
   (e) Reconstructing the final ROBDD according to a local variable ordering on each of the computing nodes.

It should be noted that joining the ROBDDs with the use of *Apply* is optional. Instead it is possible to make each node store local copies of all the ROBDDs made by other nodes. In this case, each of the nodes has an array of the ROBDDs and runs a derivation process for all the ROBDDs separately. Such approach can improve the solver's performance when the serial use of *Apply* leads to an exponential growth of the output ROBDD's size.

4   Experimental Results

We experimented on CNFs which encode a cryptanalysis of the weakened keystream generator used in the cipher A5/1. This generator is used to encrypt the traffic in GSM networks. The authors of [19] minutely described procedures for constructing a CNF encoding cryptanalysis of the generator A5/1. They also presented results on coarse-grained approach to logical cryptanalysis of the generator in a Grid system. This approach is based on the technology of decomposition of a SAT problem encoding the generator algorithm into a family of SAT problems of lower dimension. By $C(A5/1)$ we denote the CNF encoding the algorithm of the generator A5/1, and by $X(A5/1)$ we denote the set of Boolean variables appeared in $C(A5/1)$. In accordance with the technique described in [19], from $X(A5/1)$ one can select a subset of Boolean variables, each corresponds to initial contents of a cell of a register of the generator. Cardinality of this subset is d, $d \leq 64$. This set is called a decomposition set and denoted by X_d. Substituting all possible truth values for variables of X_d in $C(A5/1)$

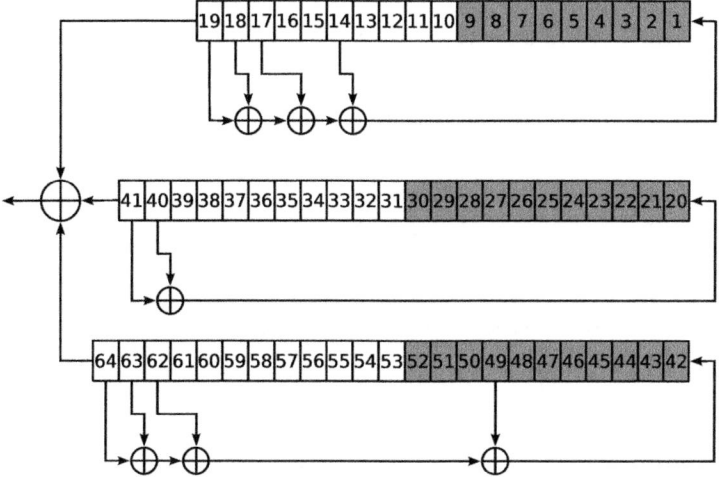

Fig. 4. Schema of the A5/1 keystream generator which consists of 3 LFSRs, given by the following connection polynomials over GF(2): LFSR 1: $X^{19} + X^{18} + X^{17} + X^{14} + 1$; LFSR 2: $X^{22} + X^{21} + 1$; LFSR 3: $X^{23} + X^{22} + X^{21} + X^8 + 1$. The algorithm of A5/1 keystream generator is encoded by CNF in accordance with the technique described in [19].

generates a decomposition family consisting of 2^d CNFs. This family forms a parallel task list that can be processed in a distributed computing environment. Inter-processor communications are extremely rare here.

The coarse-grained approach shows the best results in the case of decomposing by 31 variables. In Fig. 4 shown below the cells corresponding to this set of 31 variables are dark shaded.

In our experiments we used the decomposion set X_{20} shown in Fig. 5. Substituting all possible truth values for variables of X_{20} in $C(A5/1)$ generates a decomposition family consisting of 2^{20} CNFs. As the test material we considered 50 CNFs, chosen randomly from this decomposition family. All selected in such a way CNFs were unsatisfiable. Tests were run on a platform of Intel Xeon E5345 (4 cores, 2.33 GHz), 8 GB RAM. To evaluate efficiency of the hybrid DPLL+ROBDD derivation we used approaches listed below:

1. Coarse-grained parallelization without sharing clauses. For each of the fifty CNFs we constructed 4 simpler CNFs obtained by substituting all the possible values of two variables x_{23} and x_{45} into the original one. Thus, each of the 4 CPU cores solved its own fifty SAT problems irrespective of other cores. In this series of experiments we used the following solvers: hsat, dminisat [19] and MiniSat 2.2.0 [7].
2. The use of solvers with parallel architecture. In this series of experiments there were involved multi-threaded solvers MiraXT 1.1 [18] and ManySAT 1.1 [10], as well as mhsat, which is an MPI application.
3. Sequential solving all the considered tests by hsat using one CPU core.

We emphasize that the original versions of ManySAT and MiniSat cannot cope with tests of the set under consideration. But it is possible to solve this problem by assigning nonzero values to initial activity for those variables which correspond

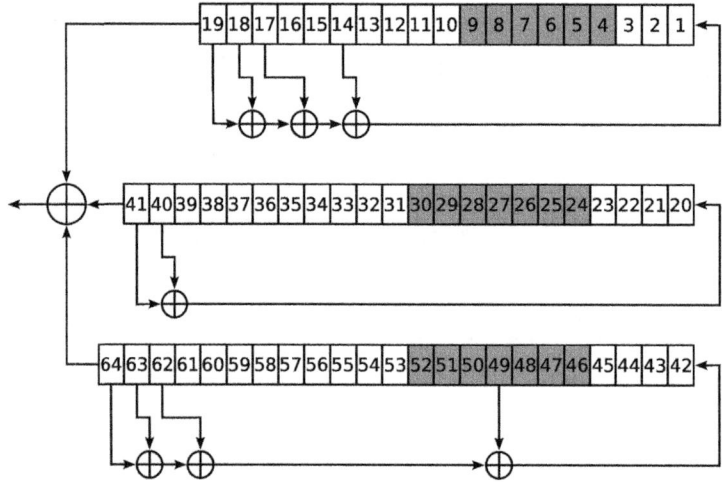

Fig. 5. Schema of the decomposition set X_{20}

Table 1. Average solving time for each of the solvers

place	solver	mode of operating	number of cores	avg. time (seconds)
1	mhsat	parallel	4	569.016
2	hsat	coarse-grained	4	644.254
3	MiraXT (mod)	parallel	4	1639.192
4	hsat	sequential	1	2385.578
5	dminisat	coarse-grained	4	2750.486
6	MiraXT (orig)	parallel	4	3214.178
7	ManySAT (mod)	parallel	4	3378.078
8	MiniSat (mod)	coarse-grained	4	5836.782

to initial contents of A5/1's registers (in Table 1 this modification is denoted by "mod"). In contrast to ManySAT and MiniSat even the original version of MiraXT can handle the considered tests. However, increasing initial activity of the same variables doubles its performance on average.

Note the fact that mhsat taking 4 CPU cores is more than 4 times faster than its sequential version (hsat).

In addition to the parallel solvers listed above, we tried to use the well-known solvers CryptoMiniSat 2.9.0 [20] and Plingeling 276 [2]. However, these solvers could not cope with the tests in a reasonable time.

5   Conclusions and Future Work

According to the experimental results we can conclude that the hybrid DPLL+ +ROBDD derivation techniques described in the paper may be useful in solving the function inversion problems that are difficult for the solvers performed better on the well-known test libraries.

We suppose that our hybrid methods have potential to be heavily improved. In particular, some improvements of the basic hsat's algorithms are expected in the near future. In addition, we also project to analyze various alternatives on inter-process sharing the arrays of conflict clauses generated by different nodes of a large-scale distributed computing environment.

Despite the interesting experimental results we realize that they are not enough to justify the efficiency of our approach to a wide class of functions. Therefore, we hope to succeed in expanding the class of tests, which can be solved by the described algorithms much more efficiently in comparison with traditional DPLL-based derivation methods.

Acknowledgements

The authors would like to thank Alexei Hmelnov,a Dmitry Bespalov and Stepan Kochemazov (ISDCT SB RAS) for their help and numerous valuable advices.

This work is supported by Russian Foundation for Basic Research (Grant No. 11-07-00377-a).

References

1. Aloul, F.A., Mneimneh, M.N., Sakallah, K.A.: ZBDD-Based Backtrack Search SAT Solver. In: Proceedings of International Workshop on Logic and Synthesis (IWLS), pp. 131–136 (2002)
2. Biere, A.: Lingeling, Plingeling, PicoSAT and PrecoSAT at SAT Race 2010. Tech. Rep. 10/1, FMV Reports Series, Institute for Formal Models and Verification, Johannes Kepler University, Altenbergerstr. 69, 4040 Linz, Austria (2010)
3. Bryant, R.E.: Graph-Based Algorithms for Boolean Function Manipulation. IEEE Transactions on Computers 35(8), 677–691 (1986)
4. Chatalic, P., Simon, L.: Zres: The old Davis-Putnam procedure meets ZBDDs. In: McAllester, D. (ed.) CADE 2000. LNCS(LNAI), vol. 1831, pp. 449–454. Springer, Heidelberg (2000)
5. Damiano, R.F., Kukula, J.H.: Checking satisfiability of a conjunction of BDDs. In: 40th Design Automation Conference, DAC 2003, pp. 818–823 (2003)
6. Dowling, W.F., Gallier, J.H.: Linear-time algorithms for testing the satisfiability of propositional horn formulae. The Journal of Logic Programming 1(3), 267–284 (1984)
7. Eén, N., Sörensson, N.: An Extensible SAT-solver. In: Giunchiglia, E., Tacchella, A. (eds.) SAT 2003. LNCS, vol. 2919, pp. 502–518. Springer, Heidelberg (2004)
8. Ganai, M., Gupta, A.: SAT-Based Scalable Formal Verification Solutions. Series on Integrated Circuits and Systems. Springer-Verlag New York, Inc., Secaucus (2007)
9. Gopalakrishnan, S., Durairaj, V., Kalla, P.: Integrating CNF and BDD based SAT solvers. In: IEEE International High-Level Design, Validation, and Test Workshop, pp. 51–56 (2003)
10. Hamadi, Y., Jabbour, S., Sais, L.: ManySAT: a Parallel SAT Solver. Journal on Satisfiability, Boolean Modeling and Computation, Special Issue on Parallel SAT Solving 6, 245–262 (2009)
11. Huang, J., Darwiche, A.: The Language of Search. Journal of Artificial Intelligence Research 29, 191–219 (2007)
12. Järvisalo, M., Junttila, T.: Limitations of restricted branching in clause learning. Constraints 14(3), 325–356 (2009)
13. Kuehlmann, A., Paruthi, V., Krohm, F., Ganai, M.K.: Robust Boolean Reasoning for Equivalence Checking and Functional Property Verification. IEEE Transactions on Computer-Aided Design 21(12), 1377–1394 (2002)
14. Lee, C.Y.: Representation of Switching Circuits by Binary-Decision Programs. Bell Systems Technical Journal 38, 985–999 (1959)
15. Lynce, I., Marques-Silva, J.: Efficient data structures for backtrack search SAT solvers. Annals of Mathematics and Artificial Intelligence 43(1), 137–152 (2005)
16. Marques-Silva, J.P., Sakallah, K.A.: GRASP: A search algorithm for propositional satisfiability. IEEE Transactions on Computers 48(5), 506–521 (1999)
17. Moskewicz, M.W., Madigan, C.F., Zhao, Y., Zhang, L., Malik, S.: Chaff: engineering an efficient SAT solver. In: Proceedings of the 38th Annual Design Automation Conference, DAC 2001, pp. 530–535. ACM, New York (2001)
18. Schubert, T., Lewis, M., Becker, B.: PaMiraXT: Parallel SAT Solving with Threads and Message Passing. Journal on Satisfiability, Boolean Modeling and Computation, Special Issue on Parallel SAT Solving 6, 203–222 (2009)

19. Semenov, A., Zaikin, O., Bespalov, D., Posypkin, M.: Parallel algorithms for SAT in application to inversion problems of some discrete functions, arXiv:1102.3563v1 [cs.DC]

20. Soos, M., Nohl, K., Castelluccia, C.: Extending SAT Solvers to Cryptographic Problems. In: Kullmann, O. (ed.) SAT 2009. LNCS, vol. 5584, pp. 244–257. Springer, Heidelberg (2009)

21. Thakur, R., Rabenseifner, R., Gropp, W.: Optimization of Collective Communication Operations in MPICH. Int'l Journal of High Performance Computing Applications 19(1), 49–66 (2005)

22. Tseitin, G.: On the complexity of derivation in propositional calculus. Studies in Constructive Mathematics and Mathematical Logic 2, 234–259 (1968)

πDD: A New Decision Diagram for Efficient Problem Solving in Permutation Space

Shin-ichi Minato*

Hokkaido University, Sapporo 060-0814, Japan
minato@ist.hokudai.ac.jp

Abstract. Permutations and combinations are two basic concepts in elementary combinatorics. Permutations appear in various problems such as sorting, ordering, matching, coding and many other real-life situations. While conventional SAT problems are discussed in combinatorial space, "permutatorial" SAT and CSPs also constitute an interesting and practical research topic.

In this paper, we propose a new type of decision diagram named "πDD," for compact and canonical representation of a *set of permutations*. Similarly to an ordinary BDD or ZDD, πDD has efficient algebraic set operations such as union, intersection, etc. In addition, πDDs hava a special Cartesian product operation which generates all possible composite permutations for two given sets of permutations. This is a beautiful and powerful property of πDDs.

We present two examples of πDD applications, namely, designing permutation networks and analysis of Rubik's Cube. The experimental results show that a πDD-based method can explore billions of permutations within feasible time and space limits by using simple algebraic operations.

1 Introduction

Permutations and combinations are two basic concepts in elementary combinatorics and discrete mathematics [4]. Permutations appear in various problems such as sorting, ordering, matching, coding and many other real-life situations. Permutations are also important in group theory since they correspond to bijective functions and generate symmetric groups. While conventional SAT problems are defined in combinatorial space, "permutatorial" SAT and CSPs also constitute an interesting research topic.

In this paper, we propose a new type of decision diagram named "πDD," for compact and canonical representation of *sets of permutations*. πDDs are based on BDDs (Binary Decision Diagrams)[1] and ZDDs (Zero-suppressed BDDs)[6]. Ordinary BDDs/ZDDs provide representations of propositional logic functions or sets of combinations, namely, they represent partial sets of combinatorial space. Data structures and algorithms on BDDs/ZDDs have been researched for more

---

* He also works for ERATO MINATO Discrete Structure Manipulation System Project, Japan Science and Technology Agency.

K.A. Sakallah and L. Simon (Eds.): SAT 2011, LNCS 6695, pp. 90–104, 2011.
© Springer-Verlag Berlin Heidelberg 2011

than twenty years, and BDD/ZDD-based SAT solving techniques have also been explored [2]. However, most DD-based methods are limited to combinatorial space, and no practical techniques for direct solving of permutational problems are known, even though they have various important applications.

πDDs are the first practical idea for efficient manipulation of sets of permutations on the basis of decision diagrams. This data structure can compress a large number of permutations into a compact and canonical representation. Similarly to ordinary BDDs/ZDDs, πDDs have efficient algebraic set operations such as union, intersection, and difference. In addition, πDDs have a special Cartesian product operation which generates all possible composite permutations (cascade of two permutations) for two given sets of permutations. This is a beautiful and powerful property for solving various problems in permutation space. For example, we can represent the primitive moves of Rubik's Cube with a small πDD, and by simply multiplying this πDD by itself k times, we can generate a single canonical πDD representing all possible positions reachable within k moves. The computation time depends on the size of the πDD, which is sometimes much smaller than the number of positions. Once we have generated πDDs for a problem, we can easily apply various analysis or testing techniques, such as counting the exact number of permutations, exploring satisfiable permutations for a given constraint and calculating the minimal or the average cost of all permutations.

The idea of πDDs provide hints about the application of state-of-the-art SAT techniques used for solving combinatorial problems in the "permutatorial world." There is a rich body of studies in group theory led by Galois and many researchers in discrete mathematics [3]. πDDs represent a new computational technique which can be applied in such research fields, and we can expect it to yield numerous exciting results in the future.

In the rest of this paper, Section 2 describes some notations and the basics of BDDs/ZDDs. In Section 3, we propose the general structure of πDDs, and Section 4 gives the algorithms of algebraic operations for πDDs, followed by Section 5, which presents experimental results for two typical problems, namely, designing permutation networks and analyzing Rubik's Cube.

2   Preliminaries

2.1   Sets of Permutations

A *permutation* is a bijective function $\pi : S \rightarrow S$, where S is a finite set $\{1, 2, 3, \ldots, n\}$. Although it is often confusing, in this paper we use the notation for permutation $\pi = (a_1, a_2, a_3, \ldots, a_n)$, in which each item k moves to a_k. For example, $\pi = (4, 2, 1, 3)$ implies $1 \rightarrow 4$, $2 \rightarrow 2$, $3 \rightarrow 1$, and $4 \rightarrow 3$. In this case, we may also use multiplicative forms, such as $1\pi = 4$, $2\pi = 2$, $3\pi = 1$, and $4\pi = 3$. A *composition* of two permutations $\pi_1 \pi_2$ simply indicates a composition of two bijective functions. For example, if $\pi_1 = (3, 1, 2)$ and $\pi_2 = (3, 2, 1)$ then $\pi_1 \pi_2 = (1, 3, 2)$ because $1\pi_1 \pi_2 = 3\pi_2 = 1$, $2\pi_1 \pi_2 = 1\pi_2 = 3$, and $3\pi_1 \pi_2 = 2\pi_2 = 2$. In general, $\pi_1 \pi_2 \neq \pi_2 \pi_1$.

In this paper, π_e denotes an *identical* permutation $(1, 2, 3, \ldots, n)$. Clearly $\pi\pi_e = \pi_e\pi = \pi$ for any π. We define the *dimension* of a permutation $dim(\pi)$ as the highest item number moved by π. For example, $dim((3, 1, 2, 4)) = 3$ as item 4 does not move. We set $dim(\pi_e) = 0$, and otherwise $dim(\pi) \geq 2$. Also, we sometimes omit items larger than $dim(\pi)$. For example, $(3,2,1,4,5)$ can be written simply as $(3,2,1)$.

The main objective of this paper is the representation of *sets of permutations*. We describe such set as $P = \{\pi_e, (2, 1), (2, 3, 1)\}$. The empty set is denoted as $\emptyset$. We also define the *dimension of a set of permutations* such that $dim(P) = max(\{dim(\pi) | \pi \in P\})$. Finaly, we set $dim(P) = 0$ iff $P = \emptyset$ or $P = \{\pi_e\}$, otherwise $dim(P) \geq 2$.

We may use a multiplicative notation between a set of permutation P and a permutation π, which is defined as follows: $P \cdot \pi = \{\pi'\pi \mid \pi' \in P\}$.

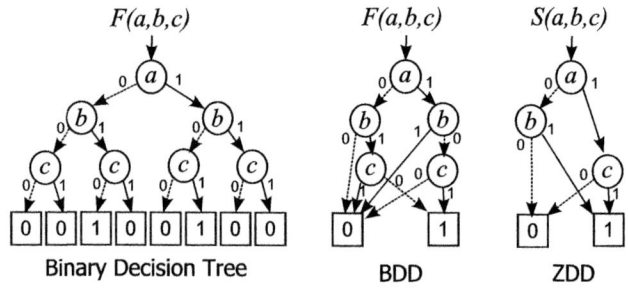

Fig. 1. Binary Decision Tree, BDD and ZDD

2.2   BDDs and ZDDs

A Binary Decision Diagram (BDD) [1] is a graph representation for a Boolean function. As illustrated in Fig. 1, it is derived by reducing a binary decision tree graph, which represents a decision making process through the input variables. If we fix the order of the input variables and apply the following two reduction rules, then we obtain a compact canonical form for a given Boolean function:

(1) Delete all redundant nodes whose both edges have the same destination, and
(2) Share all equivalent nodes having the same child nodes and the same variables.

Although the compression ratio achieved by using a BDD depends on the properties of the Boolean function to be represented, it can be between 10 and 100 times in some practical cases. In addition, we can systematically construct a BDD as a result of a binary logic operation (i.e., AND or OR) for a given pair of operand BDDs. This algorithm is based on hash table techniques, and the computation time is almost linear with respect to the size of the BDD.

A zero-suppressed BDD (ZDD) [6] is a variant of BDD customized for manipulating *sets of combinations*. ZDDs are based on special reduction rules which

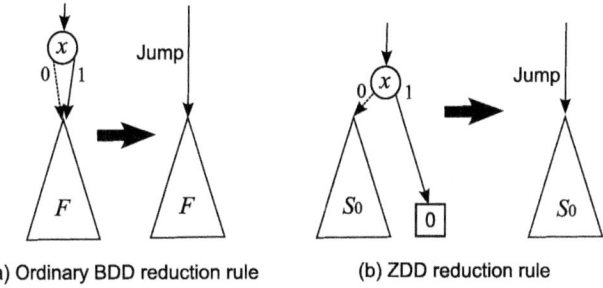

(a) Ordinary BDD reduction rule   (b) ZDD reduction rule

Fig. 2. ZDD reduction rule

differ from ordinary ones. As shown in Fig. 2, we delete all nodes whose 1-edge points directly to the 0-terminal node and do not delete the nodes that would be deleted in ordinary BDDs. Similarly to ordinary BDDs, ZDDs give compact canonical representations for sets of combinations. We can construct ZDDs by applying algebraic set operations such as union, intersection and difference, which correspond to logic operations in BDDs.

The zero-suppressing reduction rule is extremely effective for sets of sparse combinations. If the average appearance rate of each item is 1%, ZDDs are possibly up to 100 times more compact than ordinary BDDs. Such situations often appear in real-life problems, for example, in a supermarket, the number of items in a customer's basket is usually much smaller than the number of all items displayed at the supermarket. ZDDs are now widely recognized as the most important variant of BDDs (for details, see Knuth's book fascicle [5].)

3   Data Structures

3.1   Desired Properties for πDDs

Before discussing the general structure of πDDs, we list the basic properties desired for πDDs which are necessary for representing sets of permutations.

- The empty set $\emptyset$ corresponds to a 0-terminal node in a πDD since this is a zero element for union operation.
- The singleton set $\{\pi_e\}$ corresponds to a 1-terminal node since this is an identity element for composite operations.
- The form of a πDD for P does not depend on items larger than $dim(P)$. For example, $\{(3, 2, 1), (2, 1)\}$ and $\{(3, 2, 1, 4, 5), (2, 1, 3, 4, 5)\}$ should yield the same πDD.
- A πDD should provide a canonical (unique) representation for a set of permutations. This allows for efficient equivalence checking and satisfiability testing.
- Each path from the root node to a 1-terminal node should correspond to a permutation included in the set, namely, the number of paths corresponds to the cardinality of the set.

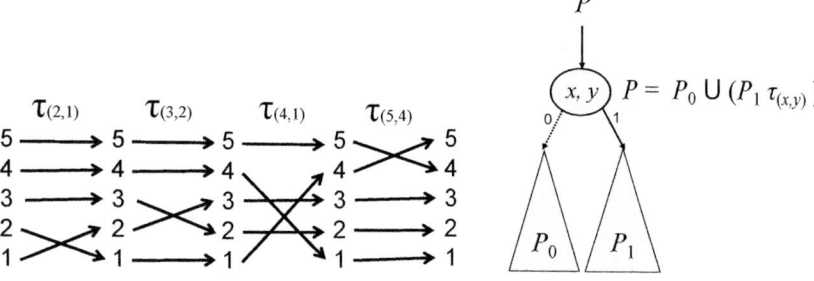

Fig. 3. Decomposition for a permutation (3,5,2,1,4)

Fig. 4. Basic structure of πDD

3.2   Decomposition of Permutations

Transposition is a basic permutation of simple swapping of two items. In this paper, $\tau_{(x,y)}$ denotes the transposition of items x and y. Clearly, $\tau_{(x,y)} = \tau_{(y,x)}$ and $(\tau_{(x,y)})^2 = \pi_e$ for any x and y. We set $\tau_{(x,x)} = \pi_e$.

The key idea behind πDDs is based on the observation that any permutation π can be decomposed into a sequence of up to $(dim(\pi) - 1)$ transpositions. For example, a permutation $(3, 5, 2, 1, 4)$ can be decomposed into $\tau_{(2,1)}\tau_{(3,2)}\tau_{(4,1)}\tau_{(5,4)}$, as illustrated in Fig. 3.

Theorem 1. *Any non-identical permutation π has a decomposition form which consists of up to $(dim(\pi) - 1)$ transpositions, and there is a way to obtain a unique decomposition form for any given permutation.*

Proof. If $dim(\pi) = 2$ then π should be a single transposition $\tau_{(2,1)}$. Next, we assume $dim(\pi) > 2$. If we let $x = dim(\pi)$ and $\pi_1 = \pi \cdot \tau_{(x,x\pi)}$, then $x\pi_1 = x$ holds. Since x is not moved by π_1, then $dim(\pi_1) < dim(\pi)$. The equation $\pi_1 = \pi \cdot \tau_{(x,x\pi)}$ can be transformed into $\pi = \pi_1 \cdot \tau_{(x,x\pi)}$, and thus π can be decomposed into a permutation π_1 followed by one transposition. In applying this procedure to π_1 recursively, the dimension decreses monotonically, and eventually we can obtain a unique decomposition form which consists of up to $(dim(\pi)-1)$ transpositions.
□

For the example shown in Fig. 3, the dimension is 5, item 5 is moved to 4, and we obtain $(3, 5, 2, 1, 4) = (3, 4, 2, 1) \cdot \tau_{(5,4)}$. Next, the dimension is 4, item 4 is moved to 1, and we obtain $(3, 4, 2, 1) = (3, 1, 2) \cdot \tau_{(4,1)}$. Similarly, we subsequently obtain $(3, 1, 2) = (2, 1) \cdot \tau_{(3,2)}$, and finally $(2, 1) = \tau_{(2,1)}$. In total, we obtain a sequence of 4 transpositions. This procedure is deterministic and the result is unique for any given permutation.

3.3   General Structure of πDDs

From the above observation, we can uniquely represent a permutation by using a combination of transpositions. Since ZDDs are efficient representations for sets

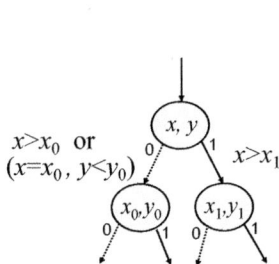

Fig. 5. Variable ordering rules in πDD

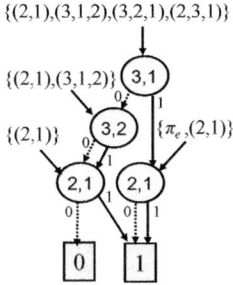

Fig. 6. Multi-rooted shared πDD

Table 1. Primitive πDD operations

$\emptyset$	Returns the empty set. (0-terminal node)
$\{\pi_e\}$	Returns the singleton set. (1-terminal node)
$P.top$	Returns the IDs (x, y) at the root node of P.
$P \cup Q$	Returns $\{\pi \mid \pi \in P \text{ or } \pi \in Q\}$.
$P \cap Q$	Returns $\{\pi \mid \pi \in P, \ \pi \in Q\}$.
$P \setminus Q$	Returns $\{\pi \mid \pi \in P, \ \pi \notin Q\}$.
$P.\tau(x, y)$	Returns $P \cdot \tau_{(x,y)}$.
$P * Q$	Returns $\{\alpha\beta \mid \alpha \in P, \ \beta \in Q\}$.
$P.cofact(x, y)$	Returns $\{\pi\tau_{(x,y)} \mid \pi \in P, \ x\pi = y\}$.
$P.count$	Returns the number of permutations.

of combinations, we might arrive at a ZDD-like data structure for representing sets of permutations.

Figure 4 shows the main idea behind πDDs. We assign a pair of item IDs (x, y) to each decision node, where $x = dim(P)$ and $x > y \geq 1$. Each decision node has the following semantics:

$$P = P_0 \cup (P_1 \cdot \tau_{(x,y)}),$$

where P_0 and P_1 represent a partition of P determined by the existence of $\tau_{(x,y)}$ in their decomposition forms. More formally, they are described as:

$$P_0 = \{\pi \mid \pi \in P, \ x\pi \neq y\}, \text{ and } P_1 = \{\pi\tau_{(x,y)} \mid \pi \in P, \ x\pi = y\}.$$

Note that $dim(P_1) < dim(P)$ holds since x has not been moved by any of the permutations in P_1. Applying this expansion recursively, we eventually obtain one of the two trivial sets of permutations, namely, the empty set $\emptyset$ (0-terminal node) or the singleton set $\{\pi_e\}$ (1-terminal node).

Similarly to ordinary ZDDs, a fixed order of variables is necessary for all $\tau_{(x,y)}$ in order to preserve the unique representation of the πDD. We use the following order from bottom to top:

$$(2, 1)(3, 2)(3, 1)(4, 3)(4, 2)(4, 1)(5, 4)(5, 3)(5, 2)(5, 1)(6, 5)(6, 4) \ldots$$

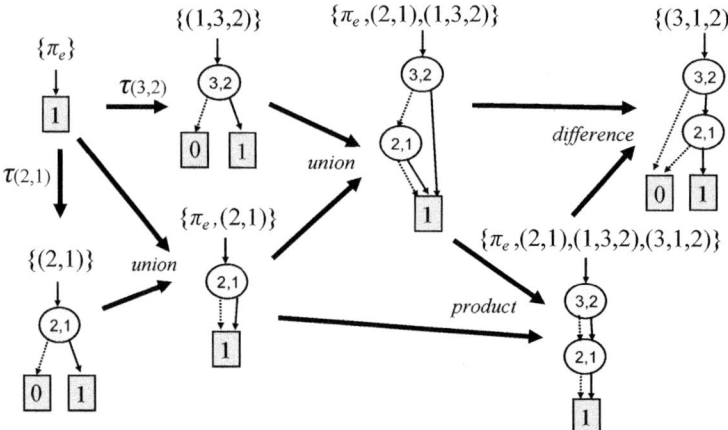

Fig. 7. Construction of πDDs by algebraic operations

Figure 5 shows the rules for variable ordering between two adjacent decision nodes in our πDDs.

In a πDD, any combination of transpositions can be represented by a unique path from the root node to a 1-terminal node.

Finally we confirm the node reduction rules in πDDs. Similarly to ordinary ZDDs, sharing of equivalent nodes is effective for πDDs as well. Note that it is necessary to check a pair of items (x, y) instead of only one decision variable in ZDDs. The zero-suppressing rule works rather well for the deletion of redundant nodes in πDDs since unnecessary transpositions are automatically deleted, and thus nodes corresponding to unmoved items never appear in πDDs.

As another similarity to BDDs/ZDDs, multiple πDDs can share their respective subgraphs with each other in a multi-rooted πDD, as shown in Fig. 6.

4   Algorithms for Algebraic Operations

In the previous section, we presented the basic structure of πDDs. However, we should consider not only compact representation but also efficient manipulation algorithms. Similarly to ordinary BDDs/ZDDs, πDDs can be constructed by applying algebraic operations, as illustrated in Fig. 7. Table 1 summarizes the primitive operations used in πDDs for manipulating sets of permutations. Here, we present a method for computing these operations efficiently. We are aiming at developing an efficient algorithm which computes in linear or small-order polynomial time with respect to the size of the relevant πDD, which is sometimes much smaller than the total number of permutations.

4.1   Binary Set Operations

First we consider the following three binary set operations: union, intersection and difference. As mentioned above, πDD is based on the expansion: $P = P_0 \cup$

$(P_1 \cdot \tau_{(x,y)})$ on each decision node. Since the two parts P_0 and $(P_1 \cdot \tau_{(x,y)})$ are disjoint, and since the τ operation is independent of the union, intersection and difference operations, we can execute those set operations in the same manner as for ordinary BDDs/ZDDs. For example, the intersection operation can be written as follows:

$$P \cap Q = (P_0 \cup (P_1 \cdot \tau_{(x,y)})) \cap (Q_0 \cup (Q_1 \cdot \tau_{(x,y)}))$$
$$= (P_0 \cap Q_0) \cup ((P_1 \cap Q_1) \cdot \tau_{(x,y)}).$$

Then, $(P_0 \cap Q_0)$ and $(P_1 \cap Q_1)$ are called recursively. Similarly to ordinary BDDs/ZDDs, we can avoid duplicate recursive calls by using cache to store previous operations and their results.

4.2   Transposition

Next, we consider the transposition operation with any pair of items for a given set of permutations. Let P be a given πDD and $P.top = (x, y)$, after which we compute $P \cdot \tau_{(u,v)}$. If $u > x$, we can simply return a decision node with items (u, v), whose 0-edge points to $\emptyset$ and whose 1-edge points to P. On the other hand, if $u \leq x$, more complex work is needed in order to traverse the internal nodes of P.

To illustrate the algorithm, we recall the example permutation $(3, 5, 2, 1, 4)$ shown in Fig. 3, and we compute $(3, 5, 2, 1, 4) \tau_{(3,1)}$. In a πDD, $(3, 5, 2, 1, 4)$ is represented by a sequence of transpositions $\tau_{(2,1)}\tau_{(3,2)}\tau_{(4,1)}\tau_{(5,4)}$, and thus we should compute $(\tau_{(2,1)}\tau_{(3,2)}\tau_{(4,1)}\tau_{(5,4)}) \tau_{(3,1)}$. Then, we can observe the following transformation:

$$\begin{aligned}
&\left(\tau_{(2,1)}\tau_{(3,2)}\tau_{(4,1)}\tau_{(5,4)}\right) \tau_{(3,1)} \\
=& \left(\tau_{(2,1)}\tau_{(3,2)}\tau_{(4,1)}\right) \left(\tau_{(5,4)}\tau_{(3,1)}\right) \\
=& \left(\tau_{(2,1)}\tau_{(3,2)}\tau_{(4,1)}\right) \left(\tau_{(3,1)}\tau_{(5,4)}\right) \\
=& \left(\tau_{(2,1)}\tau_{(3,2)}\right) \left(\tau_{(4,1)}\tau_{(3,1)}\right) \tau_{(5,4)} \\
=& \left(\tau_{(2,1)}\tau_{(3,2)}\right) \left(\tau_{(3,1)}\tau_{(4,3)}\right) \tau_{(5,4)} \\
=& \tau_{(2,1)} \left(\tau_{(3,2)}\tau_{(3,1)}\right) \tau_{(4,3)}\tau_{(5,4)} \\
=& \tau_{(2,1)} \left(\tau_{(2,1)}\tau_{(3,2)}\right) \tau_{(4,3)}\tau_{(5,4)} \\
=& \left(\tau_{(2,1)}\tau_{(2,1)}\right) \tau_{(3,2)}\tau_{(4,3)}\tau_{(5,4)} \\
=& \tau_{(3,2)}\tau_{(4,3)}\tau_{(5,4)}.
\end{aligned}$$

In this transformation, two adjacent transpositions are compared, and if the order violates the fixed order of the πDD, then the two transpositions are swapped. For example, $(\tau_{(5,4)}\tau_{(3,1)})$ is transformed into $(\tau_{(3,1)}\tau_{(5,4)})$, and $(\tau_{(4,1)}\tau_{(3,1)})$ becomes $(\tau_{(3,1)}\tau_{(4,3)})$. In this way, eventually we can obtain a normalized decomposition form of the πDD. Care should be taken since some item numbers are slightly altered in this process.

Figure 8 illustrates an example of swapping $\tau_{(x,y)}\tau_{(u,v)}$ with $\tau_{(u',v)}\tau_{(x,y')}$. In this example, u, v, and x are kept while y is changed. Here, we determine that such swapping is always possible for any pair of transpositions, and we also determine the cases in which the items should be changed.

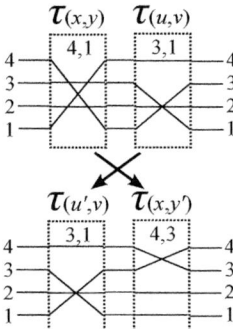

Fig. 8. Swapping of adjacent transpositions

Theorem 2. *For given positive integers x, y, u, v where $x > y > 0$ and $x \geq u > v$, a pair of cascading transpositions $\tau_{(x,y)}\tau_{(u,v)}$ can be transformed into π_e or $\tau_{(u',v)}\tau_{(x,y')}$, where u' and y' are some positive integers satisfying $u' < x$ and $x > y' > 0$.*

Proof. If there are no colliding items for $\tau_{(x,y)}$ and $\tau_{(u,v)}$, they can be swapped transparently. Next, we check all collision cases. If $y = u$, then $u' = u$ and $y' = v$. If $y = v$, then $u' = y' = u$. If $x = u$, then $u' = y' = y$. If $x = u$ and $y = v$, then $\tau_{(x,y)}\tau_{(u,v)} = \pi_e$. Otherwise, simply $u' = u$ and $y' = y$.   □

Based on this theorem, we can implement a recursive algorithm for the transposition operation. If $P.top = (x, y)$ and $u \leq x$, then $P \cdot \tau_{(u,v)}$ can be written as follows:

$$
\begin{aligned}
P \cdot \tau_{(u,v)} &= (P_0 \cup (P_1 \cdot \tau_{(x,y)})) \cdot \tau_{(u,v)} \\
&= P_0 \cdot \tau_{(u,v)} \cup (P_1 \cdot (\tau_{(x,y)}\tau_{(u,v)})) \\
&= (P_0 \cdot \tau_{(u,v)}) \cup ((P_1 \cdot \tau_{(u',v)}) \cdot \tau_{(x,y')})
\end{aligned}
$$

This formula shows that we can obtain a decision node with IDs (x, y'), whose 0-edge points to the result of $P_0 \cdot \tau_{(u,v)}$ and whose 1-edge points to the result of $P_1 \cdot \tau_{(u',v)}$. Here, it should be noted that $dim(P_1 \cdot \tau_{(u',v)})$ must be lower than x. Each sub-operation can be computed by a recursive call, and eventually we arrive at a trivial case. Similarly to other operations, we can avoid duplicate recursions by using operation cache.

4.3   Cartesian Product

The Cartesian product $P * Q = \{\alpha\beta \mid \alpha \in P, \beta \in Q\}$ computes the set of all possible composite permutations chosen from P and Q. This is the most important and useful operation in manipulating permutations.

By using transposition operations, the product $P * Q$ can be written as follows. Here, we assume $Q.top = (x, y)$.

$$
\begin{aligned}
P * Q &= P * (Q_0 \cup (Q_1 \cdot \tau_{(x,y)})) \\
&= (P * Q_0) \cup ((P * Q_1) \cdot \tau_{(x,y)})
\end{aligned}
$$

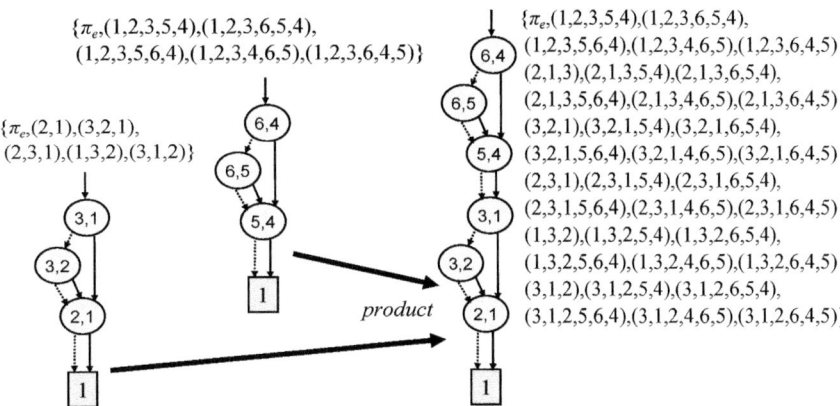

Fig. 9. Example of Cartesian product

This formula indicates that we may recursively call sub-operations $(P * Q_0)$ and $(P * Q_1)$, and we eventually arrive at a trivial operation $P * \emptyset$ or $P * \{\pi_e\}$. As in the case of other operations, we can avoid duplicate recursions by using operation cache. However, one different point here is that we cannot ensure $dim(P * Q_1) < x$, and therefore it is necessary to apply a general transposition operation for $(P * Q_1) \cdot \tau_{(x,y)}$.

Figure 9 shows an example of product operation for two πDDs whose items are disjoint. In this case, even though the number of permutations increases multiplicatively, the size of the πDD increases only additively. Since the computation time also depends on the size of the πDD, in such cases the effectiveness of the πDD-based method increses exponentially as compared to using an explicit data structure.

4.4 Cofactor

After generating a πDD for a set of permutations, it is necessary to extract a subset of permutations in order to check whether a certain property is satisfied. A *cofactor* operation $P.cofact(u, v) = \{\pi\tau_{(u,v)}|\ \pi \in P,\ u\pi = v\}$ generates a subset of permutations such that the item u is moved to v. For example,

$$\{(3, 2, 1), (2, 3, 1), (1, 3, 2), (2, 1)\}.cofact(3, 1)$$
$$= \{(3, 2, 1)\tau_{(3,1)}, (2, 3, 1)\tau_{(3,1)}\}$$
$$= \{\pi_e, (2, 1)\}.$$

Note that $P.cofact(u, u)$ can extract the permutations where u is not moved. Using cofactor and other set operations, various constraints can be specified and applied to πDDs.

Here, we discuss the method for executing the cofactor operation. If (u, v) corresponds to $P.top$, we may simply return the 1-edge of the root node. Otherwise, it is necessary to traverse the internal nodes in P. We can observe that the following equation holds.

$$P.cofact(u, v) = (P \cdot \tau_{(u,v)}).cofact(u, u),$$

Thus, the cofactor operation can be executed by using a transposition operation. Due to space limitations, we omit the details regarding the implementation of this operation.

5   Application Examples

Here, we present two application examples and the respective experimental results. We implemented a prototype version of a πDD manipulator based on our own BDD/ZDD package. The program consists of 330 lines of C++ code, newly added to the basic libraries including 6,000 lines of C/C++ code. The following experiments were performed by using a 2.4 GHz Core2Duo PC with 2 GB of RAM, SuSE 10 OS and GNU C++ compiler.

5.1   Design of Permutation Networks

A *permutation network* is an n-input and n-output network which can generate any permutation of the input items. Such circuits are often used in customized hardware of cryptographic systems and signal processing systems. Here, we consider a type of permutation networks using a set of n-bit parallel lines with a number of swapping switches X_k between any two adjacent lines, as shown in Fig. 10. We then consider an optimal layout of switches for a given permutation.

A set of permutations given by one switch can be written as $\bigcup_{i=1}^{n-1} \tau_{(i,i+1)}$. Thus, all possible permutations generated by up to k switches are described as follows.

$$\begin{aligned} P_0 &= \pi_e \\ P_1 &= P_0 \cup \left(\bigcup_{i=1}^{n-1} \tau_{(i,i+1)}\right) \\ P_k &= P_{k-1} * P_1 \qquad \text{(for } k \geq 2) \end{aligned}$$

According to this iterative formula, we can generate πDDs for $P_0, P_1, P_2, \ldots$ by increasing k, and eventually $P_{k+1} = P_k$ for any $k \geq m$. Then, m shows the minimum number of switches to required cover all permutations.

Table 2 shows the experimental results for a 10-bit permutation network. In this table, "πDD size" shows the number of decision nodes in the πDD, "# of

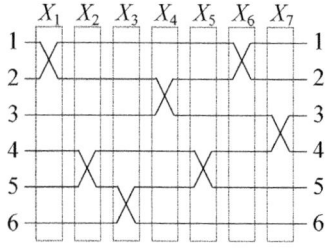

Fig. 10. A permutation network for (4,2,1,6,5,3)

Table 2. Experimental results for a 10-bit permutation network

P_k	πDD size	# of perm.	total #τ	P_k	πDD size	# of perm.	total #τ
P_0	0	1	0	P_{16}	3956	528441	3412177
P_1	9	10	9	P_{17}	4685	690778	4522462
P_2	31	54	97	P_{18}	5455	878737	5821218
P_3	63	209	546	P_{19}	6249	1089826	7296041
P_4	109	649	2152	P_{20}	7047	1319957	8915085
P_5	172	1717	6704	P_{21}	7834	1563651	10645703
P_6	261	4015	17632	P_{22}	8591	1814400	12433871
P_7	390	8504	40751	P_{23}	9293	2065149	14239194
P_8	558	16599	84985	P_{24}	9905	2308843	15996836
P_9	773	30239	162995	P_{25}	10397	2538974	17671711
P_{10}	1034	51909	291537	P_{26}	10735	2750063	19206325
P_{11}	1353	84592	491272	P_{27}	**10894**	2938022	20584666
P_{12}	1727	131635	786100	P_{28}	10857	3100359	21772380
P_{13}	2169	196524	1201963	P_{29}	10614	3236212	22773147
P_{14}	2688	282578	1764353	P_{30}	10157	3346222	23579581
P_{15}	3286	392588	2495497	P_{31}	9497	3432276	24214975

P_k	πDD size	# of perm.	total #τ
P_{32}	8655	3497165	24691907
P_{33}	7669	3544208	25039740
P_{34}	6590	3576891	25279788
P_{35}	5470	3598561	25439624
P_{36}	4374	3612201	25539440
P_{37}	3353	3620296	25598543
P_{38}	2444	3624785	25630975
P_{39}	1671	3627083	25647411
P_{40}	1055	3628151	25654943
P_{41}	602	3628591	25657983
P_{42}	305	3628746	25659023
P_{43}	136	3628790	25659303
P_{44}	59	**3628799**	25659355
P_{45}	45	**3628800**	25659360
P_{46}	45	**3628800**	25659360

perm." indicates the number of permutations included in P_k, and "total #τ" is the total number of transpositions included in all permutations in P_k. Note that the total #τ corresponds to the data size when using an explicit representation for P_k.

The result shows that P_{46} is equivalent to P_{45}, and thus we can see that $m = 45$. In other words, 45 switches are sufficient to cover all 362,880 (=10!) permutations. The number of permutations and the total number of transpositions increase monotonically in this iteration process, however, the size of the πDD reaches a peak of 10,894 at P_{27}, and consequently we require a πDD of only 45 decision nodes to represent all 10! permutations. The latter P_ks might yield more beautiful structures, and the πDD nodes are well shared, even though they include a rather large number of permutations.

We can also observe that P_{45} and P_{44} differ by only a single number of permutations by simply applying the difference set operation ($P_{45} \setminus P_{44}$), and we can confirm that the last permutation is (10,9,8,7,6,5,4,3,2,1). By applying algebraic operations for πDDs to P_ks, we can determine the minimal number of switches

Table 3. Experimental results for n-bit permutation networks

n	m	πDD size (peak)	(final)	# of perm.	total #τ	time (sec)
1	0	0	0	1	0	0.00
2	1	1	1	2	1	0.00
3	3	3	3	6	7	0.00
4	6	9	6	24	46	0.00
5	10	27	10	120	326	0.00
6	15	89	15	720	2556	0.01
7	21	292	21	5040	22212	0.02
8	28	972	28	40320	212976	0.06
9	36	3241	36	362880	2239344	0.26
10	45	10894	45	3628800	25659360	1.19
11	55	36906	55	39916800	318540960	5.77
12	66	125904	66	479001600	4261576320	27.06
13	78	435221	78	6227020800	61148511360	126.80
14	91	1520439	91	87178291200	937030429440	666.29

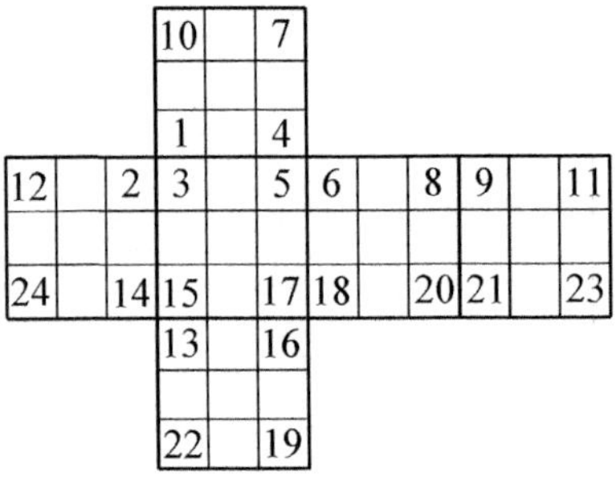

Fig. 11. Assignment of items for the corner cubes of Rubik's Cube

for any given permutation, and we can find the layout of the switches which is necessary in order to obtain this permutation.

Table 3 presents the results for n-bit permutation networks for n up to 14. We show the peak and the final size of the πDDs and their respective computation times. The number of all permutations is clearly $n!$, however, the final size of the πDD is only $n(n-1)/2$. Even though the peak size of the πDD grows exponentially, its growth rate appears to be slower than that of $n!$. Here, we can observe that the πDDs are at least 1000 times more compact than explicit representations.

5.2   Analysis of Rubik's Cube

Rubik's Cube^{TM} is one of the most popular puzzles related to permutation group theory, and πDD can be useful for analyzing it. Here, we focus only on the moves of the eight corner cubes. Figure 11 illustrates our assignment of the items to all the 24 faces of the corner cubes. Then we can describe $90°$ moves along the X-, Y- and Z-axis as follows.

$$\pi_x = \tau_{(3,5)}\tau_{(3,17)}\tau_{(3,15)}\tau_{(1,6)}\tau_{(1,16)}\tau_{(1,14)}\tau_{(2,4)}\tau_{(2,18)}\tau_{(2,13)}$$
$$\pi_y = \tau_{(2,14)}\tau_{(2,24)}\tau_{(2,12)}\tau_{(3,13)}\tau_{(3,23)}\tau_{(3,10)}\tau_{(1,15)}\tau_{(1,22)}\tau_{(1,11)}$$
$$\pi_z = \tau_{(1,10)}\tau_{(1,7)}\tau_{(1,4)}\tau_{(3,12)}\tau_{(3,9)}\tau_{(3,6)}\tau_{(2,11)}\tau_{(2,8)}\tau_{(2,5)}$$

Table 4. Experimental results for Rubik's Cube

P_k	πDD size	# of perm.	total #τ
P_0	0	1	0
P_1	63	10	72
P_2	392	64	888
P_3	1789	385	5634
P_4	6860	2232	34446
P_5	23797	12224	194406
P_6	84704	62360	1012170
P_7	290018	289896	4752582
P_8	**608666**	1159968	19087266
P_9	580574	3047716	50272542
P_{10}	18783	3671516	60540732
P_{11}	**511**	**3674160**	60579900
P_{12}	**511**	**3674160**	60579900

where all possible permutations of at most one of the primitive moves $(+90°, -90°,$ and $180°$ for each axis) are described as follows.

$$P_1 = \pi_e + \pi_x + \pi_x{}^2 + \pi_x{}^3 + \pi_y + \pi_y{}^2 + \pi_y{}^3 + \pi_z + \pi_z{}^2 + \pi_z{}^3$$

Now we can generate the set of permutations for up to k moves by using the following simple iterative formula.

$$P_k = P_{k-1} * P_1 \qquad \text{(for } k \geq 2)$$

Similarly to the case of permutation networks, we can find a fixed point m such that $P_{k+1} = P_k$ for any $k \geq m$. If we ignore all edge and center cubes, P_m contains all meaningful patterns for the eight corner cubes. Note that the cube $\{19, 20, 21\}$ is fixed to the original position in order to eliminate symmetric patterns.

Table 4 shows the result of generating πDDs for the P_k's. We can see that the number of all possible patterns of the corner cubes is 3,674,160. We confirmed that 11 moves are sufficient to generate all possible patterns, in other words, any pattern of the corner cubes can be returned to the original positions in 11 or fewer moves. As a result, this requires only 511 decision nodes of πDDs for representing all patterns, and P_8 reaches a peak at a πDD size of 608,666. The computation time for generating all πDDs was 207 seconds.

After generating the πDDs for the P_k's, we can analyze various properties of Rubik's Cube. For example, we can explore patterns where only two corner cubes are moving and the other six cubes remain at their original positions. Such patterns can be detected by cofactor operations as follows.

$$S_k = P_k.cofact(9,9).cofact(11,11).cofact(15,15)$$
$$.cofact(17,17).cofact(21,21).cofact(23,23)$$

Our experiment shows that, for $k \leq 9$, S_k only includes π_e. For $k = 10$, we discover (2,3,1,6,4,5), (3,1,2,5,6,4), (4,5,6,1,2,3) and (6,4,5,2,3,1), and by using the maximal number of moves ($k = 11$), we arrive at (6,4,5,2,3,1). After such a pattern is detected, it is not difficult to find a sequence of moves which generates it. We can apply one of the primitive moves to the final pattern in order to obtain a candidate for a preceding pattern, and we check for its existence in P_{k-1}. At least one of the candidates must be in P_{k-1}, and then we can repeat the process until we reach P_1.

Although we have considered only the corner cubes, Rokicki et al. [7] recently confirmed that all patterns of the Rubik's cube can be solved as few as 20 moves, and this is the exact minimum. They applied some mathematical pruning and used a network of PCs for massive parallel computation amounting to a total of 35 CPU years. Although the straight-forward application of πDDs to this problem might cause memory overflow, we nevertheless believe that it will be useful for accelerating such kind of problem solving.

6   Conclusion

In this paper, we proposed a new idea of decision diagrams for manipulating sets of permutations. The method of πDDs provides hints about the application of state-of-the-art SAT techniques used for solving combinatorial problems to permutational problems. There is a rich body of research in group theory led by Galois and many researchers in discrete mathematics [3]. We can expect much future work in this area, for example, developing software tools for studying group theory, considering many other practical applications, implementing various other operations for sets of permutations and considering extended models, such as sets of k-out-of-n permutations or multisets of permutations.

References

1. Bryant, R.E.: Graph-based algorithms for Boolean function manipulation. IEEE Transactions on Computers C-35(8), 677–691 (1986)
2. Chatalic, P., Simon, L.: Zres: The old davis-putnam procedure meets ZBDDs. In: McAllester, D. (ed.) CADE 2000. LNCS(LNAI), vol. 1831, pp. 449–454. Springer, Heidelberg (2000)
3. GAP Forum. GAP – Groups, Algorithms, Programming – a System for Computational Discrete Algebra (2008), http://www.gap-system.org/
4. Knuth, D.E.: Combinatorial properties of permutations. The Art of Computer Programming, vol. 3, ch. 5.1, pp. 11–72. Addison-Wesley, Reading (1998)
5. Knuth, D.E.: The Art of Computer Programming: Bitwise Tricks & Techniques; Binary Decision Diagrams. fascicle 1, vol. 4. Addison-Wesley, Reading (2009)
6. Minato, S.: Zero-suppressed BDDs for set manipulation in combinatorial problems. In: Proc. of 30th ACM/IEEE Design Automation Conference, pp. 272–277 (1993)
7. Rokicki, T., Kociemba, H., Davidson, M., Dethridge, J.: God's number is 20 (2010), http://www.cube20.org/

How to Apply SAT-Solving for the Equivalence Test of Monotone Normal Forms*

Martin Mundhenk and Robert Zeranski

Friedrich-Schiller-Universität Jena, Germany
{martin.mundhenk,robert.zeranski}@uni-jena.de

Abstract. The equivalence problem for monotone formulae in normal form MONET is in coNP, is probably not coNP-complete [1], and is solvable in quasi-polynomial time $n^{o(\log n)}$ [2].

We show that the straightforward reduction from MONET to UNSAT yields instances, on which actual SAT-solvers (SAT4J) are slower than current implementations of MONET-algorithms [3]. We then improve these implementations of MONET-algorithms notably, and we investigate which techniques from SAT-solving are useful for MONET. Finally, we give an advanced reduction from MONET to UNSAT that yields instances, on which the SAT-solvers reach running times, that seem to be magnitudes better than what is reachable with the current implementations of MONET-algorithms.

1 Introduction

The equivalence problem for Boolean formulae is one of the classical coNP-complete problems. It remains coNP-complete also if the formulae are given in normal form. For monotone formulae—i.e. formulae with conjunctions and disjunctions, but without negations—the equivalence problem is coNP-complete, too [4], but its complexity drops, if the formulae are in conjunctive or disjunctive normal form. The reason is that for every monotone formula, the minimal equivalent formula in the considered normal form is unique, and the minimal equivalent normal form formula is efficiently computable from the non-minimal normal form formula. Therefore, checking whether two monotone formulae in conjunctive normal form (resp. two monotone formulae in disjunctive normal form) are equivalent, can be done in polynomial time. The remaining case is the equivalence problem for monotone formulae, where one formula is in conjunctive normal form and the other is in disjunctive normal form. This is the problem MONET, i.e. Mo(notone) n(ormal form) e(quivalence) t(est). This problem is strongly related to dualization of monotone conjunctive normal forms and transversal hypergraph generation [5]. This means that an algorithm for MONET solves many fundamental problems in a wide range of fields, including artificial intelligence and logic, computational biology, database theory, data mining and machine learning, mobile

---

* This work was supported by the DFG under grant MU 1226/8-1 (SPP 1307/2).

K.A. Sakallah and L. Simon (Eds.): SAT 2011, LNCS 6695, pp. 105–119, 2011.
© Springer-Verlag Berlin Heidelberg 2011

communication systems, distributed systems, and graph theory (see [3] for an overview). The currently best MONET algorithms have quasi-polynomial running time $n^{o(\log n)}$, or polynomial time using $\mathcal{O}(\log^2 n)$ nondeterministic bits [6,2,1]. Thus, on the one hand, MONET is probably not coNP-complete, but on the other hand a polynomial time algorithm is not yet known. This situation turns MONET into one of the very few problems "between" P and NP- resp. coNP-hard. The exact complexity of the general problem MONET is a long standing famous open question [7].

As for evaluating the practical performance, in [8,9,10,11,12,13,14] one can find several experimental studies on known algorithms for MONET or equivalent problems. Over all, the algorithms by Fredman and Khachiyan [2], that are the MONET algorithms with the best worst-case upper-bounds, turn out to be strong practical performers [3].

In this paper, we mainly address the following two questions.

Can the performance of the algorithms by Fredman and Khachiyan be improved by using techniques from SAT*-solving?* The two algorithms by Fredman and Khachiyan basically use a technique similar to the DPLL-algorithm for SAT [15]. Both algorithms leave—more or less—open which variable to choose as the splitting variable. We added unit propagation and tried several strategies known from SAT-solving like MOMs [16], BOHM [17], and clause reduction heuristic [18]. In our experimental study we show that unit propagation and the mentioned strategies notably improve our implementations of the algorithm.

Are SAT*-solvers good for* MONET*?* Since MONET reduces to the complement of SAT, it is straightforward to use a reduction and a SAT-solver to solve MONET. Eventually, it turned out not to be that straightforward. We give a reduction from MONET to the complement of SAT that does not increase the size of the instances. Using this reduction function and a SAT-solvers reaches computation times that are very much better than what is currently possible with MONET solvers.

This paper is organized as follows. In Section 2 we introduce the basic notation and review the FK-algorithms [2]. Section 3 shows how unit propagation and strategies for choosing a splitting variable can be used in the FK-algorithms. Section 4 considers how MONET can be reduced UNSAT. Our experimental results are discussed in Section 5, and conclusions are drawn in Section 6.

2   Preliminaries

Monotone formulae and equivalence. A Boolean formula φ is called *monotone* if φ has only $\wedge$ and $\vee$ as connectives–*no negations are allowed*. Let V_φ denote the set of variables of φ. An *assignment* is denoted as a set $\mathcal{A} \subseteq V_\varphi$ of variables, where x is assigned *true* iff $x \in \mathcal{A}$. Otherwise x is assigned *false*. A *term* is a set of variables that is either interpreted as a conjunction or as a disjunction of the variables. We call a term a *monomial* if it is a conjunction and we call it a *clause* if it is a disjunction. In this paper m always denotes a monomial and c always denotes a clause. A monotone Boolean formula is a DNF (*disjunctive normal form*) if it is a disjunction of monomials, and it is a CNF (*conjunctive normal form*) if it is a conjunction of clauses. Throughout the whole paper we regard a CNF (resp. DNF) as a set of clauses (resp. monomials). A monotone DNF or CNF is called *irredundant* if it contains no two terms such that one contains the other. It is important, that every monotone Boolean formula has a unique irredundant monotone CNF and DNF [19]—and for a given monotone CNF (resp. DNF) the irredundant CNF (resp. DNF) can be obtained in quadratic time by deleting all supersets of terms. Two monotone Boolean formulae are *equivalent* if and only if they have the same irredundant monotone CNF. This paper deals with the equivalence test of monotone formulae in different normal forms.

MONET: Instance: irredundant, monotone DNF D and CNF C
                 Question: are D and C equivalent?

In this paper D always denotes a monotone DNF and C always denotes a monotone CNF. The length of a term is the number of variables in this term, and the length of a normal form D (resp. C) is the number of terms in it.

The Algorithms of Fredman and Khachyian

The algorithms with the best known worst-case upper bound for solving MONET are by Fredman and Khachiyan [2]. Both these algorithms search for a witness of non-equivalence by a depth-first search in the tree of all assignments—we call such a witness *conflict assignment*. Note that this technique is very similar to the DPLL-algorithm for SAT [15]. The FK-algorithms work as follows. In the first step the input formulae are modified to irredundant normal forms[1]. Unless the formulae are small enough to check them by brute force, the algorithm chooses a variable, sets the value of this variable to *false* and modifies the formulae due to this assignment—we call the variable chosen the *splitting variable*. Next, the equivalence of the new formulae will be tested recursively. If the recursive call does not yield a conflict assignment, the splitting variable is set to *true* and the accordingly modified formulae are tested recursively. If this second recursive call does not yield a conflict assignment, then the formulae must be equivalent.

The modifications of the formulae are as follows. If a variable is set to true in a DNF, then this variable can be deleted in each monomial. And if it is set to false, then all terms which contain this variable can be deleted. For the CNF it is dual. Let ϕ be a DNF or CNF, and let x be a splitting variable. Then ϕ_0^x denotes the formula that consists of terms of ϕ from which x is removed. Analogously, ϕ_1^x denotes the formula that consists of all terms of ϕ that do not contain x.

$$\phi_0^x = \{t - \{x\} : t \in \phi \ \text{ and } \ x \in t\} \qquad \phi_1^x = \{t : t \in \phi \ \text{ and } \ x \notin t\}$$

Thus, if x is set to true in D and C, we obtain $D_0^x \vee D_1^x$ and C_1^x. If it is set to false, we obtain D_1^x and $C_0^x \wedge C_1^x$.

---

[1] This step is necessary because the formulae will be modified in further steps and the algorithms are recursive.

Fredman and Khachiyan [2] provide necessary conditions for equivalence, that are also checked during the depth-first search, as follows.

(1) $m \cap c \neq \emptyset$ for every monomial $m \in D$ and every clause $c \in C$.
(2) D and C must contain exactly the same variables, i.e. $V_D = V_C$.
(3) $\max\{|m| : m \in D\} \leq |C|$ and $\max\{|c| : c \in C\} \leq |D|$.

FK-algorithm A [2]. The central question is about the choice of the splitting variable. Fredman and Khachiyan provide an additional necessary condition for the FK-algorithm A which ensures the existence of a frequent variable.

$$\sum_{m \in D} 2^{|V_D| - |m|} + \sum_{c \in C} 2^{|V_C| - |c|} \geq 2^{|V_D|}. \tag{4}$$

If this condition is violated the formulae are not equivalent and a conflict assignment can be computed in linear time. As splitting variable FK-algorithm A chooses a variable with frequency $\geq 1/\log(|D| + |C|)$ in either D or C.

Theorem 1. [2] *FK-algorithm A has running time $n^{\mathcal{O}(\log^2 n)}$ on input (D, C), where $n = |D| + |C|$.*

Algorithm 1 shows a pseudo-code listing of FK-algorithm A. It is shown in an experimental study in [10] that FK-algorithm A performs well in practice.

There is also a version presented in [20] which works in space polynomial in $|D|$.

FK-algorithm B [2]. What happens if the first recursive call FK-A($D_1^x, C_0^x \wedge C_1^x$) of FK-algorithm A does not yield a witness for non-equivalence? In this case we gain the information that D_1^x is equivalent to $C_0^x \wedge C_1^x$. FK-algorithm A does not use the fact that the second recursive call is performed only if the first recursive call does not yield a witness for non-equivalence. But FK-algorithm B makes use of this. The main conclusion is a restriction for the search tree when the value of the splitting variable is set to true. It then suffices to find a conflict assignment $\mathcal{A}$ for the formulae D_0^x and C_1^x with the restriction that $\mathcal{A}(C_0^x) = 0$ (cf. [2,3]). Hence, one has to check all maximal assignments not satisfying C_0^x only. Note that there are exactly $|C_0^x|$ assignments, one for every clause $c \in C_0^x$ (the resp. assignment is $V_{C_0^x} - c$).

Thus, if the first recursive call does not yield a conflict assignment it suffices to perform a recursive call for every clause $c \in C_0^x$ on the pair $(D_0^{c,x}, C_1^{c,x})$, where $D_0^{c,x}$ and $C_1^{c,x}$ denote the formulae we obtain if we set all variables in c to false. We receive a similar result if we swap the chronological order of the first and the second recursive call, cf. [2,3]. If the recursive call on the pair $(D_0^x \vee D_1^x, C_1^x)$ does not yield a witness for non-equivalence it suffices to perform a recursive call for every monomial $m \in D_0^x$ on the pair $(D_1^{m,x}, C_0^{m,x})$, where $D_1^{m,x}$ and $C_0^{m,x}$ denote the formulae we obtain if we set all variables in m to true (if we found a conflict, the resp. assignment is m).

One of the main differences to FK-algorithm A is the choice of the variable and the advanced branching. The choice of the splitting variable does not matter in [2]

Algorithm 1. The FK-algorithm A (FK-A)

Input: irredundant, monotone DNF D and CNF C
Output: $\emptyset$ in case of equivalence; otherwise, assignment $\mathcal{A}$ with $\mathcal{A}(D) \neq \mathcal{A}(C)$

 1: make D and C irredundant
 2: **if** one of conditions (1)–(4) is violated **then**
 3:     **return** conflict assignment
 4: **if** $|D| \cdot |C| \leq 1$ **then**
 5:     **return** appropriate assignment found by a trivial check
 6: **else**
 7:     choose a splitting variable x with frequency $\geq 1/\log(|D| + |C|)$ in D or C
 8:     $\mathcal{A} \leftarrow$ FK-A$(D_1^x, C_0^x \wedge C_1^x)$ // recursive call for x set to *false*
 9:     **if** $\mathcal{A} = \emptyset$ **then**
10:       $\mathcal{A} \leftarrow$ FK-A$(D_0^x \vee D_1^x, C_1^x)$ // recursive call for x set to *true*
11:       **if** $\mathcal{A} \neq \emptyset$ **then return** $\mathcal{A} \cup \{x\}$
12: **return** $\mathcal{A}$

Algorithm 2. The FK-algorithm B (FK-B)

Input: irredundant, monotone DNF D and CNF C
Output: $\emptyset$ in case of equivalence; otherwise, assignment $\mathcal{A}$ with $\mathcal{A}(D) \neq \mathcal{A}(C)$

 1: make D and C irredundant
 2: **if** one of conditions (1)–(3) is violated **then return** conflict assignment
 3: **if** $\min\{|D|, |C|\} \leq 2$ **then**
 4:     **return** appropriate assignment found by a trivial check
 5: **else**
 6:     choose a splitting variable x from the formulae
 7:     **if** x is at most μ-frequent in D **then**
 8:       $\mathcal{A} \leftarrow$ FK-B$(D_1^x, C_0^x \wedge C_1^x)$ // recursive call for x set to *false*
 9:       **if** $\mathcal{A} \neq \emptyset$ **then return** $\mathcal{A}$
10:       **for all** clauses $c \in C_0^x$ **do**
11:         $\mathcal{A} \leftarrow$ FK-B$(D_0^{c,x}, C_1^{c,x})$ // see $\langle 1 \rangle$
12:         **if** $\mathcal{A} \neq \emptyset$ **then return** $\mathcal{A} \cup \{x\}$
13:     **else if** x is at most μ-frequent in C **then**
14:       $\mathcal{A} \leftarrow$ FK-B$(D_0^x \vee D_1^x, C_1^x)$ // recursive call for x set to *true*
15:       **if** $\mathcal{A} \neq \emptyset$ **then return** $\mathcal{A} \cup \{x\}$
16:       **for all** monomials $m \in D_0^x$ **do**
17:         $\mathcal{A} \leftarrow$ FK-B$(D_1^{m,x}, C_0^{m,x})$ // see $\langle 2 \rangle$
18:         **if** $\mathcal{A} \neq \emptyset$ **then return** $\mathcal{A} \cup m$
19:     **else**
20:       $\mathcal{A} \leftarrow$ FK-B$(D_1^x, C_0^x \wedge C_1^x)$ // recursive call for x set to *false*
21:       **if** $\mathcal{A} = \emptyset$ **then**
22:         $\mathcal{A} \leftarrow$ FK-B$(D_0^x \vee D_1^x, C_1^x)$ // recursive call for x set to *true*
23:         **if** $\mathcal{A} \neq \emptyset$ **then return** $\mathcal{A} \cup \{x\}$
24: **return** $\mathcal{A}$

    $\langle 1 \rangle$: $D_1^x \equiv C_0^x \wedge C_1^x$: recursive call for all maximal non-satisfying assignments of C_0^x for x set to true
    $\langle 2 \rangle$: $D_0^x \vee D_1^x \equiv C_1^x$: recursive call for all minimal satisfying assignments of D_0^x for x set to false

---

for the theoretical upper bound, because FK-algorithm B chooses an appropriate branching with respect to the frequency of the splitting variable. Therefore, the algorithm uses a frequeny-threshold $\mu(n)$ with the property $\mu(n)^{\mu(n)} = n$. Note, that $\mu(n) \sim \log n / \log\log n$. Thus, $\mu(n) \in o(\log n)$. A pseudo-code listing of FK-algorithm B is given in Algorithm 2. There, a variable x is called *at most μ-frequent in D* (resp. C) if its frequency is at most $1/\mu(|D| \cdot |C|)$, i.e. $|\{m \in D : x \in m\}|/|D| \leq 1/\mu(|D| \cdot |C|)$.

Theorem 2. [2] *FK-algorithm B has running time $n^{o(\log n)}$ on input (D, C), where $n = |D| + |C|$.*

3   Unit Propagation and Decision Strategies

As mentioned before, the choice of the splitting variable is free in FK-algorithm B, and it is almost free in FK-algorithm A. In the old implementations in [3], the first variable in the formula is taken as splitting variable for FK-algorithm B, and the first variable that satisfies the frequency condition is taken as splitting variable in FK-algorithm A. In our new implementations, we replaced this by running unit propagation and choosing a splitting variable according to a somewhat more involved strategy.

Unit propagation for MONET. Unit propagation (UP), or also called one-literal rule, is a technique for simplifying a set of clauses in an automated theorem proving system. This technique is also used for the DPLL-algorithm [15]. A clause (resp. monomial) is a *unit clause* (resp. *unit monomial*) if it consists of one variable only. How can the FK-algorithms gain from considering unit clauses or unit monomials? There are two cases.

Case (a): *There is a unit clause in C.* Let $\{x\} \in C$, and let D and C satisfy condition (1). Then C_0^x–i.e. the set of clauses obtained from CNF C by setting x to false–is unsatisfiable, and D_1^x–i.e. the set of monomials obtained from DNF D by setting x to false–is unsatisfiable, too, because x is contained in all monomials of D (cf. condition (1)). Thus, the recursive call for setting the splitting variable x to false will yield no conflict assignment and can be left out.

Lemma 1. *Let $\{x\} \in C$. Then $D \equiv C$ if and only if $D_1^x = \emptyset$ and $D_0^x \equiv C_1^x$.*

According to Lemma 1, if C contains a unit clause $\{x\}$ then it suffices to check condition (1) on (D_0^x, C_1^x) and to do the recursive call FK-A(D_0^x, C_1^x).

Case (b): *There is a unit monomial in D.* This case is dual to case (a).

Lemma 2. *Let $\{x\} \in D$. Then $D \equiv C$ if and only if $C_1^x = \emptyset$ and $D_1^x \equiv C_0^x$.*

Note, that for formulae D and C satisfying condition (1) it is impossible that D and C contain the same unit term (excepted $D = C = \{\{x\}\}$). Thus, one can search for all unit clauses and unit monomials in the formulae and setting the variables to the resp. values. Because both formulae are irredundant we only have to delete the unit terms in the resp. formula. If there is no unit term left in

D and C we can choose a splitting variable. It is also possible to avoid checking condition (1). Thus, if we find a unit term $\{x\}$ in D (resp. C) we have to check that x is contained in all terms of C (resp D).

Decision strategies for the splitting variable. A main difference between FK-algorithm A and FK-algorithm B is the choice of the splitting variable. FK-algorithm A chooses a variable that is at least $\log(|D| + |C|)$-frequent in either D or C—one can also simply choose the most frequent variable. (If D is equivalent to C, then there exists a $\log(|D|+|C|)$-frequent variable [2]). However, FK-algorithm B is free to choose any variable as splitting variable. In general, a random choice is not a good strategy (see experiments with FKB(rMin) in Section 5). Thus, it is interesting to investigate whether strategies for choosing the splitting variable improve FK-algorithm B. We tried the following strategies.

 (i) Choose the first free variable [3]—it is related to a random choice.
 (ii) Choose the most frequent variable.
(iii) Choose the most frequent variable in the smallest terms (MOMs [16]).
(iv) Choose a variable randomly in the smallest terms.
 (v) Choose the variable with maximal occurence in small terms (BOHM [17]).
(vi) Clause reduction heuristic (CRH [18]).

4   Solving MONET Using SAT-Solvers

Since MONET is in coNP, it is polynomial-time reducible to UNSAT. Therefore, a straightforward approach to solve MONET is to use this reduction and a common SAT-solver. Clearly, (D, C) is in MONET if and only if $\neg(D \rightarrow C) \notin$ SAT and $\neg(C \rightarrow D) \notin$ SAT.

Since D is a DNF and C is a CNF, the formula $\neg(C \rightarrow D)$ is represented by the CNF $C \cup D^\neg$, where $D^\neg$ is the set of monomials in D, in which all appearances of variables are negated[2]. Similarly, the other part $\neg(D \rightarrow C)$ can be seen as a conjunction of two DNFs and can be brought into conjunctive normal form using the standard translation by Tseitin [21] that results in an equisatisfiable formula in CNF. Even though this translation enlarges the formula only linearly, our experiments with SAT-solvers on the translated formulae yielded computation times that were worse than that of the FK-algorithms (see Section 5, Table 2).

Let us consider the formula $\neg(D \rightarrow C)$ more precisely. It is satisfied by assignments that satisfy a monomial in the DNF D and falsify all clauses in the CNF C. This happens if and only if there is a monomial $m \in D$ and a clause $c \in C$ such that $m \cap c = \emptyset$. This condition can be checked in time $\mathcal{O}(|D| \cdot |C| \cdot n)$, where n denotes the number of variables. Therefore, this test can be used in the polynomial time function f that reduces MONET to UNSAT as follows.

$$f(D,C) = \begin{cases} C \cup D^\neg, & \text{if } m \cap c \neq \emptyset \text{ for all } m \in D \text{ and all } c \in C \\ true, & \text{otherwise} \end{cases}$$

---

[2] Remind that $C \cup D^\neg$ represents $C \wedge D^\neg$.

where *true* denotes a formula that is satisfied by every assignment.

Lemma 3. *The above function f is a polynomial-time function that reduces* MONET *to* UNSAT.

This reduction function can be seen as a generalization of reduction used in [5,22]. Note that the property of non-empty intersection of every clause and monomial is condition (1) from the necessary conditions for equivalence, and this is also checked in the FK-algorithms. Nevertheless, this check is not necessary for the correctness of the algorithms, but needed in the proof of the upper bound for the running time [2]. In our implementations we avoid to check condition (1), because in our experiments it seems to waste time only.

5   Experiments

We experimentally compare the following implementations of algorithms for MONET in Java. All experiments were conducted on an Intel i7-860, 2.8 GHz, 8 GB RAM running Ubuntu 10.04.

(1) The old implementations used by [3] for the FK-algorithms A and B. We call these implementations FKA(HHM) and FKB(HHM). The strategy of FKA(HHM) for choosing the splitting variable is to choose the firstly found log-frequent variable. The strategy of FKB(HHM) is to take the first variable.
(2) Our new implementations of the FK-algorithms A and B. They distinguish in the strategy how the splitting variable is chosen.

  – FKA(mf) and FKB(mf) are the FK-algorithm A and B with the strategy of choosing the most frequent variable.
  – FKA(th) is FK-algorithm A that chooses the firstly found log-frequent (threshold) variable. This is a new implementation of FKA(HHM).
  – FKB(BOHM) (resp. CRH and MOMs) denotes FK-algorithm B with BOHM (resp. CRH and MOMs) heuristic.
  – FKB(rMin) is FK-algorithm B that chooses randomly a variable in a term of minimal length.

(3) The implementation that uses our reduction to the complement of SAT and Sat4j [23] as SAT-solver. For simplicity, we call this *reduction to* SAT .

We run tests on test data that are equivalent formulae, and on those that are not equivalent. For equivalent formulae, the runtimes strongly depend on the structure of the test data. For non-equivalent formulae, this is not the case. Therefore we consider both these cases separately. We use test data $M(k)$, $TH(k)$, and $SDTH(k)$ that was used also in previous studies [10,3] and that are artificifally produced MONET instances. Additionally we have test data from the UC Irvine Machine Learning Repository [24,9] and from the Frequent Itemset Mining Implementations Repository (FIMI) [25]. Notice that an instance $(A, B) \in$ MONET if and only if $(B, A) \in$ MONET, where on the left hand side the set of terms A is read as a DNF and on the right hand side it is read as a CNF (B similarly).

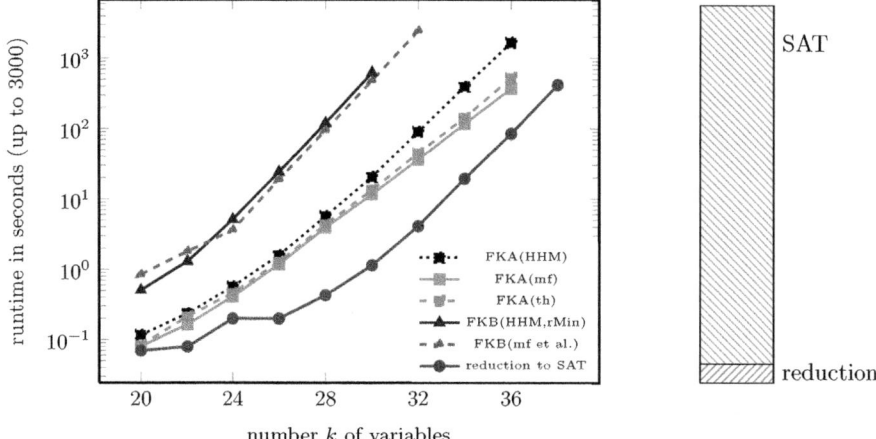

Fig. 1. Runtimes and reduction ratio on $M(k)$

Matching $M(k)$. The formula M_k of k variables (k even) consists of the terms $\{\{x_i, x_{i+1}\} : 1 \leq i < k, i \text{ is even}\}$. Thus, $|M_k| = k/2$. The formula $\widetilde{M}_k$ equivalent to M_k consists of the $2^{k/2}$ terms obtained by choosing one variable from every term of M_k. The instance $M(k)$ is the pair $(M_k, \widetilde{M}_k)$. Notice that the size of $M(k)$ is exponential in k. Simply said, the matching instances are pairs of equivalent formulae, where one formula is exponentially larger than the other.

Figure 1 shows the runtimes for the matching instances. It shows that the FKB-implementations are the slowest, the FKA-implementations are intermediate and the reduction to SAT is the fastest. For the reduction to SAT, the *reduction ratio* shows how much of the runtime was used by the reduction function and by the SAT-solver. One can see that the new FKA-implementation is better than the old one.

Threshold $TH(k)$. The formula T_k of k variables is the set of terms $\{\{x_i, x_j\} : 1 \leq i < j \leq k, j \text{ is even}\}$. Thus, $|T_k| = k^2/4$. The formula $\widetilde{T}_k$ equivalent to T_k is $\widetilde{T}_k = \{\{1, \ldots, 2t-1\} \cup \{2t+2, 2t+4, \ldots, k\} : 1 \leq t \leq k/2\} \cup \{\{2, 4, \ldots, k\}\}$. This yields $|\widetilde{T}_k| = k/2 + 1$. The instance $TH(k)$ is $(T_k, \widetilde{T}_k)$ and has size in $\mathcal{O}(k^2)$. Simply said, the threshold instances are pairs of equivalent formulae, where one formula is quadratic in the size of the other.

Figure 2 shows the runtimes for the threshold instances. It shows that the old FKA-implementation is the slowest, and the old FKB-implementation and the new FKB(rMin) are the fastest among the FK-implementations. The reason is that the choice of the variable does not really matter on these instances, and using a strategy wastes time. Again, the reduction to SAT is the fastest and seems to have the slowest slope.

Self Dual Threshold $SDTH(k)$. The formula ST_k of k variables (k even) is the set of terms $\{\{x_{k-1}, x_k\}\} \cup \{\{x_k\} \cup m : m \in T_{k-2}\} \cup \{\{x_{k-1}\} \cup m : m \in \widetilde{T}_{k-2}\}$.

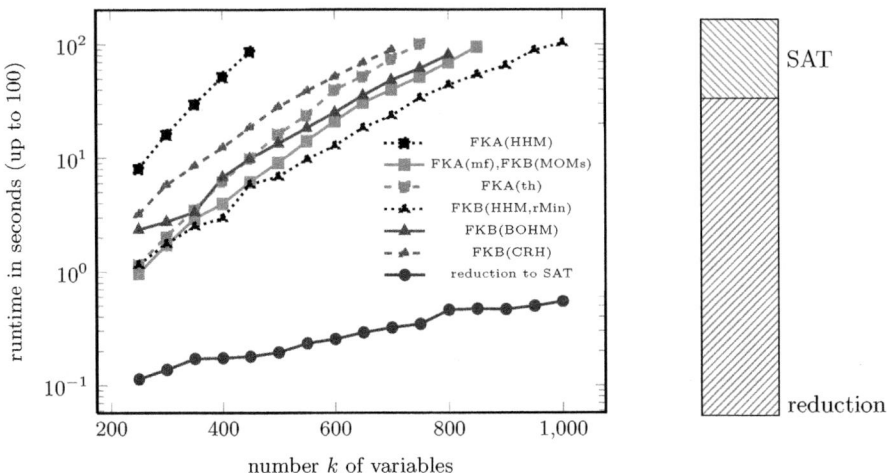

Fig. 2. Runtimes and reduction ratio on TH(k)

Note that ST_k read as DNF is equivalent to ST_k read as CNF. The instance
SDTH(k) is (ST_k, ST_k) and has size $\mathcal{O}(k^2)$. Simply said, the self dual threshold
instances are pairs of equivalent formulae of the same size.

Figure 3 shows the runtimes for the self-dual-threshold instances. It shows
that the old FKA-implementation is the slowest, but the new FKB(rMin)—that
was quite fast on the threshold instances—is very slow, too. The other new
FKB-implementations are the fastest among the FK-implementations. Again,
the reduction to SAT is the fastest.

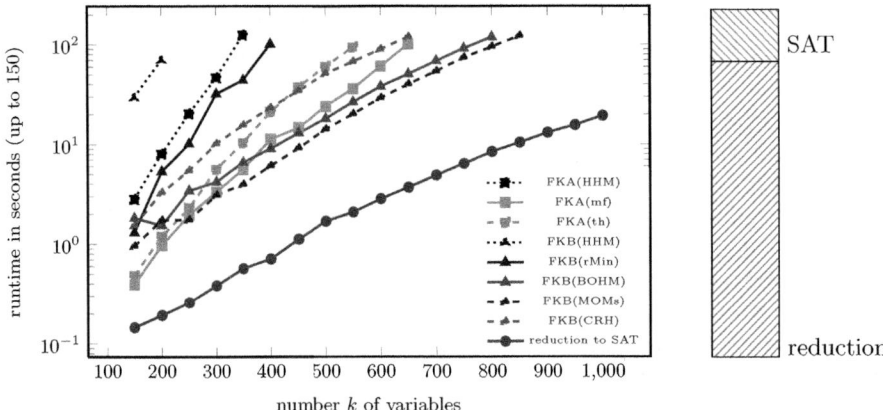

Fig. 3. Runtimes and reduction ratio on SDTH(k)

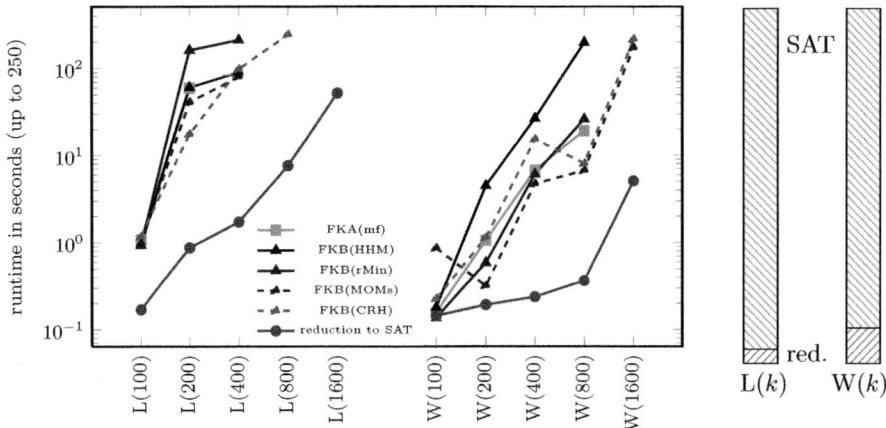

Fig. 4. Runtimes and reduction ratio on $L(r)$ and $W(r)$

Connect-4 $L(r)$ and $W(r)$. "Connect-4" is a board game. Each row of the dataset corresponds to a minimal winning (W) or losing (L) stage of the first player, and is represented as a term. A term of an equivalent formula of a set of winning stages (represented as a formula in DNF or CNF) is a minimal way to disturb winning/losing moves of the first player. To form a dataset, we take the first r rows of the minimal winning stage (called W_r) and the first r rows of the minimal losing stage (called L_r) [9,26]. To compute the equivalent formula $\widetilde{L}_r$ and $\widetilde{W}_r$ we used the DL-algorithm [9]. Thus, we have $L(r) = (L_r, \widetilde{L}_r)$ and $W(r) = (W_r, \widetilde{W}_r)$ as instances. The set of testdata are from the UC Irvine Machine Learning Repository [24]. It is used to compare algorithms that compute equivalent normal forms. The smallest formula $L(100)$ consists of 2,441 terms with 77 variables, and $L(1600)$ is the largest and has 214,361 terms with 81 variables. $W(100)$ has a size of 387 terms with 76 variables, and $W(3200)$ has 462,702 terms with 82 variables. Figure 4 shows the runtimes for the Connect-4 instances. Only few instances were solvable within the given time bound. All FK-implementations behave similar, and for sake of clarity we left some of them out in Figure 4. The new FKB(CRH) is the fastest among the FK-implementations. As before, the reduction to SAT is the fastest.

BMS-WebView-2 BMS(s) and accidents AC(s). This testdata is generated by enumerating all maximal frequent sets from datasets "BMS-WebView-2" and "accidents". For a dataset and a support threshold s, an itemset is called *frequent* if it is included in at least s members, and *infrequent* otherwise. A frequent itemset included in no other frequent itemset is called a *maximal frequent itemset*, and an infrequent pattern including no other infrequent itemset is called a *minimal infrequent itemset*. A minimal infrequent itemset is included in no maximal frequent itemset, and any subset of it is included in at

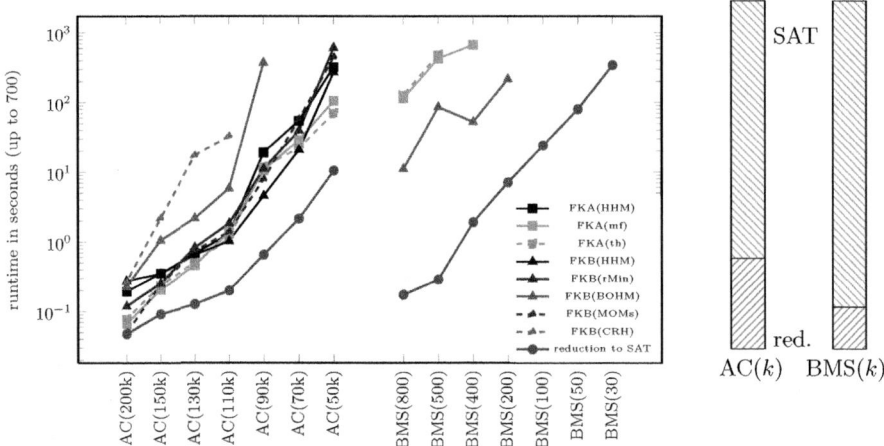

Fig. 5. Runtimes and reduction ratio on AC(s) and BMS(s)

least one maximal frequent itemset. Thus, the dual of the set of the complements of maximal frequent itemsets is the set of minimal infrequent itemsets
[26]. Note, if we want to check the correctness of enumerating all maximal frequent sets we can use MONET, because it is equivalent to this problem [2].
The problem instances are generated by enumerating all maximal frequent sets
from datasets BMS-WebView-2 BMS(s) and accidents AC(s) with threshold s,
taken from Frequent Itemset Mining Implementations Repository (FIMI) [25].
The smallest formula AC($150k$) has a size of 1,486 terms with 64 variables,
and the largest is AC($30k$) with 320,657 terms and 442 variables. BMS(500)
has a size of 17,143 terms with 3340 variables, and BMS(30) has 2,314,875
terms with 3340 variables. Figure 5 shows the runtimes for the AC and BMS
instances. The BMS instances show impressively, how good the reduction to
SAT works.

   The experiments described up to now used instances that consist of equivalent
formulae. To produce non-equivalent instances, we randomly delete variables
and terms in the above formulae. If we delete few variables or terms, we obtain
few conflict assignments. We compare some experiments on 236 non-equivalent
instances with thresholds of 60 and 360 seconds, where mf($\neg$UP) denotes the
mf-strategy without UP (see Table 1). The reduction to SAT solves all instances
within a time limit of 360 seconds, whereas our best implementation only solves
223 of 236 instances with this time. Furthermore, the experiments show that unit
propagation (UP) helps to solve more non-equivalence instances, since without
unit propagation less instances are solved. The runtimes do not depend on the
classes of test data introduced above.

   Finally, we show that in order to solve MONET using reduction to SAT with
a SAT-solver, it is much better to use the reduction function f (see Section 4)

Table 1. Non-equivalent instances (of 236) solved within 60 and 360 seconds

seconds	FKA			FKB						reduction to SAT
	mf(¬UP)	mf	HHM	mf(¬UP)	mf	MOMs	BOHM	CRH	HHM	
60	194	201	166	162	181	182	148	143	161	221
360	209	223	196	201	213	216	186	188	189	236

Table 2. Comparison of runtimes of different reductions to SAT in seconds

reduction	$M(k)$		$TH(k)$			$SDTH(k)$	
	22	24	250	500	700	250	400
using f	0.1	0.2	0.1	0.2	0.3	0.3	0.7
using Tseitin-translation	94	453	44	854	3604	539	4974
max. of FK-algorithms	1.9	5.1	8	136	563	20.5	211
	FKB(HHM)		FKA(HHM)[3]				

than the usual Tseitin-translation [21]. Table 2 shows that using the Tseitin-translation the runtimes are worse than using the FK-algorithms.

6   Conclusion

The main finding is that a good reduction function and a SAT-solver provides a more effective way for MONET than any current implementation of the FK-algorithms. It is a little surprising that it does not help to use the Tseitin translation only. Essentially, our reduction solves one direction of the equivalence test, and the SAT-solver solves the other direction. Eventually, it is not that surprising that the SAT-solvers are better than the implementations of the FK-algorithms.

On the other hand, we could improve the old implementations of the FK-algorithms [3] by using better data structures and unit propagation. Among the strategies for finding a splitting variable, it seems that MOMs is a good choice. This is not that surprising because MOMs is similar to choosing the most frequent variable, and the latter is a straightforward strategy intended in the formulation of FK-algorithm A [2].

Our next steps will be to figure out which strategies of SAT-solvers are responsible for the fast solution of reduced MONET instances and to see whether they can be integrated into the FK-algorithms. For example, clause learning seems to be useless for the FK-algorithms. Does the SAT-solver use it however for solving MONET instances? Moreover, the clauses obtained from the reduction function are easy in the sense that they are not needed for the NP-hardness of SAT—otherwise MONET would be coNP-complete. Therefore, one can assume that the "full power" of SAT-solvers is not necessary in order to solve reduced MONET instances fast. Another question is whether it makes the equivalence test easier

---

[3] Note that FKB(HHM) does not finish on SDTH(400).

if one checks both implication directions separately. The combination of reduction and SAT-solver works this way, whereas the FK-algorithms recursively make equivalence tests on decreasing formulae.

Acknowledgements. The authors thank Sebastian Kuhs for some implementations, and Markus Chimani and Stephan Kottler for helpful comments.

References

1. Kavvadias, D.J., Stavropoulos, E.C.: Checking monotone Boolean duality with limited nondeterminism. Technical Report TR2003/07/02, Univ. of Patras (2003)
2. Fredman, M.L., Khachiyan, L.: On the complexity of dualization of monotone disjunctive normal forms. Journal of Algorithms 21(3), 618–628 (1996)
3. Hagen, M., Horatschek, P., Mundhenk, M.: Experimental comparison of the two Fredman-Khachiyan-algorithms. In: Proc. ALENEX, pp. 154–161 (2009)
4. Reith, S.: On the complexity of some equivalence problems for propositional calculi. In: Rovan, B., Vojtáš, P. (eds.) MFCS 2003. LNCS, vol. 2747, pp. 632–641. Springer, Heidelberg (2003)
5. Eiter, T., Gottlob, G.: Hypergraph transversal computation and related problems in logic and AI. In: Flesca, S., Greco, S., Leone, N., Ianni, G. (eds.) JELIA 2002. LNCS (LNAI), vol. 2424, pp. 549–564. Springer, Heidelberg (2002)
6. Eiter, T., Gottlob, G., Makino, K.: New results on monotone dualization and generating hypergraph transversals. SIAM J. on Computing 32(2), 514–537 (2003)
7. Papadimitriou, C.H.: NP-completeness: A retrospective. In: Degano, P., Gorrieri, R., Marchetti-Spaccamela, A. (eds.) ICALP 1997. LNCS, vol. 1256, pp. 2–6. Springer, Heidelberg (1997)
8. Bailey, J., Manoukian, T., Ramamohanarao, K.: A fast algorithm for computing hypergraph transversals and its application in mining emerging patterns. In: Proc. of the 3rd IEEE Intl. Conference on Data Mining (ICDM 2003), pp. 485–488 (2003)
9. Dong, G., Li, J.: Mining border descriptions of emerging patterns from dataset pairs. Knowledge and Information Systems 8(2), 178–202 (2005)
10. Khachiyan, L., Boros, E., Elbassioni, K.M., Gurvich, V.: An efficient implementation of a quasi-polynomial algorithm for generating hypergraph transversals. Discrete Applied Mathematics 154(16), 2350–2372 (2006)
11. Kavvadias, D.J., Stavropoulos, E.C.: An efficient algorithm for the transversal hypergraph generation. J. of Graph Algorithms and Applications 9(2), 239–264 (2005)
12. Lin, L., Jiang, Y.: The computation of hitting sets: Review and new algorithms. Information Processing Letters 86(4), 177–184 (2003)
13. Torvik, V.I., Triantaphyllou, E.: Minimizing the average query complexity of learning monotone Boolean functions. INFORMS Journal on Computing 14(2), 144–174 (2002)
14. Uno, T., Satoh, K.: Detailed description of an algorithm for enumeration of maximal frequent sets with irredundant dualization. In: Proc. FIMI (2003)
15. Davis, M., Logemann, G., Loveland, D.W.: A machine program for theorem-proving. Commun. ACM 5(7), 394–397 (1962)
16. Pretolani, D.: Efficiency and stability of hypergraph sat algorithms. In: Proc. DIMACS Challenge II Workshop (1993)
17. Buro, M., Büning, H.K.: Report on a SAT competition (1992)

18. Kullmann, O.: Investigating the behaviour of a sat solver on random formulas. Technical Report CSR 23-2002, University of Wales (2002)
19. Quine, W.: Two theorems about truth functions. Boletin de la Sociedad Matemática Mexicana 10, 64–70 (1953)
20. Tamaki, H.: Space-efficient enumeration of minimal transversals of a hypergraph. In: Proc. SIGAL, pp. 29–36 (2000)
21. Tseitin, G.S.: On the complexity of derivation in propositional calculus. In: Slisenko, A. (ed.) Studies in Constructive Mathematics and Mathematical Logics, Part II, pp. 115–125 (1968)
22. Galesi, N., Kullmann, O.: Polynomial time SAT decision, hypergraph transversals and the hermitian rank. In: Hoos, H.H., Mitchell, D.G. (eds.) SAT 2004. LNCS, vol. 3542, pp. 89–104. Springer, Heidelberg (2005)
23. Le Berre, D., Parrain, A.: The Sat4j library, release 2.2. Journal on Satisfiability, Boolean Modeling and Computation 7, 59–64 (2010)
24. Repository, UCI (2010), http://www.archive.ics.uci.edu/ml/
25. Repository, FIMI (2010), http://www.fimi.ua.ac.be/
26. Murakami, K.: Personal communication (2010)

Enumerating All Solutions of a Boolean CSP by Non-decreasing Weight

Nadia Creignou, Frédéric Olive, and Johannes Schmidt

Laboratoire d'Informatique Fondamentale de Marseille, CNRS UMR 6166,
Aix-Marseille Université, 163, avenue de Luminy, F-13288 Marseille Cedex 9, France
{nadia.creignou,frederic.olive,johannes.schmidt}@lif.univ-mrs.fr

Abstract. We address the problem of enumerating all models of Boolean formulæ in order of non-decreasing weight in Schaefer's framework. The weight of a model is the number of variables assigned to 1. Tractability in this context amounts to enumerating all models one after the other in sorted order, with polynomial delay between two successive outputs. The question of model-enumeration has already been studied in Schaefer's framework, but without imposing a specific order. The order of non-decreasing weight changes the complexity considerably. We obtain a new dichotomous complexity classification. On the one hand, we develop new polynomial delay algorithms for Horn and 2-XOR-formulæ to enumerate the models by non-decreasing weight. On the other hand, we prove that in all other cases such a polynomial delay algorithm does not exist, unless $P = NP$.

Keywords: Enumeration, complexity, polynomial delay, generalized satisfiability, CSP.

1 Introduction

This paper is concerned with algorithmic and complexity of *enumeration*, the task of generating all solutions of a given problem. The area of enumeration algorithms has experienced tremendous growth over the last decade. This is motivated by the explosion in the size of the data that algorithms are called upon to process in everyday applications. The prime application is query answering in databases, where huge answer sets arise naturally. Computing queries incrementally and efficiently has become an increasingly important issue. For instance users of web search engines want to obtain the first results of their keyword search as quickly as possible. Other application domains include constraint solving, operations research, data mining, Web mining, bioinformatics and computational linguistics (see e.g. [17,8,1]).

Because of the amount of solutions that enumeration algorithms possibly produce, the size of their output is often much larger (e.g. exponentially larger) than the size of their input. Therefore, polynomial time complexity is not a suitable yardstick of efficiency when analyzing their performance. Actually, one would be interested in the regularity of these algorithms rather than in their total running time. For this reason, *polynomial delay* is customarily regarded as the good notion of tractability for enumeration complexity: an enumeration algorithm has polynomial delay $p(n)$ if the elapsed time between two successive outputs is polynomial in the size of the input.

K.A. Sakallah and L. Simon (Eds.): SAT 2011, LNCS 6695, pp. 120–133, 2011.

Since the seminal result of Schaefer [19], the theoretical interest of the CSP point of view on complexity questions has been largely assessed (see e.g. [5]). By offering a unified framework which has intimate connections with various problems in database, it allowed numerous classification results for a variety of computational tasks, see [7] for a survey. The present paper refers to this line of research.

In the context of non-uniform Boolean CSP we fix a constraint language Γ, which is a finite set of Boolean relations. A Γ-formula is then a conjunction of clauses where the form of the clauses is restricted by Γ. Thus the problem of enumeration can be phrased as follows: given a Γ-formula, can we efficiently enumerate all its models? Prior works handled that question. In [4], Creignou and Hébrard proved a first classification result about enumeration for Γ-formulæ: if Γ is Horn, dual-Horn, bijunctive or affine, there exists a polynomial delay algorithm that enumerates all models of any Γ-formula; otherwise, such an algorithm does not exist unless P = NP. But their result ignores an important feature in the design of enumeration algorithms: the specification of the order in which we wish the solutions to be output. This is a fundamental aspect of enumeration because in many cases we cannot afford to enumerate all the solutions, but rather we want to produce the most "important" ones in some *metric*. In other cases we need to find a solution that satisfies some other complicated side conditions and thus we generate the solutions in *order of preference* until we find an acceptable one (see some examples in [15,26]). Besides, it turns out that the order affects heavily the complexity. Johnson et al. [10] prove for instance that maximal independent sets of a graph can be enumerated in lexicographical order by a polynomial delay algorithm, while there is no such algorithm for the reverse lexicographical order, unless P = NP.

In this paper, we specify the order in which we wish the models of Γ-formulæ to be output. We deal with enumeration of models *according to their weight*, which is the number of variables they assign to 1. The weight is a natural parameter in Boolean CSPs [16,18,6], that can be seen as a cost of the assignment. Hence our approach refers to numerous works that focus on enumeration by non-decreasing cost [25,15,26]. Thus, the key problem addressed in this paper is: can one enumerate efficiently all models of a Γ-formula by non-decreasing weight? We answer this question with a dichotomous classification result: If a set of relations Γ is Horn or width-2 affine, there is a polynomial-delay algorithm that generates all models of a Γ-formula by non-decreasing weight. Otherwise such an algorithm does not exist, unless P = NP. The proof of this theorem reveals new enumeration algorithms for Boolean CSPs, different from the ones developed so far, in particular in the case of Horn formulæ.

The paper is organized as follows. In Sect. 2 we give the relevant material on Boolean CSPs and enumeration algorithms. We also state our main result, Theorem 2. The proof of this theorem is presented throughout the following sections. Section 3 deals with efficient enumeration algorithms in the case where Γ is width-2 affine or Horn. In Sect. 4 we prove the negative part of Theorem 2: for any relation Γ that is neither Horn nor width-2 affine, the existence of a polynomial delay algorithm for enumerating the models of a Γ-formula by non-decreasing weight would imply P = NP. We conclude and briefly point out open questions in Sect. 5.

2 Material

2.1 Constraint Languages and Γ-Formulæ

A *logical relation* of arity k is a relation $R \subseteq \{0,1\}^k$. By abuse of notation we do not make a difference between a relation and its predicate symbol. A *constraint*, C, is a formula $C = R(x_1, \ldots, x_k)$, where R is a logical relation of arity k and the x_i's are (not necessarily distinct) variables. If u and v are two variables, then $C[u/v]$ denotes the constraint obtained from C in replacing each occurrence of v by u. If V is a set of variables, then $C[u/V]$ denotes the result of substituting u to every occurrence of every variable of V in C. An assignment m of truth values to the variables *satisfies* the constraint C if $\big(m(x_1), \ldots, m(x_k)\big) \in R$. A *constraint language* Γ is a finite set of logical relations. A Γ-*formula* φ, is a conjunction of constraints using only logical relations from Γ and is hence a quantifier-free first order formula. With $\mathrm{Var}(\varphi)$ we denote the set of variables appearing in φ. A Γ-formula φ is satisfied by an assignment $m : \mathrm{Var}(\varphi) \to \{0,1\}$ if m satisfies all constraints in φ simultaneously (such a satisfying assignment is also called a *model* of φ). Assuming a canonical order on the variables we can regard models as tuples in the obvious way and we do not distinguish between a formula φ and the logical relation R_φ it defines, i.e., the relation consisting of all models of φ.

Throughout the text we refer to different types of Boolean relations following Schaefer's terminology [19]. We say that a Boolean relation R is *Horn* (resp. *dual Horn*) if R can be defined by a CNF formula which is Horn (resp. dual Horn). A relation R is *bijunctive* if it can be defined by a 2-CNF formula. A relation R is *affine* if it can be defined by an *affine* formula, i.e., conjunctions of XOR-clauses (consisting of an XOR of some variables plus maybe the constant 1) — such a formula may also be seen as a system of linear equations over GF[2]. A relation is *affine with width* 2 (width-2 affine, for short) if it is definable by a conjunction of clauses, each of them being either a unary clause or a 2-XOR-clause (consisting of an XOR of 2 variables plus maybe the constant 1) — such a conjunctive formula may also be seen as a system of linear equations over GF[2] with at most two variables per equation. A relation R is *0-valid* (resp., *1-valid*) if $R(0, \ldots, 0) = 1$ (resp., $R(1, \ldots, 1) = 1$). Finally, a constraint language Γ is Horn (resp. dual Horn, bijunctive, affine, width-2 affine, 0-valid, 1-valid) if every relation in Γ is Horn (resp. dual Horn, bijunctive, affine, width-2 affine). We say that a constraint language is *Schaefer* if Γ is either Horn, dual Horn, bijunctive, or affine.

There exist easy criteria to determine if a given relation is Horn, dual Horn, bijunctive, or affine. Indeed all these classes can be characterized by their polymorphisms (see e.g. [7] for a detailed description). We recall some of these properties here briefly for completeness. The operations of conjunction, disjunction, and addition applied on k-ary Boolean vectors are applied coordinate-wise.

- R is Horn if and only if $m, m' \in R$ implies $m \wedge m' \in R$.
- R is dual Horn if and only if $m, m' \in R$ implies $m \vee m' \in R$.
- R is affine if and only if $m, m', m'' \in R$ implies $m \oplus m' \oplus m'' \in R$.
- R is affine and 0-valid if and only if $m, m' \in R$ implies $m \oplus m' \in R$.

The satisfiability problem for Γ formulæ, denoted by $\mathrm{SAT}(\Gamma)$, was first studied by Schaefer [19] who obtained a famous dichotomous classification: If Γ is Schaefer or

0-valid or 1-valid, then $\text{SAT}(\Gamma)$ is in P; otherwise $\text{SAT}(\Gamma)$ is NP-complete. The complexity of finding a non-trivial solution (i.e., a solution different from all-zero and all-one), $\text{SAT}^*(\Gamma)$, was studied in [4]: If Γ is Schaefer, then $\text{SAT}^*(\Gamma)$ is in P; otherwise $\text{SAT}^*(\Gamma)$ is NP-complete. Since then and in the recent past, complexity classifications for many further computational problems for Γ-formulæ have been obtained (see [7] for a survey).

2.2 Enumeration

In this paper, we focus on enumeration by non-decreasing weight of the models of Boolean constraint formulæ, the weight of an assignment being the number of variables assigned to 1. The corresponding problem can be displayed as follows:

Problem: $\text{ENUM-SAT}_w(\Gamma)$
Input: a Γ-formula φ.
Output: generate all models of φ by non-decreasing weight.

We say that an algorithm $\mathcal{A}$ computes the enumeration problem $\text{ENUM-SAT}_w(\Gamma)$ if for a given input φ, $\mathcal{A}$ generates one by one the models of φ, by non-decreasing weight, without repetition, and stops after writing the last one.

Polynomial time complexity is not a suitable yardstick of efficiency when analyzing an enumeration algorithm since the output size is usually exponential in the size of the input. In [10] several notions of efficiency are discussed. The least we could ask is that the enumeration algorithm runs in *polynomial total time*, that is that the time required to output all solutions be polynomial in the size of the input and in the number of solutions (i.e., the size of the output). This notion is also referred to as *output polynomial*. An important feature of an enumeration algorithm is the ability to start generating configurations as soon as possible, and more generally to generate configurations in a regular way with a limited delay between two successive outputs. Hence we say that an enumeration algorithm runs in *polynomial delay* if the delay until the first solution is output and thereafter the delay between any two consecutive solutions is bounded by a polynomial $p(n)$ in the input size. It is worth noticing that such an algorithm generates the first k outputs in time $k \cdot p(n)$. This is an important property of polynomial delay algorithms, since when one has not enough time to enumerate all solutions, at least the first k solutions (the top-k-ranked in the case of an enumeration in a ranked order) can be efficiently enumerated. If $\text{ENUM-SAT}_w(\Gamma)$ is computable by a polynomial delay algorithm, we write $\text{ENUM-SAT}_w(\Gamma) \in \text{delayP}$.

For characterizing space efficiency we ignore the amount of space needed for writing the output, only the space used for storing intermediate results is measured. Enumeration algorithms are sometimes required to run in *polynomial space*, which means that the amount of space involved during the whole computation is polynomial in the size of the input. This requirement is restrictive, even for polynomial delay algorithms. Indeed there are polynomial delay algorithms, especially when a specified order is required, that build exponentially large data structures (see [10,14,24]). This is also the case for the polynomial delay algorithm for Horn formulæ, described in Sect. 3. Nevertheless, any polynomial delay algorithm runs in *incremental polynomial space*, which means that the space needed for generating the first k solutions is bounded by k times a polynomial in the input size.

2.3 Main Result

The complexity of enumerating all models of generalized Boolean formulæ, without specifying any order, has been studied in [4]. An alternative proof making use of partial polymorphisms was given later in [21].

Theorem 1. *(Model enumeration [4].) If Γ is Schaefer, then there is a polynomial-delay algorithm that generates all models of a Γ-formula. Otherwise such an algorithm does not exist unless* $\mathrm{P} = \mathrm{NP}$.

In this paper we are interested in enumerating all models by non-decreasing weight. Of course, when Γ is Schaefer, Theorem 1 enables to do that in polynomial total time: first, generate all solutions in lexicographic order with the algorithm underlying the proof of this theorem; then, sort the solutions in order to output them by non-decreasing weight. However such a procedure forbids any control on the regularity of the enumeration since for instance, the first solution is output after an exponential amount of time if there is an exponential number of models. Therefore, we have to develop specific techniques to perform efficient enumeration in this order. Sets of relations that admit a good behavior with respect to this task do not coincide with Schaefer's ones. Our main theorem details this situation, stating a new dichotomy result concerning the enumeration of models by non-decreasing weight.

Theorem 2. *(Enumeration by non-decreasing weight.) If Γ is Horn or width-2 affine, then there is a polynomial-delay algorithm that generates all models of a Γ-formula by non-decreasing weight. Otherwise such an algorithm does not exist unless* $\mathrm{P} = \mathrm{NP}$.

3 Efficient Enumeration Algorithms

The efficient enumeration algorithms proposed earlier in [4] (see Theorem 1) were based on self-reducibility ([23,13,20]). The self-reducibility property of a problem allows a "search-reduces-to-decision" algorithm to enumerate the solutions. As a consequence, the models are provided in lexicographical order. Moreover the algorithms use only polynomial space. Enumerating the solutions in order of non-decreasing weight requires new algorithms. The two classes of constraint languages under examination in this section, namely width-2 affine and Horn, invoke different algorithmic approaches. Indeed they differ in the complexity of their associated k-ONES problem, in which, given a formula and an integer k, we want to know whether there exists a model with exactly k ones. For width-2 affine formulæ this problem is in P, whereas for Horn formulæ it is NP-complete (see [6]). As a consequence for width-2 affine formulæ one can use the tractability of the k-ONES problem to get an efficient enumeration algorithm. In contrast, Horn formulæ require another strategy. It is natural to use a data structure which maintains an ordered set of elements and which supports efficient operations of insertion and extraction. The algorithm we will present for the Horn case makes use of a priority queue in order to produce the right order of the output solutions. This method was already used in e.g. [10,14]. As in these papers, our priority queue allows insertion of elements and extraction of the top element in logarithmic time in the size of the queue. Thus the size of the queue may grow exponentially whereas polynomial delay is still maintained.

Proposition 1. *If Γ is width-2 affine, then there is a polynomial-space polynomial-delay algorithm that generates all models of a Γ-formula by non-decreasing weight.*

Proof. Let Γ be width-2 affine and let φ be a Γ-formula. Without loss of generality we can suppose that φ does not contain unitary clauses. Then each clause of φ expresses either the equality or the inequality between two variables. Using the transitivity of the equality relation and the fact that in the Boolean case $a \neq b \neq c$ implies $a = c$, we can identify equivalence classes of variables such that each two classes are either independent or they must have contrary truth values. We call a pair (A, B) of classes with contrary truth values *cluster*, B may be empty. It follows easily that any two clusters are independent and thus to obtain a model of φ, we choose for each cluster (A, B) either $A = 1, B = 0$ or $A = 0, B = 1$. We suppose in the following that φ is satisfiable (otherwise, we will detect a contradiction while constructing the clusters). Let $n \geq 1$ be the number of clusters, then the number of models will be 2^n. The weight contribution of each cluster to a model is either $|A|$ or $|B|$, where $|A| = |B|$ may occur. We represent a model by an n-tuple $s \in \{0, 1\}^n$, indicating for each cluster which of the two assignments is taken. In the case $|A| \neq |B|$ we indicate by 0 the light assignment and by 1 the heavy assignment. Surely $(0, 0, \ldots, 0)$ will represent a model of minimal weight, and $(1, 1, \ldots, 1)$ will represent a model of maximal weight. For enumeration we may consider only the weight difference $||A| - |B||$ of each cluster, since we can subtract the weight of a minimal model. Setting $(w_1, \ldots, w_n)$ to these weight differences of the clusters, we reduce our problem to the following enumeration problem:

Problem:	SUBSET-SUM
Input:	A sequence of non-negative integers $(w_1, \ldots, w_n) \in \mathbb{N}^n$
Output:	generate all n-tuples $s \in \{0, 1\}^n$ by non-decreasing weight $\delta(s)$, where $\delta(s) = \Sigma_{i=1}^n s_i \cdot w_i$

To solve this enumeration problem we make use of the fact that in our case the sum of the weights $W := \Sigma_{i=1}^n w_i$ is linearly bounded by the number of variables of the original formula φ. This allows a strategy of dynamic programming to compute in polynomial time a matrix $A \in \{0, 1\}^{(n+1, W+1)}$ such that $A(i, k) = 1$ if and only if with the weights $w_1, \ldots, w_i$ one can construct the sum k, where $0 \leq i \leq n$, $0 \leq k \leq W$. The matrix A is constructed by first setting $A(0, 0) = 1$ and $A(0, k) = 0$ for all $k \geq 1$, and then filling the other fields row by row according to the rule $A(i, k) = 1$ if and only if $A(i - 1, k) = 1$ or $A(i - 1, k - w_i) = 1$. Thus the computation of A takes time $O(n \cdot W)$. After this precomputation, for each k for which there is at least one solution of weight k we enumerate all such solutions by constructing the solution strings from ϵ (the empty string) recursively.

Algorithm 1. Algorithm for SUBSET-SUM

MAIN($w_1, \ldots, w_n$)

1: compute $A \in \{0, 1\}^{(n+1, \, W+1)}$
2: **for** $k = 0$ to W **do**
3:     **if** $A(n, k) = 1$ **then**
4:         CONSTRUCTSOLUTIONS($n, \, k, \, \epsilon$)   /* enumerate all solutions of weight k */
5: **end for**

CONSTRUCTSOLUTIONS($i, \, j, \, s$)

1: **if** $i = 0$ **then**
2:     output s
3: **else**
4:     **if** $A(i - 1, j - w_i) = 1$ **then**
5:         CONSTRUCTSOLUTIONS($i - 1, \, j - w_i, \, 1 \circ s$)   /* $\circ$ stands for the concatenation operator */
6:     **if** $A(i - 1, j) = 1$ **then**
7:         CONSTRUCTSOLUTIONS($i - 1, \, j, \, 0 \circ s$)

The reader may convince himself or herself that Algorithm 1 enumerates all solutions s of the SUBSET-SUM problem by non-decreasing weight $\delta(s)$. Since both n and W are linearly bounded by the number of variables of φ, Algorithm 1 has a quadratic precomputation time and a linear delay thereafter. The translations between our original problem and SUBSET-SUM can be performed in polynomial time.   □

Proposition 2. *If Γ is Horn, then there is a polynomial-delay algorithm that generates all models of a Γ-formula by non-decreasing weight.*

Proof. Let Γ be Horn and let φ be a Γ-formula. Then φ is equivalent to a conjunction of Horn clauses. We will use a priority queue Q to respect the order of non-decreasing weight and to avoid duplicates. The command Q.enqueue(s, k) enqueues an element s with an integer key-value k (a weight). The queue sorts by non-decreasing key-value and inserts an element s only if it is not yet present in the queue.

For notational convenience we represent a model by the set of variables it sets to 1. We use the well-known fact that for Horn formulæ the intersection of all models is the unique *minimal model* which is polynomial time computable. For a satisfiable Horn formula φ we indicate the minimal model by $mm(\varphi)$. Note that for a set of variables $V \subseteq Vars(\varphi)$ the formula $\varphi \wedge V := \varphi \wedge \bigwedge_{v \in V} v$ is still representable as a Horn formula and thus, if $\varphi \wedge V$ is satisfiable, also $mm(\varphi \wedge V)$ can be computed in polynomial time.

We claim that Algorithm 2 enumerates the models of a given Horn formula with polynomial delay, by non-decreasing weight. The polynomial delay is easily seen. By definition of the priority queue and by the fact that the models m' generated out of m in line 12 are always of bigger weight than m itself, it is also easily seen that the models are output in the right order and that no model is output twice. To prove that no model is omitted, it suffices to show that for every model $m' \neq mm(\varphi)$ there exists a submodel $m \subsetneq m'$ such that in line 12 the algorithm generates m' out of m. That is,

---

Algorithm 2. Algorithm for HORN-SAT

Require: φ a Horn formula
 1: **if** φ unsatisfiable **then**
 2:     return 'no'
 3: Q = newPriorityQueue
 4: $m := mm(\varphi)$
 5: Q.enqueue($m, |m|$)
 6: **while** Q not empty **do**
 7:     $m :=$ Q.dequeue
 8:     output m
 9:     **for all** $x \in Vars(\varphi) \setminus m$ **do**
10:         **if** $\varphi \wedge m \wedge x$ satisfiable **then**
11:             $m' := mm(\varphi \wedge m \wedge x)$
12:             Q.enqueue($m', |m'|$)
13:     **end for**
14: **end while**

---

there must be an $x \in m' \setminus m$ such that $m' = mm(\varphi \wedge m \wedge x)$. Consider for this the set $H := \{m \mid m$ a model of φ and $m \subsetneq m'\}$. The set H is not empty since it contains at least the minimal model $mm(\varphi)$. A maximal element m of H fulfills our needs, since it satisfies $m' = mm(\varphi \wedge m \wedge x)$ for any $x \in m' \setminus m$.

Let us finally stress that in contrast to Algorithm 1, Algorithm 2 potentially runs in exponential space.   □

4   Hardness Results

In this section we investigate the case where Γ is neither Horn nor width-2 affine. Clearly, in order to enumerate the models of a Γ-formula by non-decreasing weight, it is a necessary condition to be able to find the lightest model efficiently. As we will prove, this is not a sufficient condition, we need also to be able to find the second one efficiently. So let us introduce the following problems.

Problem:	MIN-ONES(Γ)
Input:	a Γ-formula φ, an integer W
Question:	Is there a model of φ that assigns 1 to at most W variables?

Problem:	MIN-ONES*(Γ)
Input:	a Γ-formula φ, an integer W
Question:	Is there a model of φ different from all-0 that assigns 1 to at most W variables?

From the classification obtained in [12] for the corresponding optimization problem, one can deduce the following.

Proposition 3. *(Minimum ones satisfiability [12].) If Γ is 0-valid or Horn or width-2 affine, then* MIN-ONES(Γ) *is in* P, *otherwise* MIN-ONES(Γ) *is NP-complete.*

Our main contribution in this section is the following hardness result, which obviously proves that when Γ is neither Horn nor width-2 affine, there is no polynomial delay algorithm that enumerates all models of a Γ-formula in order of non-decreasing weight, unless P = NP.

Proposition 4. *Let Γ be a set of relations which is neither Horn nor width-2 affine. Then* MIN-ONES$^*(\Gamma)$ *is* NP-*complete.*

Proof. If Γ is not Schaefer, then SAT$^*(\Gamma)$ is NP-complete [19] and hence so is the problem MIN-ONES$^*(\Gamma)$. If Γ is not 0-valid, then, since it is neither Horn nor width-2 affine, the result follows from the NP-completeness of MIN-ONES(Γ) (Proposition 3). Therefore, it remains to study sets Γ that are Schaefer and 0-valid but that are neither Horn nor width-2 affine. There are three cases to analyse.

- Γ is bijunctive and 0-valid but neither Horn nor width-2 affine.
- Γ is affine and 0-valid but neither Horn nor width-2 affine.
- Γ is dual Horn and 0-valid but neither Horn nor width-2 affine.

Observe that a 2-CNF formula which is 0-valid is also Horn. So the first case does not occur. Besides, one can easily prove that a 0-valid affine relation which is not Horn cannot be width-2 affine. Therefore the proof of the proposition will be completed when we successively prove the NP-completeness of MIN-ONES$^*(\Gamma)$ for any set Γ such that:

1. Γ is affine and 0-valid but not Horn, or
2. Γ is dual Horn and 0-valid but neither affine nor Horn.

The NP-completness of MIN-ONES$^*(\Gamma)$ for any set Γ fulfilling the description 1 or 2 above is settled, respectively, by the forthcoming Proposition 5 and Proposition 6.   □

4.1   Affine, 0-valid, not Horn

In this section we deal with relations that are 0-valid and affine but not Horn. We will prove that for such a relation R, finding a non-all-0 model of minimal weight of an R-formula is NP-hard. In order to do so, we need some technical lemmas. The two first ones are definability results, while the third is a basic hardness result. One of the most successful techniques to obtain results on the complexity of constraints related problems (including enumeration), has been the application of tools from universal algebra. A Galois connection relates the expressive power of a constraint language to its set of so-called polymorphisms or partial polymorphisms (see e.g. [9,22]). However here it is not worth using this algebraic tool. The technical results that are needed concern only very restrictive sets of relations and can be obtained "by hand".

In the proofs of all the following lemmas, R will denote a relation of arity k and V a set of k distinct variables, say $V = \{x_1, \ldots, x_k\}$.

Lemma 1. *Let R be a relation which is 0-valid and affine but neither Horn nor 1-valid. Then there exists an R-formula equivalent to $\neg w \wedge (x \oplus y \oplus z = 0)$.*

Proof. Consider the constraint $C = R(x_1, \ldots, x_k)$. Since R is non-Horn there exist m_1 and m_2 in R such that $m_1 \wedge m_2 \notin R$. Since R is 0-valid and affine, we have $m_1 \oplus m_2 \in R$. For $i, j \in \{0, 1\}$, set $V_{i,j} = \{x \mid x \in V, m_1(x) = i \wedge m_2(x) = j\}$. Observe that $V_{0,1} \neq \varnothing$ (respectively, $V_{1,0} \neq \varnothing$), otherwise $m_1 \wedge m_2 = m_2$ (resp.,

$m_1 \wedge m_2 = m_1$), contradicting the fact that $m_1 \wedge m_2 \notin R$. Moreover $V_{1,1} \neq \varnothing$, otherwise $m_1 \wedge m_2 = \mathbf{0}$, a contradiction. Consider the $\{R\}$-constraint: $M(w, x, y, z) = C[w/V_{0,0}, x/V_{0,1}, y/V_{1,0}, z/V_{1,1}]$. According to the above remark the three variables x, y and z effectively occur in this constraint. Let us examine the set of models of M assigning 0 to w: it contains 0011 (since $m_1 \in R$), 0101 (since $m_2 \in R$), 0110 (since $m_1 \oplus m_2 \in R$) and 0000 (since R is 0-valid). But it does not contain 0001 (since by assumption $m_1 \wedge m_2 \notin R$). Thus it does not contain 0111 either. Indeed, otherwise it would contain $0011 \oplus 0101 \oplus 0111$ (since R is affine), which is equivalent to 0001, a contradiction. From this one can prove that it contains neither 0010 nor 0100 (since $0000 \oplus 0011 \oplus 0010 = 0001$ and $0110 \oplus 0101 \oplus 0100 = 0111$). Note that since R is 0-valid but not 1-valid $C[w/V] \equiv \neg w$. Hence, let us consider

$$\varphi(w, x, y, z) = C[w/V] \wedge M(w, x, y, z).$$

The R-formula φ is equivalent to $\neg w \wedge (x \oplus y \oplus z = 0)$, thus concluding the proof. $\quad\square$

Lemma 2. *Let R be a relation which is 0-valid, 1-valid, affine but not Horn. Then there exists an R-formula equivalent to $(w \oplus x \oplus y \oplus z = 0)$.*

Proof. Observe that an affine relation R which is both 0-valid and 1-valid is necessarily complementive, i.e. for all $m \in R$ we have also $\mathbf{1} \oplus m \in R$. We can mimic the analysis made in the previous lemma and consider the constraint $M(w, x, y, z) = C[w/V_{0,0}, x/V_{0,1}, y/V_{1,0}, z/V_{1,1}]$. Thus, the formula $\varphi(w, x, y, z) = M(w, x, y, z) \wedge M(w, y, z, x) \wedge M(w, z, x, y)$ verifies $\varphi(w, x, y, z) \equiv (w \oplus x \oplus y \oplus z = 0)$. $\quad\square$

Lemma 3. MIN-ONES$^*(x \oplus y \oplus z = 0)$ *and* MIN-ONES$^*(w \oplus x \oplus y \oplus z = 0)$ *are NP-complete.*

Proof. Consider a homogeneous linear system over the finite field GF(2). Finding the non-all-0 solution with minimum weight of such a system is known to be NP-hard (see [2, Theorem 4.1]). In order to prove the lemma we have to show that this problem remains hard when restricted to systems that have three (resp., four) variables by equation. Let S be a homogeneous linear system over GF(2). Suppose that S has n variables, $x_1, \ldots, x_n$. In order to reduce the number of variables in each equation we introduce auxiliary variables. If there is an equation $x_{i_1} \oplus x_{i_2} \oplus \cdots \oplus x_{i_k} = 0$ for some $k \geq 4$, we introduce a new variable y_{i_1, i_2} and replace the original equation by the two equations $y_{i_1, i_2} \oplus x_{i_1} \oplus x_{i_2} = 0$ and $y_{i_1, i_2} \oplus x_{i_3} \oplus \ldots \oplus x_{i_k} = 0$. We repeat this process until all equations have three variables. The satisfiability is preserved during this transformation. The number of auxiliary variables is bounded from above by the number of occurrences of variables in the original system. In order to keep the information on the weight of the solutions we need to introduce enough copies of the original variables, which make the auxiliary variables neglectable. Let N be the number of occurrences of variables in S. Let f be a fresh variable that will play the role of the constant 0. For each $i = 1, \ldots, n$, we introduce N copy-variables $x_i^1, \ldots, x_i^N$ of x_i and add the equations $x_i \oplus x_i^j \oplus f = 0$ for $j = 1, \ldots N$. Finally we add the equation $f \oplus f \oplus f = 0$, i.e., $f = 0$ (this will ensure that $x_i = x_i^j$ for all j). There is a one-to-one correspondence between the solutions of S and the solutions of the so-obtained system S'. Moreover S has a non-trivial solution of weight at most W if and only if S' has a non-trivial solution of weight at most $W(N+1) + N$. Since the system S' can be seen as an $(x \oplus y \oplus z = 0)$-formula we have thus proved the NP-hardness of MIN-ONES$^*(x \oplus y \oplus z = 0)$.

Let us now reduce MIN-ONES$^*(x \oplus y \oplus z = 0)$ to MIN-ONES$^*(w \oplus x \oplus y \oplus z = 0)$. Let S be a homogeneous linear system over n variables such that each equation has exactly three variables. Let w and w_i for $i = 1, \ldots, n+1$ be fresh variables. Transform S into S' as follows: transform every equation $x \oplus y \oplus z = 0$ into $w \oplus x \oplus y \oplus z = 0$ and add the $n + 1$ equations $w \oplus w \oplus w \oplus w_i = 0$ for $i = 1, \ldots, n+1$. Solutions of S' assigning 0 to w coincide with the solutions of S. Moreover any solution of S' assigning 1 to w has weight at least $n + 1$. Therefore, S has a non-trivial solution of weight at most W ($W \leq n$) if and only if S' has a non-trivial solution of weight at most W. This completes the proof.   □

Proposition 5. *If R is 0-valid and affine but not Horn, then the problem* MIN-ONES$^*(R)$ *is NP-complete.*

Proof. It follows from the three lemmas above: if R is not 1-valid, then Lemma 1 allows a reduction from MIN-ONES$^*(x \oplus y \oplus z = 0)$ to MIN-ONES$^*(R)$ (replace each constraint $(x \oplus y \oplus z = 0)$ by the R-formula equivalent to $\neg w \wedge (x \oplus y \oplus z = 0)$, where w is a fresh variable). If the relation R is 1-valid, then Lemma 2 allows a reduction from MIN-ONES$^*(w \oplus x \oplus y \oplus z = 0)$ to MIN-ONES$^*(R)$. In both cases one can conclude with Lemma 3.   □

4.2   Dual Horn, 0-valid, neither Affine nor Horn

In this section we deal with relations that are 0-valid and dual Horn but neither affine nor Horn. The method of proof is not the same as in the previous section. We will also need some intermediate lemmas.

Lemma 4. *Let R be a relation which is 0-valid, dual Horn but neither affine nor 1-valid. Then there exists an R-formula equivalent to $\neg t \wedge (u \to v)$.*

Proof. Consider the constraint $C = R(x_1, \ldots, x_k)$. Observe that since R is 0-valid but not 1-valid, $C[t/V] \equiv \neg t$. Since R is 0-valid and non-affine there exist two distinct tuples m_1 and m_2 in R such that $m_1 \oplus m_2 \notin R$. Since R is dual Horn, we have $m_1 \vee m_2 \in R$. For $i, j \in \{0, 1\}$, let $V_{i,j} = \{x \mid x \in V, m_1(x) = i \wedge m_2(x) = j\}$. Observe that $V_{1,1} \neq \varnothing$, otherwise $m_1 \vee m_2 = m_1 \oplus m_2$, a contradiction. Moreover, since $m_1 \neq m_2$ either $V_{0,1}$ or $V_{1,0}$ is nonempty. Suppose first that they are both nonempty. Consider the R-constraint $M(w, x, y, z) = C[w/V_{0,0}, x/V_{0,1}, y/V_{1,0}, z/V_{1,1}]$. The three variables x, y and z effectively appear in this constraint. Let us examine the set of models of M assigning 0 to w: it contains 0011 (since $m_1 \in R$), 0101 (since $m_2 \in R$), 0111 (since $m_1 \vee m_2 \in R$) and 0000 (since R is 0-valid), but does not contain 0110 (since by assumption $m_1 \oplus m_2 \notin R$). The membership of 0100, 0010, 0001 is open:

- If it does not contain 0100, then consider the R-formula $\varphi(t, u, v) := C[t/V] \wedge M(t, u, v, v)$. Its set of models is $\{001, 011, 000\}$ and therefore, $\varphi(t, u, v) \equiv \neg t \wedge (u \to v)$.

- If it contains 0100, then it does not contain 0010. Indeed otherwise, since R is dual-Horn it would also contain 0110, which provides a contradiction. Thus consider the R-formula $\varphi(t, u, v) := C[t/V] \wedge M(t, v, u, v)$. Its set of models is $\{001, 011, 000\}$ and therefore, $\varphi(t, u, v) \equiv \neg t \wedge (u \rightarrow v)$.

If for instance $V_{0,1} = \varnothing$, then consider $M(w, y, z) = C[w/V_{0,0}, y/V_{1,0}, z/V_{1,1}]$. In this case $\varphi(t, u, v) := C[t/V] \wedge M(t, u, v)$ is equivalent to $\neg t \wedge (u \rightarrow v)$.   □

Lemma 5. *Let R be a relation which is 0-valid, 1-valid, dual Horn but not Horn, then there exists an R-formula equivalent to $(u \rightarrow v)$.*

Proof. Since R is dual Horn but non Horn, it is non-complementive, that is, there exists an $m \in R$ such that $m \oplus 1 \notin R$. Consider the constraint $C = R(x_1, \ldots, x_k)$. For $i \in \{0, 1\}$, let $V_i = \{x \mid x \in V \wedge m(x) = i\}$. Consider the R-formula $\varphi(u, v) = C[u/V_0, v/V_1]$. Then $\varphi(u, v) \equiv \neg u \vee v \equiv u \rightarrow v$.   □

Proposition 6. *If R is 0-valid and dual Horn but neither affine nor Horn, then the problem MIN-ONES$^*(R)$ is NP-complete.*

Proof. Let R be a relation which is 0-valid and dual Horn but neither affine nor Horn. Let T be the constant unary relation $T = \{1\}$. According to Proposition 3, the problem MIN-ONES(R, T) is NP-complete. We reduce MIN-ONES(R, T) to MIN-ONES$^*(R)$. Let φ be an $\{R, T\}$-formula, $\varphi = \psi \wedge \bigwedge_{x \in V} T(x)$ where ψ is an R-formula. Let t be a fresh variable and consider

$$\varphi' = \psi[t/V] \wedge \bigwedge_{x \in \text{Var}(\varphi) \setminus V} x \rightarrow t.$$

Observe that the only solution that assigns 0 to t in φ' is the all-0 one. Therefore it is clear that φ has a solution of weight at most W ($W \geq | V |$) if and only if φ' has a non-trivial solution of weight at most $W - | V | + 1$. The two above lemmas allow to express φ' as an R-formula (modulo the introduction of an additional variable that will always take the value 0 when R is not 1-valid), thus concluding the proof.   □

5   Conclusion

We have classified the complexity of enumerating all models of a Γ-formula by non-decreasing weight. We have proved that in the case of Boolean CSPs a necessary and sufficient condition for enumerating all solutions in order of non-decreasing weight with polynomial delay is the ability to efficiently find a non-all-zero solution of minimal weight. Note that by duality, under the assumption P $\neq$ NP, one can enumerate the models of a Γ-formula by *non-increasing* weight with polynomial delay if and only if Γ is width-2 affine or dual Horn.

Another related question is: When does exist a so-called *polynomial time iterator* for the solutions' weight? That is, given a model m, when are we able to efficiently compute a model of the next weight level? In the width-2 affine case this task is tractable since k-ONES is tractable. In the Horn case this task becomes NP-hard: If it were tractable, by iteration we would be able to efficiently compute a model of maximal weight, which is NP-hard [12]. In the remaining cases, that is when enumeration by non-decreasing weight can not be done with polynomial delay unless $P = NP$, both situations may occur. A complete classification might reveal new classes of formulæ interesting in the context of enumeration.

We could also have dealt with the weighted version of the problem, i.e., we have a weight function $w: V \rightarrow \mathbb{N}$ and the weight of a model m is given by $\sum_{v \in V} w(v) \cdot m(v)$. The algorithm proposed in Proposition 2 for Horn formulæ could also tackle this variant. But the algorithm proposed in Proposition 1 for width-2 affine formulæ does not run in polynomial delay when the weights are not polynomially bounded. However, for cost of exponential space, one can also construct a polynomial delay algorithm for this case, using a priority queue in a similar way as in the Horn case.

Asking the enumeration to be performed in order of non-decreasing weight has revealed new enumeration algorithms for Boolean CSPs, different from the ones developed so far. The algorithm developed for Horn formulæ requires potentially an exponential amount of space. Ideally we would like to avoid this. Interesting open questions are to find out whether there is a space/delay trade-off and whether the exponential space requirement is inherent to the order of non-decreasing weight for Horn formulæ.

Another interesting direction of research is the study of enumeration in order of non-decreasing weight for CSPs over arbitrary finite domains. Enumeration of all solutions of such CSPs has been studied in [3,21]. As mentioned in [21] considering different orderings could be the key to discover further enumeration algorithms. Also, in [11] the authors studied the complexity of the so-called MAX-SOL(Γ) problem, which generalizes the MAX-ONES problem to arbitrary finite domains. They identified two tractable classes of constraint languages, namely injective and generalized max-closed constraint languages. These two classes can be seen as substantial and nontrivial generalizations of the tractable classes known for the MAX-ONES problem over the Boolean domain, namely width-2 affine, 1-valid and dual-Horn. It would be interesting to examine whether these classes give rise to polynomial delay algorithms for the enumeration of solutions by non-increasing weight.

Acknowledgment

We thank Arnaud Durand for providing us with the reference [2]. This work was partially supported by the project ANR ENUM (ANR 07-BLAN-0327-04).

References

1. Bagan, G., Durand, A., Grandjean, E.: On acyclic conjunctive queries and constant delay enumeration. In: Duparc, J., Henzinger, T.A. (eds.) CSL 2007. LNCS, vol. 4646, pp. 208–222. Springer, Heidelberg (2007)
2. Barg, A.: Complexity issues in coding theory. Electronic Colloquium on Computational Complexity (ECCC) 4(46) (1997)
3. Cohen, D.A.: Tractable decision for a constraint language implies tractable search. Constraints 9(3), 219–229 (2004)

4. Creignou, N., Hébrard, J.-J.: On generating all solutions of generalized satisfiability problems. Theoretical Informatics and Applications 31(6), 499–511 (1997)
5. Creignou, N., Kolaitis, P.G., Vollmer, H. (eds.): Complexity of Constraints. LNCS, vol. 5250. Springer, Heidelberg (2008)
6. Creignou, N., Schnoor, H., Schnoor, I.: Nonuniform boolean constraint satisfaction problems with cardinality constraint. ACM Trans. Comput. Log. 11(4) (2010)
7. Creignou, N., Vollmer, H.: Boolean constraint satisfaction problems: When does Post's lattice help? In: Creignou, et al [5], pp. 3–37
8. Hagen, M.: Lower bounds for three algorithms for transversal hypergraph generation. Discrete Applied Mathematics (2009)
9. Jeavons, P.G.: On the algebraic structure of combinatorial problems. Theoretical Computer Science 200, 185–204 (1998)
10. Johnson, D.S., Papadimitriou, C.H., Yannakakis, M.: On generating all maximal independent sets. Inf. Process. Lett. 27(3), 119–123 (1988)
11. Jonsson, P., Nordh, G.: Introduction to the maximum solution problem. In: Creignou, et al [5], pp. 255–282
12. Khanna, S., Sudan, M., Williamson, D.: A complete classification of the approximability of maximization problems derived from Boolean constraint satisfaction. In: Proceedings 29th Symposium on Theory of Computing, pp. 11–20. ACM Press, New York (1997)
13. Khuller, S., Vazirani, V.V.: Planar graph coloring is not self-reducible, assuming $P \neq NP$. Theoretical Computer Science 88(1), 183–189 (1991)
14. Kimelfeld, B., Sagiv, Y.: Incrementally computing ordered answers of acyclic conjunctive queries. In: Etzion, O., Kuflik, T., Motro, A. (eds.) NGITS 2006. LNCS, vol. 4032, pp. 141–152. Springer, Heidelberg (2006)
15. Kimelfeld, B., Sagiv, Y.: Efficiently enumerating results of keyword search over data graphs. Inf. Syst. 33(4-5), 335–359 (2008)
16. Krokhin, A.A., Marx, D.: On the hardness of losing weight. In: Aceto, L., Damgård, I., Goldberg, L.A., Halldórsson, M.M., Ingólfsdóttir, A., Walukiewicz, I. (eds.) ICALP 2008, Part I. LNCS, vol. 5125, pp. 662–673. Springer, Heidelberg (2008)
17. Makino, K., Uno, T.: New algorithms for enumerating all maximal cliques. In: Hagerup, T., Katajainen, J. (eds.) SWAT 2004. LNCS, vol. 3111, pp. 260–272. Springer, Heidelberg (2004)
18. Marx, D.: Parameterized complexity of constraint satisfaction problems. Computational Complexity 14(2), 153–183 (2005)
19. Schaefer, T.J.: The complexity of satisfiability problems. In: Proccedings 10th Symposium on Theory of Computing, pp. 216–226. ACM Press, New York (1978)
20. Schmidt, J.: Enumeration: Algorithms and complexity. Preprint (2009), http://www.thi.uni-hannover.de/fileadmin/forschung/arbeiten/schmidt-da.pdf
21. Schnoor, H., Schnoor, I.: Enumerating all solutions for constraint satisfaction problems. In: Thomas, W., Weil, P. (eds.) STACS 2007. LNCS, vol. 4393, pp. 694–705. Springer, Heidelberg (2007)
22. Schnoor, H., Schnoor, I.: Partial polymorphisms and constraint satisfaction problems. In: Creignou, et al [5], pp. 229–254
23. Schnorr, C.P.: Optimal algorithms for self-reducible problems. In: International Conference on Automata, Languages and Programming, pp. 322–337 (1976)
24. Strozecki, Y.: Enumeration complexity and matroid decomposition. Phd thesis (2010)
25. Vazirani, V., Yannakakis, M.: Suboptimal cuts: Their enumeration, weight and number. In: Kuich, W. (ed.) ICALP 1992. LNCS, vol. 623, pp. 366–377. Springer, Heidelberg (1992)
26. Yeh, L.-P., Wang, B.-F., Su, H.-H.: Efficient algorithms for the problems of enumerating cuts by non-decreasing weights. Algorithmica 56(3), 297–312 (2010)

A Satisfiability-Based Approach for Embedding Generalized Tanglegrams on Level Graphs

Ewald Speckenmeyer[1], Andreas Wotzlaw[1,*], and Stefan Porschen[2]

[1] Institut für Informatik, Universität zu Köln, Pohligstr. 1, D-50969 Köln, Germany
{esp,wotzlaw}@informatik.uni-koeln.de
[2] Fachbereich 4, HTW-Berlin, Treskowallee 8, D-10318 Berlin, Germany
porschen@htw-berlin.de

Abstract. A tanglegram is a pair of trees on the same set of leaves with matching leaves in the two trees joined by an edge. Tanglegrams are widely used in computational biology to compare evolutionary histories of species. In this paper we present a formulation of two related combinatorial embedding problems concerning tanglegrams in terms of CNF-formulas. The first problem is known as planar embedding and the second as crossing minimization problem. We show that our satisfiability formulation of these problems can handle a much more general case with more than two, not necessarily binary or complete, trees defined on arbitrary sets of leaves and allowed to vary their layouts.

Keywords: satisfiability, mixed Horn formula, 2-CNF, level graph, tanglegram, planar embedding, crossing minimization, graph drawing.

1 Introduction

In this paper we are interested in two combinatorial embedding problems concerning *generalized tanglegrams on level graphs*, a generalization of the well-known binary tanglegrams. A *binary tanglegram* [16] is an embedding (drawing) in the plane of a pair of rooted binary trees whose leaf sets are in one-to-one correspondence (perfect matching), such that matching leaves are connected by inter-tree edges. Clearly, the number of crossings between the inter-tree edges depends on the layout of the trees. From a practical point of view, an embedding with many crossings can hardly be analyzed. Fig. 1 shows an example of a binary tanglegram coming from phylogenetic studies done by Charleston and Perkins [5]. Thus, the first problem one can consider here consists of determining an embedding of one or both trees such that the inter-tree edges do not cross, if such an embedding exists. This problem is known as the *planar embedding* problem. If such a planar embedding is not possible, then we may want to find an embedding with as few crossing inter-tree edges as possible. This second problem, *crossing minimization*, is known in the literature also as the *tanglegram layout* problem [2,3,20]. Both problems are motivated by the desire to find

---

[*] The second author was partially supported by the DFG project under the grant No. 317/7-2

K.A. Sakallah and L. Simon (Eds.): SAT 2011, LNCS 6695, pp. 134–144, 2011.
© Springer-Verlag Berlin Heidelberg 2011

a good display of hierarchical structures, e.g., in software engineering, project management, or database design. They belong to the area of graph drawing [7]. Matching and aligning trees is also a recurrent problem in computational biology [16]. Embeddings with fewer crossings or with matching leaves close together are useful in biological analysis [20]. An embedding imposes an order among the leaves of the tree. Therefore, comparing the drawings of the trees is equivalent to comparing the permutations of the leaves. Here, prominent applications are in particular the comparisons of phylogenetic trees [5,6,8].

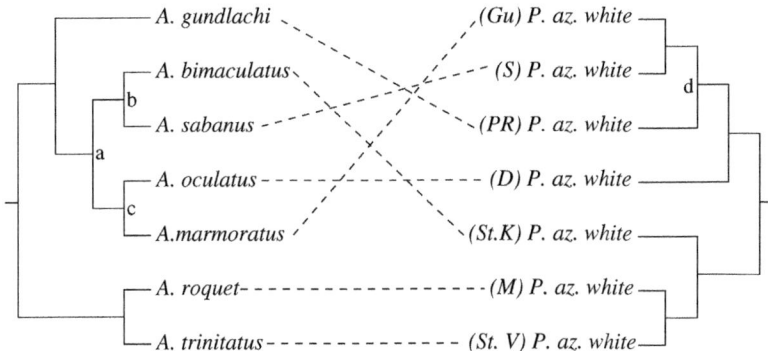

Fig. 1. A binary tanglegram from [5] showing phylogenetic trees for lizards (left tree) and strains of malaria (right tree) found in the Caribbean tropics. The dashed lines represent the host-parasite relationship. Here, the number of crossings is 7. This can be reduced to 1 by interchanging the children of nodes a, b, c, and d.

Bansal et al. [2] analyzed *generalized tanglegrams* where the number of leaves in the two binary trees may be different and a leaf in one tree may match multiple leaves in the other tree, thus no perfect matching is required here. They pointed out that such a generalization of the problem makes it possible to address not only the gene tree and species tree embedding problem, but also those problems in which the inter-tree edges between the trees can be completely arbitrary. Such general instances arise in several settings, e.g., in the analysis of host-parasite cospeciation [16].

Crossing minimization in tanglegrams has parallels to crossing minimization in graphs. Computing the minimum number of crossings in a graph is NP-complete [12]. However, it can be verified in linear time that a graph has a planar embedding [13]. The last assertion holds also for a more special case of level graphs [15,18]. Computing the minimum number of crossings is fixed-parameter tractable [3,14]. Analogously, crossing minimization in tanglegrams is NP-complete, as shown by Fernau et al. [10] by a reduction from the MAX-CUT problem [11], while the special case of planarity test can be decided in linear time [10]. Furthermore, the problem of minimizing the number of crossings where one tree is fixed and the layout of the other tree is allowed to vary can be solved efficiently. For binary trees with arbitrary topology, Fernau et al. [10] showed an $O(n \log^2(n))$ solution, further improved to $O(n \log^2(n)/ \log \log(n))$ by Bansal et al. [2]. Here, n gives the number of leaves in each tree. Venkatachalam et al. [20] provided recently an algorithm working on the integer programming formulation of the problem with the so far best-known time bound of $O(n \log(n))$. For the case of generalized tanglegrams, Bansal et al. [2] presented two algorithms with running times $O(m \log^2(m)/\log\log(m))$ and $O(mh)$, where m is the number of edges between the two trees and h is the height of the tree whose layout can change. Based on the result of Fernau et al. [10], they also showed that the existence of planar embedding can be verified in $O(m)$ time.

In our generalization of the tanglegram problem we go even further than Bansal et al. [2]. In *generalized tanglegrams on level graphs* we consider problem instances with more than two trees where every tree is defined on an arbitrary set of leaves. Notice that here the pairwise disjoint leaf sets and the corresponding inter-tree edges (no perfect matching) connecting two neighboring leaf sets constitute a level graph [18] where each level is defined by some leaf set. Thus, each tree defined on some level implies additional constraints reducing considerably the set of possible embeddings. E.g., k-ary trees with n leaves allow for at most $k!^{\frac{n-1}{k-1}}$ different leaf orders implied by different orderings of the subtrees, i.e., 2^{n-1} in case of binary trees, compared with $n!$ permutations if no restrictions are imposed on the order of the leaves. Furthermore, in our setting we do not restrict the tanglegrams only to binary trees, but consider rooted k-ary trees in which each node has not more than k children, for some fixed integer $k > 1$.

In this paper we study planar embeddability problems of generalized tanglegrams on level graphs. More specifically, we investigate the simultaneous existence of a planar embedding of the inter-tree nodes on some horizontal plane with planar embeddings of the trees on separate vertical planes, one for each tree. Our intention is to present all of them nicely on at least two orthogonal planes. To this end, we present formulations of the planarity test and the crossing minimization problem on generalized tanglegrams on level graphs in terms of CNF-formulas by incorporating ideas used already for level graphs in [18,19]. By doing this, the planarity test essentially reduces to testing satisfiability of a 2-CNF formula. The crossing minimization problem has a formulation as a PARTIAL MAX-SAT problem of a CNF formula with a mandatory part of 3- and 2-clauses that must be satisfied for the solution to be reasonable, and a second part of 2-clauses such that its truth assignment must satisfy as many of these clauses as possible. In the mandatory part, the 3-clauses reflect transitivity conditions forced by the genus of the surface, whereas the 2-clauses reflect antisymmetry conditions. These clauses have to be satisfied in order to obtain a layout. The second part of 2-clauses reflects non-crossing conditions. Each unsatisfied clause from this part represents one arc crossing. This formulation offers a simple alternative for finding reasonable approximate solutions of the crossing minimization problem. We show that the planarity test of a generalized tanglegram on a level graph having a total of n vertices and with k-ary trees defined on each level, for some fixed integer $k > 1$, can be solved in $O(n^2)$ time by an elementary 2-SAT algorithm. Finally, to the best of our knowledge, this is the first time that the generalized tanglegram problem has been treated by means of a satisfiability formulation.

The rest of the paper is organized as follows. In Section 2 we provide some basic notation and definitions of relevant computational problems for generalized tanglegrams on level graphs. The satisfiability-based formulation of the two main problems on generalized tanglegrams on level graphs is given in Section 3. Finally, in Section 4 we conclude our paper and state some open questions.

2   Preliminaries and Basic Notation

Formally, a *level graph* is a triple (G, λ, L) where $G = (V, E)$ is a directed graph, $L = \{1, ..., |L|\}$ is the set of levels, and $\lambda : V \to L$ is the level-mapping, that assigns the vertices to levels such that each arc is directed from a lower to a higher level, i.e., $\forall e = (u, v) : \lambda(v) > \lambda(u)$. For simplicity, we identify the above triple by G having the other two components in mind. Observe that there exists no arc between vertices on the same level. If in addition, for every arc $e = (u, v) \in E, \lambda(v) = \lambda(u) + 1$ holds, then the level graph is called *proper*. In the present paper we consider proper level graphs only, hence we simply will speak of level graphs. This restriction means no loss of generality since an arbitrary level graph can be turned into a proper one preserving the crossing number by simply adding dummy vertices as shown in [9,18].

Level graphs are drawn in the Euclidean x, y-plane by linear order, i.e., all vertices on the same level $j \in L$ are placed at arbitrary different positions on the line $y = j$; the x-coordinate of vertex u is denoted as $x(u)$. Arcs are represented by straight lines between the points representing their incident vertices. Often arrows at arc heads are omitted since the direction is implicitly fixed by the levels. For two vertices u, v on the same level, we simply write $u < v$ iff $x(u) < x(v)$. One is especially interested in level-graph drawings such that no two arc lines cross outside their endpoints. A level graph for which such a drawing exists is called *level-planar*. It is not hard to see that a level graph with $|E| > 2|V| - 4$ cannot be level-planar [18]. Therefore, for most level graphs all what one can hope for is to find a plane embedding such that the number of arc-crossings is minimized. Moreover, by reduction from the FEEDBACK ARC SET problem [11], Eades and Wormald [9] showed that crossing minimization in level graphs is NP-hard, even if there are only two levels with a fixed order of nodes on one level.

In generalized tanglegrams on level graphs, we define additionally on the nodes of each level $i \in L$ of a level graph G a tree T_i with nodes of level i as leaf set. Clearly, the presence of a tree on each level reduces the search space of admissible embeddings considerably. More formally, a generalized tanglegram on a level graph G is a quadruple (G, λ, L, F) where $F = \{T_1, ..., T_{|L|}\}$ is a forest of *level-trees* and $G, \lambda,$ and L are defined as above. We say that a rooted level-tree is *complete* if all its leaves have the same depth. Given a rooted, unordered tree $T \in F$, we write $V(T)$, and $E(T)$ to denote its node set, and edge set, respectively. Furthermore, for two trees T_i and T_{i+1} from F defined on two adjacent levels i and $i + 1$ of level graph G, we define the set of *inter-tree arcs* as

$$E(T_i, T_{i+1}) := \{(u, v) \in E(G) : \lambda(u) = i, \lambda(v) = i + 1\}.$$

Observe that for a proper graph G holds $E(G) = \bigcup_{i=1,...,|L|-1} E(T_i, T_{i+1})$.

For each node $v \in V(T)$, let $T(v)$ denote the subtree of T rooted at v. Given a tree T, we say that a linear order σ on the leaves of T is *compatible* with T if for each node $v \in V(T)$ the leaves in $T(v)$ form an interval (i.e., appear as a consecutive block) in σ. We write $u <_\sigma v$ to mean that leaf u appears before leaf v in the linear order σ on the leaves of T. Given compatible linear orders σ_i and σ_{i+1} on two trees T_i and T_{i+1} from F defined on two adjacent levels i and $i+1$ of level graph G, respectively, the *number of crossings* between σ_i and σ_{i+1} among the inter-tree arcs $E(T_i, T_{i+1})$ is defined as

$$\tau(\sigma_i, \sigma_{i+1}) := \left| \left\{ \{(u,a), (v,b)\} \subseteq E(T_i, T_{i+1}) : \neg\big((u <_{\sigma_i} v) \leftrightarrow (a <_{\sigma_{i+1}} b)\big) \right\} \right|.$$

Note that a pair of arcs cross at most once (see Fig. 1). Moreover, since we assume here that G is a proper level graph, only adjacent levels can induce crossings. Finally, the overall number of crossings for an instance (G, λ, L, F) and a set $S := \{\sigma_1, ..., \sigma_{|L|}\}$ of compatible orders for each level in L (tree in F) is defined as

$$\tau(G, \lambda, L, F, S) := \sum_{i=1,...,|L|-1} \tau(\sigma_i, \sigma_{i+1}).$$

Problem 1 (Planarity Test). Given an instance (G, λ, L, F), verify if there exists a planar embedding, i.e., if there exists some set S of compatible linear orders σ_i for each level $i \in L$ (tree $T_i \in F$) such that $\tau(G, \lambda, L, F, S) = 0$.

Problem 2 (Crossing Minimization). Given an instance (G, λ, L, F), find a set S of compatible linear orders σ_i for each level $i \in L$ (tree $T_i \in F$) such that $\tau(G, \lambda, L, F, S)$ is minimized.

To complete the notation, let CNF denote the set of formulas (free of duplicate clauses) in conjunctive normal form over a set $V = \{x_1, ..., x_n\}$ of propositional variables $x_i \in \{0, 1\}$. Each variable x induces a positive literal (variable x) or a negative literal (negated variable $\overline{x}$). Each formula $C \in$ CNF is considered as a clause set $C = \{c_1, ..., c_{|C|}\}$. Each clause $c \in C$ is a disjunction of different literals l_i, and is also represented as a set $c = \{l_1, ..., l_{|c|}\}$. A clause is termed a k-clause, for some $k \in \mathbb{N}$, if it contains at most k literals. The number of clauses in C is denoted by $|C|$. For $k \in \mathbb{N}$, let k-CNF denote the subset of formulas C such that each clause has length at most k. We denote by $V(C)$ the set of variables occurring in formula C. The satisfiability problem (SAT) asks, whether formula C is *satisfiable*, i.e., whether there is a truth assignment $t : V(C) \rightarrow \{0, 1\}$ setting at least one literal in each clause of C to 1. Given $C \in$ CNF, the optimization version MAX-SAT searches for a truth assignment t satisfying as many clauses of C as possible.

3   Satisfiability Formulation of Crossing Minimization

In the following we provide a formulation of the crossing minimization problem for generalized tanglegrams on level graphs in terms of propositional logic.

We proceed in two steps. Given a generalized tanglegram (G, λ, L, F), we first show the construction of CNF-formulas for the level graph (G, λ, L). In the second step, we describe a similar construction for the forest F of the generalized tanglegram.

Consider in a proper level graph G two subsequent levels i and $i + 1$ from L, as shown in Fig. 2. Let $e = (u, a)$ and $f = (v, b)$ be two arcs from $E(T_i, T_{i+1})$ directed from level i to level $i + 1$ with different tails $u \neq v$ and different heads $a \neq b$. In a drawing of G, e and f do not cross iff

$$u <_\sigma v \quad \Leftrightarrow \quad a <_\sigma b$$

for some linear order σ. Observe that arcs having the same head or tail never cross in any drawing of G.

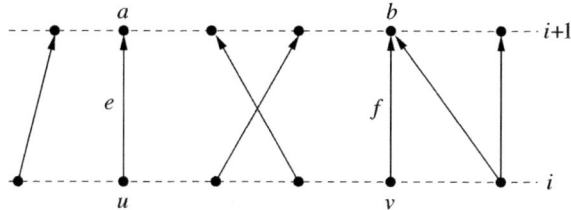

Fig. 2. Adjacent levels i and $i + 1$ of a level graph G. Arcs $e = (u, a)$ and $f = (v, b)$ have different tails and heads.

The construction of a Boolean formula C_G representing the plane embedding of G proceeds as follows:

1. For each level $i \in L$ and every pair $\{u, v\}$ of distinct vertices from level i, i.e., $\lambda(u) = \lambda(v) = i$, create a Boolean variable uv that is true iff $u <_\sigma v$ for some linear order σ.
2. Create the following Boolean subformulas:
   (i) For each level $i \in \{1, ..., |L| - 1\}$ and every two arcs $e = (u, a), f = (v, b)$ from $E(T_i, T_{i+1})$ having their tails $u \neq v$ on level i and heads $a \neq b$ on level $i + 1$, form the non-crossing preserving expression:

   $$uv \leftrightarrow ab$$

   (ii) For each level $i \in \{1, ..., |L|\}$ and each pair $\{u, v\}$ of distinct vertices on level i, form the antisymmetry expression:

   $$uv \leftrightarrow \overline{vu}$$

   (iii) For each level $i \in \{1, ..., |L|\}$ and each triple $\{u, v, w\}$ of distinct vertices on level i, form the transitivity expression:

   $$uv \wedge vw \rightarrow uw$$

Observe that the formulas resulting from (i) and (ii) yield 2-CNF formulas C_i and C_{ii} via

$$a \leftrightarrow b \equiv (\overline{a} \vee b) \wedge (\overline{b} \vee a).$$

The formula resulting from (iii) yields a Horn formula C_{iii} with clauses of length 3 via elementary equivalence

$$(a \wedge b \rightarrow c) \equiv (\overline{a} \vee \overline{b} \vee c).$$

Recall that each clause of a Horn formula contains at most one positive literal. Hence the formula $C_G = C_i \wedge C_{ii} \wedge C_{iii}$ encoding the plane embedding of a level graph G is a mixed Horn formula [17]. If G has n vertices distributed over $|L|$ levels then C_G has $|V(C_G)| \in O(n^2)$ variables. Moreover, by counting $|C_i| \in O(|E(G)|^2)$, $|C_{ii}| \in O(n^2)$, and $|C_{iii}| \in O(n^3)$. Hence the number of clauses in C_G is bounded by $O(n^3 + |E(G)|^2)$. As mentioned before, the maximal number of arcs in a level-planar graph containing $n > 2$ nodes is at most $2n - 4$. Thus, in the case we use C_G for a level planarity test, a preprocessing ensures that only $O(n^2)$ 2-clauses in C_i are generated. The following result shows that the level planarity test can be formulated as a satisfiability problem.

Proposition 1 ([18]). *A level graph G with n vertices has a level-planar embedding iff $C_G - C_{iii}$ is satisfiable. The test can be done in time $O(n^2)$.*

According to [18], the transitivity formula C_{iii} is superfluous for the level planarity test. This results in a better complexity of $O(n^2)$, since SAT for 2-CNF formulas can be decided in linear time in the number of variables and clauses in the input formula [1].

Minimizing the number of crossings of G is equivalent in terms of propositional calculus to determining a truth assignment which satisfies all clauses in C_{ii} and C_{iii} and which maximizes the number of satisfied clauses in C_i. This optimization problem is known as PARTIAL MAX-SAT [4], a variant of the MAX-SAT problem, and remains NP-hard even for (unsatisfiable) 2-CNF instances. Unfortunately, it turns out that for considering crossing minimization in terms of PARTIAL MAX-SAT, formula C_{iii} cannot be dropped in general [19].

Proposition 2 ([18]). *Let G be a level graph and $t : V(C_G) \rightarrow \{0, 1\}$ be a truth assignment satisfying all clauses of C_{ii} and C_{iii} and minimizing the number τ_G of violated clauses in C_i. Then τ_G is the minimum number of arc crossings in a level embedding of G.*

Consider now some tree T_i from F built on a level i from L. Without loss of generality assume that T_i is a complete, k-ary tree of height d, for some integers $k, d > 1$. Note that for $d = 1$ the edges of T_i never cross in any drawing of T_i and the generation of a CNF formula C_{T_i} for T_i can be omitted. Let w be some node from $V(T_i)$ such that the height of subtree $T_i(w)$ is at least 2. Note that the edges of $T_i(w)$ connecting nodes of depth 0 and 1 never cross in any drawing of $T_i(w)$. Therefore, let $e = \{u, a\}$ and $f = \{v, b\}$ be two edges from $E(T_i(w))$ with $u \neq v$ having both depth 1 and $a \neq b$ being some children of u and v, respectively, as shown in Fig. 3. In a drawing of $T_i(w)$, e and f do not cross iff

$$u <_\sigma v \quad \Leftrightarrow \quad a <_\sigma b$$

for some linear order σ.

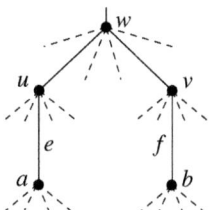

Fig. 3. Part of subtree $T_i(w)$ with two non-crossing edges e and f

We describe now the construction of a Boolean formula C_{T_i} encoding the plane embedding of T_i. We proceed as follows:

1. For each level $j = 1, ..., d$ of T_i and every pair $\{u, v\}$ of distinct vertices from level j, create a Boolean variable uv that is true iff $u <_\sigma v$ for some linear order σ.
2. Create the following Boolean subformulas:
   (iv) For each level $j = 1, ..., d - 1$ of T_i and every two edges $e = \{u, a\}$ and $f = \{v, b\}$ from $E(T_i)$ such that $u \neq v$ have depth j and a and b have depth $j + 1$ in T_i, form the non-crossing preserving expression:

$$(uv \to ab) \land (vu \to ba)$$

   (v) For each level $j = 1, ..., d$ and each pair $\{u, v\}$ of distinct vertices of depth j in T_i, form the antisymmetry expression:

$$uv \leftrightarrow \overline{vu}$$

Notice that the formulas resulting from (iv) and (v) yield after some elementary transformations 2-CNF formulas $C_{iv}^{T_i}$ and $C_v^{T_i}$, respectively, for each tree T_i. We proceed with the generation of Boolean formulas $C_{T_i} = C_{iv}^{T_i} \land C_v^{T_i}$ for all trees from F and obtain finally a Boolean formula

$$C_F = \bigwedge_{T_i \in F} C_{T_i}$$

encoding the plane embedding of F.

We shall now estimate the length of each formula C_{T_i}. The number of variables generated for each level $j = 1, ..., d$ of a k-ary tree T_i is equal to $\binom{k^j}{2}$ and thus bounded by $O(k^{2j})$. If $r_i \leq n$ is the number of vertices in level $i \in L$ of graph G, then the height of any k-ary complete tree T_i is at most $\lceil \log_k(r_i) \rceil$.

Hence, each C_{T_i} has $O\left(\frac{r_i^2-1}{k^2-1}\right)$ variables. Furthermore, the number of 2-clauses contributed to formula $C_{iv}^{T_i}$ by a level $j \in \{1, ..., \lceil \log_k(r_i) \rceil - 1\}$ of T_i is at most $2k^2\binom{k^j}{2} \in O(k^{2+2j})$, what summed up over $\lceil \log_k(r_i) \rceil - 1$ tree levels yields $|C_{iv}^{T_i}| \in O\left(\frac{r_i^2-k^2}{k^2-1}\right)$. For the number of clauses in $C_v^{T_i}$ we proceed similar as for the number of variables above and obtain that $|C_v^{T_i}| \in O\left(\frac{r_i^2-1}{k^2-1}\right)$. Thus, the number of 2-clauses in C_{T_i} is bounded by $O(r_i^2)$ for some fixed integer $k > 1$. Notice that in case of a tree T_i with r_i leaves but height greater than $\lceil \log_k(r_i) \rceil$, there must be an inner node in $V(T_i)$ with less than k children. That yields formulas $C_{iv}^{T_i}$ and $C_v^{T_i}$ with less variables and clauses than for the case of the k-ary complete tree with r_i leaves. Similar to Proposition 1, we obtain finally the following result for T_i:

Proposition 3. *For some fixed integer $k > 1$, a k-ary tree T_i built on a level i with r_i vertices has a planar embedding iff C_{T_i} is satisfiable. The test can be done in time $O(r_i^2)$.*

Since r_i is the number of vertices on level $i \in L$ in graph G and $r_1 + ... + r_{|L|} = n$, it follows that $|V(C_F)| \in O(n^2)$ and $|C_F| \in O(n^2)$.

Corollary 1. *For some fixed integer $k > 1$, a set of k-ary trees built on a level graph G with n vertices has a planar embedding iff C_F is satisfiable. The test can be done in time $O(n^2)$.*

Note that every satisfying truth assignment for C_F induces compatible linear orders σ_i on the leaves of each $T_i \in F$, and vice versa.

We are now ready to give a final satisfiability-based formulation for an instance (G, λ, L, F) of a generalized tanglegram on a level graph G. To this end, we simply generate CNF formulas C_G and C_F for (G, λ, L) and F, respectively, as described above, and combine them into a new CNF formula as follows

$$C_{GF} = C_G \wedge C_F = (C_i \wedge C_{ii} \wedge C_{iii}) \wedge \bigwedge_{T_i \in F} \left(C_{iv}^{T_i} \wedge C_v^{T_i}\right).$$

Observe that even if each T_i is planar embeddable (i.e., C_{T_i} is satisfiable) and a level graph G has a planar embedding (i.e., $C_G - C_{iii}$ is satisfiable), too, it does not imply that G plus all the T_i's together is planar embeddable. As mentioned in the introduction, in our setting we test the existence of a planar embedding of (G, λ, L, F) on at least two planes, i.e., on one horizontal plane for the level graph G and on $|L|$ vertical planes, one for each tree T_i from F.

For a level graph G with n vertices and k-ary trees F defined on its levels L, the number of clauses in C_{GF} is bounded by $O(n^3 + |E(G)|^2)$, according to the discussion above. Furthermore, C_{GF} has $O(n^2)$ variables. Note that these estimates hold only for some fixed integer $k > 1$.

Since C_{GF} contains 3-clauses, it cannot in general be solved for SAT efficiently. However, since the transitivity formula $C_{iii} \in$ 3-CNF is superfluous for the planarity test, we can remove it from C_{GF}, thus obtaining a 2-CNF formula. Similarly as for Proposition 1, we can now solve the planarity test for (G, λ, L, F)

in time $O(n^2)$ by applying the algorithm of Aspvall et al. [1]. Recall that the maximal number of arcs in a level-planar graph containing $n > 2$ nodes is at most $2n - 4$. Hence, the number of clauses $|C_{GF} - C_{iii}| \in O(n^2)$.

Proposition 4. *Let (G, λ, L, F) be an instance of a generalized tanglegram on a level graph G with n vertices and k-ary trees F, for some fixed integer $k > 1$. Then (G, λ, L, F) has a planar embedding iff $C_{GF} - C_{iii}$ is satisfiable. The test can be done in time $O(n^2)$.*

Minimizing the number of crossings of (G, λ, L, F) is equivalent to determining a truth assignment which satisfies all clauses in $C_{GF} - C_i$ and which maximizes the number of satisfied clauses in C_i, thus solving an instance of the PARTIAL MAX-SAT problem. Again, for considering crossing minimization in terms of PARTIAL MAX-SAT, formula $C_{iii} \in$ 3-CNF cannot be dropped.

Proposition 5. *Let (G, λ, L, F) be an instance of a generalized tanglegram on a level graph G with n vertices and k-ary trees F, for some fixed integer $k > 1$, and let $t : V(C_{GF}) \to \{0, 1\}$ be a truth assignment satisfying all clauses of $C_{GF} - C_i$ and minimizing the number τ of violated clauses in C_i. Then τ is the minimum number of arc crossings in an embedding of (G, λ, L, F).*

Observe that compatible linear orders σ_i for each level $i \in L$ can be extracted from a truth assignment t in time $O(n^2)$ by traversing all variables of C_{GF}.

4 Conclusion and Open Problems

We have presented a satisfiability-based formulation of the planarity test and the crossing minimization problem on generalized tanglegrams defined on level graphs. Here, the first problem essentially reduces to testing satisfiability of a 2-CNF formula and can be solved in $O(n^2)$ time for instances with n level vertices and k-ary trees defined on each level, for some fixed integer $k > 1$. Moreover, we have shown that the latter problem has a formulation as a PARTIAL MAX-SAT problem. Here, the question arises whether one could derive bounds on the approximation ratio for generalized tanglegram instances. From a practical point of view, it would be interesting to test the efficiency of our satisfiability-based approach against other techniques while solving (generalized) binary tanglegrams.

References

1. Aspvall, B., Plass, M.F., Tarjan, R.E.: A linear-time algorithm for testing the truth of certain quantified Boolean formulas. Information Processing Letters 8(3), 121–123 (1979)
2. Bansal, M.S., Chang, W., Eulenstein, O., Fernández-Baca, D.: Generalized binary tanglegrams: Algorithms and applications. In: Rajasekaran, S. (ed.) BICoB 2009. LNCS, vol. 5462, pp. 114–125. Springer, Heidelberg (2009)
3. Buchin, K., Buchin, M., Byrka, J., Nöllenburg, M., Okamoto, Y., Silveira, R.I., Wolff, A.: Drawing (complete) binary tanglegrams: Hardness, approximation, fixed-parameter tractability. In: Tollis, I.G., Patrignani, M. (eds.) GD 2008. LNCS, vol. 5417, pp. 324–335. Springer, Heidelberg (2009)

4. Cha, B., Iwama, K., Kambayashi, Y., Miyazaki, S.: Local search algorithms for partial MAXSAT. In: Proceedings of the 14th National Conference on Artificial Intelligence (AAAI/IAAI), pp. 263–268 (1997)
5. Charleston, M.A., Parkins, S.L.: Lizards, malaria, and jungles in the Caribbean. In: Tangled Trees: Phylogeny, Cospeciation and Coevolution, pp. 65–92. University of Chicago Press, Chicago (2003)
6. DasGupta, B., He, X., Jiang, T., Li, M., Tromp, J.: On the linear-cost subtree-transfer distance between phylogenetic trees. Algorithmica 25(2-3), 176–195 (1999)
7. Di Battista, G., Eades, P., Tamassia, R., Tollis, I.G.: Graph Drawing: Algorithms for Geometric Representations of Graphs. Prentice-Hall, Englewood Cliffs (1998)
8. Dufayard, J., Duret, L., Penel, S., Gouy, M., Rechenmann, F., Perriere, G.: Tree pattern matching in phylogenetic trees: automatic search for orthologs or paralogs in homologous gene sequence databases. Bioinformatics 21, 2596–2603 (2005)
9. Eades, P., Wormald, N.C.: Edge crossings in drawings of bipartite graphs. Algorithmica 11(4), 379–403 (1994)
10. Fernau, H., Kaufmann, M., Poths, M.: Comparing trees via crossing minimization. Journal of Computer and System Sciences 76(7), 593–608 (2010)
11. Garey, M.R., Johnson, D.S.: Computers and Intractability: A Guide to the Theory of NP-Completeness. W. H. Freeman & Co., New York (1979)
12. Garey, M.R., Johnson, D.S.: Crossing number is NP-complete. SIAM Journal on Algebraic and Discrete Methods 4(3), 312–316 (1983)
13. Hopcroft, J., Tarjan, R.E.: Efficient planarity testing. J. ACM 21(4), 549–568 (1974)
14. Kawarabayashi, K., Reed, B.A.: Computing crossing number in linear time. In: Proceedings of the 39th Annual ACM Symposium on Theory of Computing (STOC), pp. 382–390 (2007)
15. Leipert, S.: Level planarity testing and embedding in linear time. Ph.D. thesis, Institut für Informatik, Universität zu Köln (1998)
16. Page, R.D.M.: Tangled Trees: Phylogeny, Cospeciation, and Coevolution. University of Chicago Press, Chicago (2002)
17. Porschen, S., Speckenmeyer, E.: Satisfiability of mixed Horn formulas. Discrete Applied Mathematics 155(11), 1408–1419 (2007)
18. Randerath, B., Speckenmeyer, E., Boros, E., Hammer, P.L., Kogan, A., Makino, K., Simeone, B., Cepek, O.: A satisfiability formulation of problems on level graphs. Electronic Notes in Discrete Mathematics 9, 269–277 (2001)
19. Speckenmeyer, E., Porschen, S.: PARTIAL MAX-SAT of level graph (mixed-Horn)formulas. Studies in Logic 3(3), 24–43 (2010)
20. Venkatachalam, B., Apple, J., St. John, K., Gusfield, D.: Untangling tanglegrams: Comparing trees by their drawings. IEEE/ACM Transactions on Computational Biology and Bioinformatics 7(4), 588–597 (2010)

Minimally Unsatisfiable Boolean Circuits

Anton Belov and Joao Marques-Silva

Complex and Adaptive Systems Laboratory
School of Computer Science and Informatics
University College Dublin, Ireland
{anton.belov,jpms}@ucd.ie

Abstract. Automated reasoning tasks in many real-world domains involve analysis of redundancies in unsatisfiable instances of SAT. In CNF-based instances, some of the redundancies can be captured by computing a minimally unsatisfiable subset of clauses (MUS). However, the notion of MUS does not apply directly to non-clausal instances of SAT, particularly those that are represented as Boolean circuits. In this paper we identify certain types of redundancies in unsatisfiable Boolean circuits, and propose a number of algorithms to compute minimally unsatisfiable, that is, irredundant, subcircuits.

1 Introduction

Understanding the causes of unsatisfiability of sets of Boolean constraints is a problem of both theoretical and practical interest. Over the last decade, a large number of algorithms for identifying *minimally unsatisfiable subformulas (MUSes)* of CNF formulas have been developed. Recent accounts of practical algorithms can be found in [3,2,8], and the current theory of CNF-based MUSes in [6]. However, in many settings the original problem representation is not CNF formulas, but arbitrary Boolean formulas or circuits. For example, in hardware model checking, the next state logic can be represented as a Boolean circuit. Also, for predicate-based abstraction [10], it is necessary to compute MUSes starting from circuit structures. A fairly straightforward observation is that computing an MUS from a clausal representation of a circuit may not result in a circuit. In some contexts this is not a significant issue, but in others it can represent an important drawback. For example, circuit designers are likely to prefer to analyze a Boolean circuit than a set of apparently unrelated clauses. As a result, it is of interest to be able to compute a non-clausal Boolean formula or circuit that represents a minimal source of unsatisfiability. Early examples of work addressing minimal sources of unsatisfiability in non-clausal formulas include [6,11]. However, these early attempts are only applicable in restricted cases, and so do not provide a general solution.

This paper contains the following main contributions. First, the paper formalizes the notion of *minimally unsatisfiable circuits*. Second, the paper proposes algorithms for the computation of minimally unsatisfiable subcircuits of Boolean circuits (*circuit MUSes*). Third, the paper investigates the relationship between circuit MUSes and the recently proposed notion of group oriented MUSes [10]. Experimental results confirm the practical efficiency of the proposed algorithms, and the usefulness of dedicated techniques.

K.A. Sakallah and L. Simon (Eds.): SAT 2011, LNCS 6695, pp. 145–158, 2011.

2   Preliminaries

Propositional formulas are constructed in terms of a countably infinite set of propositional variables, logical constants F and T and a set logical connectives (in this paper, we assume this set to be $\{\neg, \vee, \wedge\}$). We denote the set of all propositional formulas by PROP, and when $\alpha \in$ PROP the set of propositional variables that occur in α by $Var(\alpha)$. A *truth-value assignment* (or simply, *assignment*) for $\alpha \in$ PROP is function h mapping $Var(\alpha) \cup \{F, T\}$ into the set $\{0, 1\}$ in such a way that $h(F) = 0$ and $h(T) = 1$. An assignment h is extended naturally to all subformulas of α. A formula α is satisfiable if there is an assignment h such that $h(\alpha) = 1$.

Propositional formulas in which the negation connective applies only to variables are said to be in the *Negation Normal Form (NNF)*. A *literal* is a propositional variable or its negation, a *clause* is a disjunction of literals. A formula is said to be in the *Conjunctive Normal Form (CNF)* if it is a conjunction of clauses. Formulas in CNF are often represented using set notation, and treated as sets of clauses – we use this representation in this paper.

For a formula $\alpha \in$ PROP, the *polarity* of a subformula α' of α is positive (resp. negative) if α' is in the scope of an even (resp. odd) number of negation connectives. We write $pol(\alpha') = 1$ (resp. $pol(\alpha') = -1$) when the polarity of α' is positive (resp. negative). Recall that if $pol(\alpha') = 1$, then for any assignment h we have $h(\alpha(\alpha'/F)) \leq h(\alpha) \leq h(\alpha(\alpha'/T))$ – the inequalities are reversed when $pol(\alpha') = -1$. Here $\alpha(\alpha'/\gamma)$ denotes the formula obtained from α by replacing the subformula α' with the logical constant γ.

Let G be a countably infinite set of *gate variables* (or simply *gates*). A *Boolean circuit* over G is a finite set C of equations of the form $g = f(g_1, \ldots, g_n)$, where $g, g_1, \ldots, g_n \in G$, and $f : \{0, 1\}^n \to \{0, 1\}$ is a Boolean function, with the additional requirements that *(i)* each $g \in G$ appears at most once as the left hand side in the equations in C, and *(ii)* the underlying directed graph $\langle G, E(C) \rangle$, where $E(C) = \{\langle g, g' \rangle \in G \times G \mid g = f(\ldots, g', \ldots) \in C\}$, is acyclic. We refer to the elements of $E(C)$ as *wires*, and to the graph $\langle G, E(C) \rangle$ as the *circuit graph* of C. If the equation $g = f(g_1, \ldots, g_n)$ is in C then g is an f-gate (or, of type f), the equation is denoted by eq_g. When no ambiguity is possible, we write $g \in C$ to denote $eq_g \in C$.

For a gate g, the set of its children (resp. parents) in the circuit graph is called the *fanin* (resp. *fanout*) of g and is denoted by $FI(g)$ (resp. $FO(g)$). A gate with the empty fanin (resp. fanout) is an *input gate* (resp. *output gate*). A gate that is neither an input nor an output is an *internal gate*. The sets of input gates and output gates in C are denoted by $Inputs(C)$ and $Outputs(C)$, respectively.

An *assignment* for a circuit C, is a function $h : Inputs(C) \to \{0, 1\}$ extended in the natural way to all gates in C – that is, for each $g = f(g_1, \ldots, g_n) \in C$, $h(g) = f(h(g_1), \ldots, h(g_n))$. Satisfiability for Boolean circuits can be defined in the following way: for each circuit C fix a designated output gate $out_C \in Outputs(C)$. Then C is *satisfiable* (with respect to out_C) if there exists an assignment h such that $h(out_C) = 1$, otherwise C is *unsatisfiable*. The polarity of gates in C with respect to the designated output out_C can be defined in terms of paths in the circuit graph and the monotonicity of functions that appear on these paths.

3   Definitions of Minimal Unsatisfiability

We begin by reviewing the well-known definition of minimal unsatisfiability for formulas in CNF, and some of the existing proposals for generalization of minimal unsatisfiability to non-clausal propositional formulas.

Definition 1. *A CNF formula F is* minimally unsatisfiable *if F is unsatisfiable, and for any clause $c \in F$, the formula $F \setminus \{c\}$ is satisfiable.*

As in [6], by MU we denote the set of minimally unsatisfiable formulas in CNF. The set of *minimally unsatisfiable subformulas* of a CNF formula F, in symbols $MUS(F)$ is defined as $MUS(F) = \{F' \mid F' \subseteq F$ and $F' \in MU\}$.

A definition of minimal unsatisfiability for propositional formulas in NNF has been proposed in [6]. Let α be an NNF formula, and T_α be the tree representation α. Consider any subtree $T_{\alpha'}$ of T_α whose root is either an $\vee$-node or a literal, and that is a successor of an $\wedge$-node. Then, formula α' represented by $T_{\alpha'}$ is called an *or-subformula* of α. The minimal unsatisfiability can be defined with respect to the elimination of or-subformulas.

Definition 2 (cf. [6]). *An NNF formula α is* minimally unsatisfiable *if α is unsatisfiable, and for any or-subformula α' of α, the formula produced by the elimination of α' from α is satisfiable.*

We denote the set of minimally unsatisfiable, according to Definition 2, NNF formulas by MU_{NNF} (it is MU^* in [6]). Note that a syntactic elimination of a subformula does not necessarily yield a well-formed formula, as such, the term "elimination" in this definition implies an additional simplification. Nevertheless, given an NNF formula α one can define a set $MUS_{\mathrm{NNF}}(\alpha)$ by analogy with CNF – this set contains all formulas in MU_{NNF} that can be obtained from α via elimination of any number of or-subformulas (including none). It is not difficult to see that on the domain of formulas in CNF, Definition 2 captures the same set of formulas as Definition 1.

A number of notions of minimal unsatisfiability for temporal formulas in LTL have been proposed in [11]. One of these notions, when specialized to the (classical) propositional logic (which is a fragment of LTL) results in the following definition:

Definition 3 (cf. [11]). *A propositional formula α is* minimally unsatisfiable *if α is unsatisfiable, and the replacement of any of its positively (resp. negatively) polarized subformula by the logical constant T (resp. F) produces a satisfiable formula.*

By MU_{PROP} we denote the set of minimally unsatisfiable, according to Definition 3, propositional formulas. Given a propositional formula α, the set $MUS_{\mathrm{PROP}}(\alpha)$ can be defined by analogy with the definition of $MUS_{\mathrm{NNF}}(\alpha)$. Note that Definition 3 captures the same set of formulas in NNF as Definition 2, that is $MU_{\mathrm{NNF}} = MU_{\mathrm{PROP}} \cap \mathrm{NNF}$.

The notions of minimal unsatisfiability presented in this section, and the related notions of MUS, both in clausal and non-clausal domains, rely on some *basic operation* with certain properties. Formally, we can describe such operation by a binary relation on a set of formulas. For example, in the case of CNF the basic operation is the removal of a single clause, and the corresponding relation $\mathcal{R}_{\mathrm{CNF}}$ on the set of CNF formulas is:

$$\langle F, F' \rangle \in \mathcal{R}_{\mathrm{CNF}} \text{ if and only if } \exists c \, (F' = F \setminus \{c\}). \tag{1}$$

Then, a CNF $F \in MU$ if F is unsatisfiable, and any $F' \in \mathcal{R}_{\mathrm{CNF}}(F)$ is satisfiable. In addition, given a CNF F, the set $MUS(F)$ is defined simply as $\mathcal{R}^*_{\mathrm{CNF}}(F) \cap MU$. The relations $\mathcal{R}_{\mathrm{NNF}}$ and $\mathcal{R}_{\mathrm{PROP}}$ on the set of propositional formulas that describe the operations used in the definitions of MU_{NNF} and MU_{PROP} can be defined analogously.

In general, given a set LE of logical entities with a defined notion of satisfiability (for example, logical formulas, or Boolean circuits), and a binary relation $\mathcal{R}$ on LE, we can define the set $MU_{LE}(\mathcal{R})$ of minimally unsatisfiable, with respect to $\mathcal{R}$, members of LE as

$$MU_{LE}(\mathcal{R}) = \{L \in \mathrm{LE} \mid L \text{ is unsatisfiable, and any } L' \in \mathcal{R}(L) \text{ is satisfiable }\},$$

and, given $L \in \mathrm{LE}$, the related set

$$MUS_{LE}(L, \mathcal{R}) = \mathcal{R}^*(L) \cap MU_{LE}(\mathcal{R}).$$

However, for some relations $\mathcal{R}$, the sets defined in this way might not capture the intuitive meaning of minimal unsatisfiability – the *irredundancy*. As such, we propose a number of characteristic properties of $\mathcal{R}$ that, albeit somewhat imprecise, aid in constructing intuitively meaningful definitions of minimal unsatisfiability and MUS.

Property 1. $\mathcal{R}$ has to be *satisfiability preserving*, that is if $L \in \mathrm{LE}$ is satisfiable, then any $L' \in \mathcal{R}(L)$ is satisfiable.

This property ensures that minimal unsatisfiability defined using $\mathcal{R}$ captures a strong notion of irredundancy – if L is in $MU_{LE}(\mathcal{R})$, then *every* L' in $\mathcal{R}^+(L)$ is satisfiable.

Property 2. For any unsatisfiable $L \in \mathrm{LE}$, $\mathcal{R}(L) \neq \emptyset$.

This property of $\mathcal{R}$ prevents definitions of minimal unsatisfiability that are vacuous – that is, elements L of $MU_{LE}(\mathcal{R})$ that are minimally unsatisfiable simply because the basic operation captured by the relation $\mathcal{R}$ cannot be applied to L.

Property 3. Every $L' \in \mathcal{R}(L)$ is in some sense "smaller" than, and is "close" to L.

This property can be made precise by defining a suitable order and a metric on LE, however for this paper we will rely on its intuitive meaning. Note that in general $L' \in \mathcal{R}(L)$ is not a necessarily a "sub-object" of L – in fact, among the definitions presented above, it is only in the case of CNF, where L' is a sub-formula of L.

In the next section we propose two relations on the set of Boolean circuits that satisfy the above requirements, and give rise to intuitively meaningful definitions of minimally unsatisfiable Boolean circuits and circuit MUSes.

4   Minimally Unsatisfiable Boolean Circuits

Consider a Boolean circuit C over a set of gates G – recall that C is a finite set of equations eq_g of the form $g = f(g_1, \ldots, g_n)$. Let out_C be the designated output of C, and assume that C is unsatisfiable, that is, for every assignment h for C, $h(out_C) = 0$. Consider the situation when there exists a gate $g \in C$ such that $C \setminus \{g\}$ is unsatisfiable. Then the gate g, or more precisely the equation eq_g, is redundant with respect to the

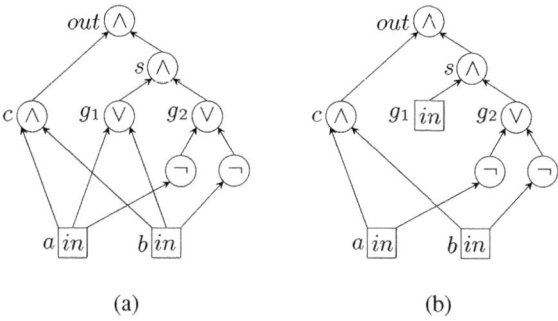

(a)   (b)

Fig. 1. (a) An example circuit C (a half-adder with both the carry c and the sum s set to 1). C is unsatisfiable. (b) The circuit $C' = C \setminus \{g_1\}$ is also unsatisfiable, and since $C' \in \mathcal{R}_g(C)$, the circuit C is not gate-minimally unsatisfiable. However, the circuit C' is, and is a gate-MUS of C.

unsatisfiability of C. This suggests a possible *gate-based* definition of minimal unsatisfiability for circuits: a circuit is unsatisfiable, and no gate equation can be removed without making it satisfiable. We will formalize this definition shortly, but first present a slightly different perspective.

Each gate equation $g = f(g_1, \ldots, g_k)$ in the circuit captures a relationship, a constraint, between the values of g and the values of gates in $FI(g)$. When the gate is redundant, the relationship of g with all these values is redundant. However, when the gate is not redundant, it does not necessarily mean that the relationship of g with *all* gates is not redundant. It is possible that only some of these relationships are important for unsatisfiability, while others can be dropped. These individual relationships corresponds to the wires in the circuit – i.e. the edges in the circuit graph – that connect the gate to the gates in its fanin. This suggests a different, *wire-based*, definition of minimal unsatisfiability for circuits – it is more refined than the gate-based, in that a minimally unsatisfiable circuit from the gate view, is not necessarily minimally unsatisfiable from the wire point of view.

With this motivation, we proceed to formalizing the two proposed types of minimal unsatisfiability.

4.1 Gate-Based Minimal Unsatisfiability

Let CIRC be a set of Boolean circuits over the set of gates G, and let $\mathcal{R}_g \subseteq \mathrm{CIRC}^2$ be defined as follows:

$$\langle C, C' \rangle \in \mathcal{R}_g \text{ if and only if } out_C = out_{C'} \text{ and } \exists g \in C \ \ C' = C \setminus \{g\}.$$

When $C' = C \setminus \{g\}$ we have $Inputs(C') = Inputs(C) \cup \{g\}$, thus the effect of the removal of eq_g from C is that g becomes an unconstrained input. As an example, consider the circuit C in Figure 1(a), and the circuit C' obtained from C by removing gate g_2 (Figure 1(b)). We now establish the basic properties of the relation $\mathcal{R}_g$, that, as argued in Section 3, will afford a meaningful definition of minimally unsatisfiable circuit and circuit MUS based on $\mathcal{R}_g$.

Proposition 1. *Let C be a satisfiable Boolean circuit. Then, for any $C' \in \mathcal{R}_g(C)$, C' is satisfiable.*

Proof. Let h be satisfying assignment for C, and let $C' = C \setminus \{g\}$. Then, $h' = h \cup \{\langle g, h(g) \rangle\}$ is a satisfying assignment for C'. □

Hence, $\mathcal{R}_g$ is satisfiability preserving (Property 1). Further, we have that for every unsatisfiable circuit C, $\mathcal{R}_g(C) \neq \emptyset$, because C must have at least one gate (Property 2). Finally, when $C' \in \mathcal{R}_g(C)$, $C' \subset C$, as such C' is smaller than C in terms of the number of gate definitions, and is close to C as the two circuits differ by exactly one gate (Property 3). Thus, paraphrasing the definitions of $MU_{\mathrm{CIRC}}(\mathcal{R}_g)$ and $MUS_{\mathrm{CIRC}}(C, \mathcal{R}_g)$ we have:

Definition 4. *A Boolean circuit C is* gate-minimally unsatisfiable, *if C is unsatisfiable and for every gate $g \in C$, the circuit $C \setminus \{g\}$ is satisfiable.*

Definition 5. *Let C be an unsatisfiable Boolean circuit. Then the circuit C' is a gate-MUS of C if $C' \subseteq C$ and C' is gate-minimally unsatisfiable.*

We denote the set $MU_{\mathrm{CIRC}}(\mathcal{R}_g)$ as MU_g, and a set $MUS_{\mathrm{CIRC}}(C, \mathcal{R}_g)$ as $MUS_g(C)$ for the rest of this paper. The circuit C' in Figure 1(b) is gate minimally unsatisfiable, and is the gate-MUS of the circuit C in Figure 1(a).

In Section 3 we have emphasized the fact that the presented definitions of minimal unsatisfiability are strict generalizations: on the domain of CNF formulas, the set MU_{NNF} coincides with the set MU, while on the domain of NNF formulas, the set MU_{PROP} coincides with the set MU_{NNF}. We now demonstrate that the proposed definition of gate-based minimal unsatisfiability for Boolean circuits, despite its intuitive appeal, is in a sense too *coarse*.

Take any $\alpha \in \mathrm{PROP}$, and let $sm : Var(\alpha) \mapsto G$ be some injective function ("subformula map"). We are going to extend sm to all subformulas of α, and, simultaneously, associate each subformula α' of α with a Boolean circuit $C_{\alpha'}$. The construction is defined inductively on the structure of α as follows:

(i) if $\alpha = p$, then let $C_\alpha = \emptyset$, and $out_C = sm(p)$;
(ii) if $\alpha = \beta \wedge \gamma$, then take a fresh $g_\alpha \in G$, and let

$$C_\alpha = C_\beta \cup C_\gamma \cup \{g_\alpha = \wedge(out_{C_\beta}, out_{C_\gamma})\},$$

and let $out_{C_\alpha} = g_\alpha$, and $sm = sm \cup \{\langle \alpha, g_\alpha \rangle\}$.
(iii) the constructions for the cases $\alpha = \beta \vee \gamma$ and $\alpha = \neg\beta$ are analogous to (ii).

Thus, the circuit graph of C_α is tree-like, with the possible exception of the inputs. Furthermore, for every gate $g \in C_\alpha$, the polarity $pol(g)$ is the same as $pol(sm^{-1}(g))$, and for every *non-variable* subformula α' of α, $pol(\alpha') = pol(sm(\alpha'))$.

Let h be an assignment to $Var(\alpha)$, then we can define define a corresponding assignment for C_α in a straightforward manner:

$$h_{sm} = \{\langle sm(p), h(sm(p)) \rangle \mid p \in Var(\alpha)\}.$$

Clearly, for any subformula α' of α, $h(\alpha') = h_{sm}(sm(\alpha'))$. Similarly, given an assignment h for circuit C_α we can define the corresponding assignment $h_{sm^{-1}}$ with the analogous property.

Theorem 1. *For every propositional formula α, if $\alpha \in MU_{PROP}$, then $C_\alpha \in MU_g$, however the converse doesn't hold.*

Proof. It is easy to see that formula α is unsatisfiable if and only if so is the circuit C_α.

Towards a contradiction, assume that $\alpha \in MU_{PROP}$, but $C_\alpha \notin MU_g$. Since α is unsatisfiable, so is C_α and we conclude that there exists $g \in C_\alpha$ such that the circuit $C' = C_\alpha \setminus \{g\}$ is unsatisfiable. Let $\alpha' = sm^{-1}(g)$, and assume $pol(\alpha') = 1$. Then the formula $\alpha(\alpha'/T)$ must be unsatisfiable, as otherwise, if h is a satisfying assignment for $\alpha(\alpha'/T)$, then $h_{sm} \cup \{\langle g, 1\rangle\}$ satisfies C'. Since $\alpha(\alpha'/T) \in \mathcal{R}_{PROP}$ we have $\alpha \notin MU_{PROP}$.

One of the reasons that the reverse implication does not hold is that the operation $\mathcal{R}_{PROP}$ allows to substitute the constants T/F for variables in the formula, while an equivalent of such operation is not captured by $\mathcal{R}_g$. Consider for example the formula $\alpha = q \wedge (\neg q \wedge r)$. Then, $\alpha \notin MU_{PROP}$ because the formula $\alpha' = q \wedge (\neg q \wedge T)$ is still unsatisfiable. However, the corresponding circuit $C_\alpha = \{out = \wedge(q, g_1), g_1 = \wedge(g_2, r), g_2 = \neg(q)\}$ is gate-minimally unsatisfiable. $\qquad\square$

Note that the formula α used as a counterexample in the above proof is a propositional representation of the CNF formula $F = \{q, \neg q, r\}$ and so $\mathcal{R}_g$ gives rise to the definition of minimal unsatisfiability that is too coarse on domain of CNF formulas as well.

The counterexample in the proof of Theorem 1 might suggest that $\mathcal{R}_g$ could be refined if it were to allow the replacement of inputs by constants according to their polarity. Unfortunately, this suggestion poses an immediate problem: in unsatisfiable circuits there must be non-polarized inputs. Selecting an arbitrary constant for non-polarized inputs results in satisfiability non-preserving operation. Further, the following, intuitively minimally unsatisfiable example circuit $C = \{out = \wedge(p, g_1), q_1 = \neg(p)\}$ would not be minimally unsatisfiable, as we could replace p with 0 (or 1) and still obtain an unsatisfiable circuit.

The real reason for the fact that in certain cases unsatisfiable formulas that are not in MU_{PROP} map to gate-minimally unsatisfiable circuits (i.e. MU_g is in this sense coarser than MU_{PROP}) is that $\mathcal{R}_{PROP}$ allows the replacement of an *individual occurrence* of a variable in the formula without affecting other occurrences of this variable. Multiple occurrences of a variable p in formula α are represented by multiple wires connecting the input $sm(p)$ in the circuit C_α, hence an operation that would allow to "break" wires in the circuit would address this weakness of $\mathcal{R}_g$. Thus, in conjunction with the discussion at the beginning of this section, we have a strong motivation for the definition of wire-based minimal unsatisfiability of Boolean circuits.

4.2   Wire-Minimal Unsatisfiable Circuits

Let I be a set of gates disjoint from G, let CIRC be a set of Boolean circuits over $G \cup I$, and let $\mathcal{R}_w \subseteq \text{CIRC}^2$ be defined as follows:

$\langle C, C'\rangle \in \mathcal{R}_w$ if and only if $out_C = out_{C'}$ and $\exists g, g_k \in G$, $\exists i \in I$ such that
$$g = f(\ldots, g_k, \ldots) \in C \text{ and } C' = C \setminus \{g\} \cup \{g = f(\ldots, i, \ldots)\}$$

In words, when $C' \in \mathcal{R}_w(C)$, the circuit C' can be obtained by replacing some wire $\langle g, g_k\rangle$ in the circuit graph of C with a wire $\langle g, i\rangle$, where i is a *fresh input gate*. Thus,

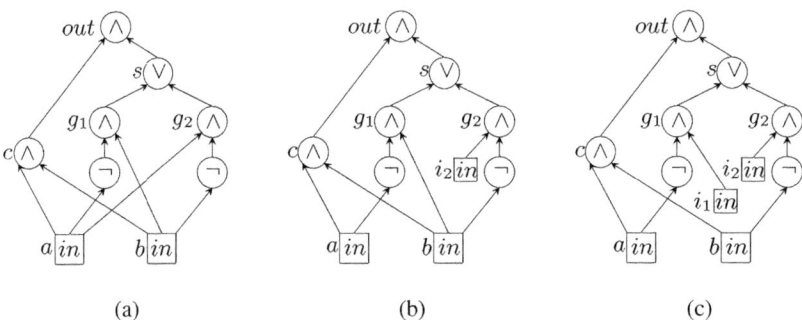

(a)   (b)   (c)

Fig. 2. (a) An example circuit C (also a half-adder with both the carry c and the sum s set to 1). C is unsatisfiable. (b) The circuit C_1 obtained from C by removing the wire $\langle g_2, a \rangle$, hence $C_1 \in \mathcal{R}_w(C)$; C_1 is also unsatisfiable. (c) The circuit C_2 obtained from C_1 by removing the wire $\langle g_1, b \rangle$, hence $C_2 \in \mathcal{R}_w(C)$. $C_2 \in MU_w$, and is a wire-MUS of C (and C_1)

effectively the operation eliminates the connection, or constraint, between the values of g_k and g in C. Note that only that wires that connect gates in G can be replaced, as such, once replaced, a wire cannot be replaced again. As such we will often refer to this operation as the *removal* of the wire $\langle g, g_k \rangle$ from C.

As an example, consider the circuit C depicted in Figure 2(a), and the circuit C_1 in Figure 2(b) that is obtained from C by removing the wire $\{g_2, a\}$. Thus, $C_1 \in \mathcal{R}_w(C)$. The circuit C_2 in Figure 2(c) is obtained from C_1 by removing the wire $\langle g_1, b \rangle$, as such $C_2 \in \mathcal{R}_w(C_1)$. We use this opportunity to point out that both C_1 and C_2 are unsatisfiable, but C is gate minimally unsatisfiable. In order to further motivate the definition of minimal unsatisfiability based on the relation $\mathcal{R}_w$, we establish the basic properties of this relation outlined in Section 3.

Proposition 2. *Let C be a satisfiable Boolean circuit. Then, for any $C' \in \mathcal{R}_w(C)$, C' is satisfiable.*

Proof. Let h be a satisfying assignment for C, and assume that C' was obtained from C by replacing some wire $\langle g, g_k \rangle$ with $\langle g, i \rangle$, where i is a fresh input gate. Then, the assignment $h' = h \cup \{\langle i, h(g_k) \rangle\}$ is satisfying for C'.   □

Thus, $\mathcal{R}_w$ is satisfiability preserving (Property 1). Further, every unsatisfiable circuit, with the exception of $C_d = \{out = 0\}$, has at least one wire, as such for every unsatisfiable $C \neq C_d$, $\mathcal{R}_w(C) \neq \emptyset$. Hence, Property 2 *almost* holds – the rather degenerate circuit C_d is the only case that violates this property, note, however, that $C_d \in MU_{\text{CIRC}}(\mathcal{R}_w)$, albeit vacuously. Finally, with respect to Property 3, when $\langle C, C' \rangle \in \mathcal{R}_w$, C' is not a subcircuit of C (but neither are the formulas related by $\mathcal{R}_{\text{NNF}}$ or $\mathcal{R}_{\text{PROP}}$). It is however, smaller than C in the sense that it has one less constraint between the values of gates. Thus, paraphrasing the definitions of $MU_{\text{CIRC}}(\mathcal{R}_w)$ and $MUS_{\text{CIRC}}(C, \mathcal{R}_w)$ we have:

Definition 6. *A Boolean circuit C is* wire-minimally unsatisfiable *if C is unsatisfiable and for any wire $\langle g, g_k \rangle$ in C, the circuit C' obtained by the replacement of this wire with $\langle g, i \rangle$ for a fresh input i is satisfiable.*

Definition 7. *Let C be an unsatisfiable Boolean circuit. Then the circuit C' is a* wire-MUS *of C, if C' can be obtained from C by removing zero or more wires, and C' is wire-minimally unsatisfiable.*

We denote the sets $MU_{\mathrm{CIRC}}(\mathcal{R}_w)$ and $MUS_{\mathrm{CIRC}}(C, \mathcal{R}_w)$ as MU_w and $MUS_w(C)$, respectively. The circuit C_2 depicted in Figure 2(c) is wire-minimally unsatisfiable, and is a wire-MUS of both the circuits C and C_1 in Figures 2(a),2(b).

The example in Figure 2 demonstrates that there are gate-minimally unsatisfiable circuits that are not wire-minimally unsatisfiable. The following theorem shows that, with the exception of circuits with constant gates, every wire-minimally unsatisfiable circuit is gate-minimally unsatisfiable. Let $\mathrm{CIRC_{nc}} \subset \mathrm{CIRC}$ be the set of Boolean circuits without constant gates. Then,

Theorem 2. *For every Boolean circuit $C \in \mathrm{CIRC_{nc}}$, if $C \in MU_w$, then $C \in MU_g$, however the converse does not hold.*

Proof. We prove the contrapositive: assume that $C \in \mathrm{CIRC_{nc}}$ is unsatisfiable and $C \notin MU_g$, we show that $C \notin MU_w$. By assumption $\exists g \in C$, such that $C_g = C \setminus \{g\}$ is unsatisfiable. Let $g = f(g_1, \ldots, g_k)$, let $\{i_1, \ldots, i_k\}$ be a set of fresh input gates, and consider the circuit $C_w = C \setminus \{g\} \cup \{g = f(i_1, \ldots, i_k)\}$. Then, C_w must be unsatisfiable, as otherwise a satisfying assignment h for C_w can be used to construct a satisfying assignment $h \cup \{\langle g, f(h(i_1), \ldots, h(i_k))\rangle\}$ for C_g. Note that $C_w \in \mathcal{R}_w^k(C)$, and therefore for some $C' \in \mathcal{R}_w(C)$, C' is unsatisfiable because $\mathcal{R}_w$ is satisfiability preserving. We conclude that $C \notin MU_w$.

The fact that the converse does not hold is demonstrated in Figure 2. □

The circuit depicted on the right margin illustrates the issue with the constant gates. While the constant gate 0 can be removed from this circuit (recall that this is equivalent to replacing it by an unconstrained input) without breaking its unsatisfiability, removing any of the wires leading from this gate will make the circuit satisfiable. Thus, by allowing the removal of wires, we get an *almost* refinement of MU_g. The following theorem shows that we also gain the equivalence with 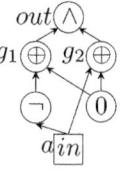 MU_{PROP}, and as such, with MU_{NNF} and MU (the sets of minimally unsatisfiable formulas in NNF and CNF, respectively).

Theorem 3. *For every propositional formula α, $\alpha \in MU_{\mathrm{PROP}}$, if and only if $C_\alpha \in MU_w$.*

Proof. We prove the contrapositives. Recall that C_α is a circuit constructed using the structure of α (Section 4.1).

Assume α is unsatisfiable and $\alpha \notin MU_{\mathrm{PROP}}$, we show that $\alpha \notin MU_w$. Without the loss of generality, let α' be a positively polarized subformula of α such that the formula $\alpha(\alpha'/T)$ is unsatisfiable. Since $pol(\alpha') = 1$, the formula $\alpha(\alpha'/F)$ is also unsatisfiable. Note that α' must be a proper subformula of α because the formula T, which is the result of substitution of α itself by T, is obviously satisfiable. As such, let β be the parent of α' in the formula tree of α. Then the circuit C' obtained from C by removing the wire $\langle sm(\beta), sm(\alpha')\rangle$ is also unsatisfiable. Hence, $C \notin MU_w$.

Assume now that C_α is unsatisfiable and $C_\alpha \notin MU_w$, we show that $\alpha \notin MU_{\text{PROP}}$. Let $\langle g_1, g_2 \rangle$ be the wire in C_α that can be removed to obtain an unsatisfiable circuit C'. If g_2 is not an input gate, then let $\alpha_2 = sm^{-1}(g_2)$, otherwise let α_2 be the occurrence of the variable $sm^{-1}(g_2)$ in the subformula $sm^{-1}(g_1)$. Then, both $\alpha(\alpha_2/T)$ and $\alpha(\alpha_2/F)$ must be unsatisfiable. Therefore, $\alpha \notin MU_{\text{PROP}}$.   □

5   Computing Circuit MUSes

In this section we propose possible solutions to the problem of computing a gate-MUS or a wire-MUS of a given unsatisfiable circuit C. For reasons of clarity, in this section we assume that C does not have constant gates, that is $C \in \text{CIRC}_{\text{nc}}$.

Most of the high-performing algorithms for computation of CNF-based MUSes are based on the identification of so called *transition clauses* [3] in unsatisfiable CNF formulas. A clause $c \in F$ is called a transition clause, if F is unsatisfiable but $F \setminus \{c\}$ is satisfiable. The key property of the transition clauses is that if c is a transition clause for F, then c belongs to *all* MUSes of F. Then, given an unsatisfiable formula F, a *deletion-based* MUS extractor picks a clause $c \in F$, and tests the formula $F' = F \setminus \{c\}$ for satisfiability. If F is satisfiable, then c is *final* – it is a part of constructed MUS. Otherwise, the algorithm continues with the formula F'. When all clauses are final, the current formula is an MUS of the initial formula F. In most cases, this basic extraction algorithm can be accelerated significantly when the underlying SAT solver supports incremental SAT solving, and is capable of producing proofs of unsatisfiability.

It is not difficult to see that in the case of Boolean circuits, the analogous concepts – *transition gates*, and *transition wires* – can be defined, and possess similar properties. As such, the existing CNF MUS algorithms, such as the deletion-based algorithm described above, can be adapted to the circuit MUS problem. It is plausible, however, that in the case of circuits the structure can be used to accelerate the circuit MUS computation – we present empirical data to support this claim in Section 6.

Unfortunately, the publicly available efficient circuit SAT solvers, such as [4], neither expose an incremental interface, nor produce proofs of unsatisfiability. Thus, it is advantageous to develop CNF-based techniques for circuit MUS extraction in order to capitalize on the continuing progress of CNF-based SAT technology.

Let C be a Boolean circuit. For each $g \in C$, let $Ts(g)$ be the set of clauses obtained by the Tseitin transformation of g to CNF [12], and thus, $Ts(C) = \{out_C\} \cup \bigcup_{g \in C} Ts(g)$ be the Tseitin encoding of C. It is tempting to compute CNF-based $M = MUS(Ts(s))$, and then "inflate" each clause in M to obtain the circuit

$$C_M = \bigcup_{c \in M} \{g \mid c \in Ts(g)\}.$$

In general, the resulting circuit C_M is not a gate-MUS of C, and so CNF MUS extractors are not applicable to circuit MUS problem. However, circuit MUSes can be computed using the tools developed for the recently proposed problem of *group oriented MUS* extraction [10]:

Definition 8 ([5]). *Given an explicitly partitioned unsatisfiable CNF formula $F = D \cup \bigcup_{G \in \mathcal{G}} G$, where $\mathcal{G} = \{G_1, \ldots, G_k\}$, and D and each G_i are disjoint sets of clauses, a group oriented MUS of F is a subset $\mathcal{G}'$ of $\mathcal{G}$ such that $D \cup \bigcup_{G \subset \mathcal{G}'} G$ is unsatisfiable, and, for every $\mathcal{G}'' \subset \mathcal{G}'$, we have that $D \cup \bigcup_{G \in \mathcal{G}''} G$ is satisfiable.*

It is not difficult to see that if we let $\mathcal{G}_C = \{Ts(g) \mid g \in C\}$, and let $D = \{out_C\}$, then the group oriented MUS of the formula $F_C = D \cup \bigcup_{G \in \mathcal{G}_C} G$ corresponds to a gate-MUS of the circuit C, and vice-versa. Thus, gate-MUSes for circuits can be computed using group oriented MUS extractors, for example SAT4J [7].

The problem of wire-MUS computation for a given circuit C, however, cannot be solved directly by computation of group oriented MUS of the formula F_C. Consider for example a gate $g \in C$ defined as $g = \wedge(g_1, g_2, g_3)$. The set $Ts(g)$ contains four clauses c_m, c_1, c_2, c_3, where

$$c_m = g \vee \neg g_1 \vee \neg g_2 \vee \neg g_3, \ c_1 = \neg g \vee g_1, \ c_2 = \neg g \vee g_2, \ c_3 = \neg g \vee g_3.$$

Assume now that we remove the wire $\langle g, g_1 \rangle$ from C, that is we replace $\langle g, g_1 \rangle$ with $\langle g, i_1 \rangle$, where i_1 is a fresh input, to obtain a circuit C'. Then, in C' the set $Ts(g)$ contains the clauses c'_m, c'_1, c_2, c_3, where

$$c'_m = g \vee \neg i_1 \vee \neg g_2 \vee \neg g_3, \ c'_1 = \neg g \vee i_1.$$

Since the variable i_1 does not appear in any other clause of $Ts(C')$ except c'_m and c'_1, from the perspective of the satisfiability the net effect of removing the wire $\langle g, g_1 \rangle$ from C on $Ts(C)$ is simply the *removal of clauses* c_m and c_1 from $Ts(C)$. Formally, the CNF formula $Ts(C')$ is satisfiable if and only if the formula $F = Ts(C) \setminus \{c_m, c_1\}$ is satisfiable. The "only if" direction is obvious, since $F \subset Ts(C')$. For the "if" direction, let h be satisfying for F – if $h(g) = 1$, then c'_m is satisfied in $Ts(C')$, and we assign 1 to i_1 to satisfy c'_1; if $h(g) = 0$, then c'_1 is satisfied, and we assign 0 to i_1 to satisfy c'_m. Similarly, the effect of removing the wire $\langle g, g_2 \rangle$ from C' amounts to removing the clause c_2 from $Ts(C')$. Finally, the subsequent removal of $\langle g, g_3 \rangle$ results in the formula with all clauses c_m, c_1, c_2, c_3 removed – note that this is equivalent to removing all clauses in $Ts(g)$ from $Ts(C)$, and as such, removing the gate g from C. Indeed, as the proof of Theorem 2 shows, removing all wires in the fanin of $g \in C$ is equivalent to removing the gate g itself.

We now point out that the problem of computing a wire-MUS of C could have been mapped to group oriented MUS directly, if in Definition 8 the groups G_i were not required to be disjoint. If this were the case, then we could set $G_1 = \{c_m, c_1\}$, $G_2 = \{c_m, c_2\}$, and $G_3 = \{c_m, c_3\}$ – note that the groups intersect on c_m. Then, removing the group G_1 from the formula F, would result in the removal the clause c_m from G_2 and G_3. As an aside, this suggests a possible generalization of the definition of the group oriented MUS problem. Meanwhile, we can still map wire-MUS problem to the group oriented MUS problem, though at the cost of adding extra variables. We demonstrate the mapping using the previous example of $g = \wedge(g_1, g_2, g_3)$ with $Ts(g) = \{c_m, c_1, c_2, c_3\}$.

The idea is to add one extra variable for every intersecting group. Let l_1, l_2 and l_3 be three fresh variables, and let G_1^*, G_2^* and G_3^* be the groups of clauses defined in the following way:

$$G_1^* = \{c_m \vee \neg l_2 \vee \neg l_3,\ c_1 \vee \neg l_1,\ l_1\}$$
$$G_2^* = \{c_m \vee \neg l_1 \vee \neg l_3,\ c_2 \vee \neg l_2,\ l_2\}$$
$$G_3^* = \{c_m \vee \neg l_1 \vee \neg l_2,\ c_1 \vee \neg l_3,\ l_3\}$$

Then, for example, the removal of group G_1^* makes the variable l_1 unconstrained, and, as such, the first clause in both groups G_2^* and G_3^* effectively becomes satisfied. It is easy to see that this has the exact effect of the removal of the group G_1 under the generalized definition of group MUS that allows non-disjoint groups. The demonstrated technique for mapping non-disjoint group MUS problem to (disjoint) group MUS problem can be applied in a general setting. We omit the formal definition of such mapping, and the proofs of its correctness from this paper.

Intuitively, it seems plausible that the structure of a given instance of group oriented MUS problem can provide additional information that allows to accelerate group-MUS extraction. In the application of group-MUS to circuit-MUS computation problem, the circuit structure can be used to deduce the relationships between groups which, in turn, can be used to guide a group-MUS extractor. We propose two such techniques, and, in the following section, demonstrate empirical evidence to their effectiveness.

One of the techniques is based on the following observation. Let g be a gate in C. By $D(g)$ let us denote the set of gates dominated by g, that is

$$D(g) = \{g' \in C \mid \text{every path from } g' \text{ to } out_C \text{ in the graph of } C \text{ includes } g \}.$$

Note that $g \in D(g)$. Then the circuit $C' = C \setminus \{g\}$ is satisfiable if and only if the circuit $C'' = C \setminus \{D(g)\}$ is satisfiable. As such, during the gate-MUS extraction, rather than testing the circuit C' for satisfiability, we can test the circuit C''. Since C'' is smaller than C' the SAT test might be faster. In addition, if C'' is unsatisfiable, we remove a potentially large set of gates at once, thus reducing the number of SAT checks. This, *domination based* optimization can be improved further by the analysis of the satisfying assignment for C'' in case it is satisfiable.

6 Empirical Study

To evaluate some of the ideas presented in this paper empirically, we implemented a prototype circuit MUS extractor ncmuser. The extractor computes gate-MUSes by mapping the gate MUS problem to group oriented MUS in the manner described in the previous section. ncmuser interfaces with the group-MUS extractor (the group-oriented version of MUSer [9]) by controlling the order in which the latter selects the groups for removal. The mapping of the wire-MUS computation problem to the group oriented MUS extraction problem described in the previous section is currently not used – instead, for wire-MUS extraction ncmuser interfaces directly with an incremental SAT solver (picosat version 935 [1]). The benchmark circuits for our experiments were selected from the following sets:

(i) unsatisfiable (i.e. correct) sequential designs from the Hardware Model Checking Competition 2010 (http://fmv.jku.at/hwmcc10/) – combinational And-Inverter-Graph (AIG) circuits were generated using aigtobmc (http://fmv.jku.at/aiger);

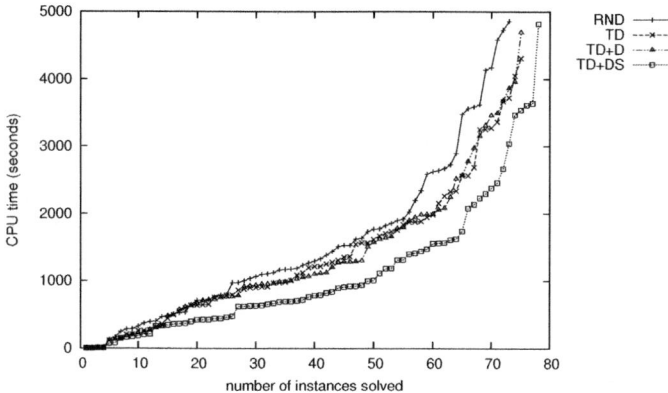

Fig. 3. Effects of gate selection strategies on gate-MUS computation times

(ii) AIGs generated using Boolector (http://fmv.jku.at/boolector/) to bit-blast QF_BV (theory of bit-vectors) instances of the SMT Competition 2009; (http://www.smtcomp.org/2009/)

(iii) unsatisfiable circuits in ISCAS format from the fvp-unsat-1.0 and fvp-unsat-2.0 benchmark suites of M. Velev (http://www.miroslav-velev.com/sat_benchmarks.html).

The objective of the first part of our empirical study was to investigate the effectiveness of the structure-based techniques for gate-MUS extraction described in Section 5. We implemented four gate selection strategies in ncmuser: the random selection (RND), the top-down traversal of the circuit (i.e. reverse topological order, TD), the top-down traversal with the domination based optimization (TD+D), and, finally, the strategy TD-D with the addition of the analysis of satisfying assignments (TD+DS). From our set of benchmarks we selected a subset of 245 instances solvable with top-down (TD) strategy given 5000 seconds of CPU time and 4 GB of RAM on HPC cluster nodes consisting of two quad-core Intel Xeon E5450's with 32 GB of RAM. From this subset we selected 75 instances that were found to have between 10% to 90% of redundant gates, and added 25 randomly selected timed-out instances. The results of the comparative evaluation of the four gate-selection strategies are presented in Figure 3. We note that the performance of gate-MUS extraction clearly improves with the amount of the circuit-based structural information used to aid the computation.

The goal of the second part of our empirical study was to find out whether the redundant wires do occur in practice. During the computation of wire-MUSes ncmuser uses the top-down circuit traversal strategy. As wire-MUS extraction may require more SAT calls than gate-MUS extraction, in wire-MUS extraction mode ncmuser was able to solve 228 instances out of 245 described above. We found that out of these 228 instances 30 had over 50%, and 70 had over 10% of redundant wires *after* all the redundant gates have been removed.

7   Conclusion

This paper addresses the problem of minimal unsatisfiability in Boolean circuits. The paper starts by formalizing the gate-based and wire-based notions of *minimally unsatisfiable circuits*, and then proposes algorithms for the computation of gate-MUSes and wire-MUSes of Boolean circuits. One key aspect is the tight relationship between circuit and group-oriented MUS extraction [10,5]. This applies both to gate-based and wire-based minimal unsatisfiability. Another key aspect is that the extraction can be accelerated by exploiting circuit structure. Experimental results, obtained on Boolean circuits from different application domains, confirm the practical efficiency of the proposed algorithms, and the usefulness of dedicated techniques. Finally, the general treatment of minimal unsatisfiability used in this paper appears to be quite convenient. Future work will investigate further the relationship between circuit and group-oriented MUS extraction.

Acknowledgements.   We thank the anonymous referees for helpful comments. This work is partially supported by SFI PI grant BEACON (09/IN.1/I2618).

References

1. Biere, A.: Picosat essentials. Journal on Satisfiability, Boolean Modeling and Computation 4, 75–97 (2008)
2. Desrosiers, C., Galinier, P., Hertz, A., Paroz, S.: Using heuristics to find minimal unsatisfiable subformulas in satisfiability problems. J. Comb. Optim. 18(2), 124–150 (2009)
3. Grégoire, É., Mazure, B., Piette, C.: On approaches to explaining infeasibility of sets of Boolean clauses. In: Int'l. Conf. on Tools with Artificial Intelligence, pp. 74–83 (2008)
4. Jain, H., Clarke, E.M.: Efficient SAT solving for non-clausal formulas using DPLL, graphs, and watched cuts. In: Proc. of the 46th Annual Design Automation Conference, pp. 563–568 (2009)
5. Järvisalo, M., Le Berre, D., Roussel, O.: Rules of the 2011 SAT Competition (2011), http://www.satcompetition.org/2011/
6. Kleine Büuning, H., Kullmann, O.: Minimal unsatisfiability and autarkies. In: Biere, A., Heule, M.J.H., van Maaren, H., Walsh, T. (eds.) Handbook of Satisfiability, ch. 11, pp. 339–401. IOS Press, Amsterdam (2009)
7. Le Berre, D., Parrain, A.: The Sat4j library, release 2.2. Journal on Satisfiability, Boolean Modeling and Computation 7, 59–64 (2010)
8. Marques-Silva, J.: Minimal unsatisfiability: Models, algorithms and applications. In: Int'l Symposium on Multiple-Valued Logic, pp. 9–14 (2010)
9. Marques-Silva, J., Lynce, I.: On improving MUS extraction algorithms. In: Sakallah, K.A., Simon, L. (eds.) SAT 2011. LNCS, vol. 6695, pp. 156–170. Springer, Heidelberg (2011)
10. Nadel, A.: Boosting minimal unsatisfiable core extraction. In: Formal Methods in Computer-Aided Design (2010)
11. Schuppan, V.: Towards a notion of unsatisfiable cores for LTL. In: Fundamentals of Software Engineering. In: Third IPM Int'l Conference, pp. 129–145 (2010)
12. Tseitin, G.S.: On the complexity of derivations in the propositional calculus. Studies in Mathematics and Mathematical Logic, Part II, pp. 115–125 (1968)

On Improving MUS Extraction Algorithms

Joao Marques-Silva[1,2] and Ines Lynce[2]

[1] University College Dublin
jpms@ucd.ie
[2] INESC-ID/IST, TU Lisbon
ines@sat.inesc-id.pt

Abstract. Minimally Unsatisfiable Subformulas (MUS) find a wide range of practical applications, including product configuration, knowledge-based validation, and hardware and software design and verification. MUSes also find application in recent Maximum Satisfiability algorithms and in CNF formula redundancy removal. Besides direct applications in Propositional Logic, algorithms for MUS extraction have been applied to more expressive logics. This paper proposes two algorithms for MUS extraction. The first algorithm is optimal in its class, meaning that it requires the smallest number of calls to a SAT solver. The second algorithm extends earlier work, but implements a number of new techniques. The resulting algorithms achieve significant performance gains with respect to state of the art MUS extraction algorithms.

1 Introduction

There has been a remarkable amount of recent work on algorithms for computing minimal explanations of unsatisfiability over the last decade (e.g. [28,16,3,15,14,9,10,11,27,12,7,13,23,25]). Most of this work is inspired by earlier work on computing explanations for inconsistencies (e.g. [5,4,1]). Algorithms for MUS extraction have often been characterized as *constructive* [12] (also referred to as insertion-based [7,23]), as *destructive* [12] (also referred to as removal-based [7], or deletion-based [23]), or as *dichotomic* [16,14]. All MUS extraction algorithms involve a number of calls to a SAT solver (or some other NP oracle). For destructive approaches, the best performing algorithms require $\mathcal{O}(m)$ calls to a SAT solver, where m is the number of clauses in the original formula. Existing constructive approaches require $\mathcal{O}(m \times k)$ calls to a SAT solver, where k is the size of the largest MUS in the original CNF formula [12]. Finally, the dichotomic approach requires $\mathcal{O}(k \log m)$ calls to a SAT solver. Recent work proposed an approach based on a weighted Maximum Satisfiability (MaxSAT) solver [7], but the function problem associated with computing a weighted MaxSAT solution is in Δ_2^P, and so unlikely to be in NP. There is also a large body of work on computing *good* approximations of MUSes (e.g. [23]). Despite the large body of work, MUS extraction algorithms are *not* industrial-strength, meaning that, with a few recent exceptions (e.g. [25]), MUS extraction algorithms are seldom evaluated on large problem instances or used in practical settings. This is demonstrated in the results section of this paper, where existing MUS extraction algorithms are shown to be in general inefficient for large complex problem instances from practical applications.

K.A. Sakallah and L. Simon (Eds.): SAT 2011, LNCS 6695, pp. 159–173, 2011.

This paper represents a first effort towards developing industrial-strength MUS extraction algorithms, and has the following main contributions. First, the paper develops a constructive algorithm for MUS extraction that requires $\mathcal{O}(m)$ calls to a SAT solver. This result implies (i) that destructive and constructive approaches have the same worst-case complexity in terms of the number of calls to a SAT solver; and (ii) that when $k = \Theta(m)$, the new algorithm represents the optimal case (as does the destructive algorithm). More importantly, this new algorithm blurs the distinction between destructive and constructive algorithms. Motivated by this observation, the paper proposes a hybrid algorithm that formally operates as a constructive algorithm, but that essentially exploits all steps of the algorithm to reduce the number of required iterations. This causes the algorithm to operate in a mostly hybrid mode, iteratively constructing the MUS, but also exploiting available information to reduce the number of iterations. Another contribution of the paper is the integration of a number of techniques that serve to simplify each SAT solver call, and to reduce the set of clauses that need to be analyzed through a call to a SAT solver. Moreover, the paper also shows that some existing techniques need not be considered for MUS extraction. Finally, the paper conducts a comprehensive evaluation of existing publicly available MUS extractors on representative industrial problem instances, obtained from well-known practical applications of SAT, where MUS extraction finds application.

2   Preliminaries

A set of variables $X = \{x_1, \ldots, x_N\}$ is assumed. A formula $\mathcal{F}$ in Conjunctive Normal Form (CNF) is defined as a set of sets of literals defined on X. A literal is either a variable or its complement. Each set of literals is referred to as a clause. Moreover, it is assumed that each clause is non-tautological. Given a clause c_i, $\{\neg c_i\}$ denotes the set of unit clauses obtained from negating c_i. Additional standard definitions can be found elsewhere (e.g. [8,24]). The focus of this paper are unsatisfiable formulas, and the characterization of the sources of unsatisfiability. Throughout the paper, $\mathcal{F}$, $\mathcal{F}' \subseteq \mathcal{F}$, $\mathcal{F}^R$, $\mathcal{F}^I$ and $\mathcal{U}$ denote CNF formulas, $\mathcal{S}$ and $\mathcal{S}'$ denote MUSes of $\mathcal{F}$, and $\mathcal{M}$ denotes a subset of an MUS $\mathcal{S}$.

Definition 1 (MUS). $\mathcal{M} \subseteq \mathcal{F}$ is a Minimally Unsatisfiable Subset (MUS) iff $\mathcal{M}$ is unsatisfiable and $\forall_{c \in \mathcal{M}}, \mathcal{M} \setminus \{c\}$ is satisfiable.

Definition 2 (MCS). $\mathcal{C} \subseteq \mathcal{F}$ is a Minimal Correction Subset (MCS) iff $\mathcal{F} \setminus \mathcal{C}$ is satisfiable and $\forall_{c \in \mathcal{C}}, \mathcal{F} \setminus (\mathcal{C} \setminus \{c\})$ is unsatisfiable.

Throughout the paper, m denotes the number of clauses in the original CNF formula $\mathcal{F}$, $m = |\mathcal{F}|$, and k denotes the number of clauses in the largest MUS $\mathcal{M}$, $k = |\mathcal{M}|$. The MUS decision problem, i.e. the problem of *deciding* whether a CNF formula $\mathcal{F}$ is an MUS is D^P-complete. In contrast, the problem of *computing* an MUS from an unsatisfiable CNF formula requires a number of calls to a SAT oracle. Over the years, three main approaches have been proposed for computing an MUS: *constructive* [5], *destructive* [4,1] and *dichotomic* [16,14]. Constructive approaches require $\mathcal{O}(m \times k)$ calls to an NP-oracle, destructive approaches require $\mathcal{O}(m)$ calls, and dichotomic approaches require $\mathcal{O}(k \times \log m)$ calls. Despite the theoretical interest of the dichotomic

---

Algorithm 1. Destructive MUS Extraction

Input   : Unsatisfiable CNF Formula $\mathcal{F}$
Output: MUS $\mathcal{M}$

1 **begin**
2 | $\mathcal{M} \leftarrow \mathcal{F}$ // MUS over-approximation
3 | **foreach** $c_i \in \mathcal{M}$ **do**
4 | | **if not** SAT($\mathcal{M} \setminus \{c_i\}$) **then**   // c_i is not transition clause
5 | | | $\mathcal{M} \leftarrow \mathcal{M} \setminus \{c_i\}$
6 | **return** $\mathcal{M}$ // Final $\mathcal{M}$ is an MUS
7 **end**

---

algorithm, the most recent implementation of MUS extraction algorithms are either destructive [2,25] or constructive [27].

Most practical MUS computation algorithms iteratively identify *transition clauses* [12]. The following definition is used throughout this paper.

Definition 3 (Transition Clause). *Let $\mathcal{F}$ be an unsatisfiable set of clauses and let $c \in \mathcal{F}$ be a clause. If $\mathcal{F} \setminus \{c\}$ is satisfiable then c is a transition clause with respect to $\mathcal{F}$.*

Lemma 1. *Let c be a transition clause of CNF formula $\mathcal{F}$. Then c is included in any MUS of $\mathcal{F}$.*

Proof. $\mathcal{F} \setminus \{c\}$ is satisfiable. Any unsatisfiable subset of $\mathcal{F}$ must include c.   □

Throughout the paper, SAT solvers are used as NP-oracles, that test the satisfiability of CNF formulas. In general, SAT($\mathcal{F}$) tests the satisfiability of a formula $\mathcal{F}$; it returns value true if the formula is satisfiable, and value false if the formula is unsatisfiable. Where necessary, SAT($\mathcal{F}$) may also return the satisfying assignment and an unsatisfiable subset. In this case, the output of the SAT solver call is represented as follows: $(\text{st}, \nu, \mathcal{U}) \leftarrow \text{SAT}(\mathcal{F})$. st is a Boolean variable assigned value *true* if the instance is satisfiable, in which case ν contains a solution to $\mathcal{F}$, or assigned value *false*, in which case $\mathcal{U} \subseteq \mathcal{F}$ is an unsatisfiable subformula. Besides the use of SAT solvers as NP-oracles, some algorithms propose the use of weighted MaxSAT solvers [7].

The standard organization of a destructive MUS extraction algorithm is shown in Algorithm 1 [12,23]. The algorithm starts with a working formula $\mathcal{M}$ equal to the original formula $\mathcal{F}$. Iteratively, the algorithm checks whether each one of the clauses $c_i \in \mathcal{M}$ is a transition clause. Non transition clauses are removed from $\mathcal{M}$. In the end, $\mathcal{M}$ is an MUS. This algorithm is studied in more detail in later sections.

Recent overviews of MUS extraction algorithms can be found in [12,7,23].

3   New Constructive Algorithm for MUS Extraction

This section develops a new constructive algorithm, that takes $\mathcal{O}(m)$ calls to a SAT oracle. This result implies that constructive and destructive approaches for MUS extraction have the same worst-case complexity in terms of the number of calls to a SAT solver, and improves known results in this area [12,23].

Algorithm 2. Constructive MUS Extraction with AtMost1 Constraint

Input : Unsatisfiable CNF Formula $\mathcal{F}$
Output: MUS $\mathcal{M}$

```
1 begin
2 │   M ← ∅                                    // M: MUS under-approximation
3 │   R ← {rᵢ | rᵢ is fresh variable for cᵢ ∈ F}    // R: relaxation variables
4 │   F^R ← {cᵢ ∪ {rᵢ} | rᵢ ∈ R ∧ cᵢ ∈ F}          // F^R: working formula
5 │   T ← CNF(∑_{rᵢ∈R} rᵢ ≤ 1)                     // ≤1 constraint
6 │   while F^R ≠ ∅ do                    // Repeat while relaxed clauses exist
7 │   │   (st, ν, U) ← SAT(F^R ∪ T ∪ M)
8 │   │   if st = true then
9 │   │   │   rᵢ ← TrueVariable(ν, R)        // Get true relaxation variable
10│   │   │   cᵢ^R ← Clause(F^R, rᵢ)          // Get clause associated with rᵢ
11│   │   │   F^R ← F^R \ {cᵢ^R}     // Remove clause cᵢ^R = cᵢ ∪ {rᵢ} from F^R
12│   │   │   M ← M ∪ {cᵢ^R \ {rᵢ}}    // Add clause cᵢ = cᵢ^R \ {rᵢ} to MUS
13│   │   else                          // If unsatisfiable, U ∩ T ≠ ∅
14│   │   │   if U ∩ F^R = ∅ then
15│   │   │   │   F^R ← ∅
16│   │   │   else
17│   │   │   │   cᵢ^R ← SelectClause(F^R ∩ U)
18│   │   │   │   F^R ← F^R \ {cᵢ^R}                  // Block one MUS
19│   return M                            // Final M is an MUS
20 end
```

Algorithm 2 shows the new constructive MUS extraction algorithm. This new algorithm borrows ideas from a number of earlier algorithms. Similarly to AMUSE [26], it adds relaxation variables to all clauses. In addition, and similarly to the use of weighted MaxSAT for MUS extraction [7], a SAT (resp. weighted MaxSAT) test is used to decide which clause to add to the MUS being built.

The operation of the algorithm is as follows. Assume the original formula $\mathcal{F}$ is unsatisfiable. The algorithm starts by creating a working formula $\mathcal{F}^R$ by relaxing all clauses in $\mathcal{F}$. An *AtMost1* constraint is created and encoded into the CNF formula $\mathcal{T}$, requiring at most one relaxation variable r_i to be assigned value true. $\mathcal{M}$ is initially an empty set and in the end is an MUS.

The outcome of the SAT solver call (see line 7) given formula $\mathcal{F}^R \cup \mathcal{T} \cup \mathcal{M}$ can either be true or false. If the outcome st is true, this means that exactly one relaxation variable was set to true. This relaxation variable r_i is associated with a clause c_i that is part of the MUS $\mathcal{M}$ being constructed. If st is false, this means that more than one relaxation variable would have to be assigned value true for the outcome to be true. This also implies the existence of more than one MUS, and so the solution is to (arbitrarily) block one MUS. This is done by simply removing a clause c_i^R from $\mathcal{F}^R$ that also occurs in the unsatisfiable formula $\mathcal{U}$ computed by the SAT solver. The process is iterated until $\mathcal{F}^R$ becomes empty (denoting that $\mathcal{M}$ is unsatisfiable), in which case $\mathcal{M}$ is an MUS.

To prove that Algorithm 2 computes an MUS of $\mathcal{F}$, the following intermediate results will be used.

Definition 4. *Throughout the execution of Algorithm 2, let $\mathcal{F}^I$ represent the clauses in $\mathcal{F}^R$ without the corresponding relaxation variables. (Observe that $\mathcal{F}^I \cap \mathcal{M} = \emptyset$.)*

Lemma 2. *Assume $\mathcal{M} \subsetneq \mathcal{S} \subseteq \mathcal{F}^I \cup \mathcal{M}$, where $\mathcal{S}$ is an MUS. Let $\mathcal{F}^R \cup \mathcal{T} \cup \mathcal{M}$ be unsatisfiable. Then $\mathcal{M}$ can be extended to strictly more than one MUS.*

Proof. Suppose that $\mathcal{M}$ can be extended to *exactly one* MUS $\mathcal{S}$. Select a clause c_i in $\mathcal{S} \setminus \mathcal{M}$, and relax clause c_i. By definition of MUS, $\mathcal{S} \setminus \{c_i\}$ must be satisfiable, and since $\mathcal{M}$ can be extended to exactly one MUS, then $\mathcal{F}^R \cup \mathcal{T} \cup \mathcal{M}$ would have to be satisfiable; a contradiction. □

Corollary 1. *Assume $\mathcal{M} \subsetneq \mathcal{S} \subseteq \mathcal{F}^I \cup \mathcal{M}$, where $\mathcal{S}$ is an MUS. Let $\mathcal{F}^R \cup \mathcal{T} \cup \mathcal{M}$ be unsatisfiable (i.e. line 13 of the algorithm), let $\mathcal{U}$ be an unsatisfiable subformula computed by the SAT solver, and let $(c_i \cup \{r_i\}) \in \mathcal{F}^R \cap \mathcal{U}$. Then there exists an MUS $\mathcal{S}'$ with $\mathcal{S}' \subseteq \mathcal{M} \cup (\mathcal{F}^I \setminus \{c_i\})$.*

Proof. $\mathcal{M} \cup (\mathcal{F}^R \setminus \{c_i \cup \{r_i\}\}) \cup \mathcal{T}$ is either satisfiable, requiring exactly one clause in $\mathcal{F}^R$ to be relaxed, or remains unsatisfiable. In either case, it still contains an MUS. □

Lemma 3. *Assume $\mathcal{M} \subsetneq \mathcal{S} \subseteq \mathcal{F}^I \cup \mathcal{M}$, where $\mathcal{S}$ is a MUS. Let $\mathcal{F}^R \cup \mathcal{T} \cup \mathcal{M}$ be satisfiable, and let c_i be a clause with an associated true relaxation variable r_i. Then, any MUS with clauses in $\mathcal{F}^I \cup \mathcal{M}$ will include c_i.*

Proof. By hypothesis, $\mathcal{F}^I \cup \mathcal{M}$ is unsatisfiable. If $\mathcal{F}^R \cup \mathcal{T} \cup \mathcal{M}$ is satisfiable, then $\mathcal{F}^R \cup \mathcal{M}$ has an MCS of size 1, which is identified by the relaxed clause c_i. Hence, by definition of MCS, c_i must be part of any MUS in $\mathcal{F}^I \cup \mathcal{M}$. □

Theorem 1. *Algorithm 2 returns an MUS of unsatisfiable CNF formula $\mathcal{F}$.*

Proof. To prove that Algorithm 2 computes an MUS of $\mathcal{F}$, the following invariants hold after each iteration of the algorithm: (i) $\mathcal{F}^I \cup \mathcal{M}$ is unsatisfiable; and (ii) there exists an MUS $\mathcal{S}$, with $\mathcal{M} \subseteq \mathcal{S} \subseteq \mathcal{F}^I \cup \mathcal{M}$. The invariants can be proved by induction on the number of iterations of the algorithm. Clearly, the invariants hold for the base case, with $\mathcal{M} = \emptyset$ and $\mathcal{F}^I$ unsatisfiable. Suppose that the invariants hold after iteration $j - 1$. Then, the objective is to analyze the invariants after iteration j. Suppose the SAT call in line 7 returns false. Hence, one clause is removed from $\mathcal{F}^I$. From Lemma 2 and Corollary 1, it is guaranteed that the resulting formula $\mathcal{F}^I \cup \mathcal{M}$ is still unsatisfiable and contains an MUS. Alternatively, suppose the SAT call in line 7 returns true. Hence, the relaxation variable is removed from the identified relaxed clause and the clause is added to $\mathcal{M}$. From Lemma 3, the identified clause is included in any MUS, and so can be added to $\mathcal{M}$. Moreover, the two invariants still hold: $\mathcal{M}$ continues to be part of an MUS and $\mathcal{F}^I \cup \mathcal{M}$ is unsatisfiable. □

Lemma 4. *The number of calls to a SAT solver by Algorithm 2 is in $\Theta(m)$.*

Proof. To prove that the number of calls is $\mathcal{O}(m)$, observe that the algorithm removes one clause from $\mathcal{F}^R$ at each iteration of the loop. Hence, there can be at most m calls to a SAT solver. To prove that the number of calls is $\Omega(m)$, consider the following CNF formula $\mathcal{F} = \{\neg x_1\} \cup_{i=1}^{N-1} \{x_i, \neg x_{i+1}\} \cup \{x_N\}$, with $|\mathcal{F}| = N + 1 = m$. $\mathcal{F}$ has a single MUS, containing all clauses. Each iteration of the algorithm will add exactly one clause to $\mathcal{M}$. Hence, the number of calls to the SAT solver is $N + 1 = m$. Thus, the number of calls to a SAT solver is in $\Omega(m)$.   □

Lemma 4 shows that deletion-based and insertion-based MUS extraction algorithms can have the same asymptotic complexity in terms of the number of calls to a SAT solver. Moreover, Algorithm 2 provides one concrete example of such algorithm. It should be noted that Algorithm 2 runs the SAT solver on a modified problem instance. However, as will be shown later, despite working on a modified problem instance, Algorithm 2 provides a few practical advantages.

4 Hybrid MUS Extraction

One of the interesting aspects of Algorithm 2 is that it blurs the distinction between constructive and destructive algorithms. On the one hand, the algorithm iteratively expands a subset of an MUS. On the other hand, the algorithm requires $\mathcal{O}(m)$ calls to a SAT solver. Similarly, one can develop a variant of Algorithm 1 that is essentially a constructive algorithm. Algorithm 3 shows this variant. As with Algorithm 2, $\mathcal{M}$ denotes a subset of an MUS, and the number of calls to a SAT solver is $\mathcal{O}(m)$. Nevertheless, Algorithm 3 also shares similarities with Algorithm 1, namely that each clause is analyzed exactly once, thus guaranteeing $\Theta(m)$ calls to a SAT solver. Besides the minor changes needed to make a constructive variant of Algorithm 1, Algorithm 3 also includes a number of key optimizations detailed below. Observe that for these techniques to be easily integrated, the algorithm *needs* to operate in constructive mode.

A first observation is that the input formula is assumed to be *trimmed*, i.e. the use of iterative identification of unsatisfiable cores was used to reduce the size of the working CNF formula. Clause set trimming is detailed in Section 4.2. To describe the techniques used to improve the performance of MUS extraction, it is convenient to isolate the clauses known to be part of an MUS (i.e. $\mathcal{M}$) from the clauses yet to be analyzed (i.e. $\mathcal{F}'$). Hence, the algorithm can be viewed as constructive. The new techniques are included in lines 7, 10, and 12.

The first technique (line 7) consists of creating a more constrained instance of SAT, by adding to the CNF formula the negation of the removed clause. It is well-known that c_i is redundant if $\mathcal{F} \setminus \{c_i\} \cup \{\neg c_i\}$ is unsatisfiable [19]. Although this technique was first proposed elsewhere [27], in the context of an $\mathcal{O}(m \times k)$ algorithm for MUS extraction, it has not been used in destructive (or hybrid) MUS extraction algorithms. In addition, its use affects the integration of other techniques, as discussed below.

Next, we analyze the technique summarized in line 12 of Algorithm 3. First, assume that the redundancy removal technique is not used, i.e. $\{\neg c_i\}$ is not added to the CNF formula given to the SAT solver. Let the outcome of the SAT solver be false. In this case, one can *refine* the working set of clauses with the unsatisfiable subformula computed by the SAT solver.

Lemma 5 (Clause Set Refinement). *Let $\mathcal{F}$, $\mathcal{F}'$, $\mathcal{M}$ and $\mathcal{U}$ be as defined in Section 2. Consider the outcome of the SAT solver on formula $\mathcal{F}' \cup \mathcal{M}$. If the result is unsatisfiable,*

Algorithm 3. Hybrid MUS Extraction

 Input : (Trimmed) Unsatisfiable CNF Formula $\mathcal{F}$
 Output: MUS $\mathcal{M}$

1 **begin**
2 $\mathcal{F}' \leftarrow \mathcal{F}$ `// Working CNF formula`
3 $\mathcal{M} \leftarrow \emptyset$ `// MUS under-approximation`
4 **while** $\mathcal{F}' \neq \emptyset$ **do**
5   $c_i \leftarrow \texttt{GetClause}(\mathcal{F}')$
6   $\mathcal{F}' \leftarrow \mathcal{F}' \setminus \{c_i\}$
7   $(\text{st}, \nu, \mathcal{U}) = \texttt{SAT}(\mathcal{M} \cup \mathcal{F}' \cup \{\neg c_i\})$ `// Add redundancy checking`
8   **if** st = true **then**   `// If SAT, `c_i` is transition clause`
9    $\mathcal{M} \leftarrow \mathcal{M} \cup \{c_i\}$
10    $(\mathcal{F}', \mathcal{M}) \leftarrow \texttt{Rotate}(\mathcal{F}', \mathcal{M}, \nu)$ `// Find more transition clauses`
11   **else if** $\mathcal{U} \subseteq \mathcal{M} \cup \mathcal{F}'$ **then** `// Equivalently, if `$\mathcal{U} \cap \{\neg c_i\} = \emptyset$
12    $\mathcal{F}' \leftarrow \mathcal{U} \setminus \mathcal{M}$ `// Clause-set refinement`
13 **return** $\mathcal{M}$ `// Final `$\mathcal{M}$` is an MUS`
14 **end**

---

with unsatisfiable subformula $\mathcal{U}$, then any MUS in $\mathcal{U}$ contains $\mathcal{M}$. Thus, the working formula $\mathcal{F}'$ can be set to $\mathcal{U} \setminus \mathcal{M}$.

Proof. By construction, $\mathcal{M}$ is composed of transition clauses, each of which is part of an MUS (see Lemma 1). Hence, any MUS in $\mathcal{U}$ must contain the clauses in $\mathcal{M}$. Since the clauses in $\mathcal{M}$ are known to be transition clauses, the working formula $\mathcal{F}'$ can be updated to $\mathcal{U} \setminus \mathcal{M}$. $\square$

A more complicated version of clause set refinement, that involves considering the resolution proof after each unsatisfiable outcome, has been described elsewhere [6,25]. Our approach considers solely the computed unsatisfiable core, and so allows using the SAT solver as a black box (provided the solver returns an unsatisfiable core).

  The integration of the redundancy removal technique (line 7) and clause set refinement is not immediate. The solution is to provide a test (line 11) to decide when the unsatisfiable core can be used as the next working CNF formula.

Proposition 1. *Let $\mathcal{U}$ be the unsatisfiable core returned by the SAT solver in line 7 of Algorithm 3. If $\mathcal{U} \cap \{\neg c_i\} = \emptyset$, then $\mathcal{U}$ contains an MUS $\mathcal{S}$ of $\mathcal{F}$.*

Finally, we analyze the technique summarized in line 10 of Algorithm 3. Let the outcome of the SAT solver be true and let ν be the computed model. This assignment *must* unsatisfy the clause removed from $\mathcal{F}'$. Similarly, *any* assignment that unsatisfies a *single* clause c_k from $\mathcal{F}'$ and satisfies all clauses in $\mathcal{M}$ proves that c_k must be part of an MUS.

Lemma 6. *Let $\mathcal{F}$, $\mathcal{F}' \subseteq \mathcal{F}$ and $\mathcal{M}$ be as defined in Section 2. Let ν be a model of $\mathcal{M} \cup \mathcal{F}' \cup \{\neg c_i\}$ (that must unsatisfy clause c_i). Then c_i is included in any MUS of $\mathcal{F}$ that contains $\mathcal{M}$.*

Proof. c_i is a transition clause. Hence, by Lemma 1, c_i is included in *any* MUS of $\mathcal{F}'$. Since $\mathcal{F}' \subseteq \mathcal{F}$, any MUS of $\mathcal{F}'$ is an MUS of $\mathcal{F}$. $\square$

Therefore, given a model ν, we can compute additional clauses to add to the MUS by selective flipping of the variable assignments in ν. The question is then how to decide which variable assignments to flip. The technique described in this paper is referred to as *model rotation*. This technique consists of analyzing changes to the computed model ν that will satisfy the single clause unsatisfied by ν. In order to keep the overhead low, only *single* literal flips are considered. This is illustrated with the following example.

Example 1 (Model Rotation). Let $\mathcal{F} = \{c_1, c_2, c_3, c_4\}$ be an unsatisfiable formula, with $c_1 = \{\neg x_1, x_2\}$, $c_2 = \{\neg x_1, \neg x_2\}$, $c_3 = \{x_1\}$, and $c_4 = \{\neg x_2, x_1, x_3\}$. Also, let $\mathcal{M} = \emptyset$. Suppose that c_1 is removed from $\mathcal{F}$. Then $\mathcal{F} \setminus \{c_1\}$ is satisfiable, with model $\nu = \{x_1, \neg x_2\}$. This means that c_1 is part of an MUS, and so it is added to $\mathcal{M}$. Observe that this model (necessarily) unsatisfies c_1. The next step is to check whether a literal flip in ν unsatisfies *exactly* another clause. For this example, flipping $\neg x_2$ to x_2 satisfies c_1 and solely unsatisfies c_2. This means that c_2 is also part of an MUS of $\mathcal{F}$. The resulting model of $\mathcal{M} \cup \mathcal{F} \setminus \{c_2\}$ is $\nu' = \{x_1, x_2\}$, and $\mathcal{M}$ is updated to $\{c_1, c_2\}$. We can now analyze ν' and check for a single flip that satisfies c_2 and unsatisfies a single clause of the remaining clauses not already in $\mathcal{M}$, namely c_3 and c_4. For example, flipping x_1 to $\neg x_1$ satisfies c_2 and unsatisfies c_3. Since c_3 is the solely unsatisfied clause, then c_3 is also part of an MUS of $\mathcal{F}$. The resulting model of $\mathcal{M} \cup \mathcal{F} \setminus \{c_3\}$ is $\nu'' = \{\neg x_1, x_2\}$, and $\mathcal{M}$ is updated to $\{c_1, c_2, c_3\}$. Observe that the model cannot be further rotated, since $\mathcal{M} = \{c_1, c_2, c_3\}$ is already unsatisfiable. This also means that c_4 is excluded from the computed MUS.

Clearly, model rotation could use more elaborate approaches for finding assignments that unsatisfy a single clause. For example, local search or even a complete SAT solver could be considered. Nevertheless, the objective of model rotation is to eliminate calls to the SAT solver, and so a simple (linear time) procedure is used instead. The analysis of computed models was first used in [27]. However, model rotation is a fundamentally different technique. Whereas the approach in [27] associates a model with each clause and requires worst-case quadratic space, model rotation simply considers single variable value changes to each computed model, so as to identify clauses that are in an MUS of the original formula.

Our results indicate that model rotation is a very effective technique, often allowing a large percentage of the satisfiable SAT calls to be skipped. Clearly, it is far more efficient to evaluate possible model rotations (in linear time), than to modify the SAT instance and call the SAT solver (in worst-case exponential time). This observation holds even if the problem instance is easy to solve.

Although the techniques described in this section are integrated in Algorithm 3, they can be applied with minor modifications to any destructive, constructive or dichotomic MUS algorithm.

4.1 Analysis of Other Techniques

Algorithm 3 integrates, adapts and extends several techniques proposed in earlier work. One additional technique could be considered, namely autarkies [17]. For example, autarkies have been successfully used in recent MUS enumeration algorithms [21]. In contrast, the use of autarkies in Algorithm 3 is less clear. First, by definition a clause is part of an autarky if and only if it is not included in *any* resolution refutation. Hence, since the proposed algorithms start by trimming the initial CNF formula, the autarkies of $\mathcal{F}$ are guaranteed to be *automatically* removed. Nevertheless, a less known observation is that, since clauses are discarded while searching for an MUS, it is possible that additional autarkies may exist with respect to $\mathcal{F}'$. Nevertheless, and similarly to clause set trimming, the use of clause set refinement also *guarantees* that autarkies are automatically eliminated, and so need not be computed. Although the previous observations suggest that identification of autarkies is unnecessary if clause set trimming and refinement are used, there are cases where autarkies *can* still find application in Algorithm 3. Observe that, due to the redundancy removal technique, clause set refinement may not be applicable after every unsatisfiable outcome. When this happens, then autarkies may exist, and can be identified. However, our experimental results indicate that the size of new autarkies does not justify their computation during the execution of the MUS extraction algorithm.

4.2 Preprocessing and Interfacing SAT Solvers

As indicated earlier, a standard technique for computing MUSes of large CNF formulas is *clause set trimming*, that consists of iteratively calling the SAT solver on computed unsatisfiable subformulas until no changes are detected in between calls to the SAT solver [28]. However, for large practical problem instances, iterating the computation of unsatisfiable subformulas until a fixed point is reached can be inefficient. A simpler alternative is to iterate the computation of unsatisfiable subformulas a constant number of times, or until the size change in the computed unsatisfiable subformulas is below a given threshold. Observe that clause set trimming can be viewed as the preprocessing step equivalent to clause set refinement described earlier in Section 4.

In MUS extraction algorithms, SAT solvers can either be used in incremental or non-incremental mode (e.g. [2]). Recent experimental results suggest that incremental mode provides significant performance gains [27,25]. Our implementation uses an incremental interface to the SAT solver, with one key change. Any clause c_i declared as being part of the MUS $\mathcal{M}$ needs not continue to be handled in incremental mode. Hence, the assumption variable used to activate c_i can be eliminated. This technique is beneficial for problem instances with large MUSes, since the overhead of the incremental interface is reduced as more clauses are added to the MUS $\mathcal{M}$.

5 Results

The algorithms described in the previous sections were implemented in the MUS extraction tool MUSer (MUS ExtratoR), built on top of the Picosat [2] SAT solver. Supported by existing experimental evidence [23], the incremental interface of Picosat was used. (Observe that other work [25] also proposes the use of the incremental interface of modern SAT solvers.) The experimental evaluation focused on the following MUS extractors: the new constructive MUS extraction algorithm based on relaxation variables (*CRV*) described in section 3; the hybrid MUS extraction algorithm (*HYB*) described in section 4; a reference destructive algorithm (*DREF*); a reference constructive algorithm [5] (*CREF*); the recent constructive algorithm from [27] (*MUNSAT*); a recent

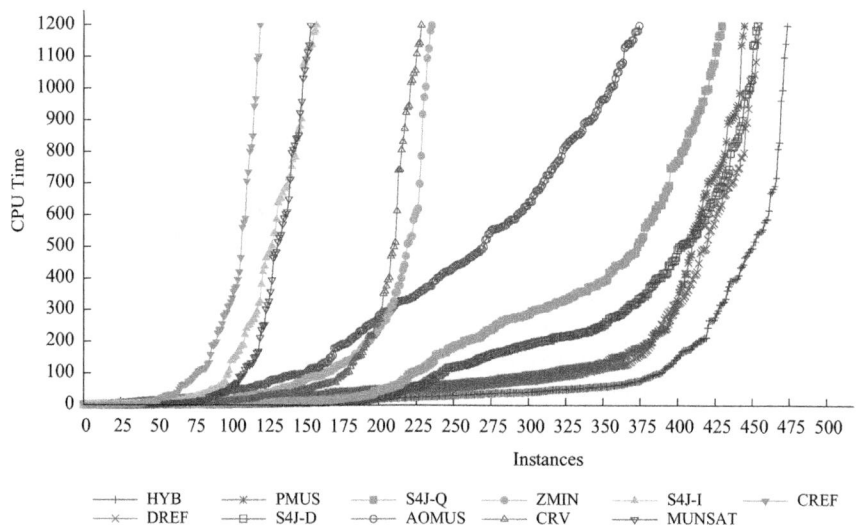

Fig. 1. Cactus plot with running times of MUS extractors

local-search-guided destructive MUS extraction algorithm from [11] (*AOMUS*); a well-known MUS extractor from [28] (ZMIN); SAT4J [18] MUS extractor in linear constructive mode (*S4J_I*), in QuickXPlain [16] mode (*S4J_Q*), and in destructive mode (*S4J_D*). Finally, a destructive MUS extraction algorithm available in the Picosat distribution [2] (PMUS). As shown by the results below, fairly recent MUS extractors [11,27,7] perform considerably worse than the most recent generation of MUS extractors, including the ones described in this paper.

The experimental evaluation focused on 500 problem instances submitted to the upcoming MUS track of the 2011 SAT Competition[1]. All problem instances were obtained from practical applications of SAT, including hardware bounded model checking, FPGA routing, hardware & software verification, equivalence checking, abstraction refinement, design debugging, function decomposition, and bioinformatics. Clause set trimming was applied to all problem instances before running *any* of the MUS extraction algorithms. Otherwise, algorithms that do not implement clause set trimming would perform poorly. All results were obtained on an HPC cluster, where each node is an 8-core CPU Xeon E5450 3GHz, with 32GByte RAM and running Linux. For each problem instance, the specified resources were a time limit of 1200 seconds and a memory limit of 4 GByte. For SAT4J, the Java virtual machine used was the Java HotSpot(TM) 64-Bit Server VM (build 19.1-b02).

Figure 1 shows a cactus plot with all MUS extractors, showing the instances solved by increasing run times. The following conclusions can be drawn. First, the new constructive algorithm based on relaxation variables (CRV) clearly outperforms all other constructive algorithms, namely MUNSAT, S4J_C and CREF. Second, and more importantly, the new hybrid algorithm *HYB* outperforms all other MUS extraction algorithms. It solves

---

[1] http://www.satcompetition.org/2011/

Table 1. Number of solved instances

Solver	CREF	MUNSAT	S4J_I	CRV	ZMIN	AOMUS	S4J_Q	PMUS	S4J_D	DREF	HYB
# Solved	112	154	158	228	235	374	429	444	453	454	**473**

Table 2. Comparison with [25]

Instance	3pipe	4pipe_1	barrel6	barrel7	barrel8	longmult6	longmult7	longmult8
Best in [25]	**167**	1528	348	700	4110	968	5099	—
HYB	194	**1143**	35	72	400	**11**	**99**	**811**
DREF	365	—	40	94	**332**	30	398	—
PMUS	—	—	68	102	701	51	283	—
S4J_S	223	—	395	829	—	152	883	—

more instances, but the plot also shows a clear performance edge with respect to all other algorithms. Third, fairly recent MUS extractors algorithms, namely MUNSAT [27] and AOMUS [11], perform significantly worse than the more recent generation of MUS extractors. Fourth, and finally, constructive algorithms perform significantly worse than destructive algorithms, the exceptions being the new algorithms *CRV* and *HYB*. However, the results confirm that constructive algorithms requiring $\mathcal{O}(m \times k)$ calls to a SAT solver simply do not scale in practice.

The cactus plot is completed with Table 1, that shows the number of solved instances. The main conclusions here are that: (i) the new algorithm *HYB* solves the largest number of instances; and (ii) recently published MUS extraction algorithms [11,27] are unable to solve many instances, many of which are easily solved by other approaches.

Finally, Figure 2 shows scatter plots comparing the run times of *HYB* with the next best MUS extraction algorithms, namely *DREF*, *S4J_D*, *PMUS*, and *AOMUS*. Again the results are clear. *HYB* clearly outperforms *DREF*, i.e. the reference implementation of destructive MUS extraction. Moreover, *HYB* clearly outperforms *PMUS*, in many cases by one order of magnitude or more. Also, *HYB* extensively outperforms *AOMUS*, in most cases by more than one order of magnitude. Finally, *HYB* also outperforms *S4J_D*, although in this case there are a number of outliers. These outliers represent problem instances with small MUSes, for which *S4J_D* performs well.

To conclude the experimental evaluation, the best performing MUS extraction tools are compared against the MUS extractor from [25], on selected problem instances. The best run times from [25] are used, since the tool is not publicly available. Moreover, the hardware where the MUS extractors were run is similar. The run times (in seconds) are shown in Table 2. As can be concluded, *HYB* performs significantly better. For the *barrel* instances, the speedup is around one order of magnitude. For the *longmult* instances, the speedup is almost two orders of magnitude. For the *pipe* instances, *HYB* performs better in one instance, and worse in another.

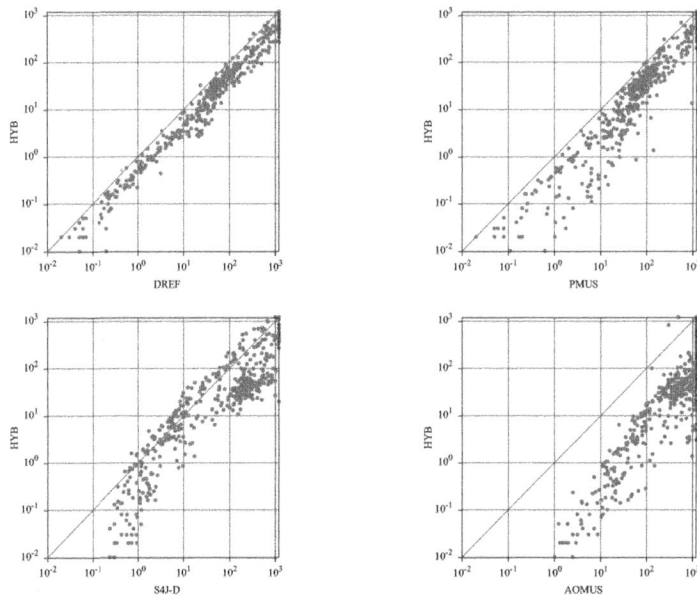

Fig. 2. Scatter plot comparing HYB with other MUS extractors

6 Related Work

To the best of our knowledge, Algorithm 2 is new. Nevertheless, the use of relaxation variables for MUS extraction has been proposed in earlier work. For example, AMUSE [26] also uses relaxation variables. However, AMUSE does not compute an MUS, and identifies instead a reduced unsatisfiable subset. The use of relaxation variables has also been considered extensively in the enumeration of MUSes [20,22], and in the use of MaxSAT for MUS extraction [7]. Although the use of relaxation variables resembles the use of selector variables [25], it is *fundamentally* different. Selector variables serve *solely* to specify clause (de)activation in incremental SAT. Relaxation variables serve to specify constraints on how many clauses can be relaxed.

Algorithm 3 is novel, even though its organization can be viewed as a (constructive) variant of Algorithm 1. Moreover, some of the techniques implemented by Algorithm 3 are novel, and their integration is also novel. Also, the implementation of these techniques requires a constructive MUS extraction algorithm. Clause set refinement was first studied in [6,25]. However, the solution proposed there is more complicated, being based on analyzing resolution proofs. In contrast, our approach simply uses the returned unsatisfiable core. The analysis of computed models for finding more than one transition clause per iteration of the algorithm was first used in [27], in the context of a constructive algorithm requiring $\Theta(m \times k)$ calls to a SAT solver. In [27], each clause is characterized by an *associated assignment*, that aims to satisfy all clauses in a working set of clauses but itself; clearly this can entail non-negligible memory requirements for large-scale problems instances. The model rotation technique proposed in this paper is novel, since computed models are only analyzed immediately after being computed, and only checked for single changes of variable values. Finally, the technique of including $\{\neg c_i\}$ in the CNF formula given to the SAT solver is standard in CNF redundancy checking [19], and was first used for MUS extraction in [27]. Our implementation follows this approach. Nevertheless, this paper proposes a new solution for integrating the redundancy removal technique and clause set refinement.

7   Conclusions

This paper develops new algorithms for the efficient extraction of MUSes from unsatisfiable CNF formulas, and has two main contributions. The first contribution is a new constructive MUS extraction algorithm. Whereas existing algorithms require $\mathcal{O}(m \times k)$ calls to a SAT oracle, the new algorithm requires $\mathcal{O}(m)$ calls. In practice, the new algorithm is shown to outperform all existing constructive algorithms. More importantly, this new algorithm shows that constructive and destructive MUS extraction algorithms share a number of important similarities. The second contribution exploits this observation, and develops a hybrid algorithm, that is organized as a constructive algorithm, but that exploits features of destructive algorithms. In addition, this algorithm integrates a number of key MUS extraction techniques, including redundancy removal, clause set refinement, and model rotation, that essentially exploit *all* of the main steps of the MUS extraction algorithm, i.e. calls to the SAT solver, and both unsatisfiable and satisfiable outcomes. Moreover, the paper also develops conditions for the integration of these techniques. Although these techniques are integrated in the new algorithm, they can be used with any MUS extraction algorithm. The resulting algorithm (*HYB*) outperforms all publicly available MUS extraction tools. The performance gains often exceed one order of magnitude when compared with state of the art MUS extraction tools. Finally, algorithm *HYB* is shown to also outperform recent non-publicly available MUS extraction algorithms [25].

The experimental results are promising and indicate that *HYB* represents the new state of the art in the area of MUS extraction algorithms. Nevertheless, practical applications of MUS extraction algorithms can gain from more efficient solutions. Envisioned research directions include better heuristics for model rotation and adapting SAT solvers to minimize computed unsatisfiable subformulas, e.g. by exploiting the AMUSE [26] heuristics.

Acknowledgement. This work is partially supported by SFI PI grant BEACON (09/IN.1/I2618), European Community FP7 project MANCOOSI (214898) and FCT through grants ATTEST (CMU-PT/ELE/0009/2009) and ASPEN (PTDC/EIA-CCO/110921/ 2009), and INESC-ID multiannual funding from the PIDDAC program funds.

References

1. Bakker, R.R., Dikker, F., Tempelman, F., Wognum, P.M.: Diagnosing and solving over-determined constraint satisfaction problems. In: International Joint Conference on Artificial Intelligence, pp. 276–281 (1993)

2. Biere, A.: PicoSAT essentials. Journal on Satisfiability, Boolean Modeling and Computation 2, 75–97 (2008)
3. Bruni, R.: On exact selection of minimally unsatisfiable subformulae. Ann. Math. Artif. Intell. 43(1), 35–50 (2005)
4. Chinneck, J.W., Dravnieks, E.W.: Locating minimal infeasible constraint sets in linear programs. INFORMS Journal on Computing 3(2), 157–168 (1991)
5. de Siqueira, J.L., Jean-Francois Puget, N.: Explanation-based generalisation of failures. In: European Conference on Artificial Intelligence, pp. 339–344 (1988)
6. Dershowitz, N., Hanna, Z., Nadel, A.: A scalable algorithm for minimal unsatisfiable core extraction. In: Biere, A., Gomes, C.P. (eds.) SAT 2006. LNCS, vol. 4121, pp. 36–41. Springer, Heidelberg (2006)
7. Desrosiers, C., Galinier, P., Hertz, A., Paroz, S.: Using heuristics to find minimal unsatisfiable subformulas in satisfiability problems. J. Comb. Optim. 18(2), 124–150 (2009)
8. Gomes, C.P., Kautz, H., Sabharwal, A., Selman, B.: Satisfiability solvers. In: Handbook of Knowledge Representation, pp. 89–134. Elsevier, Amsterdam (2008)
9. Grégoire, É., Mazure, B., Piette, C.: Extracting MUSes. In: European Conference on Artificial Intelligence, pp. 387–391 (August 2006)
10. Grégoire, É., Mazure, B., Piette, C.: Boosting a complete technique to find MSS and MUS thanks to a local search oracle. In: International Joint Conference on Artificial Intelligence, pp. 2300–2305 (January 2007)
11. Grégoire, É., Mazure, B., Piette, C.: Local-search extraction of MUSes. Constraints 12(3), 325–344 (2007)
12. Grégoire, É., Mazure, B., Piette, C.: On approaches to explaining infeasibility of sets of Boolean clauses. In: International Conference on Tools with Artificial Intelligence, pp. 74–83 (November 2008)
13. Grégoire, É., Mazure, B., Piette, C.: Using local search to find MSSes and MUSes. European Journal of Operational Research 199(3), 640–646 (2009)
14. Hemery, F., Lecoutre, C., Sais, L., Boussemart, F.: Extracting MUCs from constraint networks. In: European Conference on Artificial Intelligence, pp. 113–117 (2006)
15. Huang, J.: MUP: a minimal unsatisfiability prover. In: Asia South Pacific Design Automation, pp. 432–437 (2005)
16. Junker, U.: QUICKXPLAIN: Preferred explanations and relaxations for over-constrained problems. In: AAAI Conference on Artificial Intelligence, pp. 167–172 (2004)
17. Kullmann, O.: Lean clause-sets: generalizations of minimally unsatisfiable clause-sets. Discrete Applied Mathematics 130(2), 209–249 (2003)
18. Le Berre, D., Parrain, A.: The Sat4j library, release 2.2. Journal on Satisfiability, Boolean Modeling and Computation 7, 59–64 (2010)
19. Liberatore, P.: Redundancy in logic I: CNF propositional formulae. Artif. Intell. 163(2), 203–232 (2005)
20. Liffiton, M.H., Sakallah, K.A.: Algorithms for computing minimal unsatisfiable subsets of constraints. J. Autom. Reasoning 40(1), 1–33 (2008)
21. Liffiton, M.H., Sakallah, K.A.: Searching for autarkies to trim unsatisfiable clause sets. In: Kleine Büning, H., Zhao, X. (eds.) SAT 2008. LNCS, vol. 4996, pp. 182–195. Springer, Heidelberg (2008)
22. Liffiton, M.H., Sakallah, K.A.: Generalizing core-guided max-SAT. In: Kullmann, O. (ed.) SAT 2009. LNCS, vol. 5584, pp. 481–494. Springer, Heidelberg (2009)
23. Marques-Silva, J.: Minimal unsatisfiability: Models, algorithms and applications. In: International Symposium on Multiple-Valued Logic, pp. 9–14 (2010)
24. Marques-Silva, J., Lynce, I., Malik, S.: Conflict-driven clause learning SAT solvers. In: Biere, A., Heule, M., van Maaren, H., Walsh, T. (eds.) SAT Handbook, pp. 131–154. IOS Press, Amsterdam (2009)

25. Nadel, A.: Boosting minimal unsatisfiable core extraction. In: Formal Methods in Computer-Aided Design (October 2010)
26. Oh, Y., Mneimneh, M.N., Andraus, Z.S., Sakallah, K.A., Markov, I.L.: AMUSE: a minimally-unsatisfiable subformula extractor. In: Design Automation Conference, pp. 518–523 (2004)
27. van Maaren, H., Wieringa, S.: Finding guaranteed MUSes fast. In: Kleine Büning, H., Zhao, X. (eds.) SAT 2008. LNCS, vol. 4996, pp. 291–304. Springer, Heidelberg (2008)
28. Zhang, L., Malik, S.: Validating SAT solvers using an independent resolution-based checker: Practical implementations and other applications. In: Design, Automation and Test in Europe Conference, pp. 10880–10885 (2003)

Faster Extraction of High-Level Minimal Unsatisfiable Cores

Vadim Ryvchin[1,2] and Ofer Strichman[1]

[1] Information Systems Engineering, IE, Technion, Haifa, Israel*
[2] Design Technology Solutions Group, Intel Corporation, Haifa, Israel
rvadim@tx.technion.ac.il, ofers@ie.technion.ac.il

Abstract. Various verification techniques are based on SAT's capability to identify a small, or even minimal, unsatisfiable core in case the formula is unsatisfiable, i.e., a small subset of the clauses that are unsatisfiable regardless of the rest of the formula. In most cases it is not the core itself that is being used, rather it is processed further in order to check which clauses from a preknown set of *Interesting Constraints* (where each constraint is modeled with a conjunction of clauses) participate in the proof. The problem of minimizing the participation of interesting constraints was recently coined *high-level* minimal unsatisfiable core by Nadel [15]. Two prominent examples of verification techniques that need such small cores are 1) abstraction-refinement model-checking techniques, which use the core in order to identify the state variables that will be used for refinement (smaller number of such variables in the core implies that more state variables can be replaced with free inputs in the abstract model), and 2) assumption minimization, where the goal is to minimize the usage of environment assumptions in the proof, because these assumptions have to be proved separately. We propose seven improvements to the recent solution given in [15], which together result in an overall reduction of 55% in run time and 73% in the size of the resulting core, based on our experiments with hundreds of industrial test cases. The optimized procedure is also better empirically than the assumptions-based minimization technique.

1   Introduction

Given an unsatisfiable CNF formula φ, an unsatisfiable core (UC) is any subset of φ that is unsatisfiable. The decision problem corresponding to finding the *minimum* UC is a Σ_2-complete problem [8]. Finding a *minimal* UC (a UC such that the removal of any one of its clauses makes the formula satisfiable) is D^P-complete [17][1]. There are many works in the literature on extracting minimum [8,11], minimal [16,3,12,21], or just small cores [22,6,4] — see [15] for an extensive survey.

---

* Currently on sabbatical at the Software Engineering Institute, Pittsburgh, PA, USA
[1] D^P is the class containing all languages that can be considered as the difference between two languages in NP, or equivalently, the intersection of a language in NP with a language in co-NP.

K.A. Sakallah and L. Simon (Eds.): SAT 2011, LNCS 6695, pp. 174–187, 2011.
© Springer-Verlag Berlin Heidelberg 2011

There are many uses to the core in SAT-based verification, typically related to abstraction or decomposition. In most cases, however, it is not the core C itself that is being used, rather C is processed further in order to check which *Interesting Constraints* participate in the proof, where which constraints are interesting is given as input to the problem. Hence we can assume that in addition to the formula we are given as input a set of sets of clauses $IC = \{R_1 \ldots R_m\}$, where each R_i is a set of clauses that together encode an interesting constraint. The goal is thus to minimize the number of constraints in IC that have a non-empty intersection with C. This problem was first mentioned in [12] and recently coined *the high-level minimal unsatisfiable core* problem by Nadel [15], who observed that in his experiments with industrial problems the number of clauses that belong to interesting constraints is on average about 5% of the clause database. In fact in the verification group in Intel high-level cores are the only type of cores that are being computed, and we are not aware of any use of the general core in the EDA industry.

Two prominent examples of such techniques that are used in Intel and are described in more detail in the above reference are:

- A popular abstraction-refinement model-checking is based on iterating between a complete model checker and a SAT-based bounded model checker [14,9]. The model checker takes an abstract model, in which some of the state variables are replaced with inputs, and either proves the property or returns the depth in which it found a counterexample. In the latter case, this depth is used in a bounded-model checking run over the concrete model, which may either terminate with a concrete counterexample, or with an unsat answer. In the latter case SAT's capability to identify an unsatisfiable core is used for identifying those state variables that are sufficient for proving that there is no counterexample at that depth. All the clauses that contain a given state variable (in any time-frame) constitute a constraint in IC. Those state variables that participate in the proof define the next abstract model (these are the state variables that are *not* replaced by inputs), which is a refinement of the previous one. The process then reiterates until either the model checker is able to prove the property or the SAT solver finds a concrete counterexample.

- In formal equivalence verification (see, e.g., [10]), two similar circuits are verified to be functionally equivalent. This is done by decomposing the two circuits to 'slices' which are pair-wise verified for equivalence. The equivalence of each such pair is verified against various assumptions on the environment. In other words, rather than integrating a model of the environment with the equivalence verification condition, various properties of the environment are assumed, and added as constraints on the inputs of that condition. Then, if the equivalence is proven, it is still necessary to verify that the assumptions are indeed maintained by the environment. Each assumption is modeled with a set of clauses. The unsatisfiable core obtained when checking the equivalence is analyzed in order to find those assumptions that were used in the proof. Hence, here each constraint in IC is a set of clauses that encode an environment assumption. Here too the verification process attempts to minimize the high-level core in order to minimize the number of environment assumptions that should be verified.

We will address the question of how to minimize the core in the next section. A problem which is mostly orthogonal to minimization is how to make the SAT solver emit a core once it determines that a formula is unsatisfiable. There are two well-known approaches to solve this problem:

- **Resolution-based.** The first approach is based on the ability of many modern SAT solvers to produce a resolution proof in case the formula is unsatisfiable. The solver traverses the proof backwards from the empty clause, and reports the clauses at the leaves as the core [22,7]. This core is then intersected with the sets of clauses in IC in order to find a high-level core.
- **Assumptions-based.** A second approach is based on the *assumptions* technique, which was first implemented in an early version of Minisat [5]. Assumptions are literals that are assigned TRUE (as decisions) before any other decision. If constraint propagation leads to flipping the assignment of one of the assumptions to FALSE, it means that with these assumptions the formula is unsatisfiable. Minisat is capable of identifying which assumptions led to this conflict, which is exactly what is needed for extracting a high-level core. This can be done with *clause selectors* as follows: Let R_i be constraint in IC and let $\{c_1, \ldots, c_n\}$ be the clauses that encode it. To each clause in this set we add the literal $\neg l_i$, where l_i is a new variable. Then we add l_i to the set of assumptions. Hence setting l_i to TRUE activates this constraint, and setting it to FALSE deactivates it.

  The process of extracting the set of assumptions that led to a conflict is computationally easy. Let C be the clause that forces an assumption to its opposite value. Minisat resolves C with all its predecessors in the implication graph until a clause is generated which contains only negation of assumption literals. The negation of this clause is a conjunction of the assumptions that led to the conflict, also known as the *relevant assumptions*. The relevant assumptions constitute a high-level core.

The assumptions technique generates larger conflict clauses owing to the new selector variables, which may become significant if there are many assumptions [15,1]. The alternative of activating and deactivating constraints with unit clauses is more economic, as it simplifies and removes clauses. On the other hand, the assumptions technique does not consume memory for saving the proof, nor does it consume time to extract the core. Another difference between these two approaches, which turns out to be very important in our context, is related to clause minimization [2,20], which is a technique for shrinking conflict clauses. Whereas in resolution-based core extraction minimization of a clause may pull into the proof additional constraints, this does not happen in the assumptions-based approach. We will describe this issue in more detail in Sect. 4. The experiments in [15] showed that the assumptions-based method is on average faster than the resolution-based method, and produces slightly smaller cores. In the experiments we conducted (on a larger set of benchmarks) we witnessed similar results.

In this article we study seven improvements to the resolution-based high-level MUC problem. With these techniques, which we implemented on top of MiniSat-2.2 and ran over hundreds of industrial examples from Intel, we are able to show a 55% reduction in run time comparing to the techniques in [15], and a 28% improvement comparing to the assumptions-based technique. The configuration that achieves these improvements also reduces the core by 73% and 57%, respectively. More details on our experiments can be found in Sect. 4.

Since we take [15] as the starting point of our improvements, we begin in the next section by describing it in some detail.

2 Resolution-Based High-Level Core Minimization

The improvements we consider are relevant to resolution-based core extraction. We implemented inside Minisat 2.2 a rather standard mechanism for maintaining the resolution DAG. The resolution information is kept in a separate database, which we will call here the *resolution table*. This table maintains the indices of the parents and children of each derived clause. On top of this we implemented the reference counter technique of Shacham et al. [19]. In this technique every conflict clause has a counter, which is increased every time it resolves a new clause, and decreased when a child clause is erased. Once the counter of a clause is 0, it does not need to be maintained any longer for the purpose of later retrieving the resolution DAG. In the experiments that were reported in [19], this optimization led to a reduction by a factor of 3 to 6 in the size of the resolution table.

The unsatisfiable core is retrieved as usual by backward traversal from the empty clause to the roots. But since we are interested in minimizing the core, the story does not end here. We implemented the high-level core minimization algorithm of [15], which appears in Pseudo-code in Alg. 1. The input to this algorithm is a set of interesting constraints $IC = \{R_1 \ldots R_m\}$, each of which is a set (or a conjunction, depending on the context) of clauses, and a formula Ω, which is called the *remainder*. The formula $\Psi = \bigwedge_{j=1}^{m} R_j \wedge \Omega$ is assumed to be unsatisfiable, and the proof is available at the beginning of the algorithm. We denote the initial core by *initial_core*. The output of the algorithm is a high-level minimal unsatisfiable core with respect to IC and Ω, i.e., a subset $IC' \subseteq IC$ such that $\Psi' = \bigwedge_{R_j \in IC'} R_j \wedge \Omega$ is unsatisfiable, and no constraint can be removed of IC' without making Ψ' satisfiable.

The algorithm is rather self-explanatory, so we will be brief in describing it. In line 1 any constraint R_i that none of its clauses participated in the proof is removed together with its cone, i.e., all the clauses that were derived (transitively) from R_i clauses. The next line defines the set of candidate indices for the core, which is initiated to the indices of the constraints in IC that were not removed in the previous step. From here on the algorithm attempts to remove elements of this set. In each iteration of the loop, it removes a constraint R_k together with its cone and checks for satisfiability. If the formula is satisfiable, then R_k with its cone is returned to the formula, and R_k is added to the solution set *muc*. Otherwise, the unsatisfiability proof is checked in order to remove any constraint R_i, together with its cone, that did not participate in the proof.

Algorithm 1. Resolution-based high-level MUC extraction (Based on Alg. 2 in [15])

Input: Unsatisfiable formula of the form $\Psi = \bigwedge_{R_j \in IC} R_j \wedge \Omega$.
Output: A high-level MUC with respect to IC and Ω.

1: Remove any R_i together with its cone if it is not reachable from the empty clause;
2: $muc\_cands := \{R_i \mid R_i \cap initial\_core \neq \emptyset\}$;   ▷ MUC Candidates
3: $muc := \{\}$;
4: **while** muc_cands is non-empty **do**
5:     $R_k :=$ a member of muc_cands;
6:     Check satisfiability of the formula without R_k and its cone;
7:     **if** satisfiable **then**
8:         return R_k and its cone to the formula;
9:         $muc := muc \cup \{R_k\}$;
10:    **else**
11:        **for** $R_i \in muc\_cands$ **do**
12:            **if** $R_i \cap core = \emptyset$ **then**   ▷ *core* is the unsat core of the proof
13:                Remove R_i and its cone;
14:                $muc\_cands := muc\_cands \setminus \{R_i\}$;
15: **return** muc;

It is interesting to note that this algorithm is tailored for *high-level* core minimization, and not for general core minimization. The difference is evident by observing that the whole set of clauses associated with a constraint R_i is removed, together with their joint core. Had the object of minimization been the whole core, we would rather remove all clauses that did not participate in the proof, even if other clauses that share the same constraint *do* participate in the proof. For example, if $R_i = \{c_1, c_2\}$, and only c_1 participate in the proof, Alg. 1 retains both c_1 and c_2, because removing c_2 does not reduce the size of the high-level core, whereas it may assist in consecutive iterations. Furthermore, retaining c_2 is needed in order to guarantee minimality. Without it we may miss the fact that some other constraint can be removed.

3   Optimizations

In this section we describe seven optimizations to the basic algorithm that was presented in the previous section. We will use the following terminology: a clause is an *IC-clause* if it either belongs to one of the initial constraints in IC or is a descendant of such a clause in the resolution DAG. Other clauses are called *remainder* clauses. We say that a literal is *IC-implied* if it is implied by an IC-clause or just *implied* otherwise.

A: Maintaining Partial Resolution Proofs. In this optimization we maintain only clauses in the cone of IC-clauses in the resolution table, and the links between them. That is, we save an IC-clause, and the parents and children that are also IC-clauses. Comparing to full resolution, this reduces the amount of memory required by more than an order of magnitude in most cases, reduces the amount of time that it takes to find clauses that are in the cone of an IC (recall that in line 13 of Alg. 1 IC-clauses are removed together with their cones), and, more importantly, allows to activate a certain simplification (see below) for remainder clauses, which is normally turned off when running Alg. 1.

The simplification in point is applied in decision level 0, owing to constants. If the clause database includes a unit clause, e.g., (x), then many solvers would remove those clauses that contain x, and remove $\neg x$ from all other clauses, at decision level 0 (MiniSat is a little different in this respect: it does not remove $\neg x$ from existing clauses once x is learned, but rather it does not add $\neg x$ to new learned clauses). This simple, yet powerful simplification has to be turned off when running Alg. 1. For example, if (x) is an IC-clause associated with constraint R_1, then we cannot just remove clauses with x from the formula, since we might decide at line 13 to remove R_1, which will force us to retrieve these clauses. Empirically it is better to retain such clauses rather than keeping them in a file and then retrieving them. The same issue occurs when removing the negation of x from clauses: here too, we will need to retrieve the original clauses once R_1 is removed. One of the advantages of this optimization, therefore, is that we can turn back on this simplification for the remainder clauses.

B: Selective Clause Minimization. Clause minimization [2,20] is a technique for shrinking conflict clauses. Once a clause is learnt, each of its literals is tested: if it implies other literals in the clause, it can be removed.

Example 1. Consider the following clauses:

$$C_1 = (\neg v_1 \vee v_2) \qquad\qquad C_2 = (\neg v_2 \vee v_3)$$
$$C_3 = (\neg v_4 \vee v_5) \qquad\qquad C_4 = (\neg v_5 \vee v_6)$$
$$C_5 = (\neg v_1 \vee \neg v_3 \vee \neg v_4 \vee \neg v_6)$$

Suppose that the first decision is v_1. This decision implies v_2 (from C_1) and v_3 (from C_2). Suppose now that the next decision is v_4. This decision implies v_5 (from C_3) and v_6 (from C_4) and a conflict in clause C_5. Conflict analysis based on 1-UIP returns in this case a new clause $C = (\neg v_1 \vee \neg v_3 \vee \neg v_4)$. From C_1 and C_2 we can see that $v_1 \rightarrow v_3$, or equivalently $\neg v_3 \rightarrow \neg v_1$, which is an implication between literals in C. Clause minimization will find this implication by following the resolution DAG and remove $\neg v_3$. □

We will not present the full algorithm for clause minimization here, but rather only mention that it is based on traversing the resolution DAG backward from each literal l in the learned clause. The hope is to hit a 'frontier' of other literals from the same clause that by themselves imply l. If in this process we hit a decision variable, it means that l cannot be removed.

Example 2. Continuing the previous example, the algorithm scans each non-decision literal in C. Consider v_3: this literal was implied in C_2, and hence we progress to look at the other literal in that clause, namely v_2. This literal was implied by C_1 and hence we look at v_1. But since $v_1 \in C$, it means that we found an implication within C, and hence $\neg v_3$ can be removed. Note that the minimized clause can be resolved from the original one and the clauses that are traversed in the process. In this case $Res(C, Res(C_1, C_2)) = (\neg v_1 \vee \neg v_4)$.   □

The problem with clause minimization in our context is that it may turn a non-IC-clause C into a shorter IC-clause C'. This can happen if the minimization process uses an IC-clause: in that case C' has to be marked as an IC-clause as well. Furthermore, it can turn an IC-clause C that depends on a certain set of interesting constraints, into a shorter IC-clause that depends on *more* such constraints. This means that if that clause will participate in the proof, it will 'pull-in' more constraints into the core.

Our suggested optimization is to cancel clause minimization in any case that an IC-clause is involved. In other words, we prefer a large clause that depends on a few constraints, over a smaller one with more such dependencies. The latter may pull more constraints into the proof, and lead to other such clauses. We aspire, instead, to keep the resolution table as small as possible and with the fewest connections to IC-constraints. Ideally we should check whether using a certain IC-clause in the minimization process indeed adds dependencies, but this is simply too expensive: for this we would need to traverse the DAG backwards all the way to the roots in order to check which constraints are involved.

It is interesting to analyze the behavior of the assumptions-based method with respect to clause minimization. It turns out that it solves this problem for free, and hence in this respect it is a superior method. In fact from analyzing various cases in which it performs much better than the clause-based method (before the optimizations suggested here were added), we realized that this is the main cause for the difference in run-time, rather than the facts mentioned in the introduction (the fact that it does not need to save the resolution table, nor to extract the core in the end of each iteration). How does it solve this problem for free? Observe that with this technique all IC-clauses have as literals all the selector variables that correspond to constraints that were used in deriving that clause. For example, let R_1, R_2 be two constraints with associated selector variables l_1, l_2 respectively. If R_1 and R_2 participate in inferring C, then C must contain $\neg l_1$ and $\neg l_2$. This is implied by the fact that selector variables appear only in one phase in the formula, and hence cannot be resolved away. Hence the presence of these literals in IC-clauses is an invariant. If we falsely assume that a minimized clause C can increase its dependency on constraints, we immediately reach a contradiction: the supposedly added constraint implies that a new selector variable was added to C, which contradicts the fact that literals are only removed from C in the minimization process.

C: Postponed Propagation Over IC-Clauses. In this optimization we control the BCP order. We first run BCP over non-IC-clauses until completion.

If there is no conflict, we propagate a single implication due to an IC-clause, and run regular BCP again. We repeat this process until no more propagations are possible or reaching a conflict. The idea behind this optimization is to increase the chances of learning a remainder clause rather than an IC-clause.

D: Reclassifying IC-Clauses. When we discover that some IC-constraint R must be in the MUC (line 8 in Alg. 1), we add its clauses back as remainder clauses, together with all the clauses in its cone that do not depend on other constraints. To identify this set of constraints, we employ an algorithm in the style of a least-fix-point computation. We insert all the R clauses into a set S. Then we add all the children of those clauses that all their parents are in S. We repeat this process until reaching a fix-point.

Without this optimization R's clauses are added back as is, with their marking as IC-clauses. By adding them back as remainder clauses, we enable more simplifications, as described in the case of optimization **A**.

E: Selective Learning of IC-Clauses. When detecting a conflict, the learned clause may be an IC-clause. If all else is equal, such a clause is less preferable than a remainder clause, as it may increase the high-level core, in addition to the fact that it leads to a larger resolution table and hence longer run times. We found that learning a non-asserting remainder clause instead, combined with partial restart, improves the overall performance. The learning of the remainder clause is essential for termination, and also turns out to decrease run time. The alternative remainder clause that we learn is even closer to the conflict than the first UIP. We can learn it only if the conflicting clause is not an IC-clause; in other cases we simply revert to learning the IC-clause. Learning the remainder clause is done by reanalyzing the conflict graph *as if the IC-implications were decisions*. This optimization is only ran in conjunction with optimizations **B** and **C** above, for reasons that we will soon clarify. Alg. 2 describes the procedure for learning this clause.

Algorithm 2. An algorithm that attempts to find a remainder conflict clause by reanalyzing the conflict graph as if the IC-implications were decisions. Returns a remainder clause if one can be found, and NULL otherwise.

---

function Get_Remainder_Clause
1. If the conflicting clause is an IC-clause then return NULL.
2. Search an IC-implied literal l in the trail, starting from the latest implied literal and ending just before the 1-UIP literal.
3. Convert the implication of l into a decision, and update accordingly the decision level of all implied literals in the trail that come after it.
4. Call Analyze_Conflict() with the same conflicting clause, but while referring to the new decision levels. Let C be the resulting conflict clause.
5. Return C.

Note that the fact that we use this algorithm only when optimization **C** is active, guarantees that the literals searched and updated in steps 2 and 3 are implied by l, i.e., the fact that BCP was ran to completion on non-IC-clauses before asserting l, guarantees that the rest of the implications at that decision level depend on asserting l. Also note that the clause learnt in step 4 is necessarily a remainder clause because ANALYZE_CONFLICT() cannot cross an IC-implied literal (such implications were made into decisions), and that it corresponds to a cut in the implication graph to the right of the first UIP. The reason we activate this optimization in conjunction with optimization **B**, is that we want to refrain from a case in which we learn a remainder clause, but it then turns into an IC-clause owing to clause minimization. This is not essential for correctness, however: we could also have just compared this smaller IC-clause to the original one and choose between the two, but our experience is that it is better to give priority to minimizing the number of IC-clauses. Finally, note that there is no reason to revert the changes made to the trail, because backtracking removes this part of the trail anyway.

Example 3. Figure 1 presents an implication graph, where IC-implications are marked with dashed edges. The marked 1-UIP cut in the top drawing is calculated while considering such implications as any other implication. The suggested heuristic is to learn instead a normal clause, by considering such implications as new decisions, as depicted in the bottom drawing.   □

As mentioned earlier, learning the alternative clause is combined with a partial restart. Let dl be the level to which we would have jumped had we learned the IC-clause. We backtrack to dl, but at this point nothing is asserted because we did not learn an asserting clause. We then move to the next decision level, $dl + 1$, and decide the negation of the original 1-UIP literal. Hence instead of learning an asserting clause and implying the negation of the 1-UIP literal, we refrain from learning that clause and decide on the same value. This assignment in neither necessary or sufficient for preventing the same conflict to occur. What prevents us from entering an infinite loop in the absence of standard learning is the fact that we learn at least one clause between such partial restarts. Since the solver cannot enter a conflict state that leads to learning an existing clause, we are guaranteed not to enter an infinite loop.

Example 4. Referring again to the conflict graphs in Example 3, our solver backtracks to the end of level 3 — the same level we would have jumped with the original IC-clause — progress to level 4 and decides $\neg l_1$.   □

In our experiments we also tried other decisions (such as $\neg l_2$ in the example above), but $\neg l_1$ seems to work better in practice. We also tried different strategies of updating the scores. The best strategy we found in our experiments is to update the score according to both the original and the alternative clause.

F: Selective Chronological backtracking. Recall that optimization **E** involves a partial restart when learning an IC-clause. Different heuristics can be applied in order to choose the backtracking level. Our experiments show that if we only backtrack one level, rather than to the original backtrack level as explained above, the results improve significantly. The complete set of data, available from [18], shows that this heuristic improves the run time in most instances, and that it improves the search itself and not only reduces constants, as is evident by the fact that it reduces the number of conflicts. It seems that the reason for the success of this heuristic is related to the fact that with normal backtracking and score scheme we may lose the connection to the clause that we actually learn, i.e., the scores might divert the search from a space which is more relevant to the alternative clause that we learn.

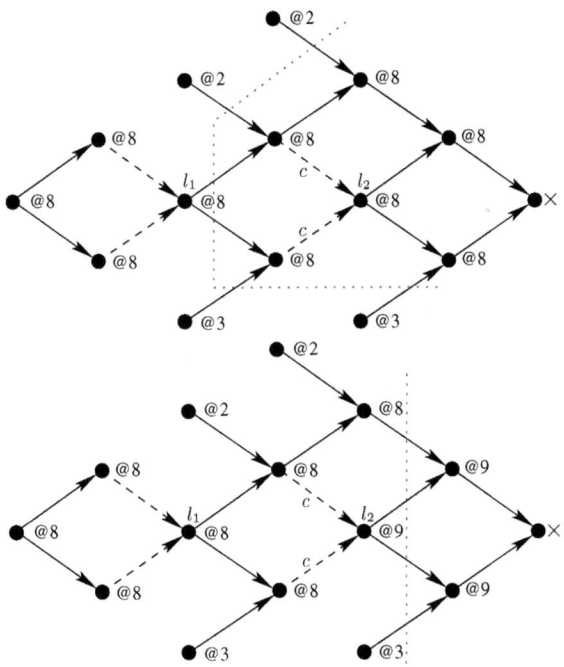

Fig. 1. In these conflict graphs, dashed arrows denote IC-implications, and the dotted lines denote 1-UIP cuts. In the top drawing, where such implications are referred to as any other implications, the learned 1-UIP clause must be marked as an IC-clause, since it is resolved from the IC-clause c. We can learn instead a normal clause by taking, for example, the 1-UIP clause in the bottom conflict graph. In that graph, c's implication are considered as decisions, which changes the decision levels labeling the nodes.

G: A removal strategy. Recall that in line 5 of Alg. 1 constraints are removed in an arbitrary order. We suggest a simple greedy heuristic instead: remove the constraint that contributed the largest number of clauses to the proof. This heuristic, as will be evident in the next section, reduces the size of the resulting core but slightly increases run time.

We also experimented with a heuristic by which we remove the constraint with the *least* number of clauses in the proof, speculating that this leaves more clauses in the formula and hence increases the chance that there will be a proof without this constraint. This option also improves performance comparing to the arbitrary order with which we started, but is not as good as the one suggested above. There is an indirect cause behind this difference: the large constraints (i.e., those that have many clauses) are typically necessary for the proof regardless of the other constraints, and hence the faster we make them remainder constraints – with optimization **D** – the faster the rest of the solution process is. This, in turn, affects the size of the core because it leads to less time-outs. As we will explain in the next section, the result of the algorithm when interrupted by a time-out is the last computed core, or, in case that even the first iteration does not terminate, the entire set of *IC*-clauses.

4   Experimental Results

Our tool HHLMUC (for Haifa's high-level MUC) was built, as mentioned earlier, on top of Minisat 2.2. It contains the algorithm from Sect. 2 and also the technique of [19] for reducing the amount of required data in the resolution table by using a reference-counter. On top of this we implemented the optimizations that were described in the previous section, and ran all possible combinations (excluding the restrictions mentioned in optimization **E**), on the set used in [15] (family 'lat-fmcad10' in the tables below), and additional nine families of harder abstraction-refinement benchmarks from Intel. We removed from the benchmark set instances that could not be solved by any of the configurations in the given time-out of one hour. This left us with 144 benchmarks, all of which are from the two application domains that were described in the introduction. This set constitute Intel's contribution to the benchmarks repository that will be used in the upcoming SAT competition dedicated to this problem. The average number of clauses per instance is 2,572,270; the average number of constraints per instance is 3804; and, finally, the average number of interesting clauses per instance is 96568 (25.3 clauses per constraint), which is approximately 6% of the clauses. All experiments were ran on Intel® Xeon® machines with 4Ghz CPU frequency and 32Gb of memory.

Table 4 shows run time results for selected configurations.[2] The second column ("Full") refers to our starting point as explained above. One may observe that the best result is achieved when combining the first six optimizations, whereas the seventh slightly increases the overall run-time.

We also compared our results to assumptions-based minimization. We tried both a simple scheme, and the improvement suggested in [15]. In the simple scheme, a constraint is added to the MUC (line 8 in Alg.1) by setting its associated selector variable to true; In the improved method the same effect is achieved by adding a unit clause asserting this literal to TRUE. Similarly, in the simple scheme an environment assumption is removed from the formula (line 13 in Alg.1) by setting its associated selector to FALSE; In the improved method the

---

[2] The tool and the full set of results, including a comparison to MUC tools (which does not appear here) can be downloaded from [18].

same effect is achieved by adding a unit clause asserting this literal to FALSE. The improved method is better empirically apparently because the unit clause invokes a simplification step in decision level 0, which removes the selector variable and erases some clauses. The results we witnessed with the two methods appear in the last two columns of the table. Overall the combination of optimizations achieve a reduction of 55% in run time comparing to our starting point, and a reduction of 28% comparing to the assumptions-based method.

All the presented methods can be affected by the order in which constraints are removed in line 5. We therefore tried three different arbitrary removal orders in each case. Empirically this hardly had an effect on the average run-time when using the resolution-based methods, whereas it had some effect when using the assumption-based methods. The table below represents the best overall run times among the different orders we tried (i.e., we present the results that together have the minimum run-time). Regarding the size of the resulting core, the different arbitrary orders had inconsistent effect, as expected, but the order referred to in optimization **G** had a non-negligible positive effect on the size of the core, as will be shown momentarily.

Table 1. Summary of run-time results by family (144 instances all together)

Benchmark	Resolution-based								Assumptions-based
family	Full	A	AB	ABC	ABCE	A–E	A–F	A–G		units
latch1	2001	1604	660	465	570	575	425	423	819	798
gate1	3747	1403	705	636	620	579	490	477	856	855
latch2	9113	5915	6636	6116	5685	5656	2424	2370	8153	8043
latch3	348	293	274	274	283	275	262	200	236	236
latch4	769	529	506	457	467	455	443	379	504	521
latch5	1103	820	735	657	678	630	632	625	747	689
lat-fmcad10	785	457	445	451	435	435	400	394	417	425
latch6	8868	5456	5329	5188	5007	5006	4948	4943	5322	5279
latch7	9956	7050	5719	5244	5094	5096	5302	5286	5688	5652
latch8	8223	7946	5673	6133	5459	5420	5127	5587	8004	5534
Total	44913	31473	26682	25621	24298	24127	**20453**	20684	30746	28032

Next, we consider the size of the resulting high-level MUC. The configuration that achieves the best run-time (A–F) achieves the second smallest high-level core, whereas the second best configuration in terms of run time (A–G) achieves the smallest core. If a solver timed-out in our experiments, we considered its latest computed core, i.e., the set $muc \cup muc\_cands$. If a solver did not finish even the first iteration, then we considered the entire set of clauses in IC as its achieved core. This policy, which reflects the way such cores are used, explains the different results of strategies that are supposed to be equivalent with respect to the size of the core. For example, the partial-resolution proof optimization (**A**) does not remove more clauses than 'Full', but since the latter is generally slower, it times-out more times and hence its core count is larger. The 'TO' row contains the number of such time-outs with each configuration.

Table 2. Summary of the size of the high-level core by family. The 'TO' row indicates the number of time-outs.

Benchmark family	Resolution-based								Assumptions-based
	Full	A	AB	ABC	ABCE	A–E	A–F	A–G	units
latch1	41	41	41	41	42	42	41	42	52	45
gate1	1143	1210	1089	568	1029	1029	870	901	618	1192
latch2	5887	2851	127	3040	2851	2851	131	129	3782	4165
latch3	168	202	202	199	211	211	208	123	140	132
latch4	236	237	248	236	238	238	237	162	177	217
latch5	224	266	266	206	206	206	220	222	222	223
lat-fmcad10	577	456	456	489	540	540	453	454	457	450
latch6	2550	2502	2502	2490	2490	2490	2480	2480	2463	2502
latch7	2578	322	585	253	154	154	211	204	304	287
latch8	5591	615	2867	393	344	344	371	373	2887	2877
TO	8	5	3	3	2	2	2	2	6	5
Total	18995	8702	8383	7915	8105	8105	5222	**5090**	11102	12090

5   Summary and Future Work

The recently introduced problem of finding a *high-level* minimal unsatisfiable core has various applications in the industry. Until [15] the standard practice was to minimize the core itself, and only then to find the interesting part of it. Our experiments show that this approach cannot compete with a solver that focuses on the high-level core. In this article we introduced seven techniques that reduce both the run time and the resulting high-level core.

A straight-forward direction for future research is to migrate some of the suggested optimizations to the assumptions-based approach. Related SAT problems may also benefit from these methods. First - it is possible that general SAT solving can be improved with some combination of optimizations **E** and **F**. Second, the same techniques can potentially expedite other methods in which the SAT component needs to extract only partial information from the resolution proof, like interpolation-based model checking [13]. In interpolation only a small part of the proof is necessary in order to generate the interpolant, and we want to explore possibilities to minimize that part and decrease the overall run time with variants of the methods suggested here.

References

1. Asín, R., Nieuwenhuis, R., Oliveras, A., Rodríguez-Carbonell, E.: Efficient generation of unsatisfiability proofs and cores in SAT. In: Cervesato, I., Veith, H., Voronkov, A. (eds.) LPAR 2008. LNCS (LNAI), vol. 5330, pp. 16–30. Springer, Heidelberg (2008)
2. Beame, P., Kautz, H., Sabharwal, A.: Towards understanding and harnessing the potential of clause learning. Journal of Artificial Intelligence Research 22, 319–351 (2004)

3. Dershowitz, N., Hanna, Z., Nadel, A.: A scalable algorithm for minimal unsatisfiable core extraction. In: Biere, A., Gomes, C.P. (eds.) SAT 2006. LNCS, vol. 4121, pp. 36–41. Springer, Heidelberg (2006)
4. Desrosiers, C., Galinier, P., Hertz, A., Paroz, S.: Using heuristics to find minimal unsatisfiable subformulas in satisfiability problems. J. Comb. Optim. 18(2), 124–150 (2009)
5. Eén, N., Sörensson, N.: Temporal induction by incremental SAT solving. Electr. Notes Theor. Comput. Sci. 89(4) (2003)
6. Gershman, R., Koifman, M., Strichman, O.: An approach for extracting a small unsatisfiable core. J. on Formal Methods in System Design, 1–27 (2008)
7. Goldberg, E., Novikov, Y.: Verification of proofs of unsatisfiability for CNF formulas. In: Proceedings of Design, Automation and Test in Europe Conference and Exhibition (DATE 2003), pp. 886–891 (2003)
8. Gupta, A.: Learning Abstractions for Model Checking. PhD thesis, Carnegie Mellon University (2006)
9. Gupta, A., Ganai, M.K., Yang, Z., Ashar, P.: Iterative abstraction using sat-based bmc with proof analysis. In: ICCAD, pp. 416–423 (2003)
10. Khasidashvili, Z., Kaiss, D., Bustan, D.: A compositional theory for post-reboot observational equivalence checking of hardware. In: FMCAD, pp. 136–143 (2009)
11. Liffiton, M.H., Mneimneh, M.N., Lynce, I., Andraus, Z.S., Marques-Silva, J., Sakallah, K.A.: A branch and bound algorithm for extracting smallest minimal unsatisfiable subformulas. Constraints 14(4), 415–442 (2009)
12. Liffiton, M.H., Sakallah, K.A.: Algorithms for computing minimal unsatisfiable subsets of constraints. J. Autom. Reasoning 40(1), 1–33 (2008)
13. McMillan, K.: Interpolation and SAT-based model checking. In: Hunt Jr., W.A., Somenzi, F. (eds.) CAV 2003. LNCS, vol. 2725, pp. 1–13. Springer, Heidelberg (2003)
14. McMillan, K., Amla, N.: Automatic abstraction without counterexamples. In: Garavel, H., Hatcliff, J. (eds.) TACAS 2003. LNCS, vol. 2619, pp. 2–17. Springer, Heidelberg (2003)
15. Nadel, A.: Boosting minimal unsatisfiable core extraction. In: Bloem, R., Sharygina, N. (eds.) FMCAD (2010)
16. Oh, Y., Mneimneh, M.N., Andraus, Z.S., Sakallah, K.A., Markov, I.L.: Amuse: a minimally-unsatisfiable subformula extractor. In: DAC 2004, pp. 518–523 (2004)
17. Papadimitriou, C.H., Wolfe, D.: The complexity of facets resolved. J. Comput. Syst. Sci. 37(1), 2–13 (1988)
18. Ryvchin, V.: Benchmarks + results, http://ie.technion.ac.il/~ofers/sat11.html
19. Shacham, O., Yorav, K.: On-the-fly resolve trace minimization. In: DAC, pp. 594–599 (2007)
20. Sörensson, N., Biere, A.: Minimizing learned clauses. In: Kullmann, O. (ed.) SAT 2009. LNCS, vol. 5584, pp. 237–243. Springer, Heidelberg (2009)
21. van Maaren, H., Wieringa, S.: Finding guaranteed mUSes fast. In: Kleine Büning, H., Zhao, X. (eds.) SAT 2008. LNCS, vol. 4996, pp. 291–304. Springer, Heidelberg (2008)
22. Zhang, L., Malik, S.: Extracting small unsatisfiable cores from unsatisfiable boolean formulas. In: Sixth International Conference on Theory and Applications of Satisfiability Testing (SAT 2003), S. Margherita Ligure (2003)

On Freezing and Reactivating Learnt Clauses[*]

Gilles Audemard, Jean-Marie Lagniez, Bertrand Mazure, and Lakhdar Saïs[**]

Université Lille-Nord de France
CRIL - CNRS UMR 8188
Artois, F-62307 Lens
{audemard,lagniez,mazure,sais}@cril.fr

Abstract. In this paper, we propose a new dynamic management policy of the learnt clause database in modern SAT solvers. It is based on a dynamic freezing and activation principle of the learnt clauses. At a given search state, using a relevant selection function, it activates the most promising learnt clauses while freezing irrelevant ones. In this way, clauses learned at previous steps can be frozen at the current step and might be activated again in future steps of the search process. Our strategy tries to exploit pieces of information gathered from the past to deduce the relevance of a given clause for the remaining search steps. This policy contrasts with all the well-known deletion strategies, where a given learned clause is definitely eliminated. Experiments on SAT instances taken from the last competitions demonstrate the efficiency of our proposed technique.

1 Introduction

The SAT problem, i.e. the problem of checking whether a set of Boolean clauses is satisfiable or not, is central to many domains of computer science and artificial intelligence (theorem proving, planning, non-monotonic reasoning, VLSI correctness checking or knowledge-base verification and validation). During the last two decades, SAT has gained considerable audience with the advent of a new generation of SAT solvers that are able to solve large instances encoding real-world applications. These solvers, called CDCL (Conflict Driven, Clause Learning) [11,5], are based on a nice combination of (i) clause learning [9,10,15], (ii) VSIDS heuristics [11] and (iii) restart policies [6,7], enhanced with efficient data structures (eg. Watched literals). On the theoretical side, K. Pipatsrisawat and A. Darwiche [13] proved that modern SAT solvers formalized as a proof system are equivalent in strength to general resolution, if the search is restarted at each conflict. This result shows that resolution-based clause learning is an important component of modern SAT solvers, since it pushes forward DPLL-like procedures from tree-like to general resolution, a more powerful proof system. On the practical side, as the set of clauses that can be derived from conflicts is of exponential size in the worst case, several strategies have been designed to cope with this combinatorial explosion problem. To maintain a learnt clause database of polynomial size - and consequently perform unit propagation with reasonable cost - all these strategies dynamically reduce the learnt database by deleting clauses considered to be irrelevant to the next search

---

[*] Nominated as Best Paper candidate.

[**] This work is (partially) supported by ANR UNLOC project: ANR 08-BLAN-0289-01.

K.A. Sakallah and L. Simon (Eds.): SAT 2011, LNCS 6695, pp. 188–200, 2011.
© Springer-Verlag Berlin Heidelberg 2011

steps. The most popular strategy considers a learnt clause as irrelevant if its activity or its involvement in recent conflict analysis is marginal. In [2], a static measure called literal block distance (LBD, corresponding to the number of different levels involved in a given learnt clause) is used to quantify the quality of learnt clauses. Clauses with smaller LBD are considered as more relevant. Theoretically, the first unique implication point (UIP) scheme is shown to be optimal among schemes that learn an asserting clause in terms of LBD measure [1]. The main drawback of these cleaning strategies is that they cannot avoid the elimination of relevant learnt clauses. Their irreversible elimination makes it possible that the same clause will be derived repeatedly.

The problem of determining what is a useful learnt clause in advance remains very challenging and computationally hard. In this paper, we propose a new dynamic management policy of the learnt clause database in modern SAT solvers. It is based on a dynamic freezing and activation principle of the learnt clauses. At a given search state, it activates the most promising learnt clauses while freezing irrelevant ones. In this way, previously learned clauses can be discarded for the current step, but may be activated again in future steps of the search process. Our policy tries to exploit pieces of information gathered from the past to deduce the relevance of a given clause for the remaining search steps. This policy contrasts with all well-known deletion strategies, where a given learned clause is definitely eliminated. In this way, a clause can be useless at a given step and relevant at another step of the search process. The ideal is to freeze a learnt clause when it is not used and just to reactivate it at the time when it could play a role in the proof.

The next part of the paper is organized as follows: section 2 introduces necessary background. In section 3, we introduce a new relevance measure of learnt clauses, whereas in section 4, we present our dynamic freezing and activation strategy of learnt clauses. Before concluding, we present in section 5, an experimental comparison of our new dynamic learnt clauses management policy with the well known state-of-the-art reduction policies as well as state of the art solvers.

2   Definitions, Notations and Technical Background

In this section, after some preliminary definitions and notations, we introduce the most salient computational features of modern SAT solvers.

A CNF *formula* Σ is a conjunction (interpreted as a set) of *clauses*, where a clause is a disjunction (interpreted as a set) of *literals*. A literal is a positive (x) or negative ($\neg x$) Boolean variable. The two literals x and $\neg x$ are called *complementary*. A *unit* clause is a clause with only one literal (called *unit literal*). An *empty clause*, is interpreted as false, while an *empty* CNF *formula*, is interpreted as true. A set of literals is *complete* if it contains one literal for each variable occurring in Σ and *fundamental* if it does not contain complementary literals. An *interpretation* $\mathcal{I}$ of a boolean formula Σ associates a value $\mathcal{I}(x)$ to some of the variables x appearing in Σ. An interpretation can be represented by a fundamental set of literals, in the obvious way. A *model* of a formula Σ is an interpretation $\mathcal{I}$ that satisfies the formula, *i.e.* that satisfies all clauses of the formula. Finally, SAT is the problem of deciding whether a given CNF formula Σ admits a model or not.

Algorithm 1. CDCL solver

Input: a CNF formula Σ
Output: SAT or UNSAT
1 $\Delta = \emptyset$;   /* learnt clause database */
2 **while** *(true)* **do**
3   **if** *(!propagate())* **then**
4   **if** *((c = analyzeConflict()) == $\emptyset$)* **then return** UNSAT;
5        $\Delta = \Delta \cup \{c\}$;
6   **if** *(timeToRestart())* **then** backtrack to level 0;
7   **else**
8   backtrack to the assertion level of c;
9   **else**
10   **if** *((l = decide()) == null)* **then** return SAT;
11   assert l in a new decision level;
12   **if** *(timeToReduce())* **then** clean(Δ);

Let us now briefly describe the basic components of CDCL based SAT solvers [11,5]. To be exhaustive, these solvers incorporate unit propagation (enhanced by efficient and lazy data structures), variable activity based heuristic, literal polarity phase, clause learning, restarts and a learnt clause database reduction policy.

These main components are depicted by the general scheme given in Algorithm 1. At each step of the main loop, the algorithm performs unit propagation (line 3). In case of conflict (lines 4-8), a new asserting clause is derived by conflict analysis (line 4). If such a clause is empty, then the formula is answered unsatisfiable, otherwise it is added to the learnt clause database (line 5). If it is not time to restart, the algorithm backjumps to the assertion level of the learnt clause, *i.e.* the level where the learnt clause becomes unit (line 8), otherwise it backjumps to the root of the search. When the formula is closed under unit propagation without generating the empty clause, a new decision literal - if it exists - is selected and asserted in a new decision level (line 11), otherwise a model is found and the formula is answered to be satisfiable (line 10).

Finally, when it is time to reduce, the learnt clause database is cleaned (line 12). This component, usually omitted in the description of CDCL solvers, is clearly crucial to the solvers' performance. Indeed, keeping too many learnt clauses will slow down the unit propagation process, while deleting too many of them will break the overall learning benefit. Consequently, identifying good learnt clauses - relevant to the proof derivation - is clearly an important challenge. The first proposed quality measure follows the success of the activity based VSIDS heuristic. More precisely, a learnt clause is considered relevant to the proof, if it is involved more often in recent conflicts, *i.e.* usually used to derive asserting clauses. Clearly, this deletion strategy supposes that a useful clause in the past could be useful in the future. More recently, a more accurate measure called LBD is used to estimate the quality of a learnt clause leading to a better cleaning strategy than the previous one [2]. This new measure is based on the number of different decision levels appearing in a learnt clause and is computed when the clause is learnt.

Extensive experiments demonstrates that clauses with small LBD values are used more often than those with higher LBD ones.

Another feature of CDCL solvers recently proposed in [12] concerns the literal polarity to be chosen when the next decision variable is selected thanks to the VSIDS heuristic. Usually, a default polarity (e.g. false) is defined and used each time a decision literal is assigned. Based on the observation that restarts and backjumping might lead to repetitive solving of same subformulas, Pipatsrisawat and Darwiche [12] proposed to dynamically save for each variable the last used polarity. This literal polarity based heuristic, called progress saving, prevents the solver from solving the same satisfiable subformulas several times. These memorized polarities can be represented as a complete interpretation $\mathcal{P}$. Each time a decision variable is chosen, its assignment polarity is selected from $\mathcal{P}$. Consequently, $\mathcal{P}$ gives us at least the polarities of the decision literals. And each time a literal is assigned by the solver, its associated polarity is set in $\mathcal{P}$.

In the next section, we exploit $\mathcal{P}$ (progress saving) to approximate the usefulness that one can expect in the near future from a learnt clause, in other words to measure the likelihood for a given clause to be part of the implication graph.

3   A New Measure for Identifying Relevant Learnt Clauses

As mentioned above a CDCL-based SAT solver can be formulated as a resolution proof system [13,3]. In practice, the main problem behind resolution-based techniques arises from their exponential space complexity. Consequently, the practical incarnation of modern SAT solvers can be seen as a resolution-based procedure with a deletion strategy. As a consequence, the completeness of modern SAT solvers is heavily connected to both the chosen deletion and restart policies. For example, if we use a restart with a static cutoff value and an aggressive deletion strategy, we cannot guarantee the completeness of the solver. For this reason one needs to be careful when designing a deletion strategy. Consequently, defining what is a relevant clause before completing the proof itself is of a great importance for the efficiency of the solver. However answering such a question is computationally hard and it is related to finding a proof of minimal size.

In this section we define a simple measure to identify the relevance of a given learnt clause and we experimentally show its effectiveness. Our measure is based on the progress saving polarity [12] introduced in the previous section. This *progress saving based quality measure*, in short *psm*, is defined as follows: given a clause c and a complete set of literals $\mathcal{P}$ representing the current set of saved literals polarities, we define $psm_{\mathcal{P}}(c) = |\mathcal{P} \cap c|$. This measure can be related to another one proposed in [11]. In this paper, a learnt clause was tagged useless, in the goal to delete it, if its number of unassigned literals has reached a predefined threshold.

First let us note that the *psm* measure is highly dynamic. Since the set $\mathcal{P}$ of saved literals polarities will evolve during search, the *psm* of a given clause will also evolve consequently. For example, when a clause is learnt, its *psm* value is equal to zero and becomes one after backjumping to the assertion level. It is also important to note that when a given learnt clause is at the origin of unit propagation, its *psm* value is also equal to one. These preliminary remarks suggest that clauses with small *psm* value are the most relevant to the near future of the search. Let $\mathcal{I}$ be the current partial interpretation and $\mathcal{P}$ be the current complete interpretation representing the current saved literals polarities, and c a learnt clause. As $\mathcal{I} \subset \mathcal{P}$, $psm_{\mathcal{P}}(c)$ represents the number of literals that are assigned to true by $\mathcal{I}$ or that would be assigned to true by $\mathcal{P} \backslash \mathcal{I}$. Consequently, a clause with a small *psm* value has a lot of chance to be unit propagated or to be falsified. On the contrary, a clause with a big *psm* value has a lot of chance to be satisfied by more than one literal and then to be irrelevant for the subsequent part of the search.

To analyze and to validate this assumption, experiments are conducted on some SAT instances. Figure 1 shows, for a sample of instances, the average number of times clauses with a given *psm* value are used during the unit propagation process. In this experiment, we consider a time sequence t_k with $k > 0$ (the search starts at t_0) corresponding to the successive steps of the search where the learnt database is classically reduced. Let $\mathcal{P}_{t_k}$ and $\mathcal{P}_{t_{k+1}}$ be the progress saving literal polarities at the steps t_k and t_{k+1} respectively. Let us consider the time window between t_k and t_{k+1}, when a given clause c from the learnt database is used for unit propagation, we compute $psm = psm_{\mathcal{P}_{t_k}}(c)$ and then $\alpha(psm)$ the number of times a clause with such *psm* value is used for propagation is increased by one. The average number of times a clause with a given *psm* value (x-$axis$) is used in unit propagation (y-$axis$), corresponds to $\alpha(psm)$ divided by the total number of times a learnt database is reduced.

As we can observe from Figure 1, learnt clauses with small *psm* value are used more often in the unit propagation process than those with higher *psm* value. If we look closer, we can see that the most used clauses are those with *psm* value around 10. Based on extensive experiments, we observed that on the majority of instances the distribution of *psm* values looks like those represented in the two upper curves of Figure 1.

This first experiment illustrates the relevance of clauses with small *psm* value. To compare it with previous learnt clauses quality measure, we integrate our *psm* measure to the learnt clauses reduction policy (clean(Δ) - line 12) of MINISAT 2.2 which is the latest version of the well known solver MINISAT [5]. Similarly to previous approaches, each time a reduction is performed, the set of clauses is sorted according to the increasing order of *psm* value. When two clauses admit the same *psm* value, the one with the greatest activity (VSIDS) is preferred. Then the learnt database is reduced by half. Like other strategies, we keep the binary clauses in the learnt database.

In the sequel, all our experiments are conducted on a Quad-core Intel XEON X5550 with 32Gb of memory, using the 292 application instances of the SAT 2009 competition. The CPU time limit is set to 900 seconds.

For each solver, we indicate the number of solved instances (#Solved) with the number of satisfiable (#SAT) and unsatisfiable instances (#UNSAT) in brackets. We also give the average time in seconds (avg time) necessary to solve these instances.

Table 1 summarises the results obtained by MINISAT^d [5], MINISAT^d + LBD [2] and MINISAT^d + *psm* using the default time sequence (noted MINISAT^d) of MINISAT. As we can see, MINISAT^d + *psm* obtains the best overall results and is the best on satisfiable instances. This first experiment shows the efficiency of our new measure *psm* using the default time reduction sequence of MINISAT.

To make a fair comparison between these three approaches, we also present in Table 2 the results obtained using an aggressive cleaning policy as presented in [2]

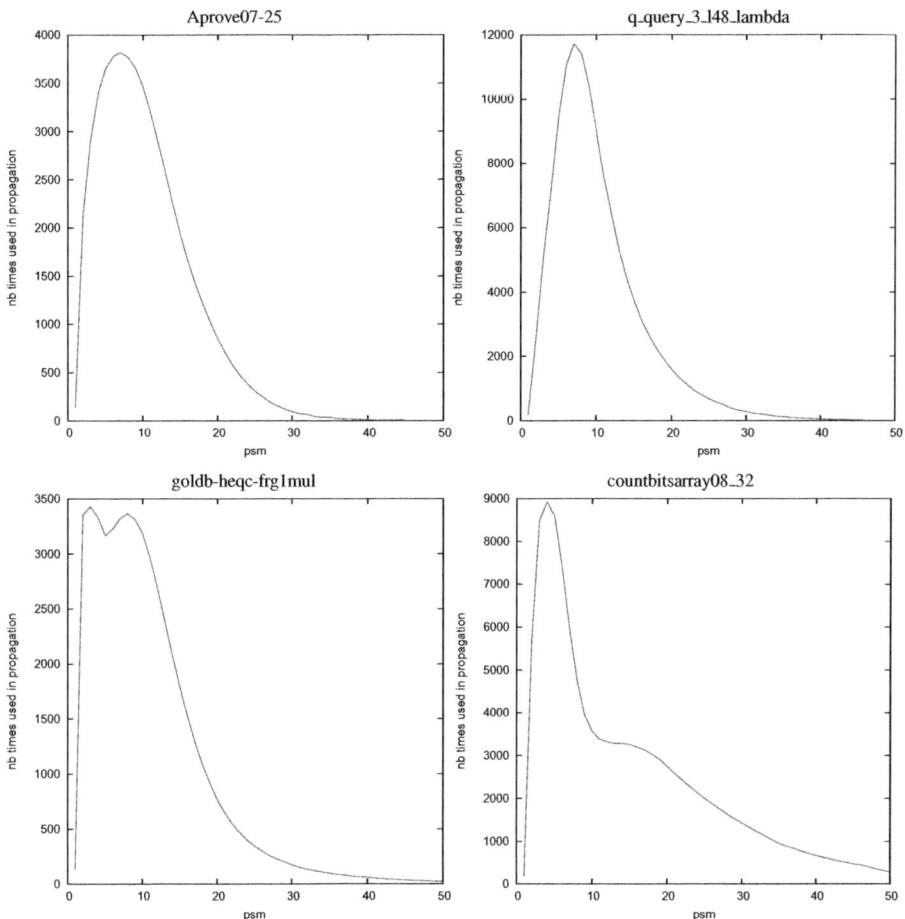

Fig. 1. Progress saving measure / relevance with respect to UP

Table 1. Results with the MINISAT default time cleaning sequence

Solver	#Solved (#SAT- #UNSAT)	avg time
MINISAT^d	174 (68 - **106**)	142
MINISAT^d + *psm*	**177** (**73** - 104)	130
MINISAT^d + LBD	173 (71 - 102)	132

(noted MINISAT^a). In this experiment the learnt database Δ is reduced using the following time sequence, $t_0 = 4000$ conflicts and $t_k = t_{k-1} + 300$ conflicts for $k > 0$. Using aggressive (more frequent) cleaning time sequence, the result obtained by the LBD measure are better than those obtained by VSIDS like criterion and the *psm* measure.

As a summary, considering the classical reduction and deletion strategies, these first experiments clearly show that our measure is competitive with the two other

Table 2. Results with an aggressive time cleaning sequence

Solver	#Solved (#SAT- #UNSAT)	avg time
MINISAT^{a}	162 (68 - 94)	136
MINISAT^{a} + psm	163 (70 - 93)	140
MINISAT^{a} + LBD	**168 (72 - 96)**	128

well-known measures using both aggressive and less aggressive cleaning policy. This measure will be used in next section in order to design a dynamic management policy of learnt clauses.

4 Freeze and Reactivate: A Dynamic Management Policy

In section 3, we defined a new measure based on progress saving [12] for identifying relevant learnt clauses. In this section, we describe our dynamic management policy of the learnt clause database. Our proposed framework is based on two important key points. First, the progress saving based measure is highly dynamic and evolves during search. Consequently, a clause might be considered irrelevant (high *psm* value) at a given step of the search and could become relevant (small *psm* value) in the future steps of the search. Secondly, determining if a given learnt clause will be involved again in the resolution proof is a computationally hard task. All the well-known management policies are not safe from regularly eliminating relevant learnt clauses. For both reasons, our proposed approach introduces an additional and new concept of frozen learnt clauses. A learnt clause considered as irrelevant at a given step can be frozen and reactivated when it is considered as useful again. More precisely, freezing (respectively activating) a clause means that the clause is disconnected (respectively attached) to the learnt database, and then it is not used during the search (respectively used).

This kind of management strategy cannot be defined using the other known measures such as activity and LBD-based ones. Indeed, the LBD value of a given clause is definitely set at the time of its generation and does not change during search, while the activity (VSIDS-based) measure is dynamic but can only be used to update the activity of learnt clauses currently in the database.

Let us now formally describe our new learnt clause management policy. First, as the *psm* value of a given clause is highly dynamic, we introduce a notion of deviation between two successive sets of progress saving polarities. Let V_{t_k} be the set of variables assigned by the solver between two consecutive time sequences (as defined in previous section) t_{k-1} and t_k. The deviation d_{t_k} is defined as follows: $d_{t_k} = \frac{h(\mathcal{P}_{t_k}, \mathcal{P}_{t_{k-1}})}{|V_{t_k}|}$, where h is the usual hamming distance.

This deviation defined as a normalized hamming distance, gives us an outline of the evolution of progress saving polarities between two successive cleanings of the learnt database. A deviation tending to zero indicates that the solver explores around the same part of the search space whereas a value close to one indicates that the solver explores different part of the search space.

To obtain a more precise view of the search behavior, we introduce another notion of minimal deviation $d_{t_k}^m = min\{d_{t_i} | 0 \leq i \leq k\}$ at time step t_k.

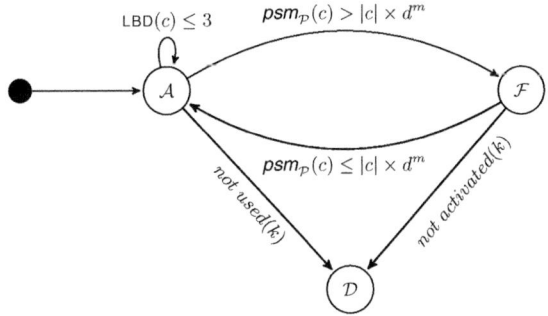

Fig. 2. State diagram of a learnt clause

Using this minimal deviation, we can now refine our *psm* measure. Indeed, let c be a clause to be evaluated at time step t_k, if $psm_{\mathcal{P}_{t_k}}(c) > d^m_{t_k} \times |c|$ then the clause c is likely to be satisfied in a near future, otherwise it is likely to be involved in the propagation process.

Our approach depicted in Figure 2 is represented as a state diagram. At each cleaning t_k, learnt clauses can move from a state to another one following some conditions.

First, a learnt clause c can be in one of the three following states:

1. *Active state $\mathcal{A}$*: c is active and watched.
2. *Frozen state $\mathcal{F}$*: c is frozen i.e. c is not watched
3. *Dead state $\mathcal{D}$*: c is deleted.

Let us describe these different transitions:

- Each time a clause is learnt it enters the state $\mathcal{A}$.
- A clause $c \in \mathcal{A}$ with a short LBD ($lbd(c) \le 3$ in the figure) remains in the state $\mathcal{A}$ until the end of the search process.
- A clause $c \in \mathcal{A}$ such that $\dfrac{psm_{\mathcal{P}_{t_k}}(c)}{|c|} > d^m_{t_k}$ enters the frozen state $\mathcal{F}$.
- A clause $c \in \mathcal{F}$ such that $\dfrac{psm_{\mathcal{S}_{n_i}}(c)}{|c|} \le d^m_{n_i}$ enters the active state $\mathcal{A}$.
- A clause $c \in \mathcal{F}$ not activated after k time steps is deleted. Similarly, a clause $c \in \mathcal{A}$ remaining active more than k steps without participating to the search is also deleted. In both cases, it enters the state $\mathcal{D}$ after $k = 7$ time steps in our experiments.

One of the main advantages of our approach comes from the fact that we can perform frequent cleaning of the learnt clause database without taking care of removing relevant clauses. So we choose a very aggressive policy. We set $t_0 = 500$ conflicts, and $t_k = t_{k-1} + 100$ conflicts.

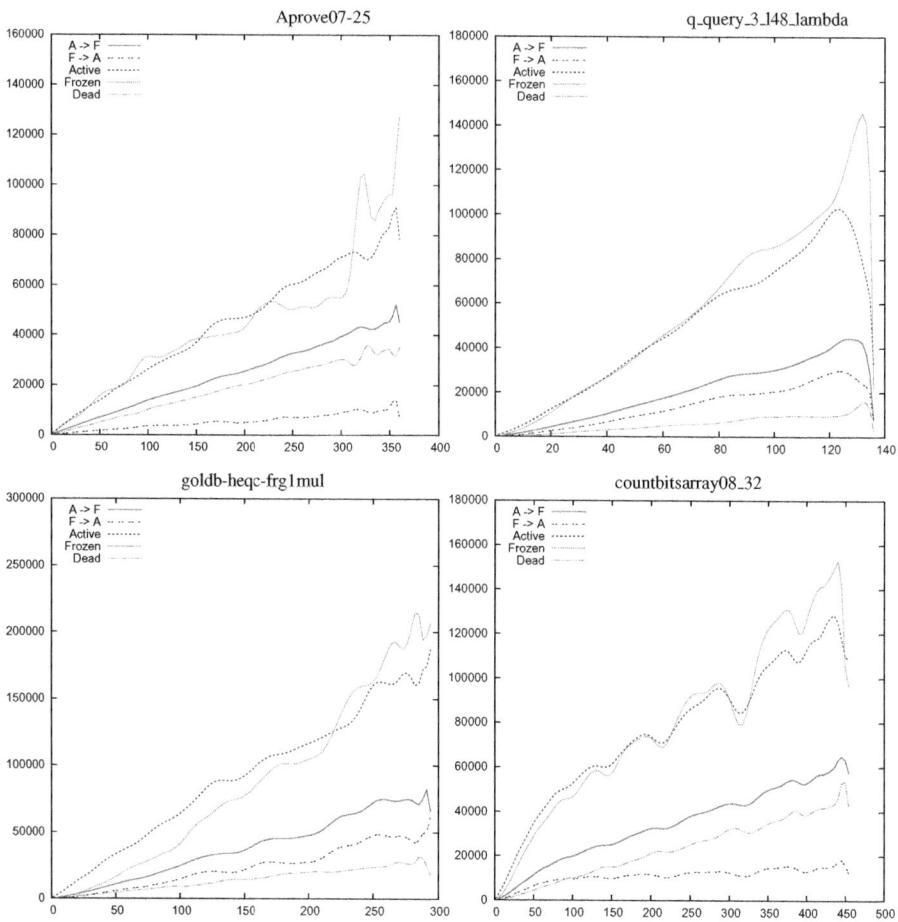

Fig. 3. Evolution of the number of clauses in different states and number of state transfers

We conducted some experiments to analyse the transfer of the clauses from the state $\mathcal{A}$ to the state $\mathcal{F}$ and *vice versa*. Figure 3 shows, for the same sample of instances as in the Figure 1, the number of deleted clauses, the number of transitions to the frozen state, the number of transitions to the active state, the number of active (or watched) learnt clauses and finally the number of frozen clauses. These data are represented by the y-axis, whereas the x-axis represents the cleaning operated at the time step t_k. For clarity reasons, all curves have been smoothed. For all instances, the number of frozen clauses ($Frozen$) and the number of active clauses ($Active$) are relatively similar. The curve representing the number of clauses becoming active ($\mathcal{F} \rightarrow \mathcal{A}$) is dominated by those representing the number of clauses becoming frozen ($\mathcal{A} \rightarrow \mathcal{F}$). However, the two curves evolve similarly and they are closer on some instances (e.g. $q\_query\_3\_l48\_lambda$) than on others (e.g. $Aprove07 - 25$) . Finally, we can also observe that, at each cleaning time step, some clauses are definitively deleted ($Dead$).

5   Empirical Evaluation

This section is divided in two parts. In the first, we compare our dynamic management policy (psm_{dym}) against the classical reduction approach with different quality measures (LBD, VSIDS like, *psm*). In the second experiment, we compare it with three state-of-the-art solvers: GLUCOSE which embeds LBD measure, a dynamic restart policy and some other features [2], LINGELING which also embeds more powerful reasoning like blocked clause elimination [8], and finally, CRYPTOMINISAT which adds many other features (e.g. vivification, reasoning on xor clauses...). Descriptions of these solvers are available on the SATRACE 2010 website http://baldur.iti.uka.de/sat-race-2010. Except for LINGELING and CRYPTOMINISAT which embed preprocessing inside, the other solvers use SatElite for preprocessing [4].

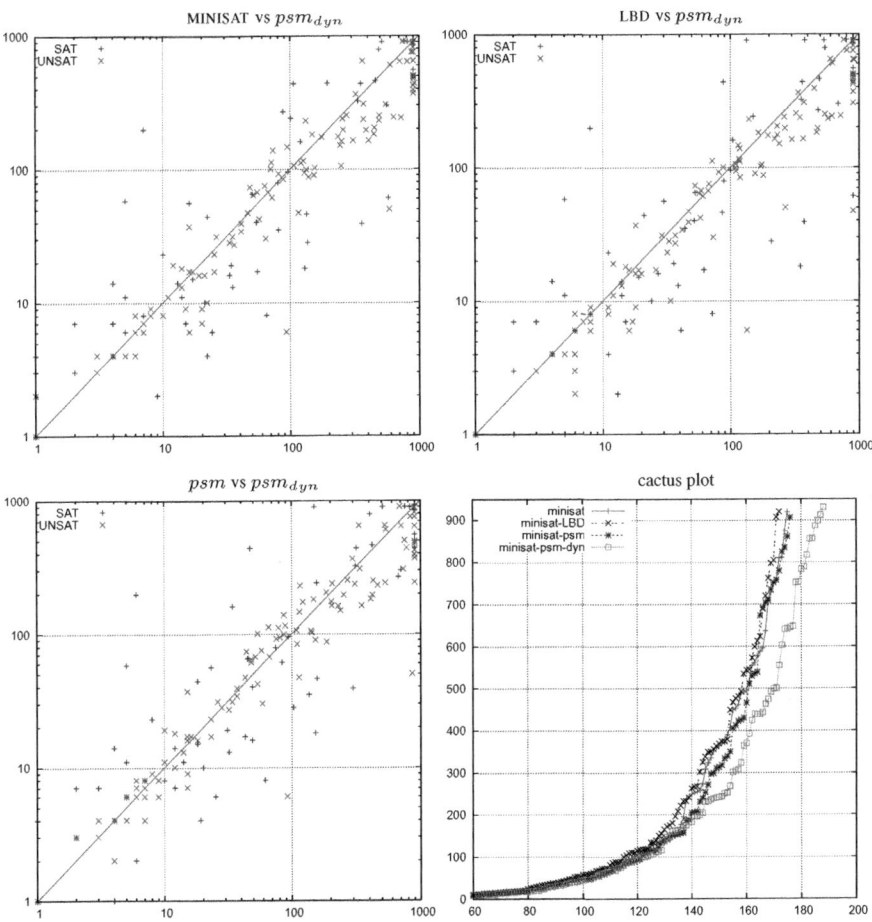

Fig. 4. Comparison with different learnt clauses quality measures

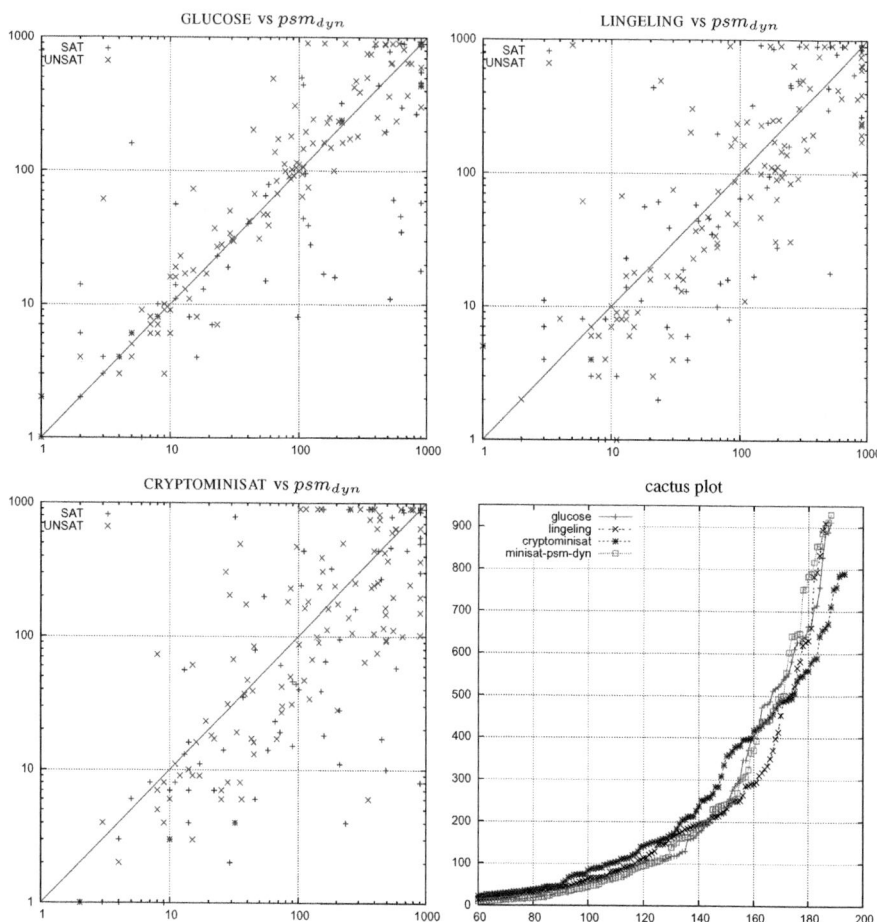

Fig. 5. Comparison with state of the art solvers: GLUCOSE, LINGELING and CRYPTOMINISAT

In the first experiment, we use the same solver and the only difference is in the learnt clause management policy. In the second experiment, our aim is to compare our learnt clause management approach integrated in MINISAT 2.2 (MINISAT-psm_{dyn}) with the state-of-the-art SAT solvers. Source code and extensive experiments can be found at http://www.cril.fr/~lagniez/ressource.html.

5.1   Comparison with Different Quality Measures

We compare our dynamic policy, called MINISAT-psm_{dyn} with the classic MINISAT, and MINISAT with learnt database reduction based on *psm* (MINISAT-*psm*) and on LBD (MINISAT-LBD) (like in section 3). Figure 4 summarizes the results. It contains three scatter plots corresponding to the comparison of MINISAT-psm_{dyn} with the 3 others solvers. In such a plot, each dot corresponds to a given instance, the x-axis corresponds to the cpu time needed by the MINISAT, LBD or *psm* to solve the instance, whereas the-y axis corresponds to the cpu time needed by psm_{dyn} to solve it. So, dots below the diagonal correspond to instances solved faster by MINISAT-psm_{dyn} (SAT and UNSAT instances are differentiated). Figure 4 also contains a cactus plot related to the comparison of the 4 solvers.

It is quite clear that our freezing strategy outperforms the other strategies. It solves 189 instances (76 SAT and 113 UNSAT), which is significantly better than the other solvers (see Table 1). Furthermore, as we can see on the scatter plots, MINISAT-psm_{dyn} solves instances faster than the others solvers.

5.2 Comparison with State of the Art Solvers

Figure 5 summarizes the comparison with state of the art solvers. It is structured as figure 4. Let us detail the number of solved instances by each solver: LINGELING solves 187 instances (77 SAT, 110 UNSAT), GLUCOSE 189 (70 SAT and 119 UNSAT) and CRYP-TOMINISAT 194 (74 SAT, 120 UNSAT). These results and the plots of Figure 5 show that our dynamic management policy is really competitive with state-of-the-art solvers (remember, it solves 189 instances (76 SAT and 113 UNSAT)). It does not even embed sophisticated components such as dynamic restart, etc.

6 Conclusion

In this paper, we introduced a new measure for identifying relevant learnt clauses. The main advantage of this measure is that it is dynamic (unlike the LBD measure) and it can be computed even if clauses do not participate in the search process (unlike the VSIDS like measure). Thanks to this property, a new learnt clause database management framework has been proposed. It exploits a novel dynamic policy that activates the most promising learnt clauses while freezing irrelevant ones. This is in contrast with all the well-known deletion strategies, where a given learned clause is definitely eliminated. Experiments on SAT instances taken from the last competitions demonstrate the effectiveness of our approach.

As future work, we plan to exploit the evolution of the set of progress saving literal polarities in order to decide if cleaning has to be performed. Considering the connection between restarts and clause learning [14], we plan to exploit this connection to improve our proposed leant database management approach.

Acknowledgments. We would like to thank the anonymous reviewers for insightful comments.

References

1. Audemard, G., Bordeaux, L., Hamadi, Y., Jabbour, S., Saïs, L.: A Generalized Framework for Conflitcs Analysis. Technical Report MSR-TR-2008-34, Microsoft Research (2008)
2. Audemard, G., Simon, L.: Predicting learnt clauses quality in modern SAT solvers. In: Proceedings of IJCAI, pp. 399–404 (2009)
3. Beame, P., Kautz, H., Sabharwal, A.: Towards understanding and harnessing the potential of clause learning. Journal of Artificial Intelligence Research 22, 319–351 (2004)

4. Eén, N., Biere, A.: Effective preprocessing in SAT through variable and clause elimination. In: Bacchus, F., Walsh, T. (eds.) SAT 2005. LNCS, vol. 3569, pp. 61–75. Springer, Heidelberg (2005)

5. Eén, N., Sörensson, N.: An extensible SAT-solver. In: Giunchiglia, E., Tacchella, A. (eds.) SAT 2003. LNCS, vol. 2919, pp. 502–518. Springer, Heidelberg (2004)

6. Gomes, C., Selman, B., Kautz, H.: Boosting combinatorial search through randomization. In: Proceedings of AAAI, pp. 431–437 (1998)

7. Huang, J.: The effect of restarts on the efficiency of clause learning. In: Proceedings of IJCAI, pp. 2318–2323 (2007)

8. Järvisalo, M., Biere, A., Heule, M.: Blocked clause elimination. In: Esparza, J., Majumdar, R. (eds.) TACAS 2010. LNCS, vol. 6015, pp. 129–144. Springer, Heidelberg (2010)

9. Bayardo Jr., R.J., Schrag, R.: Using csp look-back techniques to solve real-world sat instances. In: Proceedings of AAAI, pp. 203–208 (1997)

10. Marques-Silva, J., Sakallah, K.: GRASP - A New Search Algorithm for Satisfiability. In: Proceedings of ICCAD, pp. 220–227 (1996)

11. Moskewicz, M., Madigan, C., Zhao, Y., Zhang, L., Malik, S.: Chaff: Engineering an efficient SAT solver. In: Proceedings of DAC, pp. 530–535 (2001)

12. Pipatsrisawat, K., Darwiche, A.: A lightweight component caching scheme for satisfiability solvers. In: Marques-Silva, J., Sakallah, K.A. (eds.) SAT 2007. LNCS, vol. 4501, pp. 294–299. Springer, Heidelberg (2007)

13. Pipatsrisawat, K., Darwiche, A.: On the power of clause-learning SAT solvers with restarts. In: Gent, I.P. (ed.) CP 2009. LNCS, vol. 5732, pp. 654–668. Springer, Heidelberg (2009)

14. Pipatsrisawat, K., Darwiche, A.: Width-based restart policies for clause-learning satisfiability solvers. In: Kullmann, O. (ed.) SAT 2009. LNCS, vol. 5584, pp. 341–355. Springer, Heidelberg (2009)

15. Zhang, L., Madigan, C., Moskewicz, M., Malik, S.: Efficient conflict driven learning in boolean satisfiability solver. In: Proceedings of ICCAD, pp. 279–285 (2001)

Efficient CNF Simplification Based on Binary Implication Graphs[*]

Marijn J.H. Heule[1], Matti Järvisalo[2], and Armin Biere[3]

[1] Department of Software Technology, Delft University of Technology, The Netherlands
[2] Department of Computer Science, University of Helsinki, Finland
[3] Institute for Formal Models and Verification, Johannes Kepler University Linz, Austria

Abstract. This paper develops techniques for efficiently detecting redundancies in CNF formulas. We introduce the concept of *hidden literals*, resulting in the novel technique of *hidden literal elimination*. We develop a practical simplification algorithm that enables "*Unhiding*" various redundancies in a unified framework. Based on time stamping literals in the binary implication graph, the algorithm applies various binary clause based simplifications, including techniques that, when run repeatedly until fixpoint, can be too costly. *Unhiding* can also be applied during search, taking learnt clauses into account. We show that *Unhiding* gives performance improvements on real-world SAT competition benchmarks.

1 Introduction

Applying reasoning techniques (see e.g. [1,2,3,4,5,6,7]) to simplify Boolean satisfiability (SAT) instances both before and during search is important for improving state-of-the-art SAT solvers. This paper develops techniques for efficiently detecting and removing redundancies from CNF (conjunctive normal form) formulas based on the underlying *binary clause structure* (i.e., the binary implication graph) of the formulas.

In addition to considering known simplification techniques (hidden tautology elimination (HTE) [6], hyper binary resolution (HBR) [1,7], failed literal elimination over binary clauses [8], equivalent literal substitution [8,9,10], and transitive reduction [11] of the binary implication graph [10]), we introduce the novel technique of *hidden literal elimination* (HLE) that removes so-called *hidden literals* from clauses without affecting the set of satisfying assignments. We establish basic properties of HLE, including conditions for achieving confluence when combined with equivalent literal substitution.

As the second main contribution, we develop an efficient and practical simplification algorithm that enables "*Unhiding*" various redundancies in a unified framework. Based on time stamping literals via randomized depth-first search (DFS) over the binary implication graph, the algorithm provides efficient approximations of various binary clause based simplifications which, when run repeatedly until fixpoint, can be too costly. In particular, while our *Unhiding* algorithm is linear time in the total number of literals (with an at most logarithmic factor in the length of the longest clause), notice as an example that fixpoint computation of failed literals, even just on the binary implication

---

[*] The 1st author is financially supported by Dutch Organization for Scientific Research (grant 617.023.611), the 2nd author by Academy of Finland (grant 132812) and the 1st and 3rd author are supported by the Austrian Science Foundation (FWF) NFN Grant S11408-N23 (RiSE).

K.A. Sakallah and L. Simon (Eds.): SAT 2011, LNCS 6695, pp. 201–215, 2011.

graph, is conjectured to be at least quadratic in the worst case [8]. *Unhiding* can be implemented without occurrence lists, and can hence be applied not only as a preprocessor but also *during search*, which allows to take learnt clauses into account. Indeed, we show that, when integrated into the state-of-the-art SAT solver Lingeling [12], *Unhiding* gives performance improvements on real-world SAT competition benchmarks.

On related work, Van Gelder [8] studied exact and approximate DFS-based algorithms for computing equivalent literals, failed literals over binary clauses, and implied (transitive) binary clauses. The main differences to this work are: (i) *Unhiding* approximates the additional techniques of HTE, HLE, and HBR; (ii) the advanced DFS-based time stamping scheme of *Unhiding* detects failed and equivalent literals *on-the-fly*, in addition to *removing* (instead of adding as in [8]) transitive edges in the binary implication graph; and (iii) *Unhiding* is integrated into a clause learning (CDCL) solver, improving its performance on real application instances (in [8] only random 2-SAT instances were considered). Our advanced stamping scheme can be seen as an extension of the BinSATSCC-1 algorithm in [13] which excludes (in addition to cases (i) and (iii)) transitive reduction. Furthermore, while [13] focuses on simplifing the binary implication graph, we use reachability information obtained from traversing it to simplify larger clauses, including learnt clauses, in addition to extracting failed literals.

As for more recent developments, CryptoMiniSAT v2.9.0 [14] caches implied literals, and updates the cache after top-level decisions. The cache can serve a similar purpose as our algorithms, removing literals and clauses. Yet, the cache size is quadratic in the number of literals, which is also the case for using the cache for redundancy removal for the whole CNF. Thus, at least from a complexity point of view, the cache of CryptoMiniSAT does not improve on the quadratic algorithm [8]. In contrast, *Unhiding* requires only a single sweep over the binary implication graph and the other clauses.

After preliminaries (CNF satisfiability and known CNF simplification techniques, Sect. 2), we introduce hidden literal elimination and establish its basic properties (Sect. 3). We then explain the *Unhiding* algorithm: basic idea (Sect. 4) and integration of simplification techniques (Sect. 5). Then we develop an advanced version of *Unhiding* that can detect further redundancies (Sect. 6), and present experimental results (Sect. 7).

2   Preliminaries

For a Boolean variable x, there are two *literals*, the positive literal x and the negative literal $\bar{x}$. A *clause* is a disjunction of literals and a CNF formula a conjunction of clauses. A clause can be seen as a finite set of literals and a CNF formula as a finite set of clauses. A truth assignment for a CNF formula F is a function τ that maps literals in F to $\{0, 1\}$. If $\tau(x) = v$, then $\tau(\bar{x}) = 1 - v$. A clause C is satisfied by τ if $\tau(l) = 1$ for some literal $l \in C$. An assignment τ satisfies F if it satisfies every clause in F.

Two formulas are *logically equivalent* if they are satisfied by exactly the same set of assignments. A clause is a *tautology* if it contains both x and $\bar{x}$ for some variable x. The length of a clause is the number of literals in the clause. A clause of length one is a *unit clause*, and a clause of length two is a *binary clause*. For a CNF formula F, we denote the set of binary clauses in F by F_2.

Binary Implication Graphs. For any CNF formula F, we associate a unique directed *binary implication graph* $\mathsf{BIG}(F)$ with the edge relation $\{\langle \bar{l}, l' \rangle, \langle \bar{l'}, l \rangle \mid (l \vee l') \in F_2\}$. In other words, for each binary clause $(l \vee l')$ in F, the two implications $\bar{l} \rightarrow l'$ and $\bar{l'} \rightarrow l$, represented by the binary clause, occur as edges in $\mathsf{BIG}(F)$. A node in $\mathsf{BIG}(F)$ with no incoming arcs is a *root* of $\mathsf{BIG}(F)$ (or, simply, of F_2). In other words, literal l is a root in $\mathsf{BIG}(F)$ if there is no clause of the form $(l \vee l')$ in F_2. The set of roots of $\mathsf{BIG}(F)$ is denoted by $\mathsf{RTS}(F)$.

2.1   Known Simplification Techniques

BCP and Failed Literal Elimination (FLE). For a CNF formula F, *Boolean constraint propagation* (BCP) (or *unit propagation*) propagates all unit clauses, i.e. repeats the following until fixpoint: if there is a unit clause $(l) \in F$, remove from $F \setminus \{(l)\}$ all clauses that contain the literal l, and remove the literal $\bar{l}$ from all clauses in F, resulting in the formula $\mathsf{BCP}(F)$. A literal l is a *failed literal* if $\mathsf{BCP}(F \cup \{(l)\})$ contains the empty clause, implying that F is logically equivalent to $\mathsf{BCP}(F \cup \{(\bar{l})\})$. FLE removes failed literals from a formula, or, equivalently, adds the complements of failed literals as unit clauses to the formula.

Equivalent Literal Substitution (ELS). The strongly connected components (SCCs) of $\mathsf{BIG}(F)$ describe equivalent classes of literals (or simply equivalent literals) in F_2. *Equivalent literal substitution* refers to substituting in F, for each SCC G of $\mathsf{BIG}(F)$, all occurrences of the literals occurring in G with the representative literal of G. ELS is confluent, i.e., has a unique fixpoint, modulo variable renaming.

Hidden Tautology Elimination (HTE). [6] For a given CNF formula F and clause C, (*hidden literal addition*) $\mathsf{HLA}(F, C)$ is the *unique* clause resulting from repeating the following clause extension steps until fixpoint: if there is a literal $l_0 \in C$ such that there is a clause $(l_0 \vee l) \in F_2 \setminus \{C\}$ for some literal l, let $C := C \cup \{\bar{l}\}$. Note that $\mathsf{HLA}(F, C) = \mathsf{HLA}(F_2, C)$. Further, for any $l \in \mathsf{HLA}(F, C) \setminus C$, there is a path in $\mathsf{BIG}(F)$ from l to some $l_0 \in C$. For any CNF formula F and clause $C \in F$, $(F \setminus \{C\}) \cup \{\mathsf{HLA}(F, C)\}$ is logically equivalent to F [6]. Intuitively, each extension step in computing HLA is an application of self-subsuming resolution [2,15,16] in reverse order. For a given CNF formula F, a clause $C \in F$ is a *hidden tautology* if and only if $\mathsf{HLA}(F, C)$ is a tautology. *Hidden tautology elimination* removes hidden tautologies from CNF formulas. Note that *distillation* [4] is more generic than HTE [6] (and also more generic than HLE as defined in this paper). However, it is rather costly to apply, and is in practice restricted to irredundant/original clauses only.

Transitive reduction of the binary implication graph (TRD). A directed acyclic graph G' is a *transitive reduction* [11] of the directed graph G provided that (i) G' has a directed path from node u to node v if and only if G has a directed path from node u to node v, and (ii) there is no graph with fewer edges than G' satisfying condition (i). It is interesting to notice that, by applying FLE restricted to the literals in F_2 before HTE, HTE achieves a transitive reduction of $\mathsf{BIG}(F)$ for any CNF formula F purely on the clausal level [6].

3   Hidden Literal Elimination

In this section we present a novel redundancy elimination procedure exploiting the binary clause structure of a CNF formula. We call the technique *hidden literal elimination*.

For a given CNF formula F and literal l, we denote by $\mathrm{HL}(F, l)$ the *unique* set of *hidden literals* of l w.r.t F. $\mathrm{HL}(F, l)$ is defined as follows. First, let $L = \{l\}$. Then repeat the following steps until fixpoint: if there is a literal $l_0 \in L$ such that there is a clause $(l_0 \vee l') \in F_2$ for some literal l', let $L := L \cup \{\bar{l}'\}$. Now, let $\mathrm{HL}(F, l) := L \setminus \{l\}$. In other words, $\mathrm{HL}(F, l)$ contains the complements of all literals that are reachable from $\bar{l}$ in $\mathrm{BIG}(F)$, or, equivalently, all literals from which l is reachable in $\mathrm{BIG}(F)$. Notice that $\mathrm{HL}(F, l) = \mathrm{HL}(F_2, l)$. Also, HL captures failed literals in F_2 in the sense that by definition, for any literal l in F_2, there is a path from l to $\bar{l}$ in $\mathrm{BIG}(F)$ if and only if $\bar{l} \in \mathrm{HL}(F, l)$.

Proposition 1. *For any CNF formula F, a literal l in F_2 is failed iff $\bar{l} \in \mathrm{HL}(F, l)$.*

For a given formula F, *hidden literal elimination* (HLE) repeats the following: if there is a clause $C \in F$ and a literal $l \in C$ such that $C \cap \mathrm{HL}(F, l) \neq \emptyset$, let $F := (F \setminus \{C\}) \cup \{C \setminus \mathrm{HL}(F, l)\}$. In fact, the literals in $\mathrm{HL}(F, l)$ can be removed from all clauses that contain l.

Proposition 2. *For every CNF formula F, any result of applying HLE on F is logically equivalent to F.*

Proof. For any CNF formula F and two literals l and l', if $l' \in \mathrm{HL}(F, l)$, then $F \cup \{(l')\}$ logically implies l by the definition of HL. Hence, for any clause $C \in F$ with $l, l' \in C$, for any satisfying assignment τ for F with $\tau(l') = 1$ we have $\tau(l) = 1$, and hence τ satisfies $(F \setminus \{C\}) \cup \{C \setminus \mathrm{HL}(F, l)\}$. □

A relevant question is how many literals HLE eliminates relative to other literal elimination techniques. One example is self-subsuming resolution (SSR) [2] that replaces clauses that have a resolvent that subsumes the clause itself with the resolvent (essentially eliminating from the clause the literal not in the resolvent).

Proposition 3. *There are CNF formulas from which HLE can remove more literals from clauses than SSR.*

Proof. Consider the formula $F = (a \vee b) \wedge (\bar{b} \vee c) \wedge (a \vee \bar{c} \vee d)$. Since $\mathrm{HL}(F, a) = \{\bar{b}, \bar{c}\}$, HLE can remove literal $\bar{c}$ from the last clause in contrast to SSR. □

HLE can also strengthen formulas by increasing possibilities for unit propagation.

Proposition 4. *Removal of hidden literals can increase BCP.*

Proof. Consider the formula $F = (a \vee b) \wedge (\bar{b} \vee c) \wedge (a \vee \bar{c} \vee d)$. Since $\mathrm{HL}(F, a) = \{\bar{b}, \bar{c}\}$, HLE removes literal $\bar{c}$ from the last clause. When d is assigned to 0 after eliminating literal $\bar{c}$, BCP will infer a. □

In general, HLE does not have a unique fixpoint.

Proposition 5. *Applying* HLE *until fixpoint is not confluent.*

Proof. Consider the formula $F = (a \vee b) \wedge (\bar{a} \vee \bar{b}) \wedge (a \vee \bar{b} \vee c)$. Since $\text{HL}(F, a) = \{\bar{b}\}$ and $\text{HL}(F, \bar{b}) = \{a\}$, HLE can remove either $\bar{b}$ or a from $(a \vee \bar{b} \vee c)$. A fixpoint is reached after removing one of these two literals.   □

In the example the non-confluence is due to a and $\bar{b}$ being equivalent literals. In fact, assume that all clauses in F_2 are kept even in the case HLE turns a binary clause into a unit clause (i.e., in such cases HLE will introduce new unit clauses into F). Then HLE can be made confluent (modulo variable renaming) by substituting equivalent literals.

Theorem 1. *For any CNF formula F, assuming that all clauses in the original F_2 are kept, alternating* ELS *and* HLE *(until fixpoint) until fixpoint is confluent modulo variable renaming.*

Proof sketch. ELS is confluent modulo variable renaming. Now consider HLE. Assume that we do not change F_2. Take any clause C with $l, l' \in C$ and $l' \in \text{HL}(F, l)$. The only possible source of non-confluence is that $l \in \text{HL}(F, l')$. Then there is a cycle in F_2, and hence l and $\bar{l}'$ are equivalent literals. This is handled by ELS afterwards. Now assume a binary clause is added to F_2 by HLE shortening a clause of length > 2. Newly produced cycles are handled by ELS afterwards.   □

4   Unhiding Redundancies Based on Time Stamping

In this section we present an efficient algorithm for detecting several kinds of redundancies in CNF formulas, focusing on techniques which exploit binary clauses.

For a given CNF formula F, our algorithm, referred to as *Unhiding* (see Fig. 1, details explained in the following), consists in essence of two phases. First, a depth-first search (DFS) over the binary implication graph $\text{BIG}(F)$ is performed. During the DFS, each literal in $\text{BIG}(F)$ is assigned a time stamp; we call this process *time stamping*. In the second phase, these time stamps are used for discovering the various kinds of redundancies in F, which are then removed.

In the following, we will first describe a *basic time stamping procedure* (Sect. 4.1). Then we will show how redundancies can be detected and eliminated based on the time stamps (Sect. 5). After these, in Sect. 6 we describe a more *advanced time stamping procedure* that embeds additional simplifications that are captured *during* the actual depth-first traversal of $\text{BIG}(F)$.

4.1   Basic Time Stamping

The basic time stamping procedure implements a depth-first search on the binary implication graph $\text{BIG}(F)$ of a given CNF formula F. The procedure associates a *discovered-finished interval* (or a *time stamp*) with each literal in $\text{BIG}(F)$ according to the depth-first traversal order. For any depth-first traversal of a graph G, a node in G is *discovered* (resp. *finished*) the first (resp. last) time it is encountered during search. For a given depth-first traversal, the *discovery* and *finish times* of a node v in G, denoted by $\text{dsc}(v)$ and $\text{fin}(v)$, respectively, are defined as the number of steps taken at the time of discovering and finishing, respectively, the node v. The important observation here is that, according to the well-known "parenthesis theorem", for two nodes u and v with discovered-finished intervals $[\mathrm{dsc}(u), \mathrm{fin}(u)]$ and $[\mathrm{dsc}(v), \mathrm{fin}(v)]$, respectively, we know that v is a descendant of u in the DFS tree if and only if $\mathrm{dsc}(u) < \mathrm{dsc}(v)$ and $\mathrm{fin}(u) > \mathrm{fin}(v)$, i.e., if the time stamp (interval) of u *contains* the time stamp (interval) of v. These conditions can be checked in constant time given the time stamps.

Pseudo-code for the main unhiding procedure *Unhiding* and the time stamping procedure *Stamp* is presented in Fig. 1. The main procedure *Unhiding* (left) initializes the attributes and calls the recursive stamping procedure (right) for each root in $\mathrm{BIG}(F)$ in a random order. When there are no more roots, we pick a literal not visited yet as the next starting point until all literals have been visited.[1] *Stamp* performs a DFS in $\mathrm{BIG}(F)$ from the given starting literal, assigns for each literal l encountered the discovery and finish times $\mathrm{dsc}(l)$ and $\mathrm{fin}(l)$ according to the traversal order, updates *stamp* (initially 0), and for each literal l, defines its DFS parent $\mathrm{prt}(l)$ and the root $\mathrm{root}(l)$ of the DFS tree in which l was discovered.

In the following, we say that a given time stamping *represents the implication* $l \to l'$ if the time stamp of l contains the time stamp of l'.

	Unhiding (formula F)		*Stamp* (literal l, integer *stamp*)
1	$stamp := 0$	1	$stamp := stamp + 1$
2	**foreach** literal l in $\mathrm{BIG}(F)$ **do**	2	$\mathrm{dsc}(l) := stamp$
3	$\quad \mathrm{dsc}(l) := 0; \mathrm{fin}(l) := 0$	3	**foreach** $(\bar{l} \vee l') \in F_2$ **do**
4	$\quad \mathrm{prt}(l) := l; \mathrm{root}(l) := l$	4	$\quad$ **if** $\mathrm{dsc}(l') = 0$ **then**
5	**foreach** $r \in \mathrm{RTS}(F)$ **do**	5	$\quad\quad \mathrm{prt}(l') := l$
6	$\quad stamp := \textit{Stamp}(r, stamp)$	6	$\quad\quad \mathrm{root}(l') := \mathrm{root}(l)$
7	**foreach** literal l in $\mathrm{BIG}(F)$ **do**	7	$\quad\quad stamp := \textit{Stamp}(l', stamp)$
8	$\quad$ **if** $\mathrm{dsc}(l) = 0$ **then**	8	$stamp := stamp + 1$
9	$\quad\quad stamp := \textit{Stamp}(l, stamp)$	9	$\mathrm{fin}(l) := stamp$
10	**return** $\textit{Simplify}(F)$	10	**return** $stamp$

Fig. 1. The *Unhiding* algorithm. Left: the main procedure. Right: the basic stamping procedure.

Example 1. Consider the formula

$$E = (\bar{a} \vee c) \wedge (\bar{a} \vee d) \wedge (\bar{b} \vee d) \wedge (\bar{b} \vee e) \wedge (\bar{c} \vee f) \wedge (\bar{d} \vee f) \wedge (\bar{f} \vee h) \wedge (\bar{g} \vee f) \wedge$$
$$(\bar{g} \vee h) \wedge (\bar{a} \vee \bar{e} \vee h) \wedge (\bar{b} \vee \bar{c} \vee h) \wedge (a \vee b \vee c \vee d \vee e \vee f \vee g \vee h).$$

The formula contains several redundant clauses and literals. The clauses $(\bar{a} \vee \bar{e} \vee h)$, $(\bar{g} \vee h)$, and $(\bar{b} \vee \bar{c} \vee h)$ are hidden tautologies. In the last clause, all literals except e and h are hidden. The binary implication graph $\mathrm{BIG}(E)$ of E, as shown in Fig. 2, consists of two components. A partition of $\mathrm{BIG}(E)$ produced by the basic time stamping procedure is shown in Fig. 3. The nodes are visited in the following order: g, f, h, $\bar{e}$, $\bar{b}$, b, e, d, $\bar{h}$, $\bar{g}$, $\bar{f}$, $\bar{d}$, $\bar{a}$, $\bar{c}$, a, c. $\mathrm{BIG}(E)$ consists of 30 implications including the transitive ones. However, the trees and time stamps in the figure explicitly represent only 16 of them, again including transitive edges such as $\bar{h} \to \bar{a}$. The implications $b \to f$, $\bar{f} \to \bar{b}$, $b \to h$,

---

[1] Thus, BIG need not be acyclic. Note that eliminating cycles in BIG by substituting variables might shorten longer clauses to binary clauses, which in turn could introduce new cycles. This process cannot be bounded to be linear and is not necessary for our algorithms.

and $\bar{h} \to \bar{b}$ are not represented by this time stamping. Note that the implication $\bar{f} \to \bar{c}$ is represented, and thus implicitly $c \to f$ as well. Using contraposition this way the four transitive edges mentioned above are not represented, the other 26 edges are.

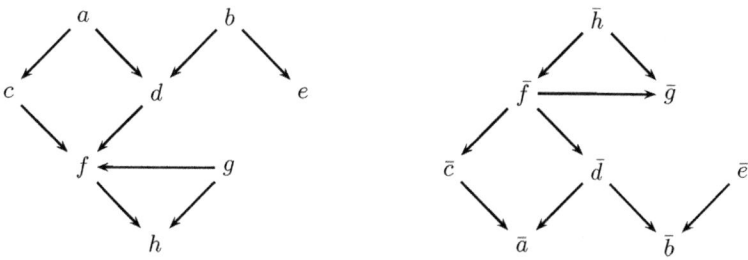

Fig. 2. BIG(E). The graph has five root nodes: a, b, $\bar{e}$, g, and $\bar{h}$.

The order in which the trees are traversed has a big impact on the quality, i.e. the fraction of implications that are represented by the time stamps. The example shows that randomized stamping may not represent all implications in BIG. Yet, for this formula, there is a DFS order that produces a stamping that represents *all* implications: start from the root $\bar{h}$ and stamp the tree starting with literal $\bar{f}$. Then, by selecting a as the root of the second tree, regardless of the order of the other roots and literals, the time stamps produced by stamping will represent all implications.   ■

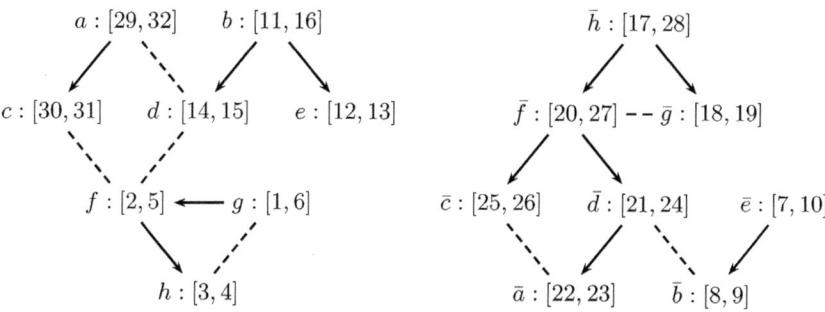

Fig. 3. A partition of BIG(E) into a forest with discovered-finished intervals [dsc(v), fin(v)] assigned by the basic time stamping routine. Dashed lines represent implications in BIG(E) which are not used to set the time stamps.

5   Capturing Various Simplifications

We now explain how one can remove hidden literals and hidden tautologies, and furthermore perform hyper binary resolution steps based on a forest over the time stamped literal nodes produced by the main DFS procedure. The main procedure *Simplify* for this second phase, called by the main *Unhiding* procedure after time stamping, is shown in Fig. 4. For each clause C in the input CNF formula F, *Simplify* removes C from F.

Simplify (formula F)

1   **foreach** $C \in F$
2       $F := F \setminus \{C\}$
3   **if** *UHTE*(C) **then continue**
4       $F := F \cup \{UHLE(C)\}$
5   **return** F

Fig. 4. Procedure for applying HTE and HLE based on time stamps

Then, it first checks whether the *UHTE* procedure detects that C is a hidden tautology. If not, literals are (possibly) eliminated from C by the *UHLE* procedure (using hidden literal elimination). The resulting clause is added to F.

Notice that the simplification procedure visits each clause $C \in F$ only once. The invoked sub-procedures, *UHTE* and *UHLE*, exploit the time stamps, and use two sorted lists: (i) $S^+(C)$, list of the literals in C sorted according to increasing discovery time, and (ii) $S^-(C)$, list of the complements of the literals in C, sorted according to increasing discovery time. We will now explain both of these sub-procedures in detail.

5.1 Hidden Literals

Once literals are stamped using the unhiding algorithm, one can cheaply detect (possibly a subset of) hidden literals. In this context, literal $l \in C$ is hidden if there is (i) an implication $l \to l'$ with $l' \in C$ that is represented by the time stamping, or (ii) an implication $\bar{l}' \to \bar{l}$ with $l' \in C$ that is represented by the time stamping.

We check for such implications as follows using the *UHLE* procedure shown in Fig. 5. For each input clause C, the procedure returns a subset of C with some hidden literals removed from C. For this procedure, we use $S^+(C)$ in reverse order, denoted by $S^+_{rev}(C)$. In essence, we go through the lists $S^+_{rev}(C)$ and $S^-(C)$, and compare the finish times of two successive elements in the lists. In case an implication is found, a hidden literal is detected and removed.

Lines 1-4 in Fig. 5 detect implications of the form $l \to l'$ with $l, l' \in C$ that are represented by the time stamping. Recall that in $S^+_{rev}(C)$ literals are ordered with decreasing discovering time. Let l' be located before l in $S^+_{rev}(C)$. If fin(l) > fin(l') we found the implication $l \to l'$, and hence l is a hidden literal (in the code *finished* = fin(l')). Line 3 checks whether the next element in $S^+_{rev}(C)$ is a hidden literal, and if so, the literal is removed. Lines 5-8 detect implications $\bar{l}' \to \bar{l}$ with $l, l' \in C$. In $S^-(C)$ literals are ordered with increasing discovering time. Now, $\bar{l}'$ be located before $\bar{l}$ in $S^-(C)$ and *finished* = fin($\bar{l}'$). On Line 7 we check that fin($\bar{l}$) < fin($\bar{l}'$) or, equivalently, fin($\bar{l}$) < *finished*. In that case l is a hidden literal and is hence removed.

Example 2. Recall the formula E from Example 1. All literals except e and h in the clause $C = (a \lor b \lor c \lor d \lor e \lor f \lor g \lor h) \in E$ are hidden. In case the literals in $RTS(E)$ are stamped with the time stamps shown in Figure 3, the *UHLE* procedure can detect them all. Consider first the sequence $S^+_{rev}(C) = (c, a, d, e, b, h, f, g)$. Since fin($c$) < fin($a$), a is removed from C. Similarly, fin(e) < fin(b) and fin(f) < fin(g), and hence b and g are removed from C. Second, consider the complements of the literals in the reduced clause: $S^-(C) = (\bar{e}, \bar{h}, \bar{f}, \bar{d}, \bar{c})$. Now, fin($\bar{h}$) > fin($\bar{f}$), fin($\bar{d}$), fin($\bar{c}$), and hence f, d, and c are removed. ∎

UHLE (clause C)

1   *finished* := finish time of first element in $S_{\text{rev}}^+(C)$
2   **foreach** $l \in S_{\text{rev}}^+(C)$ starting at second element
3   **if** $\text{fin}(l) > \textit{finished}$ **then** $C := C \setminus \{l\}$
4   **else** *finished* := $\text{fin}(l)$
5   *finished* := finish time of first element in $S^-(C)$
6   **foreach** $\bar{l} \in S^-(C)$ starting at second element
7   **if** $\text{fin}(\bar{l}) < \textit{finished}$ **then** $C := C \setminus \{l\}$
8   **else** *finished* := $\text{fin}(\bar{l})$
9   **return** C

Fig. 5. Eliminating hidden literals using time stamps

5.2 Hidden Tautologies

Fig. 6 shows the pseudo-code for the *UHTE* procedure that detects hidden tautologies based on time stamps. Notice that if a time stamping represents an implication of the form $\bar{l} \rightarrow l'$, where both l and l' occur in a clause C, then the clause C is a hidden tautology.

The *UHTE* procedure goes through the sorted lists $S^+(C)$ and $S^-(C)$ to find two literals $l_{\text{neg}} \in S^-(C)$ and $l_{\text{pos}} \in S^+(C)$ such that the time stamping represents the implication $l_{\text{neg}} \rightarrow l_{\text{pos}}$, i.e., it checks if $\text{dsc}(l_{\text{neg}}) < \text{dsc}(l_{\text{pos}})$ and $\text{fin}(l_{\text{neg}}) > \text{fin}(l_{\text{pos}})$. The procedure starts with the first literals $l_{\text{neg}} \in S^-(C)$ and $l_{\text{pos}} \in S^+(C)$, and loops through the literals in $l_{\text{pos}} \in S^+(C)$ until $\text{dsc}(l_{\text{neg}}) < \text{dsc}(l_{\text{pos}})$ (Lines 4–6). Once such a l_{pos} is found, if $\text{fin}(l_{\text{neg}}) > \text{fin}(l_{\text{pos}})$ (Line 7), we know that C is a hidden tautology, and the procedure returns true (Line 10). Otherwise, we loop through $S^-(C)$ to select a new l_{neg} for which the condition holds (Lines 7–9). Then (Lines 4–6), if $\text{dsc}(l_{\text{neg}}) < \text{dsc}(l_{\text{pos}})$, C is a hidden tautology. Otherwise, we select a new l_{pos}. Unless a hidden tautology is detected, the procedure terminates once it has looped through all literals in either $S^+(C)$ or $S^-(C)$ (Lines 5 and 8).

One has to be careful while removing binary clauses based on time stamps. There are two exceptions in which time stamping represents an implication $l_{\text{neg}} \rightarrow l_{\text{pos}}$ with $l_{\text{neg}} \in S^-(C)$ and $l_{\text{pos}} \in S^+(C)$ for which C is not a hidden tautology. First, if $l_{\text{pos}} = \bar{l}_{\text{neg}}$, then l_{neg} is a failed literal. Second, if $\text{prt}(l_{\text{pos}}) = l_{\text{neg}}$, then C was used to set the time stamp of l_{pos}. Line 7 takes both of these cases into account.

Example 3. Recall again the formula E from Example 1. E contains three hidden tautologies: $(\bar{g} \lor h)$, $(\bar{a} \lor \bar{e} \lor h)$, and $(\bar{b} \lor \bar{c} \lor h)$. In the time stamping in Fig. 3, $\bar{h} : [17, 28]$ contains $\bar{g} : [18, 19]$. However, $\text{prt}(\bar{g}) = \bar{h}$, and hence $(\bar{g} \lor h)$ cannot be removed. On the other hand, $\bar{g} : [1, 6]$ contains $\bar{h} : [3, 4]$, and $\text{prt}(h) \neq g$, and hence $(\bar{g} \lor h)$ is identified as a hidden tautology. We can also identify $(\bar{a} \lor \bar{e} \lor h)$ as a hidden tautology because $\bar{h} : [17, 28]$ contains $\bar{a} : [22, 23]$. This is not the case for $(\bar{b} \lor \bar{c} \lor h)$ because the implications $b \rightarrow h$ and $\bar{h} \rightarrow \bar{b}$ are not represented by the time stamping. ∎

Proposition 6. *For any Unhiding time stamping, UHTE detects all hidden tautologies that are represented by the time stamping.*

UHTE (clause C)

1    $l_{\text{pos}} :=$ first element in $S^+(C)$
2    $l_{\text{neg}} :=$ first element in $S^-(C)$
3   **while** *true*
4   **if** $\text{dsc}(l_{\text{neg}}) > \text{dsc}(l_{\text{pos}})$ **then**
5   **if** l_{pos} is last element in $S^+(C)$ **then return** false
6            $l_{\text{pos}} :=$ next element in $S^+(C)$
7   **else if** $\text{fin}(l_{\text{neg}}) < \text{fin}(l_{\text{pos}})$ **or** $(|C| = 2$ **and** $(l_{\text{pos}} = \bar{l}_{\text{neg}}$ **or** $\text{prt}(l_{\text{pos}}) = l_{\text{neg}}))$ **then**
8   **if** l_{neg} is last element in $S^-(C)$ **then return** false
9            $l_{\text{neg}} :=$ next element in $S^-(C)$
10   **else return** true

Fig. 6. Detecting hidden tautologies using time stamps

Proof sketch. For every $l_{\text{neg}} \in S^-(C)$, *UHTE* checks if time stamping represents the implication $l_{\text{neg}} \to l_{\text{pos}}$ for the first literal in $l_{\text{pos}} \in S^+(C)$ for which $\text{dsc}(l_{\text{neg}}) < \text{dsc}(l_{\text{pos}})$ holds. The key observation is that if there is a $l_{\text{neg}} \in S^-(C)$ and a $l_{\text{pos}} \in S^+(C)$ such that time stamping represents the implication $l_{\text{neg}} \to l_{\text{pos}}$, then the stamps also represent $l_{\text{neg}} \to l'_{\text{pos}}$ with l'_{pos} being the first literal in $S^+(C)$ for which $\text{dsc}(l_{\text{neg}}) < \text{dsc}(l_{\text{pos}})$ holds. □

If a clause C is a hidden tautology, then $\text{HLA}(F, C)$ is a hidden tautology due to $\text{HLA}(F, C) \supseteq C$. However, it is possible that, for a given clause C, *UHTE*(C) returns true, while *UHTE*(*UHLE*(C)) returns false. In other words, *UHLE* could in some cases disrupt *UHTE*. For instance, consider the clause $(a \lor b \lor c)$ and the following time stamps: $a : [2, 3]$, $\bar{a} : [9, 10]$, $b : [1, 4]$, $\bar{b} : [5, 8]$, $c : [6, 7]$, $\bar{c} : [11, 12]$. Now *UHLE* removes literal b which is required for *UHTE* to return true. Therefore *UHTE* should be called before *UHLE*, as is done in our *Simplify* procedure (recall Fig. 4).

5.3   Adding Hyper Binary Resolution

An additional binary clause based simplification technique that can be integrated into the unhiding procedure is *hyper binary resolution* [1] (HBR). Given a clause of the form $(l_1 \lor \cdots \lor l_k)$ and $k - 1$ binary clauses of the form $(l' \lor \bar{l}_i)$, where $2 \leq i \leq k$, the hyper binary resolution rule allows to infer the clause $(l_1 \lor l')$ in one step.

For HBR in the unhiding algorithm we only need the list $S^-(C)$. Let C be a clause with k literals. We find a hyper binary resolvent if (i) all literals in $S^-(C)$, except the first one l_1, have a common ancestor l', or (ii) all literals in $S^-(C)$, except the last one l_k, have a common ancestor l''. In case (i) we find $(l_1 \lor \bar{l'})$, and in case (ii) we find $(l_k \lor \bar{l''})$. It is even possible that all literals in $S^-(C)$ have a common ancestor l''' which shows that l''' is a failed literal, in which case we can learn the unit clause $(\bar{l'''})$.

While *UHBR*(C) could be called in *Simplify* after Line 4, our experiments show that applying *UHBR*(C) does not give further gains w.r.t. running times, and can in cases degrade performance. We suspect that this is because *UHBR*(C) may add transitive edges to BIG(F). Consider the formula $F = (a \lor b \lor c) \land (\bar{a} \lor d) \land (\bar{b} \lor d) \land (c \lor e) \land (c \lor f) \land (d \lor \bar{e})$. Assume that the time stamping DFS visits the literals in the order $\bar{f}$,

$c, a, d, \bar{d}, \bar{e}, \bar{a}, \bar{b}, \bar{c}, f, e, b.$ *UHBR*$((a \vee b \vee c))$ can learn $(c \vee d)$, but it cannot check that this binary clause adds a transitive edge to BIG(F).

5.4 Some Limitations of Basic Stamping

As already pointed out, time stamps produced by randomized DFS may not represent all implications of F_2. In fact, the fraction of implications represented can be very small in the worst case. Especially, consider the formula $F = (a \vee b \vee c \vee d) \wedge (\bar{a} \vee \bar{b}) \wedge (\bar{a} \vee \bar{c}) \wedge (\bar{a} \vee \bar{d}) \wedge (\bar{b} \vee \bar{c}) \wedge (\bar{b} \vee \bar{d}) \wedge (\bar{c} \vee \bar{d})$ that encodes that exactly one of a, b, c, d must be true. Due to symmetry, there is only one possible DFS traversal order, and it produces the time stamps $a : [1, 8], \bar{b} : [2, 3], \bar{c} : [4, 5], \bar{d} : [6, 7], b : [9, 12], \bar{a} : [10, 11], c : [13, 14], d : [15, 16]$. Only three of the six binary clauses are represented by the time stamps. This example can be extended to n variables, in which case only $n - 1$ of the $n(n-1)/2$ binary clauses are represented. In order to capture as many implications (and thus simplification opportunities) as possible, in practice we apply multiple repetitions of *Unhiding* using randomized DFS (as detailed in Sect. 7).

6 Advanced Stamping for Capturing Additional Simplifications

In this section we develop an advanced version of the DFS time stamping procedure. Our algorithm can be seen as an extension of the BinSATSCC-1 algorithm in [13]. The advanced procedure, presented in Fig. 7, enables performing additional simplifications *on-the-fly during* the actual time stamping phase: the on-the-fly techniques can perform some simplifications that cannot be done with *Simplify*(F), and, on the other hand, enlarging the time stamps of literals may allow further simplifications in *Simplify*(F). Although not discussed further in this paper due to the page limit, we note that, additionally, all simplifications by *UHTE*, *UHLE*, and *UHBR* which only use binary clauses could be performed on-the-fly within the advanced stamping procedure.

Here we introduce the attribute obs(l) that denotes the latest time point of observing l. The value of obs(l) can change frequently during *Unhiding*. Each line of the advanced stamping procedure (Fig. 7) is labeled. The line labeled with OBS assigns obs(l) for literal l. The label BSC denotes that the line originates from the basic stamping procedure (Fig. 1). Lines with the other labels are techniques that can be performed on-the-fly: transitive reduction (TRD / Sect. 6.1), failed literal elimination (FLE / Sect. 6.2), and equivalent literal substitution (ELS / Sect. 6.3). The technique TRD depends on FLE and both techniques use the obs() attribute while ELS is independent of obs().

6.1 Transitive Reduction

Binary clauses that represent transitive edges in BIG are in fact hidden tautologies [6]. Such clauses can already be detected in the stamping phase (i.e., before *UHTE*), as shown in the advanced stamping procedure on Line 6 with label TRD.

A binary clause $(\bar{l} \vee l')$ can only be observed as a hidden tautology if dsc$(l') > 0$ during *Stamp*$(l, stamp)$. Otherwise, prt$(l') := l$, which satisfies the last condition on Line 7 of *UHTE*. If dsc$(l') >$ dsc(l) just before calling *Stamp*$(l', stamp)$, then $(\bar{l} \vee l')$

is a hidden tautology. When transitive edges are removed on-the-fly, *UHTE* can focus on clauses of size ≥ 3, making the last check on Line 7 of *UHTE* redundant.

Transitive edges in $\mathsf{BIG}(F)$ can hinder the unhiding algorithm by reducing the time stamp intervals. Hence as many transitive edges as possible should be removed. Notice that in case $0 < \mathrm{dsc}(l') < \mathrm{dsc}(l)$, *Stamp*$(l, stamp)$ cannot detect that $(\bar{l} \vee l')$ is a hidden tautology. Yet by using $\mathrm{obs}(l')$ instead of $\mathrm{dsc}(l')$ in the check (Line 14 of Fig. 7), we can detect additional transitive edges. For instance, consider the formula $F = (\bar{a} \vee b) \wedge (b \vee \bar{c}) \wedge (b \vee \bar{d}) \wedge (\bar{c} \vee d)$ where $(b \vee \bar{c})$ is a hidden tautology. If *Unhiding* visits the literals in the order $a, b, c, d, \bar{b}, \bar{a}, \bar{c}, \bar{d}$, then this hidden tautology is not detected using $\mathrm{dsc}(l')$. However, while visiting d in advanced stamping, we assign $\mathrm{obs}(b) := \mathrm{dsc}(d)$. Now, using $\mathrm{obs}(l')$, *Stamp*$(c, stamp)$ can detect that $(b \vee \bar{c})$ is a hidden tautology.

6.2 Failed Literal Elimination over F_2

Detection of failed literals in F_2 can be performed on-the-fly during stamping. If a literal l in F_2 is failed, then all ancestors of l in $\mathsf{BIG}(F)$ are also failed. Recall that there is a strong relation between HLE restricted to F_2 and failed literals in F_2 (Prop. 1).

To detect a failed literal, we check for each observed literal l' whether $\bar{l}'$ was also observed in the current tree, or $\mathrm{dsc}(\mathrm{root}(l)) \leq \mathrm{dsc}(\bar{l}')$. In that case the lowest com-

---

Stamp (literal l, integer $stamp$)

1	BSC	$stamp := stamp + 1$	
2	BSC/OBS	$\mathrm{dsc}(l) := stamp; \mathrm{obs}(l) := stamp$	
3	ELS	$flag := \mathbf{true}$	// l represents a SCC
4	ELS	$S.\mathrm{push}(l)$	// push l on SCC stack
5	BSC	**for each** $(\bar{l} \vee l') \in F_2$	
6	TRD	**if** $\mathrm{dsc}(l) < \mathrm{obs}(l')$ **then** $F := F \setminus \{(\bar{l} \vee l')\};$ **continue**	
7	FLE	**if** $\mathrm{dsc}(\mathrm{root}(l)) \leq \mathrm{obs}(\bar{l}')$ **then**	
8	FLE	$\quad l_{\mathrm{failed}} := l$	
9	FLE	$\quad$ **while** $\mathrm{dsc}(l_{\mathrm{failed}}) > \mathrm{obs}(\bar{l}')$ **do** $l_{\mathrm{failed}} := \mathrm{prt}(l_{\mathrm{failed}})$	
10	FLE	$\quad F := F \cup \{(\bar{l}_{\mathrm{failed}})\}$	
11	FLE	$\quad$ **if** $\mathrm{dsc}(\bar{l}') \neq 0$ **and** $\mathrm{fin}(\bar{l}') = 0$ **then continue**	
12	BSC	**if** $\mathrm{dsc}(l') = 0$ **then**	
13	BSC	$\quad \mathrm{prt}(l') := l$	
14	BSC	$\quad \mathrm{root}(l') := \mathrm{root}(l)$	
15	BSC	$\quad stamp := $ **Stamp**$(l', stamp)$	
16	ELS	**if** $\mathrm{fin}(l') = 0$ **and** $\mathrm{dsc}(l') < \mathrm{dsc}(l)$ **then**	
17	ELS	$\quad \mathrm{dsc}(l) := \mathrm{dsc}(l'); flag := \mathbf{false}$	// l is equivalent to l'
18	OBS	$\mathrm{obs}(l') := stamp$	// set last observed time attribute
19	ELS	**if** $flag = \mathbf{true}$ **then**	// if l represents a SCC
20	BSC	$\quad stamp := stamp + 1$	
21	ELS	$\quad$ **do**	
22	ELS	$\quad\quad l' := S.\mathrm{pop}()$	// get equivalent literal
23	ELS	$\quad\quad \mathrm{dsc}(l') := \mathrm{dsc}(l)$	// assign equal discovered time
24	BSC	$\quad\quad \mathrm{fin}(l') := stamp$	// assign equal finished time
25	ELS	$\quad$ **while** $l' \neq l$	
26	BSC	**return** $stamp$	

Fig. 7. Advanced literal time stamping capturing failed and equivalent literals

mon ancestor in the current tree is a failed literal. Similar to transitive reduction, the number of detected failed literals can be increased by using the $\mathrm{obs}(\bar{l}')$ attribute instead of $\mathrm{dsc}(\bar{l}')$. We compute the lowest common ancestor l_{failed} of l' and $\bar{l}'$ (Lines 8–9 in Fig. 7). Afterwards the unit clause $(\bar{l}_{\mathrm{failed}})$ is added to the formula (Line 10).

At the end of on-the-fly FLE (Line 11), the advanced stamping procedure checks whether to stamp l' after finding a failed literal. In case we learned that $\bar{l}'$ is a failed literal, then we have the unit clause (l'). Then it does not make sense to stamp l', as all implications of l' can be assigned to true by BCP. This check also ensures that binary clauses currently used in the recursion are not removed by transitive reduction.

6.3 Equivalent Literal Substitution

In case $\mathrm{BIG}(F)$ contains a cycle, then all literals in that cycle are equivalent. In the basic stamping procedure all these literals will be assigned a different time stamp. Therefore, many implications of F_2 will not be represented by any of the resulting time stampings. To fix this problem, equivalent literals should be assigned the same time stamps.

A cycle in $\mathrm{BIG}(F)$ can be detected after calling $Stamp(l', stamp)$, by checking whether $\mathrm{fin}(l')$ still has the initial value 0. This check can only return true if l' is an ancestor of l. We implemented ELS on-the-fly using a variant of Tarjan's SCC decomposition algorithm [17] which detects all cycles in $\mathrm{BIG}(F)$ using any depth-first traversal order. We use a local boolean *flag* that is initialized to true (Line 3). If true, *flag* denotes that l represents a SCC. In case it detects a cycle, *flag* is set to false (Lines 16–17). Additionally, a global stack S of literals is used, and is initially empty. At each call of $Stamp(l, stamp)$, l is pushed on the stack (Line 4). At the end of the procedure, if l is still the representative of a SCC, all literals in S being equivalent to l, all literals in S are assigned the same time stamp (Lines 19–25).

7 Experiments

We have implemented *Unhiding* in our state-of-the-art SAT solver Lingeling [12] (version 517, source code and experimental data at http://fmv.jku.at/unhiding) as an additional preprocessing or, more precisely, *inprocessing* technique applied during search. Batches of randomized unhiding rounds are interleaved with search and other already included inprocessing techniques. The number of unhiding rounds per unhiding phase and the overall work spent in unhiding is limited in a similar way as is already done in Lingeling for the other inprocessing. The cost of *Unhiding* is measured in the number of recursive calls to the stamping procedure and the number of clauses traversed. Sorting clauses (in *UHTE* and *UHLE*) incurs an additional penalty. In the experiments *Unhiding* takes on average roughly 7% of the total running time (including search), which is more than twice as much as standard failed literal probing (2%) and around half of the time spent on SatElite-style variable elimination (16%).

The cluster machines used for the experiments, with Intel Core 2 Duo Quad Q9550 2.8-GHz processors, 8-GB main memory, running Ubuntu Linux version 9.04, are around twice as fast as the ones used in the first phase of the 2009 SAT competition. For the experiments we used a 900 s timeout and a memory limit of 7

GB. Using the set of all 292 application instances from SAT Competition 2009 (http://satcompetition.org/2009/), a comparison of the number of solved instances for different configurations of *Unhiding* and the baseline (up-to-date version of Lingeling without *Unhiding*) is presented in Table 1. Note that we obtained similar results also for the SAT Race 2010 instances, and also improved performance on the crafted instances of SAT Competition 2009.

The three main observations are: (i) *Unhiding* increases the number of solved satisfiable instances already when using the basic stamping procedure; (ii) using the advanced stamping scheme, the number of solved instances increases notably for both satisfiable and unsatisfiable instances; and (iii) the *UHBR* procedure actually degrades the performance (in-line with the discussion in Sect. 5.3). Hence the main advantages of *Unhiding* are due to the combination of the advanced stamping procedure, *UHTE*, and *UHLE*.

8   Conclusions

The *Unhiding* algorithm efficiently (close to linear time) approximates a combination of binary clause based simplifications that is conjectured to be at least quadratic in the worst case. In addition to applying known simplification techniques, including the recent hidden tautology elimination, we introduced the novel technique of hidden literal elimination, and implemented it within *Unhiding*. We showed that *Unhiding* improves the performance of a state-of-the-art CDCL SAT solver when integrated into the search procedure for inprocessing formulas (including learnt clauses) during search.

Table 1. Comparison of different configurations of *Unhiding* and the baseline solver Lingeling. The 2nd to 4th columns show the number of solved instances (sol), resp. solved satisfiable (sat) and unsatisfiable (uns) instances. The next three columns contain the average percentage of total time spent in unhiding (unhd), all simplifications through inprocessing (simp), and variable elimination (elim). Here we also take unsolved instances into account. The rest of the table lists the number of hidden tautologies (hte) in millions, the number of hidden literal eliminations (hle), also in millions, and finally the number of unhidden units (unts) in thousands which includes the number of unhidden failed literals. We also include the average percentage (stp) of hidden tautologies resp. derived units during stamping, and the average percentage (red) of redundant/learned hidden tautologies resp. removed literals in redundant/learned clauses. A more detailed analysis shows that for many instances, the percentage of redundant clauses is very high, actually close to 100%, both for HTE and HLE. Note that "unts" is not precise as the same failed literal might be found several times during stamping since we propagate units lazily after unhiding.

configuration	sol	sat	uns	unhd	simp	elim	hte	stp	red	hle	red	unts	stp
adv.stamp (no uhbr)	188	78	110	7.1%	33.0%	16.1%	22	64%	59%	291	77.6%	935	57%
adv.stamp (w/uhbr)	184	75	109	7.6%	32.8%	15.8%	26	67%	70%	278	77.9%	941	58%
basic stamp (no uhbr)	183	73	110	6.8%	32.3%	15.8%	6	0%	52%	296	78.0%	273	0%
basic stamp (w/uhbr)	183	73	110	7.4%	32.8%	15.8%	7	0%	66%	288	76.7%	308	0%
no unhiding	180	74	106	0.0%	28.6%	17.6%	0	0%	0%	0	0.0%	0	0%

References

1. Bacchus, F.: Enhancing Davis Putnam with extended binary clause reasoning. In: Proc. AAAI, pp. 613–619. AAAI Press, Menlo Park (2002)
2. Eén, N., Biere, A.: Effective preprocessing in SAT through variable and clause elimination. In: Bacchus, F., Walsh, T. (eds.) SAT 2005. LNCS, vol. 3569, pp. 61–75. Springer, Heidelberg (2005)
3. Gershman, R., Strichman, O.: Cost-effective hyper-resolution for preprocessing CNF formulas. In: Bacchus, F., Walsh, T. (eds.) SAT 2005. LNCS, vol. 3569, pp. 423–429. Springer, Heidelberg (2005)
4. Han, H., Somenzi, F.: Alembic: An efficient algorithm for CNF preprocessing. In: Proc. DAC, pp. 582–587. IEEE, Los Alamitos (2007)
5. Järvisalo, M., Biere, A., Heule, M.J.H: Blocked clause elimination. In: Esparza, J., Majumdar, R. (eds.) TACAS 2010. LNCS, vol. 6015, pp. 129–144. Springer, Heidelberg (2010)
6. Heule, M.J.H., Järvisalo, M., Biere, A.: Clause elimination procedures for CNF formulas. In: Fermüller, C.G., Voronkov, A. (eds.) LPAR-17. LNCS, vol. 6397, pp. 357–371. Springer, Heidelberg (2010)
7. Marques-Silva, J.P.: Algebraic simplification techniques for propositional satisfiability. In: Dechter, R. (ed.) CP 2000. LNCS, vol. 1894, pp. 537–542. Springer, Heidelberg (2000)
8. Van Gelder, A.: Toward leaner binary-clause reasoning in a satisfiability solver. Annals of Mathematics and Artificial Intelligence 43(1), 239–253 (2005)
9. Li, C.M.: Integrating equivalency reasoning into Davis-Putnam procedure. In: Proc. AAAI, pp. 291–296 (2000)
10. Brafman, R.: A simplifier for propositional formulas with many binary clauses. IEEE Transactions on Systems, Man, and Cybernetics, Part B 34(1), 52–59 (2004)
11. Aho, A., Garey, M., Ullman, J.: The transitive reduction of a directed graph. SIAM Journal on Computing 1(2), 131–137 (1972)
12. Biere, A.: Lingeling, Plingeling, PicoSAT and PrecoSAT at SAT Race 2010. FMV Report Series Technical Report 10/1, Johannes Kepler University, Linz, Austria (2010)
13. del Val, Á.: Simplifying binary propositional theories into connected components twice as fast. In: Nieuwenhuis, R., Voronkov, A. (eds.) LPAR 2001. LNCS (LNAI), vol. 2250, pp. 392–406. Springer, Heidelberg (2001)
14. Soos, M.: Cryptominisat 2.5.0, sat race 2010 solver description (2010)
15. Korovin, K.: iProver – an instantiation-based theorem prover for first-order logic (System description). In: Armando, A., Baumgartner, P., Dowek, G. (eds.) IJCAR 2008. LNCS (LNAI), vol. 5195, pp. 292–298. Springer, Heidelberg (2008)
16. Groote, J.F., Warners, J.P.: The propositional formula checker HeerHugo. J. Autom. Reasoning 24(1/2), 101–125 (2000)
17. Tarjan, R.: Depth-first search and linear graph algorithms. SIAM J. Computing 1(2) (1972)

Between Restarts and Backjumps

Antonio Ramos, Peter van der Tak, and Marijn J.H. Heule*

Department of Software Technology, Delft University of Technology, The Netherlands

Abstract. This paper introduces a novel technique that significantly reduces the computational costs to perform a restart in conflict-driven clause learning (CDCL) solvers. Our technique exploits the observation that CDCL solvers make many redundant propagations after a restart. It efficiently predicts which decisions will be made after a restart. This prediction is used to backtrack to the first level at which heuristics may select a new decision rather than performing a complete restart.

In general, the number of conflicts that are encountered while solving a problem can be reduced by increasing the restart frequency, even though the solving time may increase. Our technique counters the latter effect. As a consequence CDCL solvers will favor more frequent restarts.

1 Introduction

Restarts are used in satisfiability (SAT) solvers to avoid heavy-tail behavior [4]. Restart strategies [7,14] have been a crucial feature in conflict-driven clause learning (CDCL) solvers [8] to tackle hard industrial problems. These solvers favor frequent restarts in recent years [5].

CDCL solvers select decision variables based on their involvement in emerged conflicts [10]. In case of frequent restarts, only several new conflicts have been hit between two succeeding restarts. As a consequence, CDCL solvers tend to select the same variables in a similar order after succeeding restarts. Additionally, phase-saving [12] ensures that decision variables are assigned to the same truth value as the value they were assigned to before a restart. Due to these heuristics, CDCL solvers generally do not perform a full restart, but effectively they perform a *partial restart*.

This paper capitalizes on this observation by introducing two techniques to reduce the computational costs to perform a restart. In case the solver wants to restart, we show how to efficiently predict the first level at which the heuristics may select a different decision variable. The solver can perform a partial restart by backtracking to this level, rather than perform a more costly full restart.

Additionally, by reducing the restart costs, it appears that restarting even more frequently improves the performance of CDCL solvers. We implemented our techniques in MiniSAT 2.2 [2]. Experiments show that the enhanced version with rapid restarts solves more real-world SAT instances from the SAT 2009 application suite than the original version.

---

* Supported by Dutch Organization for Scientific Research under grant 617.023.611.

K.A. Sakallah and L. Simon (Eds.): SAT 2011, LNCS 6695, pp. 216–229, 2011.

The remainder of this paper is structured as follows: the next section provides some background information about CDCL solvers and corresponding terminology. In Section 3 we motivate our work and Section 4 presents two novel techniques to reduce the computational costs to perform a restart. Experimental results are described in Section 5. Finally, we offer suggestions for future work in Section 6 and we draw conclusions in Section 7.

2   Conflict-Driven Clause Learning Solvers

The strategy used by conflict-driven clause learning (CDCL) solvers is to make a series of decisions (heuristically chosen assignments) and to propagate assignments that can be derived from these decisions by means of unit propagation (satisfying the remaining literal in every unit clause). The solver will continue to make decisions and propagate information until either a satisfying assignment is found for the problem, or a conflict emerges.

A conflict emerges if the solver finds a *conflicting clause* – a clause for which all literals are false. When this occurs, the solver analyzes the reason for the conflict. This is captured in a so-called *learned clause* [9,10] , which intuitively can be considered a clause that will avoid recurrence of the same combination of assignments that led to the conflict. Now, the solver unassigns variables until the learned clause becomes unit and continues to make decisions and apply other unit propagations as before.

The terminology introduced in this section is used in the remainder of this paper. Fig. 1 graphically shows the most important terms. This figure will also be used as a running example throughout the paper.

2.1   Heuristics

In addition to the general process described above, most CDCL solvers use the Variable State Independent Decaying Sum (VSIDS) heuristic [10] to determine the order in which decisions should be made. This heuristic stores an activity value for each variable, which is increased by 1 whenever a variable appears in a learned clause. After incrementing the activity value, the value of every variable is decreased by multiplying them with a constant factor δ[1], called the variable decay. This decay factor δ has a value in interval $(0, 1)$. In general, CDCL solvers use $\delta = 0.95$. The lower the value of δ, the more VSIDS prefers to select variables that were involved in recent conflicts. When no more information can be propagated, a new decision is made by selecting the unassigned variable with the highest activity value.

After a decision variable is selected by the solver, it must be assigned a value. A commonly used method is phase-saving [12], which stores for each variable the last value to which it was assigned by unit propagation. Decision variables are assigned to that value. By assigning variables to their last implied value, the solver picks up where it left off and continues its search in a similar part of the search space after a restart. Therefore, phase-saving facilitates frequent restarts.

---

[1] In practice, VSIDS is implemented by multiplying the incremental value by $\frac{1}{\delta}$ instead.

2.2   Decision Levels and Backjumping

Each decision introduces a new *decision level*. A decision level consists of the sequence of assignments of a decision variable and all variables that are implied by that decision. Decision levels are numbered incrementally, where 0 is the level where no decisions have yet been made – also known as the *restart level*. Decision level 1 is the first level that involves an actual decision. A decision is the first assignment in each level (denoted in Fig. 1 by the rectangles), other assignments, if any, are caused by unit propagation.

The decision levels form a *trail* of assignments. This trail can be seen as a list of variable assignments at a certain moment in time. The trail comprises both decisions and unit propagations, where each decision starts a new decision level. Finally, the *backjump level* [3] is the level to which the solver backtracks whenever a conflict is found. This is the level at which the learned clause is a unit clause. Notice that backjumping could be seen as performing a partial restart.

2.3   Restart Strategies

Modern solvers use restarts to avoid spending too much time searching for a solution in the same region without finding useful information. By restarting, CDCL solvers try to avoid heavy-tail behavior [4]. When a restart is performed, the solver will undo every assignment on the trail and make a new series of decisions and propagations. Because the learned clauses and the VSIDS heuristic will have changed since the previous run, the new run may perform decisions in a different order. This could reduce the total number of decisions necessary to solve a problem [6].

A commonly used restart strategy in recent years is based on a sequence of restart sizes suggested by Luby et al. [7]. In their work the authors show that the suggested sequence is log optimal when the runtime distribution of the problems is unknown. In this strategy the length of restart i is $u \cdot t_i$ when u is a constant unit run and

$$t_i = \begin{cases} 2^{k-1}, & \text{if } i = 2^k - 1 \\ t_{i-2^{k-1}+1}, & \text{if } 2^{k-1} \le i < 2^k - 1. \end{cases}$$

Since unit runs are commonly short, solvers using the Luby restart strategy exhibit frequent restarts. The solvers Rsat [12] and TiniSat [6] use a unit run of 512 conflicts, while MiniSAT 2.2 [2] and precoSAT [1] use a shorter unit run of 100 conflicts.

In this paper, we propose *partial* restarts which can be combined with these *full* restart strategies. An alternative partial restart strategy that has been proposed is *random jump* [15]. This strategy randomly backtracks to a level between the restart level and the backjump level. In [11] a technique is proposed to partially restart based on the learned clause if certain conditions are met.

$$F = (\neg x_1 \lor x_2 \lor \neg x_7) \land (\neg x_1 \lor \neg x_4) \land (x_1 \lor x_5) \land (\neg x_2 \lor x_6 \lor \neg x_8) \land$$
$$(\neg x_2 \lor x_4 \lor x_7) \land (x_3 \lor \neg x_5 \lor \neg x_6) \land (\neg x_3 \lor x_9) \land (x_6 \lor x_8 \lor \neg x_9)$$

trail before restart   **trail after restart**

─────────────────────── restart level ───────────────────

decision level 1 $\Big\{$ $\boxed{x_1 = 1}$ $\quad$ $\boxed{x_1 = 1}$ $\Big\}$ decision level 1
$\quad$ $x_4 = 0$ $\qquad$ $x_4 = 0$

──────────────────── MATCHINGTRAIL level ────────────────

decision level 2 $\Big\{$ $\boxed{x_7 = 1}$ $\quad$ $\boxed{x_2 = 1}$ $\Big\}$ decision level 2
$\quad$ $x_2 = 1$ $\qquad$ $x_7 = 1$

decision level 3 $\Big\{$ $\boxed{x_5 = 0}$ $\quad$ $\boxed{x_5 = 0}$ $\Big\}$ decision level 3

──────────────────── PERMUTEDTRAIL level ────────────────

decision level 4 $\Big\{$ $\boxed{x_3 = 1}$ $\quad$ $\boxed{x_9 = 1}$ $\Big\}$ decision level 4
$\quad$ $x_9 = 1$ $\qquad$ $x_6 = 1$

─────────────────────── backjump level ──────────────────

decision level 5 $\Big\{$ $\boxed{x_6 = 0}$ $\qquad\qquad$ $\vdots$
$\quad$ $x_8 = 1$

conflict : $(\neg x_2 \lor x_6 \lor \neg x_8)$

learned : $(\neg x_2 \lor x_6 \lor \neg x_9)$

VSIDS	VSIDS
$x_1 : 5.42$	$x_1 : 5.42$
$x_7 : 4.11$	$x_2 : 4.51$
$x_5 : 3.96$	$x_7 : 4.11$
$x_2 : 3.51$	$x_5 : 3.96$
$x_3 : 3.19$	$x_9 : 3.91$
$x_4 : 3.02$	$x_6 : 3.84$
$x_9 : 2.91$	$x_3 : 3.19$
$x_6 : 2.84$	$x_4 : 3.02$
$x_8 : 2.55$	$x_8 : 2.55$

Fig. 1. Visualization of our running example. Example of the outcome of MATCHING-TRAIL and PERMUTEDTRAIL. In both trails, the first five assigned variables are x_1, x_2, x_4, x_5, and x_7, albeit in different order. Therefore, backtracking to decision level 3 – right after the five matching assignments – causes the state of the solver to be equivalent to the state after restarting to decision level 0 and assigning the first five variables.

3   Motivation

The main contributions of this paper are two techniques to reduce the computational costs of performing a restart. In this section, we motive our work. First, we argue that CDCL solvers actually perform a partial restart and we indicate how to capitalize on that observation (Section 3.1). Second, we show that CDCL solvers traverse a smaller part of the search space when they restart more frequently. However, due to the restart costs, this may not always translate into improved performance (Section 3.2).

3.1   Partial Restart

The first observation that inspired the work below is that modern CDCL solvers usually make partial restarts rather than a full one. Yet in practice a full restart is performed, followed by the setup of a similar trail to the one that was just removed. By avoiding the redundant propagations, the cost of a restart can be reduced significantly.

Consider our example formula F and the two trails shown in Fig. 1. At the bottom of this figure the activities (VSIDS scores) are shown. Due to the learned clause $(\neg x_2 \vee x_6 \vee \neg x_9)$ the activity of the corresponding variables is increased by 1, which slightly changes the order. Recall that the variable with the highest activity that is not yet assigned is always selected as the next decision. The left part of Fig. 1 shows the assignments before the restart; the right shows the assignments after.

Let us compare the two trails. The first similarity is that decision level 1 is exactly the same before and after the restart, because variable x_1 still has the highest VSIDS score after the restart. Due to this similarity, the solver actually performs a partial restart. Yet this observation is not exploited by current solvers. As a result, they perform redundant propagations. Because the second decision after the restart is x_2 (instead of x_7), the trails no longer match. We denote by MATCHINGTRAIL, the last level at which the trails before and after a restart completely match. We show how to compute this level efficiently in Section 4.1.

A second similarity can be observed between the two trails. Notice that (i) the first five variables in both trails are the same and (ii) that these variables are assigned to the same values in the new trail as in the former trail. This is not a coincidence. The reason for (i) is that CDCL solvers restart frequently. Therefore, only a few clauses are learned between two restarts. This changes the VSIDS order of the variables only slightly. Additionally, (ii) is ensured by the phase-saving heuristic which is used by most CDCL solvers.

Since we know that there are no new propagations before the backjump level, the only difference in the trail is that the order of variables are permuted. We denote the last level at which both (i) and (ii) hold by PERMUTEDTRAIL. Notice that at the PERMUTEDTRAIL level the reduced formula is exactly the same before and after the restart. Therefore, performing a partial restart to the PERMUTEDTRAIL level is similar to performing a full restart. Section 4.2 shows how to compute this level efficiently.

3.2   Restart Frequency

Another observation regarding restarts in modern CDCL solvers was presented in [5] showing that restarting with shorter unit runs reduces the size of the search space the solver explores to tackle a problem. More specifically, more frequent restarts reduce the number of conflicts encountered during the search.

We computed the effect of Luby-based restarts with various unit runs on the average number of conflicts. The only difference between our experiment and [5] is the use of the latest version of MiniSAT (2.2) [2]. The results for the SAT 2009 application suite are shown in Table 1.

Table 1. Average number of conflicts for several unit runs of the Luby sequence. The numbers between brackets denote the number of solved instances within 1200 seconds. We used three seeds to initialize the VSIDS scores to obtain a more stable image.

Strategy	SAT	UNSAT	SOLVED	UNSOLVED	ALL
Luby-1	190525 (64)	428450 (102)	336830 (166)	2470522	1241357
Luby-2	236141 (67)	609316 (101)	460046 (168)	2547307	1323072
Luby-4	235651 (67)	626690 (101)	471207 (168)	2708730	1401559
Luby-8	209926 (68)	725730 (102)	519003 (170)	2834041	1465453
Luby-16	252346 (67)	729354 (102)	539303 (169)	2939230	1526033
Luby-32	249255 (69)	835857 (102)	599158 (171)	3062220	1594681
Luby-64	297142 (70)	764207 (97)	569364 (167)	3130413	1640186
Luby-128	264409 (69)	770363 (96)	559378 (165)	3147708	1662650
Luby-256	222895 (68)	688930 (94)	492907 (162)	3277263	1708840
Luby-512	238800 (68)	725555 (93)	520394 (161)	3186994	1687999

First consider the number of solved instances shown in the first three columns. Although the Luby unit run is incrementally increased by a factor two, the number of instances solved remains quite comparable. The biggest differences are on the satisfiable instances. This was expected because CDCL solvers are not very stable on those formulas. When comparing the average number of conflicts, we observe that the longer the unit run, the higher this average. For the longest unit runs, we do not observe this pattern. These averages have been influenced by the lack of solved hard unsatisfiable instances within the timeout.

The last two columns show a clearer pattern regarding the average number of conflicts. Both columns are almost strictly increasing. Based on the data in these columns we can estimate the number of conflicts per second for different unit runs. Both the averages shown in the UNSOLVED and ALL columns indicate that the long unit runs handle about 35% more conflicts per second compared to the short unit runs. This difference is likely to be caused by restart costs.

By restarting with a short unit run, the solver encounters on average fewer conflicts while solving a problem. However, due to the costs of restarting frequently, using shorter unit runs does not result in solving more instances. In fact, both effects appear to cancel each other out since the various settings solve practically the same number of instances. We aim to reduce the costs of restarts which should in turn favor solvers that restart more frequently.

4   Reducing Restart Costs

This section describes the two algorithms we propose to compute the level to which to backtrack, MATCHINGTRAIL and PERMUTEDTRAIL. The algorithms rely on phase-saving, VSIDS ordering, and the absence of random decisions – all default in e.g. the latest MiniSAT 2.2. Furthermore, they should have access to the assignment type of each variable (Decision, Implication, Unassigned) and the decision level at which the variable was assigned. In the algorithms these are denoted by $AssignmentType[x]$ and $DecisionLevel[x]$ respectively, where x is a variable.

4.1   Matching Trail

Fig. 2 shows the pseudo-code of how to compute the MATCHINGTRAIL level. The algorithm increases $MTLevel$ for every decision that will be made at the same level in the current trail and the trail after the restart. The algorithm loops through variables in descending order of activity. If the variable is not currently assigned, the next decision level after the restart will be different and the algorithm will terminate (Line 4). If the variable is already assigned a value at $MTLevel$ or before, it will be an implied variable in both trails and can be ignored (Line 5). Finally, if it is a decision variable, it will be the next decision in the trail after the restart. Therefore, if the variable matches the decision at $MTLevel$, a match is found and $MTLevel$ is incremented (Line 7). If not, the decisions at the next level will be different, and the algorithm returns the last level at which they were the same (Line 9).

Example. Again consider the example in Fig. 1. The algorithm starts with $MTLevel = 0$ and considers x_1. It detects that both trails will have matching decisions at level 1, and increments $MTLevel$ to 1. Next, variable x_2 is found to become the decision at level 2 after the restart, but it does not match decision variable x_7 at the same level of the current trail. Therefore, the algorithm terminates and returns $MTLevel = 1$.

MatchingTrail (DecisionLevel, AssignmentType, VSIDS order)

1   MTLevel $\leftarrow$ 0
2   **forever do**
3        $x \leftarrow$ Next variable with highest activity
4   **if** AssignmentType[x] = Unassigned **then break**
5   **if** DecisionLevel[x] $\leq$ MTLevel **then continue**
6   **if** AssignmentType[x] = Decision **and** DecisionLevel[x] = MTLevel + 1 **then**
7   MTLevel $\leftarrow$ MTLevel + 1
8   **else break**
9   **return** MTLevel

Fig. 2. Pseudo-code of the MATCHINGTRAIL algorithm. This algorithm returns the last level at which all decisions will occur in the exact same order after the restart.

4.2 Permuted Trail

The PERMUTEDTRAIL algorithm (Fig. 3) aims to compute the last level at which the partial assignment (and hence the reduced formula under this assignment) is exactly the same before and after a restart (recall Section 3.1). Like MATCHING-TRAIL, PERMUTEDTRAIL loops through variables in descending order of activity. For each variable, it determines at which level it was assigned, and stores the running maximum in $MinimalLevel$ (Line 7). This value represents the level at which all variables that have been processed so far have been assigned. Also, it counts how many of these are currently decision variables, and stores this value in $MatchCount$ (Line 9). Any variable that is currently unassigned terminates the algorithm, since this variable will become a decision that can never be part of a permutation of the current trail (Line 6).

Now consider what happens when $MatchCount = MinimalLevel$. By definition of $MinimalLevel$, the set of variables that the algorithm has processed so far is a subset of the variables that are assigned up to $MinimalLevel$. Because this set includes $MatchCount$ decision variables, it must include each decision variable up to $MinimalLevel$. Since at least the same decisions are made, unit propagation will ensure that any currently implied variable is also assigned after the restart. Since the algorithm is performed at the backtrack level, no additional unit clauses may appear in the trail after the restart up to this point, which means that both trails must contain the exact same variables. Therefore the algorithm indicates that a partial restart is possible at this level in $PTLevel$ (Line 10).

PermutedTrail (DecisionLevel, AssignmentType, VSIDS order)

```
1      PTLevel ← 0
2      MinimalLevel ← 0
3      MatchCount ← 0
4   forever do
5          x ← Next variable with highest activity
6          if AssignmentType[x] = Unassigned then break
7          if DecisionLevel[x] > MinimalLevel then MinimalLevel ← DecisionLevel[x]
8          if AssignmentType[x] = Decision then
9              MatchCount ← MatchCount + 1
10             if MatchCount = MinimalLevel then PTLevel ← MatchCount
11  return PTLevel
```

Fig. 3. Pseudo-code of the PERMUTEDTRAIL algorithm. This algorithm returns the decision level at which all decisions occur in the trail after the restart (so that there are no intermediate decisions), but possibly in a different order.

Example. Consider how the algorithm will find the PERMUTEDTRAIL level for the running example in Fig. 1. The algorithm starts considering x_1, which is set in decision level 1, so that $MinimalLevel$ is set to 1. Since it is also a decision variable, $MatchCount$ is incremented to 1. The values match, and hence the algorithm finds that a partial restart is possible at $PTLevel = 1$. Next, x_2 has the highest activity. It is a propagation in level 2, and it updates $MinimalLevel = 2$ and $MatchCount = 1$. Next, x_7 is a decision in level 2, so that $MinimalLevel = MatchCount = 2$. Both values match, and $PTLevel = 2$ is another possible backtrack level for a partial restart. Note that this is detected even though x_2 became a decision variable and x_7 became an implied variable. Now x_5 is considered, leading to $MatchCount = MinimalLevel = 3$, which means that $PTLevel = 3$ is the best candidate so far. For x_9, $MinimalLevel = 4$ and $MatchCount = 3$, so that $PTLevel = 3$ remains unchanged. Finally, x_6 is currently unassigned because the algorithm runs after backtracking to the back-jump level. The algorithm thus terminates with $PTLevel = 3$.

4.3   Discussion

The MATCHINGTRAIL technique has the nice feature that solvers will explore the search space exactly the same as when performing a full restart. Yet although the reduced formula before and after a restart is exactly the same at the PERMUTEDTRAIL level, the solver may explore the search space differently when this technique is applied. This is caused by the so-called *reason clauses* [2]. The reason clause for an implied variable is the one that assigned its truth value (the first to become unit). Reason clauses are used to compute learned clauses. By making decisions in a different order, the reason clauses may be different, which in turn could make the conflict clauses different. This may influence the way the search space is explored.

Ideally one wants to backtrack to the last level at which the partial assignment is exactly the same before and after a restart. Although the PERMUTEDTRAIL algorithm is designed to do that, it may return a "subprime" level.

To illustrate this, let LASTSAMEASSIGNMENT be the ideal backtrack level (i.e. the last level where the partial assignment is the same before and after the restart). Let y be the decision variable at the LASTSAMEASSIGNMENT level. Now, assume that there is a variable x which is a decision variable before the LASTSAMEASSIGNMENT level in the current trail, and which has a lower activity than y after the restart. Because the partial assignment is the same, x is assigned in the trail after the restart, and because it has a lower activity than y it must be an implied variable. However, using the PERMUTEDTRAIL algorithm, we cannot detect that x is implied by the assignments in the new trail. Therefore it will not return the LASTSAMEASSIGNMENT level. During our experiments we observed that in practice this does not occur often, so that there is not much difference between the level returned by PERMUTEDTRAIL and the LASTSAMEASSIGNMENT level.

5   Experimental Results

We implemented MATCHINGTRAIL and PERMUTEDTRAIL in MiniSAT 2.2 and configured it to facilitate our analysis. There are three main requirements for using the proposed algorithms: phase-saving, the VSIDS heuristic, and the absence of random selection of decision variables. Each of these is default in MiniSAT 2.2, therefore the implementation of the algorithms was easy and straightforward.

We used the (292) application instances of the SAT 2009 competition[2]. Each instance was run with a time limit of 1200 seconds using different configurations on a server of 20 Intel Xeon X5570 CPUs running on 2.9 GHz with 32 GB of memory.

5.1   Matching Trail

It turned out that the MATCHINGTRAIL algorithm was much less effective than the PERMUTATEDTRAIL algorithm. Fig. 4 shows a typical distribution of the MATCHINGTRAIL, PERMUTEDTRAIL, and backjump levels. The distributions of the PERMUTEDTRAIL and backjump level are quite comparable. However, the MATCHINGTRAIL levels are generally much lower – which explains why it is less successful in reducing the restart costs. Therefore, we focused our experiments on the usefulness of the PERMUTEDTRAIL algorithm.

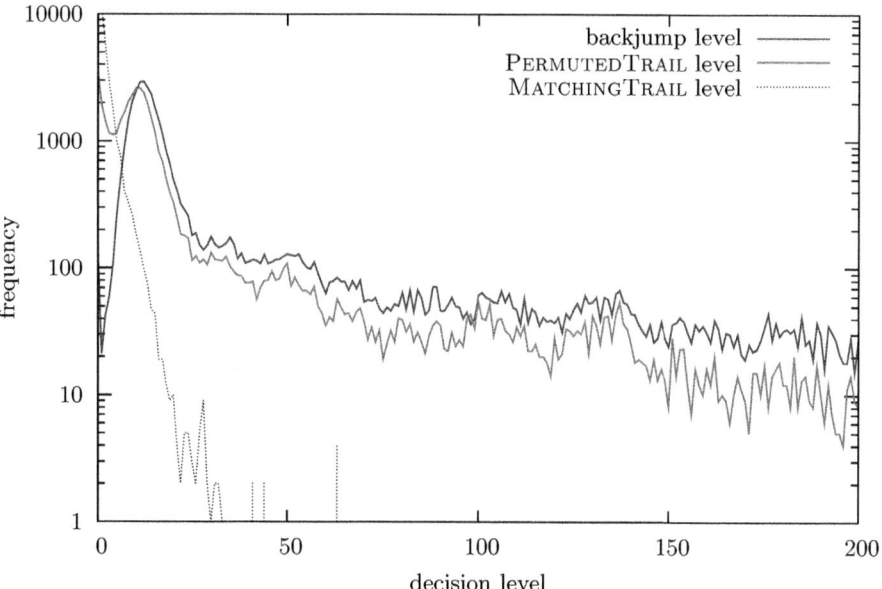

Fig. 4. Distribution of the MATCHINGTRAIL, PERMUTEDTRAIL, and backjump levels while solving the ACG-15-10p1.cnf benchmark of the SAT 2009 competition using MiniSAT 2.2 with a Luby-1 restart strategy

---

[2] Available from http://www.satcompetition.org/

5.2   Permuted Trail

In this section we compare the performance of MiniSAT 2.2 with and without PERMUTEDTRAIL. Comparing the performance of SAT solvers is hard, because a small change, for example to the order in which unit propagation is applied, can have a huge impact on the performance. Therefore, our focus in this section is not only on solving time, but also on the number of conflicts per second. The latter seems a bit more robust (recall also Section 3.2).

We experimented with three restart strategies. First, the default strategy of MiniSAT 2.2, which uses the Luby sequence with a unit run of 100 (in short Luby-100). Second, because we expect that the PERMUTEDTRAIL technique is especially useful for short unit runs, we added the strategy with the shortest unit run 1 (Luby-1). Third, we included a radical strategy that restarts before every decision (Const-1). This strategy would profit most from PERMUTEDTRAIL, showing thereby the maximum one could gain using this technique. Notice that a CDCL solver with the Const-1 restart strategy is still complete [13].

Fig. 5 shows the number of conflicts per second (left) and the solving time (right) for MiniSAT using the three restart strategies with and without partial restarts to the PERMUTEDTRAIL level. The Const-1 and Luby-1 strategies can clearly process more conflicts per second when PERMUTEDTRAIL is enabled. For the Luby-100 strategy no real improvement is observable, as expected. For some instances, the PERMUTEDTRAIL actually had a negative effect. For these benchmarks the performance greatly depends on the seed[3].

In our last experiment, we wanted to see whether the use of PERMUTED-TRAIL would make a rapid restart strategy preferred over the default Luby-100. In the tests above we used the default variable decay of MiniSAT $\delta = 0.95$ (see Section 2.1). However, preliminary tests showed that when using rapid restarts such as Const-1 and Luby-1, a lower value of the δ results in improved performance. Notice that a lower variable decay will make PERMUTEDTRAIL itself a bit less effective because variables will go up and down faster in the VSIDS order. We found that $\delta = 0.75$ results in strong performance. Therefore, in this last experiment we combined PERMUTEDTRAIL with $\delta = 0.75$ (denoted by an asterisk, e.g. Luby-1*).

Figure 6 shows the results. Combining PERMUTEDTRAIL with $\delta = 0.75$ increases the number of instances solved for each restart strategy, especially for Const-1 (156 vs 168 instances solved) and Luby-1 (173 vs 187 instances solved). The impact on Luby-100 is hardly visible. This is expected since this strategy restarts much less frequently, therefore the cost reduction of restarts hardly influences the performance. Luby-1* performed best during our experiments. This shows that PERMUTEDTRAIL reduced the restart costs to such level that the benefits of encountering fewer conflicts to solve a problem can be exploited to the point where it solves 10 instances more than the default configuration of MiniSAT.

---

[3] MiniSAT has the option to randomly initialize the VSIDS scores. For many benchmarks the seed used for the initialization has a huge impact on the performance.

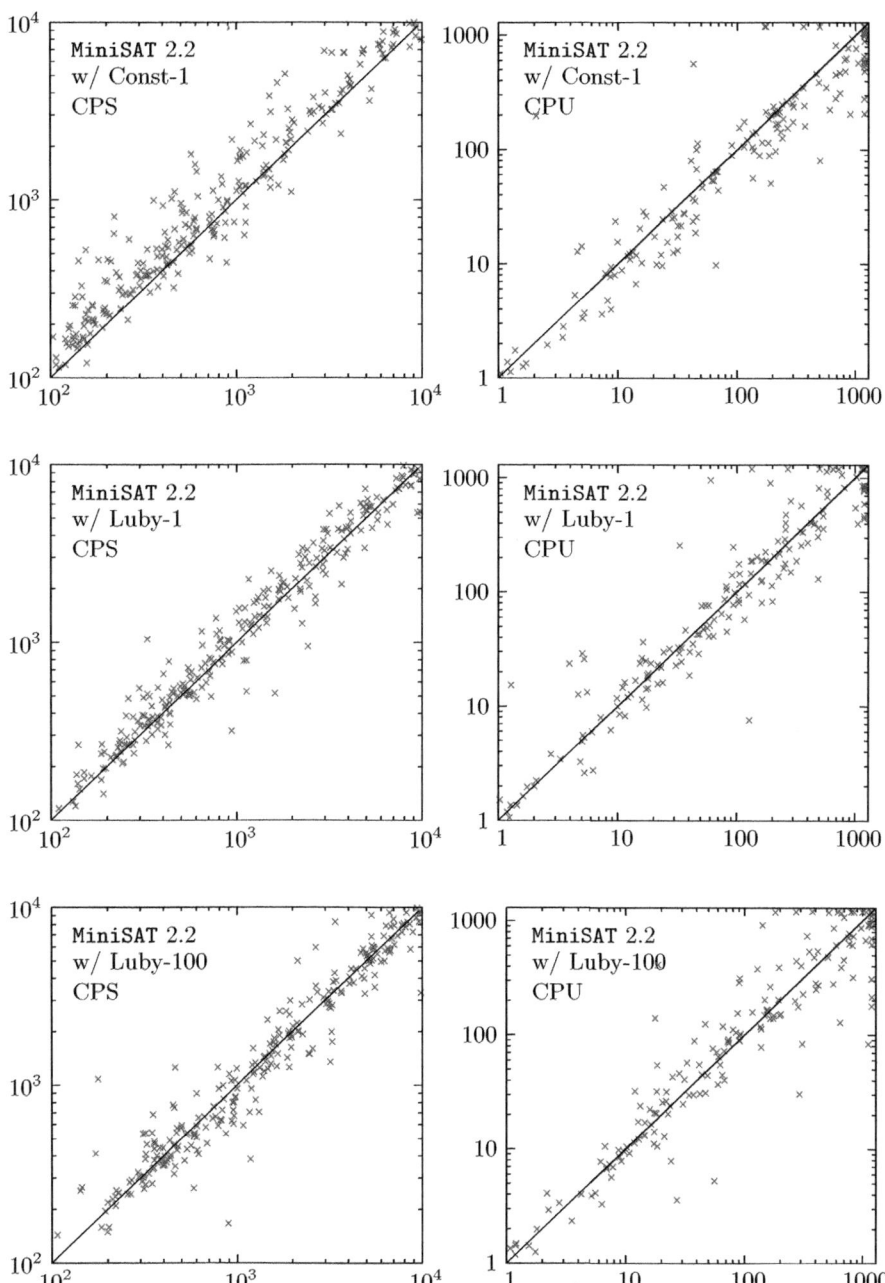

Fig. 5. Comparison of the number of conflicts per second (left) and solver runtime (right) between full restarts and PERMUTEDTRAIL for the SAT 2009 application benchmarks. PERMUTEDTRAIL propagates more conflicts per second above the diagonal (left) and solves instances faster below the diagonal (right).

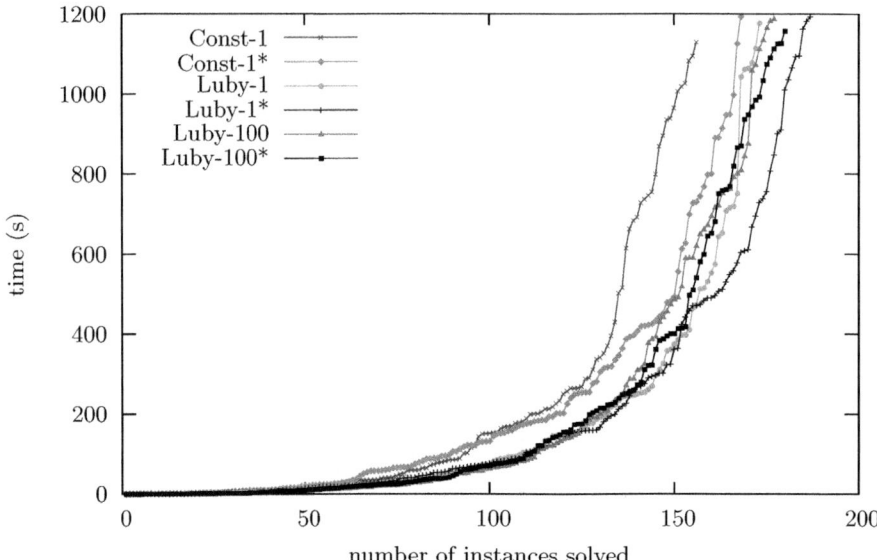

Fig. 6. Cactus plot showing the number of instances solved versus the time required to do so for three restart strategies and two configurations. PERMUTEDTRAIL with $\delta = 0.75$ – denoted in the legend with an asterisk – improves the performance of MiniSAT.

6   Suggestions for Future Work

Although our algorithm finds a reasonably high partial restart level, the solver still performs redundant work sometimes when a decision variable becomes an implied variable after a restart (recall Section 4.3). The current algorithms will not always detect that, and therefore may not return the optimal backtrack level. Although we have not seen this happen frequently in practice, it is possible that for some instances this occurs often, in which case it might be interesting to further analyze this issue and to develop efficient solutions.

We expect that the performance improvements are mainly caused by the reduced restart costs. Yet, the PERMUTEDTRAIL algorithm has also an important side effect. After a full restart, the reason clauses of implied variables may change, while after a partial restart the reason clauses stay the same. We want to study whether this effect influences the performance positively or negatively.

7   Conclusion

In this work, we implemented and tested two performance enhancements that reduce restart costs for CDCL solvers. We implemented both techniques in the latest MiniSAT solver. We show how to reduce the redundant work that is introduced by a restart by predicting the trail that will occur after a restart.

By applying a partial restart based on this prediction and by restarting more frequently, the performance of CDCL solvers can be improved.

References

1. Biere, A.: P{re,i}coSAT@SC 2009. In: SAT 2009 competitive events booklet: preliminmary version, pp. 41–44 (2009)
2. Eén, N., Sörensson, N.: An extensible SAT-solver. In: Giunchiglia, E., Tacchella, A. (eds.) SAT 2003. LNCS, vol. 2919, pp. 502–518. Springer, Heidelberg (2004)
3. Ginsberg, M.L.: Dynamic backtracking. J. Artif. Int. Res. 1, 25–46 (1993)
4. Gomes, C.P., Selman, B., Crato, N.: Heavy-tailed distributions in combinatorial search. In: Smolka, G. (ed.) CP 1997. LNCS, vol. 1330, pp. 121–135. Springer, Heidelberg (1997)
5. Haim, S., Heule, M.J.H.: Towards ultra rapid restarts, Technical report, UNSW and TU Delft (2010),
   http://www.st.ewi.tudelft.nl/~marijn/publications/rapid.pdf,
6. Huang, J.: The Effect of Restarts on the Efficiency of Clause Learning. In: Proceedings of the International Joint Conference on Artificial Intelligence, pp. 2318–2323 (2007)
7. Luby, M., Sinclair, A., Zuckerman, D.: Optimal speedup of las vegas algorithms. In: ISTCS, pp. 128–133 (1993)
8. Marques-Silva, J.P., Lynce, I., Malik, S.: Conflict-Driven Clause Learning SAT Solvers. In: Frontiers in Artificial Intelligence and Applications, vol. 185, ch. 4, pp. 131–153. IOS Press, Amsterdam (February 2009)
9. Marques-Silva, J., Sakallah, K.: Grasp: a search algorithm for propositional satisfiability. IEEE Transactions on Computers 48(5), 506–521 (1999)
10. Moskewicz, M.W., Madigan, C.F., Zhao, Y., Zhang, L., Malik, S.: Chaff: engineering an efficient sat solver. In: Proceedings of the 38th annual Design Automation Conference, DAC 2001, pp. 530–535. ACM, New York (2001)
11. Nadel, A., Ryvchin, V.: Assignment stack shrinking. In: Strichman, O., Szeider, S. (eds.) SAT 2010. LNCS, vol. 6175, pp. 375–381. Springer, Heidelberg (2010)
12. Pipatsrisawat, K., Darwiche, A.: A lightweight component caching scheme for satisfiability solvers. In: Marques-Silva, J., Sakallah, K.A. (eds.) SAT 2007. LNCS, vol. 4501, pp. 294–299. Springer, Heidelberg (2007)
13. Pipatsrisawat, K., Darwiche, A.: On the power of clause-learning SAT solvers with restarts. In: Gent, I.P. (ed.) CP 2009. LNCS, vol. 5732, pp. 654–668. Springer, Heidelberg (2009)
14. Walsh, T.: Search in a small world. In: IJCAI 1999: Proceedings of the Sixteenth International Joint Conference on Artificial Intelligence, pp. 1172–1177 (1999)
15. Zhang, H.: A complete random jump strategy with guiding paths. In: Biere, A., Gomes, C. (eds.) SAT 2006. LNCS, vol. 4121, pp. 96–101. Springer, Heidelberg (2006)

Abstraction-Based Algorithm for 2QBF

Mikoláš Janota[2] and Joao Marques-Silva[1,2]

[1] University College Dublin, Ireland
[2] INESC-ID, Lisbon, Portugal

Abstract. Quantified Boolean Formulas (QBFs) enable standard representation of PSPACE problems. In particular, formulas with two quantifier levels (2QBFs) enable representing problems in the second level of the polynomial hierarchy (Π_2^P, Σ_2^P). This paper proposes an algorithm for solving 2QBF satisfiability by counterexample guided abstraction refinement (CEGAR). This represents an alternative approach to 2QBF satisfiability and, by extension, to solving decision problems in the second level of polynomial hierarchy. In addition, the paper presents a comparison of a prototype implementing the presented algorithm to state of the art QBF solvers, showing that a larger set of instances is solved.

1 Introduction

The *Quantified Boolean Formula* (QBF) decision problem represents the paradigmatic PSPACE-complete decision problem [21]. Restrictions of QBF have also been used to characterize the polynomial hierarchy [21]. The QBF problem is important not only from a theoretical perspective, but also from an applied one, with many applications being easily modeled as instances of QBF [14]. There has been renewed interest in QBF solving over the last decade, in part motivated by the practical success of SAT solvers [23]. Most modern QBF algorithms build on the success of SAT solvers, but implement dedicated techniques [14].

One of the most successful approaches for symbolic model checking is *counterexample-guided abstraction refinement* (CEGAR) [6,7], having been applied in BDD-based and SAT-based model checking [5,7,8]. The success of CEGAR motivated its use with more expressive logics [1,27,12]. Moreover, recent work has applied the CEGAR paradigm in handling quantification on a number of different settings with promising results, including propositional circumscription [20], quantified bit-vector formulas [40], and linear real arithmetic [26]. Although the previous approaches for handling quantification could be used for solving QBF, it is expected that dedicated solutions will result in more effective algorithms.

This paper develops a CEGAR approach for solving QBF with 2 levels of quantifiers. The proposed algorithm generalizes the algorithm from [20] to the 2QBF case. Although the two algorithms exhibit similar abstraction refinement loops, the actual implementation of the key steps of the algorithms differs substantially. These differences are detailed in this paper. Experimental results, obtained on a wide range of problem instances, shows that the new algorithm outperforms the best QBF solvers from the most recent QBF evaluation [28]. Moreover, the new algorithm also outperforms the encoding of 2QBF to propositional circumscription [20], thus confirming that reduction of 2QBF to other domains is unlikely to result in efficient algorithms.

K.A. Sakallah and L. Simon (Eds.): SAT 2011, LNCS 6695, pp. 230–244, 2011.
© Springer-Verlag Berlin Heidelberg 2011

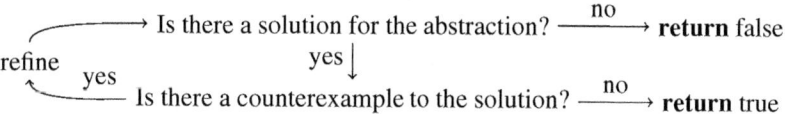

Fig. 1. CEGAR loop

2 Preliminaries

Lowercase letters from the end of the alphabet are used for Boolean variables (x_1, y, etc.); capital letters from the end of the alphabet (X, Y, etc.) are used to denote vectors of variables. Quantified Boolean formulas (QBF) are assumed to be in the *prenex* form $Q_1 z_1 \ldots Q_n z_n . \phi$ where $Q_i \in \{\forall, \exists\}$, z_i are distinct variables, and ϕ is a propositional formula using only the variables z_i and the constants 0 (false), 1 (true). The sequence of quantifiers in a QBF is called the *prefix* and the propositional formula the *matrix*. If a prefix contains a subsequence $\forall x_1 \ldots \forall x_n$, resp. $\exists x_1 \ldots \exists x_n$, we denote it by $\forall X$, resp. $\exists X$, for the variable vector $X = \{x_1, \ldots, x_n\}$.

The Greek letters ν and μ are used to denote vectors of the constants 0 and 1. For a Boolean formula ϕ and vectors $X = \{x_1, \ldots, x_n\}$, $\nu = \{a_1, \ldots, a_n\}$ we write $\phi[X/\nu]$ for the simultaneous substitution of occurrences of x_i by a_i. Further, $\phi[X/\nu]$ assumes that the formula has been partially evaluated ($x \vee 0 \equiv x$, etc.). If X are all the variables in ϕ, we treat the value vector ν as a variable valuation and $\phi[X/\nu]$ is the formula's value under that valuation. We write $\mathcal{B}$ for the set $\{0, 1\}$ and $\mathcal{B}^n$ for the set of vectors of the values 0,1 of length n.

A Boolean formula in *conjunctive normal form (CNF)* is a conjunction of *clauses*, where a clause is a disjunction of *literals*, and a literal is either a variable or a complement. Whenever convenient, a CNF formula is treated as a set of clauses and a clause is treated as a set of literals. For a literal l we write $\mathrm{var}(l)$ for the variable in l, i.e. $\mathrm{var}(\neg x) = \mathrm{var}(x) = x$.

Some of the heuristics proposed in the paper use a *partial MAX-SAT* problem formulation. The partial MAX-SAT problem is specified with two sets of clauses: a set of *hard* clauses, and a set of *soft* clauses. A solution to the problem is a variable valuation that satisfies all the hard clauses and maximizes the number of satisfied soft clauses [22].

2.1 Counterexample Guided Abstraction Refinement (CEGAR)

Counterexample Guided Abstraction Refinement (CEGAR) was designed for tackling problems whose implicit representation is infeasible to solve and thus an abstract representation is tackled instead. Here we present a informal description of the approach necessary for the understanding of the article. For more details see [7].

In CEGAR-based algorithm we talk about *concrete* and *abstract* representation of the problem. Then we talk about *abstract solutions*, which are solutions to the abstract version, and *concrete solutions* which are solutions to the actual problem. The goal of the CEGAR approach is to get to a concrete solution via abstract solutions. The relation between these two representations is characterized by the following properties:

1. If the abstraction does not have a solution, the concrete problem does not have a solution either.
2. If an abstract solution is not a concrete solution, a *counterexample* is produced to demonstrate this fact.
3. If there are no counterexamples to an abstract solution, it is also a concrete solution.

The search for a solution of the concrete problem is carried out in the following loop (see Figure 1). First, a solution for the abstraction is computed. If such does not exist, the search terminates unsuccessfully due to property 1. If the abstraction has a solution, it needs to be checked whether it is also a concrete solution. If there are no counterexamples to the abstract solution, it is also a concrete solution and the search terminates successfully due to property 3. Otherwise, the abstraction needs to be *refined* where the obtained counterexample is used to guide the refinement. Observe that an abstract solution is in fact a *candidate* for a solution to the given problem, the second step of the iteration then checks if it is really a solution.

We should note that conceptually, here we are looking for a solution to the problem. However, often, especially in verification, the goal of the CEGAR loop is to show *unsatisfiability* of the problem, i.e., lack of solutions. Finding a solution then corresponds to finding an error in the modeled system. Algorithmically, these goals are identical but the pertaining terminology in literature may differ.

3   Problems

This article focuses on the satisfiability of formulas with two levels on quantifiers. In particular, we focus on the following two problems.

Name: 2QBF PROBLEM
Given: $\exists X \forall Y. \phi$, where ϕ is a propositional formula
Question: Is there value a vector ν such that $\forall Y. \phi[X/\nu]$?

Name: 2QCNF PROBLEM
Given: $\exists X \forall Y. (\neg \phi')$, where ϕ' is a CNF
Question: Is there value a vector ν such that $\forall Y. \neg \phi'[X/\nu]$?

In both problems, a vector ν satisfying the condition is called a *solution* of $\exists X \forall Y. \phi$. While deciding the satisfiability of a QBF is PSPACE complete, the above problems are Σ_2^P-complete [24,21].

Here we make several notes on the specific form of the problems that we chose for this article. While 2QCNF PROBLEM is a special case of 2QBF PROBLEM, we single out 2QCNF PROBLEM as the uniformity of the format can be exploited for efficiency. The satisfiability of a formula of the form $\forall X \exists Y. \phi$ can be decided by negating to $\exists X \forall Y. \neg \phi$ and negating the response, which is why we consider only the latter form. Conceptually, $\exists X \forall Y. \neg \phi$ can be thought of as an attempt to *refute* $\forall X \exists Y. \phi$.

While for CNF ϕ' the satisfiability of $\forall X \exists Y. \phi'$ is in Π_2^P, $\exists X \forall Y. \phi'$ is an NP problem. Hence, we consider only the satisfiability of $\exists X \forall Y. (\neg \phi')$, which corresponds to refuting $\forall X \exists Y. \phi'$.

4   Algorithm for the 2QBF PROBLEM

This section presents a CEGAR-based algorithm for the 2QBF PROBLEM. The algorithm relies on a SAT oracle (a SAT solver) and hence we begin by observing the relation of 2QBF to Boolean satisfiability. First let us observe that the 2QBF PROBLEM can be expressed as a Boolean satisfiability problem if the universal quantifier is expanded into conjunctions.

4.1   Algorithm

Observation 1. *A value vector ν is a solution to $\exists X \forall Y.\phi$ iff ν is a satisfying assignment of the following formula:*

$$\bigwedge_{\mu \in \mathcal{B}^{|Y|}} \phi[Y/\mu] \tag{1}$$

Consequently, $\exists X \forall Y.\phi$ has a solution iff (1) is satisfiable.

Hence, a naïve approach to solving the 2QBF problem would be to perform the expansion outlined above and invoke a SAT solver. However, this is infeasible since the formula grows exponentially. We continue by observing that the question whether a certain value vector is a solution can be formulated as a Boolean satisfiability question[1].

Observation 2. *A value ν is a solution to $\exists X \forall Y.\phi$ iff the following formula is unsatisfiable:*

$$\neg\phi[X/\nu] \tag{2}$$

Example 1. Expanding the formula $Q = \exists x \forall y.x \rightarrow y$ yields $x \rightarrow 0 \wedge x \rightarrow 1$, which is equivalent to the formula $\neg x$. Hence, according to Observation 1, $\{x = 0\}$ is a solution of the formula Q. In contrast, Observation 2 tells us that the value vector $\{x = 1\}$ is not a solution since $1 \wedge \neg y$ is satisfiable.

The two observations above motivate the following abstraction-based approach. Instead of considering the full expansion of the given problem (see (1)), we consider only a partial expansion, which will serve as the abstraction in the approach.

Definition 1 (W-abstraction). *Let $W \subseteq \mathcal{B}^{|Y|}$, then the W-abstraction of $\exists X \forall Y.\phi$ is the following formula.*

$$\bigwedge_{\mu \in W} \phi[Y/\mu] \tag{3}$$

A satisfying valuation of a W-abstraction is not necessarily a solution for the given problem. Recall that a solution ν must satisfy $\forall Y.\phi[X/\nu]$. Hence, if ν is not a solution, then there must exist a valuation μ of the variables Y for which $\phi[X/\nu]$ does not hold, i.e. $\phi[X/\nu][Y/\mu] = 0$. This valuation μ is used as the counterexample in the approach.

Definition 2 (counterexample). *If ν and μ are value vectors and μ is a satisfying valuation of $\neg\phi[X/\nu]$ then μ is called a counterexample to ν.*

---

[1] This observation is in fact a consequence of the fact that the problem is in Σ_P^2.

Algorithm 1. CEGAR loop for 2QBF

---

input : $\exists X \forall Y. \phi$
output: (true, ν) if there exists ν s.t. $\forall Y \phi[X/\nu]$,
       $(\text{false}, -)$ otherwise

1 $\omega \leftarrow 1$
2 **while** true **do**
3      $(\text{outc}_1, \nu) \leftarrow \text{SAT}(\omega)$ `// find a candidate solution`
4   **if** $\text{outc}_1 = \text{false}$ **then**
5         $\lfloor$ **return** $(\text{false}, -)$ `// no candidate found`
6      $(\text{outc}_2, \mu) \leftarrow \text{SAT}(\neg\phi[X/\nu])$ `// find a counterexample`
7   **if** $\text{outc}_2 = \text{false}$ **then**
8         $\lfloor$ **return** (true, ν) `// candidate is a solution`
9      $\omega \leftarrow \omega \wedge \phi[Y/\mu]$ `// refine`

Proposition 1. *If a W-abstraction is unsatisfiable then the corresponding 2QBF* PROBLEM *has no solutions. If for a value vector ν there are no counterexamples, then ν is a solution to the* 2QBF PROBLEM.

Proof (sketch). For any $W \subseteq \mathcal{B}^{|Y|}$, the W-abstraction ω is weaker than (1), hence if ω is unsatisfiable, the given problem does not have a solution due to Observation 1. There are no counterexamples to ν iff $\neg\phi[X/\nu]$ is unsatisfiable, which is true only if ν is a solution (Observation 2).

Algorithm 1 shows a pseudo-code representation of the 2QBF CEGAR loop using the notion of abstraction and counterexample defined above. The pseudo-code assumes a satisfiability oracle SAT which for a Boolean formula returns whether it is satisfiable or not. If it is satisfiable, it also returns a satisfying valuation; this information is returned as a pair with the first element representing satisfiability and the second element the valuation (if applicable).

    The algorithm maintains the W-abstraction in the variable ω and it starts with $W = \emptyset$, i.e. $\omega = 1$ (line 1). Each iteration of the loop begins by looking for a solution for the abstraction, this solution is called the *candidate*. If the abstraction is unsatisfiable—there are no candidates—the given problem is unsatisfiable and hence the loop terminates (line 5). If a candidate was found, the algorithm checks whether the candidate is indeed a concrete solution to the given problem or not. If the candidate is a solution, then the loop terminates successfully (line 8). If the candidate is not a concrete solution, the abstraction is refined according to the counterexample. The refinement consists in adding the counterexample μ to the set W, which corresponds to conjoining $\phi[Y/\mu]$ to ω (line 9). Observe that the set W monotonically increases from $\emptyset$ to $\mathcal{B}^{|Y|}$ with one iteration adding one element to it.

Example 2. Let $\phi = (x_1 \vee y_1) \wedge (x_2 \vee y_2)$ then $\neg\phi = (\neg x_1 \wedge \neg y_1) \vee (\neg x_2 \wedge \neg y_2)$ and the following is a possible run of Algorithm 1. Initial $\omega_1 = 1$, yields a candidate

$\text{SAT}(\omega_1) = (\text{true}, \nu_1)$ with $\nu_1 = \{x_1 = 0, x_2 = 0\}$. In turn we obtain a counterexample $\text{SAT}(\neg\phi[X/\nu_1]) = (\text{true}, \mu_1)$ with $\mu_1 = \{y_1 = 1, y_2 = 0\}$. The counterexample yields the refinement $\omega_2 = \omega_1 \wedge x_2 = x_2$. The second iteration yields a candidate $\text{SAT}(\omega_2) = (\text{true}, \nu_2)$ with $\nu_2 = \{x_1 = 0, x_2 = 1\}$ and a counterexample $\text{SAT}(\neg\phi[X/\nu_2]) = (\text{true}, \mu_2)$ with $\mu_2 = \{y_1 = 0, y_0 = 1\}$. The corresponding refinement is $\omega_3 = \omega_2 \wedge x_1 = x_2 \wedge x_1$. The candidate in the third iteration is inevitably $\nu_3 = \{x_1 = 1, x_2 = 1\}$, which is a solution as there are no counterexamples to it ($\text{SAT}(\neg\phi[X/\nu_3]) = (\text{false}, -)$).

4.2 Properties

Here we discuss the correctness and some other properties of the algorithm. A crucial observation is that no counterexample can appear in two distinct iterations of the CEGAR loop, which is stated by the following lemma.

Lemma 1. *Let μ_i and μ_k be counterexamples found in the i-th and k-th iterations of the loop, respectively, where $i < k$. Then $\mu_i \neq \mu_k$.*

Proof (sketch). For contradiction assume that $\mu = \mu_i = \mu_k$ and let ν be a candidate found in the k-th step. The candidate ν satisfies the current abstraction, which is of the form $\omega' \wedge \phi[Y/\mu]$ since the abstraction was refined with μ in the step i. Hence, μ is a model of $\phi[X/\nu]$, which is a contradiction since it is also a model of $\neg\phi[X/\nu]$.

The above lemma ensures that the CEGAR loop will have a finite number of iterations since there is only a finite number of possible counterexamples. However, the algorithm has even a stronger property. Once a counterexample μ is found, all candidates to which μ is a counterexample are eliminated from the space of possible candidates. This is stated formally in the following lemma.

Lemma 2. *Let μ_i be a counterexample found in the i-th iteration of the loop and ν_k be a candidate found in the k-th iteration of the loop, where $k > i$. Then μ_i is not a counterexample to ν_k. In particular, no candidate can appear more than once.*

Proof (sketch). In the i-th iteration of the loop the abstraction has been refined as $\omega = \omega' \wedge \phi[Y/\mu_i]$. Since ν_k must satisfy ω, and therefore also $\phi[Y/\mu_i]$. Consequently, μ_i and ν_k cannot together satisfy $\neg\phi$.

Lemma 1 and Lemma 2 tell us that neither candidates nor counterexamples can repeat in the iteration loop, which yields the following upper bound on the total number of iterations.

Proposition 2. *Let $k = min(|X|, |Y|)$, then Algorithm 1 performs at most 2^k iterations of the loop and requires $O(|\phi| * 2^k)$ space.*

Proof (sketch). Immediate consequence of Lemma 1, Lemma 2, and the fact that there are $2^{|Z|}$ different value assignments to a set of variables Z.

Algorithm 2. CEGAR loop for 2QCNF

---

input : $\exists X \forall Y.(\neg\phi')$
output: (true, ν) if there exists ν s.t. $\forall Y \neg\phi'[X/\nu]$,
          $(\text{false}, -)$ otherwise

1 $\omega \leftarrow 1$
2 **while** true **do**
3      $(\text{outc}_1, \nu) \leftarrow \text{SAT}(\omega)$ `// find a candidate solution`
4   **if** $\text{outc}_1 = \text{false}$ **then**
5 | **return** $(\text{false}, -)$ `// no candidate found`
6      $(\text{outc}_2, \mu) \leftarrow \text{SAT}(\phi'[X/\nu])$ `// find a counterexample`
7   **if** $\text{outc}_2 = \text{false}$ **then**
8 | **return** (true, ν) `// candidate is a solution`
       `// refine`
9      $C \leftarrow \{c \mid c' \in \phi' \wedge c = c'[Y/\mu] \wedge c \neq 1\}$ `// substitute`
10   let z_c be a fresh variable for each $c \in C$
11     $\omega \leftarrow \omega \cup \{\neg z_c \vee \neg l \mid c \in C \wedge l \in c\}$
12     $\omega \leftarrow \omega \cup \{\bigvee_{c \in C} z_c\}$

While the theoretical upper bound given by Proposition 2 is rather crude, we can observe that Lemma 2 gives us some further insight. A refinement according to a counterexample μ prevents the algorithm from finding any candidates to which μ is also a counterexample. In other words, the space of possible candidates is diminished more if the counterexample is a counterexample to many possible candidates. This is illustrated by the following example.

Example 3. Let $\Phi = \exists xy \forall q. ((x \wedge q) \vee (x \wedge \neg q)) \wedge ((y \wedge q) \vee (y \wedge \neg q))$ and consider the following run of the algorithm. The first candidate $\nu_1 = \{x = 0, y = 0\}$ yielding the counterexample $\mu_1 = \{q = 1\}$. The corresponding refinement is $\omega_2 = x \wedge y$. Inevitably, the second candidate $\nu_1 = \{x = 1, y = 1\}$ is a solution to the problem. Observe that μ_1 is a counterexample to all candidates that are not solutions.

More generally, we hypothesize that the algorithm is likely to work well for problems where *one counterexample is a counterexample to many potential candidates*.

5   Algorithm for the 2QCNF PROBLEM

This section looks in more detail at the CNF formulation of the problem. Recall that in 2QCNF PROBLEM the input formula ϕ is of the form $\neg\phi'$ where ϕ' is in CNF. The structure of the algorithm remains the same but we make several observations that enable more efficient implementation. The pseudo-code is presented by Algorithm 2.

First, observe that $\neg\phi[X/\nu] = \phi'[X/\nu]$ since $\phi = \neg\phi'$. Hence, a search for a counterexample to the candidate ν is simplified to $\text{SAT}(\phi'[X/\nu])$ (instead of $\text{SAT}(\neg\phi[X/\nu])$). Second, the refinement $\omega \leftarrow \omega \wedge \phi[Y/\mu]$ now has the form $\omega \leftarrow \omega \wedge \neg\phi'[Y/\mu]$. To maintain ω in CNF, we perform a variant of Plaisted-Greenbaum transformation [30][2]. For each clause $c \in \phi'[Y/\mu]$ introduce a fresh variable z_c and add to ω the clauses $\neg z_c \vee \neg l$ for each literal $l \in c$. Finally, add the clause $\bigvee_{c \in \phi'[Y/\mu]} z_c$. Intuitively, a variable z_c represents that the clause c is false and their disjunction represents that at least one of them is false, thus enforcing $\neg \phi'[Y/\mu]$.

The size of the refinement is trimmed by omitting those clauses that are immediately satisfied by the counterexample μ (this happens whenever μ satisfies at least one literal in the clause). Further, at the implementation level, the incremental interface of the SAT solver is used as ω is gradually strengthened; the variables z_c are reused if the clause c appears in multiple iterations of the CEGAR loop.

5.1 Heuristics

The CEGAR loop relies on two calls to a SAT solver and either of these two calls may yield different models for the same abstraction or candidate, respectively. While the correctness of the algorithm is not affected by which of these models is returned, the overall efficiency of the algorithm may be affected. Here we propose heuristics that determine which candidates and counterexamples are better.

Candidate heuristic. The objective of the heuristic used in computing a candidate is to find such candidates that there likely to be solutions to the original problem $\exists X \forall Y. (\neg \phi')$. Since a solution must satisfy $\forall Y. \neg \phi'$, we propose a heuristic that maximizes the number of unsatisfied clauses in ϕ'. In particular, the satisfiability problem $\mathrm{SAT}(\omega)$ is replaced by the following MAX-SAT problem:

$$
\begin{aligned}
&\{hard\ c \mid c \in \omega\} \\
&\{hard\ \neg z_c \vee \neg l \mid z_c \text{ is a fresh variable } \wedge c \in \phi' \wedge l \in c \wedge \mathrm{var}(l) \notin Y\} \qquad (4)\\
&\{soft\ z_c \mid c \in \phi'\}
\end{aligned}
$$

Counterexample heuristic. In the refinement step we need to consider only those clauses that are not satisfied by the counterexample μ, i.e. $c \in \phi \wedge c[Y/\mu] \neq 1$. Hence, the clause $\bigvee z_c$, added in line 12, has less literals the more clauses are satisfied by μ. Since, in general, short clauses represent stronger constraints then long clauses, we propose the heuristic to look for those counterexamples that maximize the number of satisfied clauses in ϕ'. In particular, the satisfiability problem $\mathrm{SAT}\ (\phi'[X/\nu])$ is replaced by the following MAX-SAT problem:

$$
\begin{aligned}
&\{hard\ c \mid c' \in \phi' \wedge c = c[X/\nu]\} \\
&\{soft\ c \mid c' \in \phi' \wedge c = \{l \mid (l \in c') \wedge \mathrm{var}(l) \notin X\}\}
\end{aligned} \qquad (5)
$$

Implementing heuristics. In both of the aforementioned heuristics the corresponding SAT problem is transformed into a MAX-SAT problem. Solving these MAX-SAT problems in each iteration of the CEGAR loop is not feasible because typically a large number of iterations is required (up to hundreds of thousands) and MAX-SAT is significantly more time-consuming than SAT. Hence, in the implementation we compute

---

[2] As opposed to Tseitin transformation [38], implications in only one direction are introduced.

Table 1. Numbers of solved instances

	struqs	QuBE7.1	qbf2circ	AReQS	AReQS-H
2qbf '10 pre (114)	30	93	37	**101**	**101**
circ pre (117)	6	113	117	**117**	**117**
icore pre (140)	30	23	33	**62**	**62**
robots pre (999)	516	921	647	974	**975**
noprepro (232)	15	47	18	51	**55**
total (1602)	597	1197	852	1305	**1310**

an approximate solution to the MAX-SAT problems by skewing the default decision polarity and variable activity of a SAT solver. Hard clauses are given to the SAT solver as standard clauses without any change. Each soft clause c is represented by the clause $r_c \vee c$ where r_c is a fresh variable. The polarity of the variable r_c is set to 0 and the activity increased. This instructs the SAT solver to set r_c to 0 as soon as possible in the search for a satisfying valuation, which then enforces c to be satisfied. While this approach does not guarantee the optimum, it is commonly used in modern MAX-SAT and PB solvers and has been successfully applied to SAT solving with preference [35].

6   Experimental Results

A prototype implementing Algorithm 2 was developed using MiniSat2.2 as the underlying SAT solver [10]. In the following text we refer to the prototype as *AReQS* (**A**bstraction **Re**finement **Q**BF **S**olver). Two versions of AReQS were evaluated: one that does not use any heuristics (denoted *AReQS*) and the second that uses the heuristics described in Section 5.1 (denoted *AReQS-H*).

For comparison, two QBF solvers were chosen: *struqs* [37] and *QuBE7.1* [16], which are the official and unofficial[3] winner, respectively, of the 2QBF track of the 2010 QBF evaluation [31]. Besides comparing to these two solvers, AReQS was compared to our own tool *qbf2circ*. The tool utilizes a transformation from 2QBF to propositional circumscription [11], and invokes a dedicated *propositional circumscription solver* [20]. Since the dedicated solver is based on similar ideas presented in this paper, the purpose of this translation was to investigate whether a dedicated QBF solver pays off.

A variety of benchmarks was chosen for the empirical evaluation. The sources for the benchmarks were: QBF library [32], QBF evaluation [31], and two well-known Σ_2^P and Π_2^P complete problems. From the QBF library [32] we chose the Robots2D benchmarks, from QBF evaluation the set of problems used in 2010 2QBF track. Entailment in propositional circumscription is a well-known Π_2^P problem and instances from product configuration were used [19]; implicates core is the problem of deciding for a given clause c, a constant k, and a CNF ϕ whether there exists a clause $c' \subseteq c$, s.t. $|c'| < k$ and $\phi \rightarrow c'$, the problem is well known to be Σ_2^P-complete [39][4]. Only problems of

---

[3] QuBE7 was disqualified because of discrepancies, which are already fixed in QuBE7.1.

[4] The problem is usually presented for an implicant rather than implicate, which is easily convertible to the implicate problem by negating the input formula.

the form $\forall\exists$ were considered from the QBF library (this was true for all the problems in the 2QBF track of the QBF-Evaluation); the implicant core problem was directly generated in its negated form (again producing the $\forall\exists$ form). All experimental results were obtained on an Intel Xeon 5160 3GHz, with 4GB of memory, and running Linux. The experiments were obtained with a 1000 seconds time limit and 2GB memory limit.

Our initial experiments showed that all the tested solvers perform extremely poorly when the input problem is not preprocessed. Hence, the preprocessor *sQueezeBF* [13] (part of QuBE7.1[5]) was first applied on the instances (discarding instances solved completely by the preprocessor). A random subset of the aforementioned problems were chosen for the evaluation without the processing (noprepro). Therefore, all the sets instances except for noprepro consist of instances already simplified by sQueezeBF.

Table 1 shows the number of solved instances for each set of benchmarks and solver. The new tool AReQS solves ca. 10% more instances than QuBE7.1 and more than double the instances solved by struqs. The gains achieved with AReQS are uniformly distributed among the classes of problem instances considered. The tool qbf2circ solves more instances than struqs, but less than QuBE7.1. Figure 2 shows a more detail overview of the runtimes with cactus and scatter plots. Both versions of AReQS consistently outperform all the other approaches; QuBE7.1 comes second; struqs performs slightly worse than qbf2circ, and both perform significantly worse than QuBE7.1.

The first scatter plot compares AReQS-H and QuBE7.1 on all the instances combined. This plot shows that AReQS-H not only solves more instances but the majority is solved faster. The last two scatter plots compare the heuristic approach to the non-heuristic approach. In the first of the to scatter plots the times are compared and in the second the number of iterations of the CEGAR loop. The heuristics yield an overall improvement, both in time and iteration count; in a number of instances the improvement is in orders of magnitude.

The experimental results suggest that CEGAR is a promising approach for developing dedicated algorithms for 2QBF. Although the AReQS tool is still a prototype, it consistently outperforms state of the art QBF solvers on several classes of problem instances. Nevertheless, the importance of preprocessing should be noted, and any approach needs preprocessing for achieving good overall performance.

7   Related Work

QBF is a well-known PSPACE-complete problem (e.g. [21]), with a wide range of practical applications [14]. Restrictions on the number of alternations have been used to characterize the polynomial hierarchy [21]. QBF algorithms have been the subject of significant improvements over the last decade [14,33,29]. Examples of recent work can be found in [28,15].

Counterexample guided abstraction refinement was successfully applied in model checking [6,7] and since then it has appeared in various forms. In satisfiability modulo theories (SMT) solvers, CEGAR has been used to abstract first order theories as propositional theories with the use of a SAT solver and decision procedures [1,27,12]. The refinement in these works consists in blocking the abstract solution just found.

---

[5] QuBE itself was then run with the no processing option.

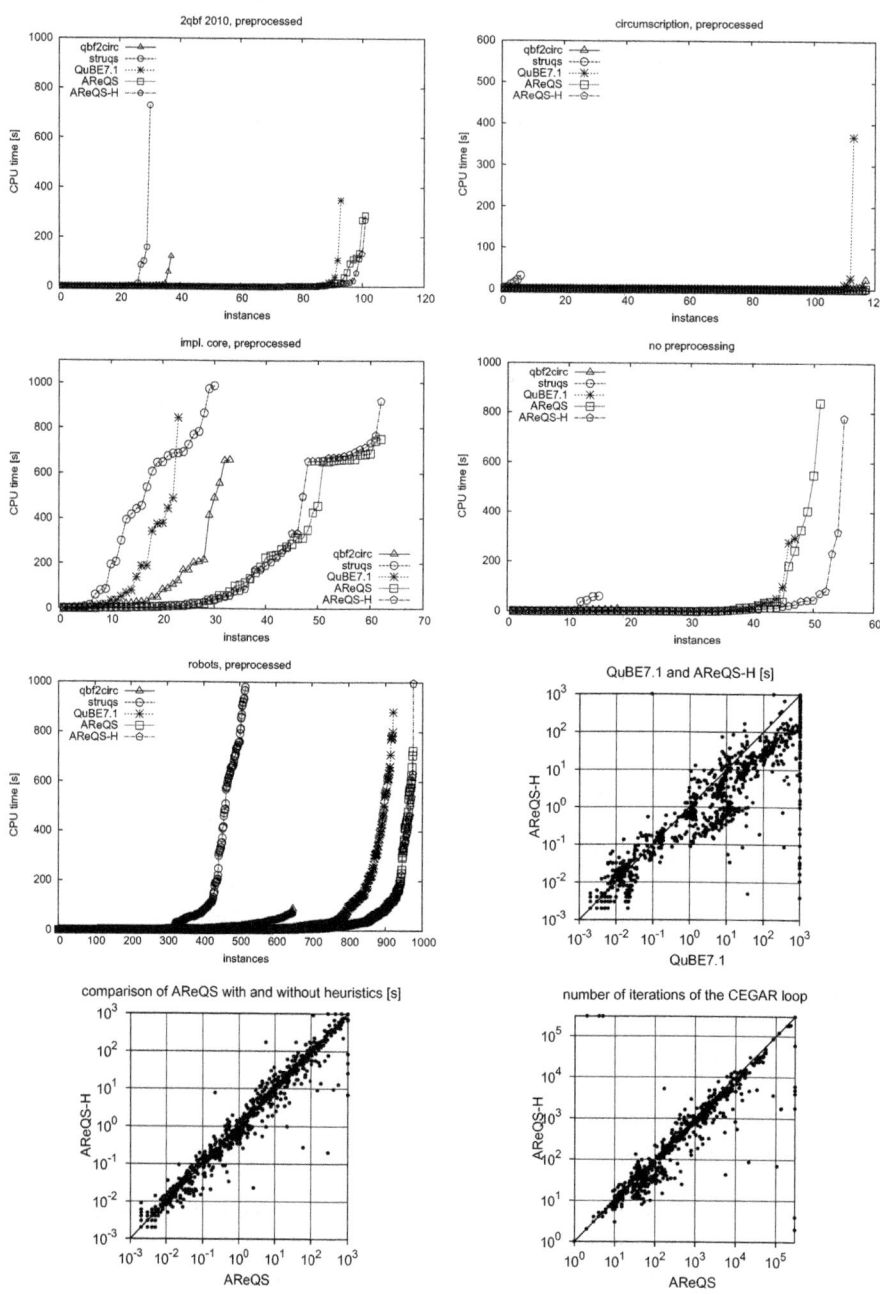

Fig. 2. Overview of the experimental results

Rintanen uses an idea similar to the W-abstraction in a technique called *inversion of quantifiers* in the context of search-based QBF solving [34]. This abstraction is populated randomly (whereas in our case it is determined by counterexamples); the abstraction does not drive the algorithm, instead, it is used as a simplification technique in a search-based algorithm. SAT-based algorithms were used for solving 2QBF formulas in the context of bounded model checking [9] and in planning [4]. However, these algorithms are highly specialized for the problems in question and it is not clear how they could be generalized for arbitrary 2QBF.

A variation of the CEGAR approach used in SMT [1,27,12] was applied to certain special forms of 2QBF. Mneimneh and Sakallah compute the *vertex eccentricity* of a transition system (also known as *the diameter*) [25]. Browning and Remshagen tackle the validity of *Q-ALL SAT* [3]. Besides the fact that the algorithms presented in these articles are specialized to subsets of 2QBF, there is also an important difference in the refinement they use. Once a candidate is found, it is simply blocked so that it is not found again. That means that the set of possible candidates is explored one by one. In contrast, in our approach multiple candidates are removed upon each iteration (see Lemma 2). The one-by-one iteration over candidates not only affects the theoretical upper bound for number of iterations (Proposition 2) but also is likely to lead to an unmanageable number of iterations, especially for unsatisfiable instances where *all* possible candidates need to be considered. Browning and Remshagen address this problem by a heuristic for decreasing the size of the blocking clause. This heuristic is computationally expensive since it requires additional calls to the solver, does not provide any theoretical guarantee, and it is unclear how it could be generalized for arbitrary QBF.

More recently, there has been increased interest in the CEGAR approach in the context of quantification [20,40,26]. Out of these works, our own work on *propositional circumscription entailment* is probably the most similar [20]. Although the algorithm based on propositional circumscription can be used to solve 2QBF, e.g. by using the well-known reduction from [11], the new dedicated 2QBF algorithm is shown to outperform this approach. The dedicated algorithm exploits the problem representation, and this provides a natural performance edge.

The work described in [26] is for quantified linear real arithmetic. Although this work could be used on QBF formulas, the key techniques do not aim Boolean formulas. Finally, the work reported in [40] is solving a computationally harder decision problem, namely quantified bit-vector formulas. This means that [40] can be used to solve arbitrary QBFs, but it is also unlikely to scale as well as a dedicated algorithm.

8 Conclusions

This paper develops a new algorithm for the 2QBF and 2QCNF problems. The algorithm exploits the counterexample-guided abstraction refinement paradigm [6,7], and is shown to outperform the best peforming QBF solvers from the most recent QBF Evaluation [28]. Although the work builds on recent work on using counterexample-guided abstraction for handling quantification [20,40,26], the algorithm exploits the natural properties of the problem formulation, and is shown to outperform approaches based on mapping QBF to another domain [20]. Refining the abstraction in some sense corresponds to traversing the search space with the use of learned clauses [23,17]. However, there are some important differences. Learned clauses can be removed without affecting the correctness of the algorithm, which is not the case for the abstraction refinements. This has the adversary effect that the abstraction algorithm requires exponential space. One the other hand, the CEGAR-based search does not require traversal in any particular order, which enables us to focus on *likely solutions*. This advantage is demonstrated by the heuristics developed for the approach (Section 5.1). Further, we hypothesize that the approach will work well on certain types of problems (Section 4.2).

The promising experimental results motivate extending the work to arbitrary levels of the polynomial hierarchy and to general QBF. Nevertheless, many interesting applications lie in the second level of the polynomial hierarchy and this paper suggests that dedicated algorithm may in general represent the best approach for achieving the most efficient solutions.

Acknowledgement. This work is partially supported by SFI PI grant BEACON (09/IN.1/I2618), EC FP7 project MANCOOSI (214898), FCT grant ATTEST (CMU-PT/ELE/0009/2009), and INESC-ID multiannual PIDDAC program funds.

References

1. Barrett, C.W., Dill, D.L., Stump, A.: Checking satisfiability of first-order formulas by incremental translation to SAT. In: Brinksma, E., Larsen, K.G. (eds.) CAV 2002. LNCS, vol. 2404, pp. 236–249. Springer, Heidelberg (2002)
2. Biere, A., Heule, M., van Maaren, H., Walsh, T. (eds.): Handbook of Satisfiability, Frontiers in Artificial Intelligence and Applications, vol. 185. IOS Press, Amsterdam (2009)
3. Browning, B., Remshagen, A.: A SAT-based solver for Q-ALL SAT. In: Menezes, R. (ed.) ACM Southeast Regional Conference, pp. 30–33. ACM, New York (2006)
4. Castellini, C., Giunchiglia, E., Tacchella, A.: Improvements to SAT-based conformant planning. In: European Conference on Planning (2001)
5. Chauhan, P., Clarke, E.M., Kukula, J.H., Sapra, S., Veith, H., Wang, D.: Automated abstraction refinement for model checking large state spaces using SAT based conflict analysis. In: Aagaard, M.D., O'Leary, J.W. (eds.) FMCAD 2002. LNCS, vol. 2517, pp. 33–51. Springer, Heidelberg (2002)
6. Clarke, E.M., Grumberg, O., Jha, S., Lu, Y., Veith, H.: Counterexample-guided abstraction refinement. In: Emerson, E.A., Sistla, A.P. (eds.) CAV 2000. LNCS, vol. 1855, pp. 154–169. Springer, Heidelberg (2000)
7. Clarke, E.M., Grumberg, O., Jha, S., Lu, Y., Veith, H.: Counterexample-guided abstraction refinement for symbolic model checking. J. ACM 50(5), 752–794 (2003)
8. Clarke, E.M., Gupta, A., Strichman, O.: SAT-based counterexample-guided abstraction refinement. IEEE Trans. on CAD of Integrated Circuits and Systems 23(7), 1113–1123 (2004)
9. Dershowitz, N., Hanna, Z., Katz, J.: Bounded model checking with QBF. In: Bacchus, F., Walsh, T. (eds.) SAT 2005. LNCS, vol. 3569, pp. 408–414. Springer, Heidelberg (2005)
10. Eén, N., Sörensson, N.: An extensible SAT-solver. In: Giunchiglia, E., Tacchella, A. (eds.) [18]
11. Eiter, T., Gottlob, G.: Propositional circumscription and extended closed-world reasoning are Π_2^P-complete. Theor. Comput. Sci. 114(2), 231–245 (1993)
12. Flanagan, C., Joshi, R., Ou, X., Saxe, J.B.: Theorem proving using lazy proof explication. In: Hunt Jr., W.A., Somenzi, F. (eds.) CAV 2003. LNCS, vol. 2725, pp. 355–367. Springer, Heidelberg (2003)

13. Giunchiglia, E., Marin, P., Narizzano, M.: An effective preprocessor for QBF pre-reasoning. In: 2nd International Workshop on Quantification in Constraint Programming, QiCP (2008)
14. Giunchiglia, E., Marin, P., Narizzano, M.: Reasoning with quantified boolean formulas. In: Biere, et al (eds.) [2], pp. 761–780
15. Giunchiglia, E., Marin, P., Narizzano, M.: sQueezeBF: An effective preprocessor for QBFs based on equivalence reasoning. In: Strichman, O., Szeider, A. (eds.) [36], pp. 85–98
16. Giunchiglia, E., Narizzano, M., Tacchella, A.: QuBE++: An Efficient QBF Solver. In: Hu, A.J., Martin, A.K. (eds.) FMCAD 2004. LNCS, vol. 3312, pp. 201–213. Springer, Heidelberg (2004)
17. Giunchiglia, E., Narizzano, M., Tacchella, A.: Clause/term resolution and learning in the evaluation of quantified boolean formulas. J. Artif. Intell. Res (JAIR) 26, 371–416 (2006)
18. Giunchiglia, E., Tacchella, A. (eds.): SAT 2003. LNCS, vol. 2919. Springer, Heidelberg (2004)
19. Janota, M., Botterweck, G., Grigore, R., Marques-Silva, J.: How to complete an interactive configuration process? In: van Leeuwen, J., Muscholl, A., Peleg, D., Pokorný, J., Rumpe, B. (eds.) SOFSEM 2010. LNCS, vol. 5901, pp. 528–539. Springer, Heidelberg (2010)
20. Janota, M., Grigore, R., Marques-Silva, J.: Counterexample guided abstraction refinement algorithm for propositional circumscription. In: Janhunen, T., Niemelä, I. (eds.) JELIA 2010. LNCS, vol. 6341, pp. 195–207. Springer, Heidelberg (2010)
21. Kleine-Büning, H., Bubeck, U.: Theory of quantified boolean formulas. In: Biere, et al (eds.) [2]
22. Li, C.M., Manyà, F.: Maxsat, hard and soft constraints. In: Biere, et al (eds.) [2], pp. 613–631
23. Marques-Silva, J., Lynce, I., Malik, S.: Conflict-driven clause learning SAT solvers. In: Biere, et al (eds.) [2], pp. 131–153
24. Meyer, A.R., Stockmeyer, L.J.: The equivalence problem for regular expressions with squaring requires exponential space. In: Symposium Switching and Automata Theory (October 1972)
25. Mneimneh, M.N., Sakallah, K.A.: Computing vertex eccentricity in exponentially large graphs: QBF formulation and solution. In: Giunchiglia, E., Tacchella, A. (eds.) [18], pp. 411–425
26. Monniaux, D.: Quantifier elimination by lazy model enumeration. In: Touili, T., Cook, B., Jackson, P. (eds.) CAV 2010. LNCS, vol. 6174, pp. 585–599. Springer, Heidelberg (2010)
27. de Moura, L.M., Rueß, H., Sorea, M.: Lazy theorem proving for bounded model checking over infinite domains. In: Voronkov, A. (ed.) CADE 2002. LNCS (LNAI), vol. 2392, pp. 438–455. Springer, Heidelberg (2002)
28. Peschiera, C., Pulina, L., Tacchella, A., Bubeck, U., Kullmann, O., Lynce, I.: The seventh QBF solvers evaluation (QBFEVAL'10). In: Strichman, O., Szeider, S. (eds.) [36], pp. 237–250
29. Plaisted, D.A., Biere, A., Zhu, Y.: A satisfiability procedure for quantified boolean formulae. Discrete Applied Mathematics 130(2), 291–328 (2003)
30. Plaisted, D.A., Greenbaum, S.: A structure-preserving clause form translation. J. Symb. Comput. 2(3), 293–304 (1986)
31. QBF solver evaluation portal, http://www.qbflib.org/index_eval.php
32. The Quantified Boolean Formulas satisfiability library, http://www.qbflib.org/
33. Ranjan, D.P., Tang, D., Malik, S.: A comparative study of 2QBF algorithms. In: SAT (2004)
34. Rintanen, J.: Improvements to the evaluation of quantified Boolean formulae. In: Dean, T. (ed.) IJCAI, pp. 1192–1197. Morgan Kaufmann, San Francisco (1999)
35. Rosa, E.D., Giunchiglia, E., Maratea, M.: Solving satisfiability problems with preferences. Constraints 15(4), 485–515 (2010)
36. Strichman, O., Szeider, S. (eds.): SAT 2010. LNCS, vol. 6175. Springer, Heidelberg (2010)

37. STRUQS: A Structural QBF Solver, www.qbflib.org/DESCRIPTIONS/struqs.pdf
38. Tseitin, G.S.: On the complexity of derivation in propositional calculus. Studies in Constructive Mathematics and Mathematical Logic 2(115-125), 10–13 (1968)
39. Umans, C.: The minimum equivalent DNF problem and shortest implicants. J. Comput. Syst. Sci. 63(4), 597–611 (2001)
40. Wintersteiger, C.M., Hamadi, Y., de Moura, L.: Efficiently solving quantified bit-vector formulas. In: Proceedings of Formal Methods in Computer Aided Design FMCAD (October 2010)

Transformations into Normal Forms
for Quantified Circuits[*]

Hans Kleine Büning[1], Xishun Zhao[2], and Uwe Bubeck[1]

[1] Computer Science Institute, University of Paderborn, Germany
kbcsl@upb.de , bubeck@upb.de
[2] Institute of Logic and Cognition, Sun Yat-sen University, PR China
hsszxs@mail.sysu.edu.cn

Abstract. We consider the extension of Boolean circuits to quantified Boolean circuits by adding universal and existential quantifier nodes with semantics adopted from quantified Boolean formulas (QBF). The concept allows not only prenex representations of the form $\forall x_1 \exists y_1 ... \forall x_n \exists y_n\ c$ where c is an ordinary Boolean circuit with inputs $x_1, ..., x_n, y_1, ..., y_n$. We also consider more general non-prenex normal forms with quantifiers inside the circuit as in non-prenex QBF, including circuits in which an input variable may occur both free and bound. We discuss the expressive power of these classes of circuits and establish polynomial-time equivalence-preserving transformations between many of them. Additional polynomial-time transformations show that various classes of quantified circuits have the same expressive power as quantified Boolean formulas and Boolean functions represented as finite sequences of nested definitions (NBF). In particular, universal quantification can be simulated efficiently by circuits containing only existential quantifiers if overlapping scopes of variables are allowed.

1 Introduction

Boolean circuits are a powerful concept to store propositional formulas. On the one hand, they can suitably represent important structural information, and on the other hand, they allow sharing of common subexpressions. For example, a formula like $(\alpha_1 \vee \beta_1) \wedge (\alpha_1 \vee \beta_2) \wedge (\alpha_2 \vee \beta_1) \wedge (\alpha_2 \vee \beta_2)$ with arbitrary subformulas $\alpha_1, \alpha_2, \beta_1, \beta_2$ can be represented by a circuit in which the value of each of those subformulas is computed only once and then reused multiple times by fanout. Fig. 1 shows a simple circuit for the formula $(\neg x \vee (x \wedge \neg y)) \wedge ((x \wedge \neg y) \vee z)$ in which $(x \wedge \neg y)$ is shared. Under favorable circumstances, sharing can lead to significantly shorter encodings of formulas as circuits. Accordingly, circuits have been used successfully also for SAT and QBF solvers, e.g. in [9, 10, 11, 13].

Similar to other representations like [14], Boolean circuits can be extended by allowing quantifiers. That is, a circuit can contain universal and existential quantifier nodes in addition to propositional logic gates. Typically, existing work on

[*] Research partially supported by DFG grant KL 529/QBF and NSFC grant 60970040.

K.A. Sakallah and L. Simon (Eds.): SAT 2011, LNCS 6695, pp. 245–258, 2011.

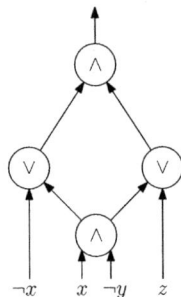

Fig. 1. Circuit in negation normal form

quantified circuits, e.g. [2], has focused on decision problems for circuits in prenex form $\forall x_1 \exists y_1 ... \forall x_n \exists y_n \ c(x_1, y_1, ..., x_n, y_n)$, where c is an ordinary quantifier-free Boolean circuit and quantification is applied subsequently on its input variables. In this paper, we want to investigate the expressive power of quantified circuits not only in prenex form, but also in arbitrary negation normal form where quantifiers are allowed inside the circuit. The most general case that we consider covers circuits in which an input variable may occur both free and bound. In QBF, it is easy to obtain unique variable names and prenex representations by renaming bound variables. We will see that such renamings are problematic for quantified circuits due to the sharing of subcircuits, but this paper shows that there still exist efficient transformations between many classes of quantified circuits. Another main contribution is the interesting result that universal quantification can be simulated in linear time by negation normal form circuits containing only existential quantifiers if overlapping variable scopes are allowed. For QBF, this appears to be possible only for special classes such as quantified Horn or 2-CNF formulas [3, 5], unless the polynomial hierarchy collapses.

2   Boolean Circuits and Propositional Formulas

We begin our discussion with a brief review of basic relationships between circuits and propositional formulas. A *Boolean circuit* is a directed acyclic graph with one outgoing edge (the *sink*) and multiple input nodes labeled with Boolean variables. The other nodes are AND-, OR-, and NOT-gates that each have two (AND and OR) or one (NOT) incoming edges and an arbitrary number of outgoing edges. We let $\mathcal{C}$ be the class of circuits in *negation normal form* (NNF), that means circuits in which the inner nodes are only AND- and OR-gates and the inputs are variables x and negated variables $\neg x$. Fig. 1 shows an example of a circuit in NNF. The *length* or *size of a circuit* is the number of gates, including negations associated with input variables. By the laws of De Morgan and the elimination of double negations, any circuit can be transformed in linear time in NNF, although in the worst case the size of the circuit might double. Subsequently, we consider only circuits in NNF, because we later want to avoid quantifier nodes being negated. The negation of a quantifier essentially inverts its meaning, which is in particular problematic in combination with subcircuit sharing, because one quantifier might then be shared in its original and in its negated form, so a variable would be existentially and universally quantified in the same subcircuit. For clarity, we sometimes use the equivalence operator (written as =). But since operations such as equivalence, and also implication or XOR, implicitly contain negation and therefore cause the same problems with quantifiers, we consider such operations only as abbreviations which must be expanded in the actual representation of the circuit.

For propositional formulas, we allow the same operators $\wedge$ (AND), $\vee$ (OR) and $\neg$ (NOT) and propositional variables. Similar to circuits, a formula is in negation normal form if the negation symbols occur only directly in front of variables. The *length of a formula* is the number of variable occurrences.

Definition 1. *(Propositional Formulas and Circuits)*
1. BF $:= \{\phi \mid \phi$ *is propositional formula in negation normal form*$\}$
2. $\mathcal{C} := \{c \mid c$ *is a Boolean circuit in negation normal form*$\}$

Let α and β be two formulas or two circuits or a mixed pair of one formula and one circuit, and let $z_1, ..., z_r$ be the union of all variables which occur in α or β inside a formula or in a circuit input. Then α and β are *(logically) equivalent*, in symbols $\alpha \approx \beta$, if and only if for every truth assignment to $z_1, ..., z_r$ it holds that α and β evaluate to the same truth value after substituting the assigned truth values for $z_1, ..., z_r$. This is significantly stronger than *satisfiability equivalence*, which requires that if one of the formulas or circuits is satisfied by some assignment to $z_1, ..., z_r$, the other one must also have some (possibly different) satisfying assignment to $z_1, ..., z_r$. Satisfiability equivalence is sometimes too weak: for example, replacing a term inside a larger formula with a different term is in general only sound if both terms are logically equivalent.

For circuit or formula classes A and B, we write $A \leq_p B$ to express that there are polynomial-time transformations from A to B. That means $A \leq_p B$ if and only if there is a poly-time mapping T, such that $T(a) \in B$ and $T(a) \approx a$ for each $a \in A$. $A =_p B$ is an abbreviation for $A \leq_p B$ and $B \leq_p A$.

There is obviously a close relationship between Boolean circuits and propositional formulas: Every propositional formula can be considered as a circuit with fanout 1, and vice versa. Fanout 1 means that every node has exactly one outgoing edge and there is no sharing of subcircuits. On the other hand, an arbitrary Boolean circuit can be encoded as a formula when we label the edges of the circuit with new auxiliary variables and describe the gates by propositional clauses over these auxiliary variables [1]. For example, we obtain $y = x_1 \wedge x_2$ for an AND-node having incoming edges labeled with x_1 and x_2 and output edges labeled with y. This can be performed in linear time, but the resulting formula is in general only satisfiability equivalent to the circuit, because adding new variables typically destroys logical equivalence. To achieve full logical equivalence, the auxiliary variables must be bound by existential quantifiers, which are formally introduced in the following section.

3   Quantified Boolean Formulas

Quantified Boolean formulas (QBF) extend propositional logic with universal ($\forall$) and existential ($\exists$) quantifiers over variables. For example, $\forall x(\neg x \vee (\exists y(y \vee x \vee z)))$ is a quantified Boolean formula. Variables on which a quantifier is applied are called *bound* variables, and variables which are not bound by a quantifier are *free*. In the example, x and y are bound, and z is free. $\forall x \; \phi(x)$ is defined to be true if and only if $\phi(0)$ is true *and* $\phi(1)$ is true, and $\exists y \; \phi(y)$ means that $\phi(0)$ *or* $\phi(1)$ is true. To save parentheses, we assume that the logical connectives have a higher binding priority than the quantifiers, so we could also write the previous example as $\forall x \; \neg x \vee (\exists y \; y \vee x \vee z)$. As for propositional formulas, the length of a quantified Boolean formula is the number of variable occurrences, but now including occurrences with quantifiers. Accordingly, the example has length 6.

A formula $\Phi(z_1, ..., z_r)$ with free variables $z_1, ..., z_r$ is *satisfiable* if and only if there exists a truth assignment τ to the free variables such that $\Phi(\tau(z_1), ..., \tau(z_r))$ is true. Here, $\Phi(\tau(z_1), ..., \tau(z_r))$ denotes the substitution of the truth values in τ for the free variables in Φ. Two quantified Boolean formulas $\Phi(z_1, ..., z_r)$ and $\Psi(w_1, ..., w_s)$ with free variables $z_1, ..., z_r$ and $w_1, ..., w_s$ are logically equivalent if and only if for every truth assignment τ to the free variables $z_1, ..., z_r, w_1, ..., w_s$ both formulas evaluate to the same truth value. This means that the bound variables are not directly considered when checking for logical equivalence, which makes them local to the respective formula. The ability to introduce new local variables without losing full equivalence is a powerful advantage over ordinary propositional calculus. A propositional formula can be considered as a special case of a quantified Boolean formula in which all variables are free. Similarly, all input variables in a Boolean circuit can be treated as free variables. Logical equivalence between QBF formulas is then a generalization of the equivalence criterion presented in the previous section and can naturally be extended to mixed pairs from all three representations.

Negation normal form is defined for QBF as it is for propositional formulas and can be achieved with the additional equivalences $\neg(\forall x \; \Phi) \approx \exists x \neg \Phi$ and $\neg(\exists x \; \Phi) \approx \forall x \neg \Phi$. A QBF formula Φ is in *prenex form* if $\Phi = Q_1 v_1 ... Q_k v_k \; \phi$ with quantifiers $Q_i \in \{\forall, \exists\}$ and a propositional formula ϕ. We call $Q := Q_1 v_1 ... Q_k v_k$ the *prefix* and ϕ the *matrix* of Φ.

Definition 2. *(Quantified Boolean Formulas)*
1. QBF := $\{\Phi \mid \Phi$ *is a quantified Boolean formula in negation normal form*$\}$
2. $\exists$BF := $\{\Phi \mid \Phi \in$ QBF *and* Φ *contains only existential quantifiers*$\}$
3. pQBF := $\{\Phi \mid \Phi \in$ QBF *in prenex form*$\}$

An arbitrary QBF can be transformed in linear time into prenex form. The result is in general not unique, and different prenexing strategies have been widely investigated, e.g. in [8]. Following the notation $\leq_p$ and $=_p$, we write $\leq_{linear}$ and $=_{linear}$ for linear-time transformations.

Proposition 1. QBF $=_{linear}$ pQBF

4   Nested Boolean Functions

Before we introduce quantified circuits, we briefly consider another representation of Boolean functions which appears rather different at first glance, but will soon turn out to be closely related to both quantified circuits and QBF. Every Boolean function can be defined by a suitable set of initial functions, for example AND, OR and NOT, and a composition of these functions. Instead of a fixed set of starting functions, we can also allow arbitrary propositional formulas as starting functions. In order to illustrate the idea, we present a short example:

Let $f_1(x, y) := x \vee \neg y$ and $f_2(z, w) := z \wedge w$ be two initial functions, and let
$f_3(x_1, x_2) := f_2(0, f_1(x_1, x_2))$ and
$f_4(x, y, z) := f_2(f_3(x, z), f_1(z, y))$ be compound functions.
Then $f_4(x, y, z)$ is equivalent to the propositional formula $(0 \wedge (x \vee \neg z)) \wedge (z \vee \neg y)$.

Such definition schemes for Boolean functions have been introduced in [6] as *Boolean Programs*. But this name is also used for different concepts in other fields. To avoid confusion, we use the name *Nested Boolean Functions* (NBF).

Definition 3. *A nested Boolean function (NBF) is a finite sequence $D(f_k) = (f_1, ..., f_k)$ of definitions of Boolean functions. For fixed $t \in \{1, ..., k\}$, it contains*

- *initial functions $f_1, ..., f_t$, which are each defined by $f_i(\mathbf{x}^i) := \alpha_i(\mathbf{x}^i)$ for a propositional formula α_i over variables $\mathbf{x}^i := (x^{i,1}, ..., x^{i,n_i})$, and*
- *compound functions $f_{t+1}, ..., f_k$ of the form $f_i(\mathbf{x}^i) := f_{j_0}(f_{j_1}(\mathbf{x}_1^i), ..., f_{j_r}(\mathbf{x}_r^i))$ for previously defined functions $f_{j_0}, ..., f_{j_r} \in \{f_0, ..., f_{i-1}\}$. The arguments $\mathbf{x}_1^i, ..., \mathbf{x}_r^i$ are tuples containing variables in $\mathbf{x}^i$ or Boolean constants, such that the arity of $\mathbf{x}_l^i$ matches the arity of f_{j_l} and r is the arity of f_{j_0}.*

We call f_k the defined Boolean function. *The* length *of a NBF $D(f_k)$ is $|D(f_k)|$: $= |f_1| + ... + |f_k|$, where $|f_i|$ is the total number of occurrences of constants, variables and function symbols on the right hand side of the defining equation of f_i.*

A Boolean circuit $c(\mathbf{x})$ with input variables $\mathbf{x}$ can be represented as a finite sequence of definitions $D(f_c(\mathbf{x})) \in$ NBF with $c(\mathbf{x}) \approx f_c(\mathbf{x})$. The initial functions are $f_{id}(x) := x, f_\neg(x) := \neg x, f_\wedge(x, y) := x \wedge y$ and $f_\vee(x, y) := x \vee y$. Bottom-up, we assign a function to each edge in the circuit. For an input x or $\neg x$, we simply use $f_{id}(x)$ or $f_\neg(x)$. For the output edges of an AND-node over incoming edges associated with $g(\mathbf{x})$ and $h(\mathbf{y})$, we choose $f(\mathbf{x} \cap \mathbf{y}) := f_\wedge(g(\mathbf{x}), h(\mathbf{y}))$. Here, $\mathbf{x} \cap \mathbf{y}$ is the tuple of all variables occurring in g and h, without multiple occurrences and in arbitrary order. Analogously, we assign functions to the OR-nodes and NOT-nodes (for non-NNF circuits). It is not difficult to see that the function associated with the outgoing edge of the circuit is equivalent to the circuit. We have already mentioned a linear-time encoding of circuits as existentially quantified Boolean formulas (Section 2). Now, the number of function symbols in the resulting NBF is again linear in the size of the circuit, but the length of the NBF also includes occurrences of variables as arguments. We obtain a time and space bound of $O(|\mathbf{x}| \cdot |c(\mathbf{x})|)$, where $|\mathbf{x}|$ is the number of input variables and $|c(\mathbf{x})|$ is the circuit size. Subsequently, we use the term *v-linear* for $O(|v|\cdot|A|)$, where $|A|$ is the length of an expression A and $|v|$ the number of variables in A.

Proposition 2. $\mathcal{C} \leq_{v\text{-}linear}$ NBF

For an arbitrary NBF $D(f_k) = (f_1, ..., f_k)$, the problem of deciding whether $f_k(a_1, ..., a_{n_k}) = 1$ for given arguments $a_1, ..., a_{n_k} \in \{0, 1\}$ has been shown to be PSPACE-complete [6]. This immediately implies also the PSPACE-completeness of the NBF satisfiability problem, i.e. the problem of determining whether there exists a choice of arguments for which the defined function is true. This PSPACE-completeness suggests a close relationship to quantified Boolean formulas. In fact, it is not difficult to encode QBF formulas as NBFs. For example, let $\Phi(\mathbf{z}) = \forall x \exists y \, \phi(x, y, \mathbf{z})$ be a pQBF formula with matrix ϕ and free variables $\mathbf{z}$. First, we define the initial functions $f_\wedge(x_1, x_2) := x_1 \wedge x_2$, $f_\vee(x_1, x_2) := x_1 \vee x_2$ and $f_1(x, y, \mathbf{z}) := \phi$. Then we simulate the existential quantifier by means of the $\vee$ function, using the equivalence $\exists y \, \phi(x, y, \mathbf{z}) \approx \phi(x, 0, \mathbf{z}) \vee \phi(x, 1, \mathbf{z})$. This leads to the definition $f_2(x, \mathbf{z}) := f_\vee(f_1(x, 0, \mathbf{z}), f_1(x, 1, \mathbf{z}))$, and analogously $f_3(\mathbf{z}) := f_\wedge(f_2(0, \mathbf{z}), f_2(1, \mathbf{z}))$ to simulate the universal quantifier by the $\wedge$ function. By construction, we obtain $f_3(\mathbf{z}) \approx \Phi(\mathbf{z})$. Because of the need to count the variables in the arguments, the length of the resulting definition is in general not linear in the length of the formula, but only v-linear.

The inverse direction from NBF to QBF is less intuitive. By a general argument, every Boolean function defined by a NBF can be simulated by a poly-space Turing machine [6]. And a poly-space Turing machine can be encoded as a QBF of polynomial length [12]. Recently, a linear-time equivalence-preserving transformation from NBF to pQBF has been found [4].

Lemma 1. NBF $\leq_{linear}$ pQBF, QBF *and* pQBF, QBF $\leq_{v\text{-}linear}$ NBF.

Whether there exists a linear-time transformation from (p)QBF to NBF is an open question and closely related to the question whether for circuits there is a linear-time transformation to NBF.

5   Quantified Circuits

Similar to quantified Boolean formulas being an extension of propositional formulas, we now introduce *quantified circuits* as Boolean circuits which may in addition contain nodes labeled with $\forall x$ or $\exists x$ for a propositional variable x. Each of these *quantifier nodes* has exactly one incoming edge and an arbitrary number of outgoing edges. Examples of quantified circuits are given in Fig. 2. We say that a variable x occurs *bound* in a quantified circuit if there is a path from the input x or $\neg x$ to the sink which passes through a quantifier node $\exists x$ or $\forall x$. On the other hand, x occurs *free* if there is a path from x or $\neg x$ to the sink which contains no node $\exists x$ or $\forall x$. In the left circuit in Fig. 2, x is bound, z is free, and y occurs free and bound.

We borrow the semantics of quantified circuits from quantified Boolean formulas by mapping to each edge of a quantified circuit a QBF formula in the

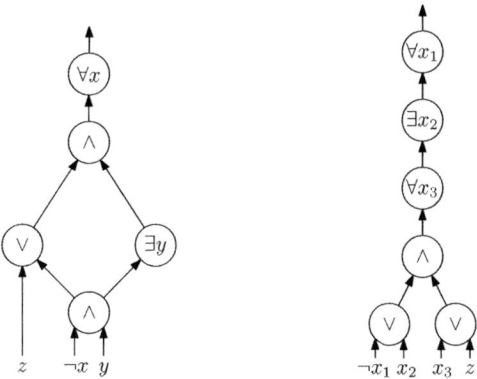

Fig. 2. Quantified circuits in non-prenex form (left) and prenex form (right)

following bottom-up manner: The formula for an input node x or $\neg x$ is x or $\neg x$ itself. The output edge of an AND-node over incoming edges associated with formulas $\alpha(\mathbf{x})$ and $\beta(\mathbf{y})$ is then labeled with $\alpha(\mathbf{x}) \wedge \beta(\mathbf{y})$. OR- and NOT-nodes are treated analogously. When we encounter a quantifier node $\exists x$ that has an incoming edge labeled with $\alpha(x, \mathbf{y})$, we label its outgoing edges with $\exists x\, \alpha(x, \mathbf{y})$, and analogously $\forall x\, \alpha(x, \mathbf{y})$ for nodes $\forall x$. Finally, the interpretation of the circuit is defined as the value of the formula associated with its output when the variables are assigned as determined by the circuit inputs. For example, the left circuit in Fig. 2 is equivalent to $\forall x\ ((z \vee (\neg x \wedge y)) \wedge (\exists y(\neg x \wedge y)))$.

This semantics means that when we ignore the direction of the edges, a quantified circuit can be understood as the extension of a syntax tree of a quantified Boolean formula with additional sharing of subformulas. That is also the reason why we prefer to draw those circuits with the sink on the top.

When we construct the associated QBF formula as in the semantics definition, it is obvious that the length of the formula can be super-polynomial in the length (size) of the circuit, since we lose the sharing of subcircuits. It turns out that this can be avoided easily when we represent the quantified circuit as NBF. The encoding is the same as the one for ordinary Boolean circuits from Section 4, with the following extension: For an $\exists x$ node (a $\forall x$ node, respectively) that has an incoming edge labeled with $f(x, \mathbf{y})$, we define a new function symbol $g(\mathbf{y}) := f_{\vee}(f(0, \mathbf{y}), f(1, \mathbf{y}))$ $(g(\mathbf{y}) := f_{\wedge}(f(0, \mathbf{y}), f(1, \mathbf{y})),$ respectively).

Proposition 3. Quantified Circuits $\leq_{v\text{-}linear}$ NBF

5.1   Normal Forms

For QBF, various normal forms are well known. We now define analogous normal forms for quantified circuits and analyze them in the following subsections.

Definition 4. *Let c be a quantified circuit. Then we say:*

1. c is in negation normal form (NNF) *if each inner node is either an AND-node, an OR-node or a quantifier node, and the inputs are variables x or negated variables $\neg x$. That means only inputs can be negated.*

2. *c is in* prenex form *if every successor node of a quantifier node is a quantifier node, too. That means quantifiers are only allowed in front of the sink.*
3. *c is* pure *if no variable has both bound and free occurrences in c.*

The quantified circuits shown earlier in Fig. 2 are both in negation normal form. The circuit on the left is neither in prenex form nor pure, while the one on the right is in prenex form and thus also pure. Subsequently, we assume that all quantified circuits are in negation normal form, unless stated otherwise.

Proposition 4. *A quantified circuit c is in negation normal form (in prenex form, or pure, respectively) if and only if the associated QBF formula is in negation normal form (in prenex form, or pure, respectively).*

In QBF, it is easy to transform NNF formulas into equivalent pure or even prenex formulas by renaming and shifting of quantifiers. Consider the example $\Phi = \forall x \, ((z \vee (\neg x \wedge y)) \wedge (\exists y \, (\neg x \wedge y)))$ where y occurs free and bound. Then $\Phi \approx \forall x \, ((z \vee (\neg x \wedge y)) \wedge (\exists u \, (\neg x \wedge u))) \approx \forall x \exists u \, ((z \vee (\neg x \wedge y)) \wedge (\neg x \wedge u))$. But such renamings are problematic for non-prenex quantified circuits due to shared subcircuits. Consider again the left circuit in Fig. 2. If we rename $\exists y$ into $\exists u$, the resulting term $(\neg x \wedge u)$ can no longer be shared with $(\neg x \wedge y)$, which we still need to keep, because the free occurrence of y cannot be renamed without losing equivalence. Thus, we need to have two copies of the previously shared subcircuit. In general, this can cause exponential growth. That observation motivates separate investigations of prenex and non-prenex circuits.

5.2   Quantified Circuits in Prenex Form

Definition 5. *(Quantified Circuits in Prenex Form)*
1. $QC := \{c \mid c \text{ is a circuit over } \wedge, \vee, \neg, \forall \text{ and } \exists \text{ in NNF and prenex form}\}$
2. $\exists C := \{c \mid c \text{ is a circuit over } \wedge, \vee, \neg \text{ and } \exists \text{ in NNF and prenex form}\}$

Because of the previously mentioned linear transformation of Boolean circuits into existentially quantified Boolean formulas by labeling edges with auxiliary variables as in Section 2, we immediately get the following relationships:

Proposition 5. $\exists C =_{linear} \exists BF$ *and* $QC =_{linear} QBF, pQBF$.

5.3   Pure Quantified Circuits in Non-prenex Form

Now, we investigate quantified circuits which are not in prenex form, but still pure. In the next section, we will also drop the purity restriction.

Definition 6. *(Pure Quantified Circuits in Negation Normal Form)*
1. $C_{\forall, \exists} := \{c \mid c \text{ is a pure circuit over } \wedge, \vee, \neg, \forall \text{ and } \exists \text{ in NNF}\}$
2. $C_{\exists} := \{c \mid c \text{ is a pure circuit over } \wedge, \vee, \neg \text{ and } \exists \text{ in NNF}\}$

In order to discuss the expressive power of $C_{\exists}$, we further refine this class. The idea is to restrict the number of occurrences of $\exists x$ nodes for a single variable x.

Definition 7. *For $k \geq 1$, we let*
$$\mathcal{C}_{\exists(k)} \doteq \{c \in \mathcal{C}_\exists \mid each\ variable\ occurs\ in\ at\ most\ k\ \exists\ nodes\ of\ the\ pure\ circuit\ c\}$$

When $k = 1$, all $\exists$ nodes have distinct variable names. Furthermore, no variable occurs both free and bound, since the circuits are pure. That allows us to move all $\exists$ nodes in front of the sink, just like prenexing in QBF. As the circuits are also in NNF, we obtain an equivalent circuit in $\exists\mathcal{C}$. In the other direction, every $\exists\mathcal{C}$ circuit is trivially equivalent to a $\mathcal{C}_{\exists(1)}$ circuit, because if a prenex circuit has quantifier nodes with duplicate names, all but the innermost can be dropped.

Proposition 6. $\mathcal{C}_{\exists(1)} =_{linear} \exists\mathcal{C}$

For $k = 2$, we shall see that the expressive power jumps to the full power of quantified circuits. That means $\mathcal{C}_{\exists(2)}$ circuits can compactly encode universal quantifiers using only existential quantifiers. This is quite simple if we waive the NNF requirement and use the equivalence $\neg\exists x\phi \approx \forall x\neg\phi$. But if we only consider circuits and QBF formulas in NNF, this is not possible. In fact, a poly-time simulation of universal quantifiers in negation normal form QBF formulas would lead to the collapse of the polynomial hierarchy. Accordingly, the following idea is specific to circuits and uses structure sharing by fanout: It is well known that quantifiers can be expanded, e.g. by the equivalence $\forall x\,\alpha(x, \mathbf{y}) \approx \alpha(0, \mathbf{y}) \wedge \alpha(1, \mathbf{y})$ with some free variables $\mathbf{y}$. Repeated application clearly causes exponential growth due to the duplication of α. Unfortunately, sharing by fanout only works for subexpressions which are exactly the same, but α has different arguments here. Our trick is to use the equivalence $\forall x\,\alpha(x, \mathbf{y}) \approx (\exists x(x \wedge \alpha(x, \mathbf{y})) \wedge (\exists x(\neg x \wedge \alpha(x, \mathbf{y}))$ instead. Now, we have two identical occurrences of $\alpha(x, \mathbf{y})$, which can be shared as shown in Fig. 3. By repeated expansion of universal nodes in this way, any quantified circuit in NNF can be transformed into a representation that uses only $\exists$ nodes. Obviously, the size of the resulting circuit remains linear in the size of the initial circuit.

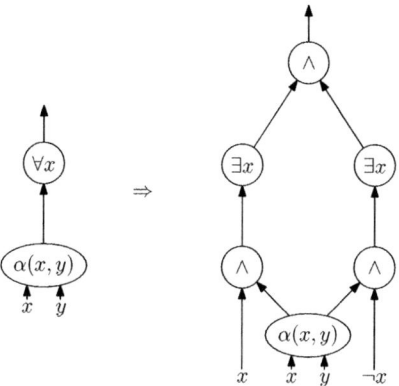

Fig. 3. $\forall$-simulation by $\exists$ nodes and sharing of subcircuits

We can now summarize our expressiveness results in the following theorem.

Theorem 1. $QC =_p C_{\forall,\exists} =_p C_\exists =_p C_{\exists(2)} =_p NBF =_p QBF =_p pQBF$

Proof. We show $QBF \leq_p QC \leq_p C_{\exists(2)} \leq_p C_\exists \leq_p C_{\forall,\exists} \leq_p NBF \leq_p QBF$.

1. $QBF \leq_p QC$ according to Proposition 5.
2. $QC \leq_p C_{\exists(2)}$ by encoding a universal node with two existentials (Fig. 3).
3. $C_{\exists(2)} \leq_p C_\exists \leq_p C_{\forall,\exists}$: $C_{\exists(2)}$ is a subset of $C_\exists$, which in turn is a subset of $C_{\forall,\exists}$.
4. $C_{\forall,\exists} \leq_p NBF$ follows from Proposition 3.
5. $NBF \leq_p QBF$ by Lemma 1.   □

5.4   Non-pure Quantified Circuits

We now consider quantified circuits in which a variable may occur both free and bound. In the analogous QBF case, we can rename such bound variables in consideration of their scope, so that finally no variable is both free and bound. For example, $(\exists x\, \alpha(x, y)) \wedge \alpha(x, y) \approx (\exists x'\, \alpha(x', y)) \wedge \alpha(x, y)$. We have already pointed out that such renamings are problematic for circuits, since they make direct sharing of subcircuits impossible. In the example, α can no longer be shared, so we need two copies $\alpha(x', y)$ and $\alpha(x, y)$. Can we avoid such copying?

Clearly, non-pure circuits can be made pure by an indirect transformation: We know that all quantified circuits, including non-pure ones, can be encoded as v-linear NBFs, which in turn correspond to linear-size pure quantified circuits.

But there is also a direct linear-time transformation from non-pure to pure circuits: for every variable x which has both free and bound occurrences in a given quantified circuit c, let x' be a new variable which does not yet occur in c. Then we substitute x' for all occurrences of x in c, including occurrences in quantifier nodes. We call the resulting circuit $c[x/x']$. To make it equivalent to the original one, we bind x' with an existential quantifier and require x' to be true if and only if x is true. We obtain a circuit which represents $\exists x'((x' = x) \wedge c[x/x'])$, as shown in Fig. 4 for the above example $(\exists x\, \alpha(x, y)) \wedge \alpha(x, y)$.

Essentially, this construction turns a free variable into a bound variable, so instead of having a free variable named like a bound variable, we now have bound variables with duplicate names (in Fig. 4, we get two quantifier nodes $\exists x'$). For the previously mentioned indirect transformation from non-pure to pure circuits via NBF, the situation is similar, because it requires universal quantifier nodes, and their simulation by existential ones as in Fig. 3 also introduces existential quantifier nodes with duplicate names. This is not unexpected, since pure circuits with distinct quantifier names, previously denoted $C_{\exists(1)}$, seem to be significantly weaker ($C_{\exists(1)} =_p \exists BF$) than those with duplicate names ($C_{\exists(2)} =_p QBF$). Interestingly, we can now show that the full expressiveness of QBF can be achieved by existentially quantified circuits with distinct quantifier names if we allow non-purity. Formally, the non-pure counterpart of $C_{\exists(1)}$ is defined as follows:

Definition 8. $C^*_{\exists(1)} := \{c \mid c$ *is a circuit over* $\wedge, \vee, \neg$ *and* $\exists$ *in negation normal form, every quantified variable occurs exactly once in the set of* $\exists$ *nodes* $\}$

In contrast to $C_{\exists(1)} =_{linear} \exists BF$, we now prove $C^*_{\exists(1)} =_{v-linear} QBF$.

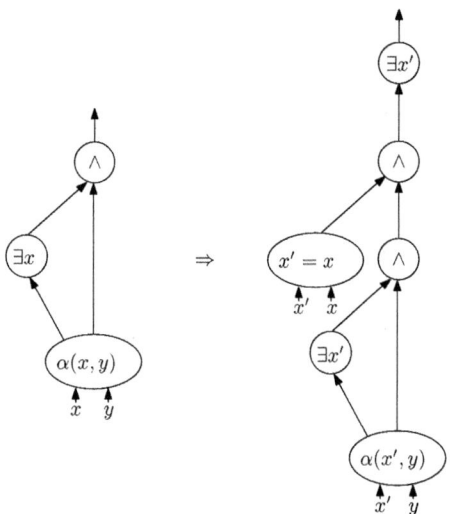

Fig. 4. Transformation of a non-pure into a pure circuit

Theorem 2. $\mathcal{C}^*_{\exists(1)} =_{v-linear}$ QBF

Proof. 1. $\mathcal{C}^*_{\exists(1)} \leq_{v-linear}$ QBF is obvious from the fact that any quantified circuit, and thus also a $\mathcal{C}^*_{\exists(1)}$ circuit, can be transformed into a v-linear NBF (Proposition 3), followed as before by a linear-time transformation to QBF.

2. We show QBF $\leq_{linear} \mathcal{C}^*_{\exists(1)}$ with a similar approach as for the transformation from QBF to $\mathcal{C}_{\exists(2)}$ from Theorem 1: we bring the formula into prenex form (with uniquely named variables) and use the obvious linear mapping from pQBF into a corresponding quantified circuit in prenex form. We then simulate the universal quantifier nodes using only existential nodes, but with a new procedure which is different from the one shown in Fig. 3.

The new procedure uses the equivalence $\forall x\ \alpha(x, \mathbf{y}) \approx \alpha(x, \mathbf{y}) \wedge \alpha(\neg x, \mathbf{y})$. Notice that x is now free in the right-hand formula. Typically, equivalent formulas need to have the same free variables, but it is possible to have additional ones which occur only on one side, as long as their actual values do not influence the truth value of the formula. This is the case here with x: no matter whether $x = 0$ or $x = 1$, the formula on the right represents $\alpha(0, \mathbf{y}) \wedge \alpha(1, \mathbf{y})$, which is just the definition of universal quantification.

Now, we need to express this equivalence by a quantified circuit which contains only one copy of α. Sharing of subcircuits by fanout only works if both instances of α have exactly the same arguments, say $\alpha(x, \mathbf{y})$. Our idea is to first introduce a new existential variable x' which abbreviates $\neg x$:

$$\forall x\ \alpha(x, \mathbf{y}) \approx \alpha(x, \mathbf{y}) \wedge \exists x'((x' = \neg x) \wedge \alpha(x', \mathbf{y}))$$

In order to turn $\alpha(x', \mathbf{y})$ into $\alpha(x, \mathbf{y})$, we then perform another renaming, but this time, the new existential variable is named x, just like the free variable:

$$\forall x\ \alpha(x, \mathbf{y}) \approx \alpha(x, \mathbf{y}) \wedge \exists x'((x' = \neg x) \wedge \exists x((x = x') \wedge \alpha(x, \mathbf{y})))$$

We end up with two copies of $\alpha(x, \mathbf{y})$ that can be shared by fanout, as shown in Fig. 5. Everything else added to the resulting circuit has constant size, so repeated application of this procedure will lead to a circuit of linear size. It is clearly in $\mathcal{C}^*_{\exists(1)}$, since all quantified variables have distinct names.   □

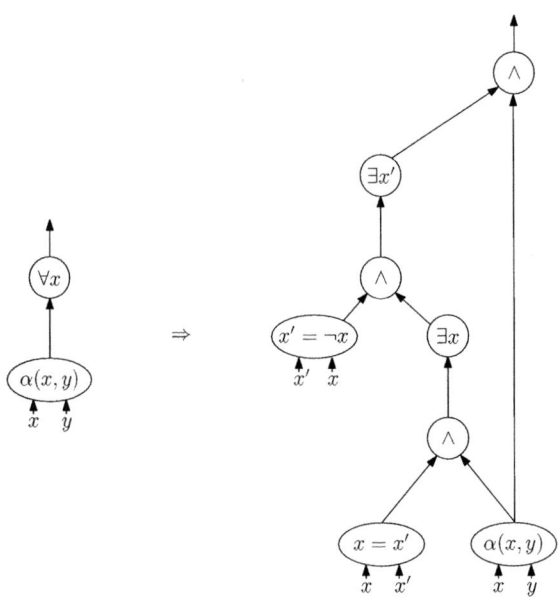

Fig. 5. ∀-simulation with unique quantifier names

We now have a second method to simulate universal quantifiers. Both of them require the ability to express overlapping scopes of variables with identical names, but the theorem shows that it does not matter whether that overlapping is between two bound variables or between a bound and a free variable.

The expressive power of $\mathcal{C}^*_{\exists(1)}$ is also evident from the following observation: In QBF, it is possible to compress conjunctions of formula instantiations for different variable names with the following equivalence [7]:

$$\alpha(x', y) \wedge \alpha(x, y) \approx \forall u \, (((u = x') \vee (u = x)) \rightarrow \alpha(u, y))$$

For the dual case $\alpha(x', y) \vee \alpha(x, y)$, we can use existential quantification:

$$\alpha(x', y) \vee \alpha(x, y) \approx \exists u \, (\alpha(u, y) \wedge ((u = x') \vee (u = x)))$$

In both cases, only one copy of α is needed. The second equivalence can obviously be applied exactly the same in $\mathcal{C}^*_{\exists(1)}$. An equivalence of the first form could be translated into $\mathcal{C}^*_{\exists(1)}$ by the above simulation of universal quantifiers, but it turns out that the same degree of compression can be achieved by a direct encoding without universal quantifiers. The idea is to introduce a local renaming of x':

$$\alpha(x, y) \wedge \alpha(x', y) \approx \alpha(x, y) \wedge \exists x((x = x') \wedge \alpha(x, y))$$

Now, both instances of $\alpha(x, y)$ can be computed by one subcircuit (Fig. 6).

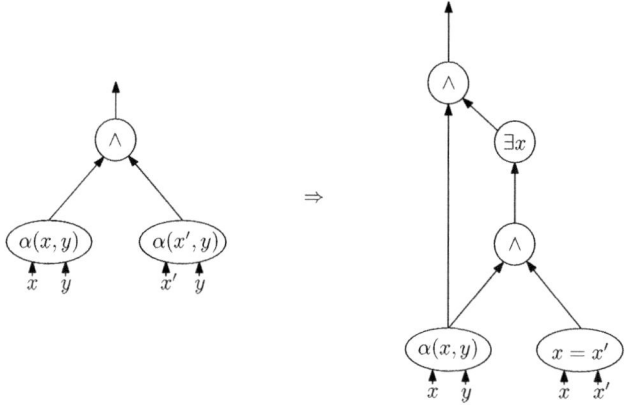

Fig. 6. Compression of renamed instantiations

6   Conclusion

We have studied quantified circuits, an extension of Boolean circuits with additional quantifier nodes. Such a circuit can be understood as syntax tree of a quantified Boolean formula with additional sharing of subformulas. In general, quantified circuits have the same expressive power under polynomial-time equivalence reductions as quantified Boolean formulas and nested Boolean functions:

$$\text{Quantified Circuits} =_p \text{QBF} =_p \text{NBF}$$

We could, however, prove the interesting result that every quantified circuit can be transformed into a polynomial-size equivalent circuit containing only existential quantifier nodes, which implies:

$$\mathcal{C}_\exists =_p \text{QBF}$$

This simulation of universal quantifiers does require the ability to express overlapping scopes of variables with identical names, be it pairs of bound variables or bound and free variables having the same names:

$$\mathcal{C}_{\exists(2)} =_p \mathcal{C}^*_{\exists(1)} =_p \text{QBF}$$

Such overlapping makes it possible to rename variables, and in combination with the subformula sharing ability of the underlying circuit structure allows compact encodings of subformula instantiations with different names, such as e.g. $\alpha(x, y) \wedge \alpha(x', y)$. With the construction from Fig. 6, this can be expressed with one copy of the circuit $\alpha(x, y)$ and without the need to use universal quantifiers as in the QBF encoding $\alpha(x', y) \wedge \alpha(x, y) \approx \forall u \, (((u = x') \vee (u = x)) \rightarrow \alpha(u, y))$. Ordinary Boolean circuits and purely existentially quantified Boolean formulas ($\exists$BF) seem to be lacking exactly that renaming ability and appear to be limited to abbreviate exact repetitions of subformulas. Nevertheless, it remains a long-standing open problem to show that QBF is indeed exponentially more powerful than ∃BF. Perhaps, quantified circuits might provide a new perspective onto that problem, especially since they allow focusing only on one kind of quantifier and instead to consider renaming as the crucial feature.

References

[1] Anderaa, S., Börger, E.: The Equivalence of Horn and Network Complexity for Boolean Functions. Acta Informatica 15, 303–307 (1981)

[2] Borchert, B., Stephan, F.: Looking for an Analogue of Rice's Theorem in Circuit Complexity Theory. In: Gottlob, G., Leitsch, A., Mundici, D. (eds.) KGC 1997. LNCS, vol. 1289, pp. 114–127. Springer, Heidelberg (1997)

[3] Bubeck, U., Kleine Büning, H.: Models and Quantifier Elimination for Quantified Horn Formulas. Discrete Applied Mathematics 156(10), 1606–1622 (2008)

[4] Bubeck, U., Kleine Büning, H.: Encoding Nested Boolean Functions as Quantified Boolean Formulas. In: 3rd Guangzhou Symposium on Satisfiability in Logic-based Modeling, Zhuhai, China (2010)

[5] Bubeck, U., Kleine Büning, H.: Rewriting (Dependency-)Quantified 2-CNF with arbitrary free literals into existential 2-HORN. In: Strichman, O., Szeider, S. (eds.) SAT 2010. LNCS, vol. 6175, pp. 58–70. Springer, Heidelberg (2010)

[6] Cook, S., Soltys, M.: Boolean Programs and Quantified Propositional Proof Systems. The Bulletin of the Section of Logic 28(3), 119–129 (1999)

[7] Dershowitz, N., Hanna, Z., Katz, J.: Bounded model checking with QBF. In: Bacchus, F., Walsh, T. (eds.) SAT 2005. LNCS, vol. 3569, pp. 408–414. Springer, Heidelberg (2005)

[8] Egly, U., Seidl, M., Tompits, H., Woltran, S., Zolda, M.: Comparing Different Prenexing Strategies for Quantified Boolean Formulas. In: Giunchiglia, E., Tacchella, A. (eds.) SAT 2003. LNCS, vol. 2919, pp. 214–228. Springer, Heidelberg (2004)

[9] Ganai, M., Zhang, L., Ashar, P., Gupta, A., Malik, S.: Combining Strengths of Circuit-based and CNF-based Algorithms for a High-performance SAT Solver. In: Proc. 39th Design Automation Conference (DAC 2002), pp. 747–750. ACM, New York (2002)

[10] Goultiaeva, A., Bacchus, F.: Exploiting circuit representations in QBF solving. In: Strichman, O., Szeider, S. (eds.) SAT 2010. LNCS, vol. 6175, pp. 333–339. Springer, Heidelberg (2010)

[11] Kuehlmann, M., Ganai, M., Paruthi, V.: Circuit-based Boolean Reasoning. In: Proc. 38th Design Automation Conference (DAC 2001), pp. 232–237. ACM, New York (2001)

[12] Meyer, A., Stockmeyer, L.: Word Problems Requiring Exponential Time. In: Proc. 5th ACM Symp. on Theory of Computing (STOC 1973), pp. 1–9 (1973)

[13] Pigorsch, F., Scholl, C.: Exploiting Structure in an AIG based QBF Solver. In: Proc. Intl. Conf. on Design, Automation and Test in Europe (DATE 2009), pp. 1596–1601 (2009)

[14] Williams, P., Biere, A., Clarke, E., Gupta, A.: Combining Decision Diagrams and SAT Procedures for Efficient Symbolic Model Checking. In: Emerson, E.A., Sistla, A.P. (eds.) CAV 2000. LNCS, vol. 1855, pp. 124–138. Springer, Heidelberg (2000)

Failed Literal Detection for QBF

Florian Lonsing and Armin Biere[*]

Institute for Formal Models and Verification
Johannes Kepler University, Linz, Austria
http://fmv.jku.at/

Abstract. Failed literal detection (FL) in SAT is a powerful approach
for preprocessing. The basic idea is to assign a variable as assumption. If
boolean constraint propagation (BCP) yields an empty clause then the
negated assumption is necessary for satisfiability. Whereas FL is common
in SAT, it cannot easily be applied to QBF due to universal quantification. We present two approaches for FL to preprocess prenex CNFs. The
first one is based on abstraction where certain universal variables are
treated as existentially quantified. Second we combine QBF-specific BCP
(QBCP) in FL with Q-resolution to validate assignments learnt by FL.
Finally we compare these two approaches to a third common approach
based on SAT. It turns out that the three approaches are incomparable.
Experimental evaluation demonstrates that FL for QBF can improve the
performance of search- and elimination-based QBF solvers.

1 Introduction

The logic of quantified boolean formulae (QBF) is an extension of propositional
logic (SAT) where variables are existentially or universally quantified. Whereas
this often allows for more succinct encodings of problems, it also causes PSPACE-
completeness of the decision problem of QBF.

A standard and common input format of QBF solvers is prenex conjunctive
normal form (PCNF). The conversion of an arbitrary QBF encoding into PCNF
might hide relevant structural properties like variable dependencies. This in turn
can influence solver performance negatively. In order to overcome this drawback,
several approaches for preprocessing PCNFs have been suggested. Some were
ported from SAT to QBF such as binary clause reasoning, equivalence detection
and variable elimination by resolution or expansion [2,5,6,11,21,25,26].

This work focuses on the detection of failed literals (FL) in QBF. The idea is
to make an assumption, i.e. a trial assignment of a variable, and apply boolean
constraint propagation with QBF-specific inference rules (QBCP, see also Section
2). If the empty clause is discovered then the negated assumption is a necessary
assignment for satisfiability and can be added to the formula as a learnt unit
clause. In contrast to SAT where application of FL is straightforward, complications arise in the context of QBF.

---

[*] The 2nd author is financially supported by the Austrian Science Foundation (FWF)
NFN Grant S11408-N23 (RiSE).

K.A. Sakallah and L. Simon (Eds.): SAT 2011, LNCS 6695, pp. 259–272, 2011.

Example 1. The satisfiable PCNF $\psi := \forall x \exists y. \; (x \lor \neg y) \land (\neg x \lor y)$ expresses equivalence of x and y. When assuming y in FL, then the first clause becomes empty due to universal reduction (see Section 2). But adding unit clause $(\neg y)$, which corresponds to the negated assumption, produces an unsatisfiable formula. The problem pointed out in Example 1 is that y depends on x. That is, y must not be assigned as assumption without considering the value of x as was erroneously done by FL. This might destroy satisfiability of PCNFs due to violations of the quantifier ordering: x is outermost but y was assumed first. Hence that ordering must be respected in preprocessing just as in QBF solving in general.

The objective of our work is to apply *sound* variants of FL to PCNFs for preprocessing, i.e. where preprocessing by FL produces an equivalent formula. In the following we introduce and evaluate three approaches to detect necessary assignments of QBFs in PCNF. First we reconsider a known technique based on SAT solving (Section 3). A SAT solver is used to check if a unit clause is implied by the plain (i.e. ignoring quantifiers) CNF of some PCNF. Such unit clauses can then be added to the PCNF as well.

Second we introduce FL with respect to an abstraction of a PCNF (Section 4). A unit clause learnt by FL on the abstraction can also be added to the original PCNF. An abstraction is obtained separately for each assumption. Variables are treated as existentially quantified if they are smaller in the quantifier ordering than the current assumption.

Third we combine FL on the original PCNF with Q-resolution (Section 5). If an empty clause is discovered in FL then we try to derive the unit clause corresponding to the negated assumption by Q-resolution. The choice of Q-resolution candidates is heuristically guided by the current run of QBCP in FL.

A major contribution of our work is a comparison (Sections 6 and 7). We provide examples pointing out that the three aforementioned approaches are incomparable with respect to effectiveness. There are formulae where particular unit clauses can be learnt with one approach but not with another. This observations also raise questions whether conflict-driven clause learning (CDCL) in QBF solvers could be improved in this respect. Moreover, our abstraction-based approach can be regarded as a polynomial-time[1] alternative to the common SAT-based approach. Hence FL could be applied dynamically in search-based QBF solvers when interleaved with the search. This idea is similar to optimizations carried out in many SAT solvers when backtracking to the topmost decision level, e.g. after restarting or if a unit clause was learnt.

We implemented the three approaches for FL in our novel preprocessor QxBFand evaluated their effectiveness in combination with various search- and elimination-based QBF solvers (Section 7). Our experiments confirm observations regarding incomparability. Although the vast performance improvement is observed for elimination-based solvers, we conjecture that search-based solvers could benefit from dynamic applications of various approaches of FL.

Related Work. FL for SAT originated in [9] and is an integral part of look-ahead SAT solvers [14]. A comprehensive treatment of FL for preprocessing is

---

[1] We interpret QBCP as the polynomial-time procedure defined in Section 2.

given in [1,18], which also includes inferences of unit clauses based on complementary assumptions [13]. We consider that future work and refrain from discussion in this work.

FL was applied to QBF in [22] but with a special treatment of empty clauses and QBCP lacking universal reduction and pure literal detection. Thus the full propagation power of QBCP was not exploited. In contrast to this, our FL approaches presented in Sections 4 and 5 are an improvement.

A theoretical foundation of QBF preprocessing was given in [26] in terms of QBF models. Additionally binary clause reasoning for QBF was introduced in [25] where QBCP was used for detecting binary clause inferences. We combine QBCP in FL with Q-resolution to find derivations of unit clauses (Section 5).

In [21] a SAT solver was applied to detect necessary assignments of the CNF part of PCNFs. We consider the same approach mainly for reference (Section 3) but also show that it is incomparable to our FL approaches (Section 6).

A SAT solver was integrated into a search-based QBF solver in [24]. It was observed that the two solvers are capable of learning clauses the other one can not learn. We make similar observations and provide examples regarding the detection of unit clauses by SAT solving and by our FL approaches (Section 6).

2   Preliminaries

For a set of propositional variables V, a *literal* is either a variable $x \in V$ or its negation $\neg x$ where $v(x) = x$ and $v(\neg x) = x$ denotes the variable of a literal. A *clause* C_i is a disjunction over literals where $\{x, \neg x\} \not\subseteq C_i$ for all $x \in V$. For clauses C_i, a propositional formula $\phi := C_1 \wedge \ldots \wedge C_n$ is in *conjunctive normal form (CNF)*. For a CNF ϕ and a literal l, the set of *occurrences* of l is $O(l) := \{C \mid C \in \phi, l \in C\}$.

A quantified boolean formula (QBF) $\psi := Q_1 S_1 \ldots Q_n S_n. \phi$ in *prenex conjunctive normal form (PCNF)* consists of a CNF ϕ over a set of variables V and a *quantifier prefix* $Q_1 S_1 \ldots Q_n S_n$. The quantifier prefix is a linearly ordered set of *scopes* S_i forming a partition on V. A scope S_i is *existential* ($Q_i = \exists$) if it is associated with an existential quantifier and *universal* ($Q_i = \forall$) otherwise. For scopes S_i and S_{i+1}, $Q_i \neq Q_{i+1}$ for $1 \leq i < n$. For a literal x with $v(x) \in S_i$, $q(x) := Q_i$ is the *quantifier type* of (the variable of) x. For clause C and $Q \in \{\forall, \exists\}$, $L_Q(C) := \{l \in C \mid q(l) = Q\}$. For literals l, k with $v(l) \in S_i$ and $v(k) \in S_j$, $l \leq k$ if, and only if $i \leq j$ for $1 \leq i, j \leq n$.

Given CNF ϕ, an *assignment* is a mapping $A : V \rightarrow \{true, false\}$ from variables in ϕ to truth values. An assignment m is a *CNF-model* of ϕ, written as $m \models \phi$, if every clause in ϕ is satisfied under m.

We introduce QBF semantics based on tree-like models as in [26]. This allows for a simpler definition of necessary assignments (see below) and proofs. In this framework, a necessary assignment is a property of all models of a QBF. This cannot easily be expressed within standard QBF semantics based on recursive evaluation like e.g. in [7]. Further, relying on tree-like models, our results can naturally be generalized to QBF solving using dependency schemes [23].

Given a PCNF $\psi := Q_1 S_1 \ldots Q_n S_n.\phi$. An *assignment tree* T is a tree of assignments according to the following restrictions. Every node N in T except the root represents a truth assignment to a variable v in V. A node has at most (exactly) one sibling if it assigns a truth value to an existential (universal) variable. Two siblings altogether denote assignments *true* and *false*. Every path P from the root to a leaf of T corresponds to an assignment A for variables in CNF ϕ. A node N for variable v is an ancestor of another node N' for variable v' in P if and only if $v \leq v'$. That is, assignments along every path P respect the quantifier ordering. An assignment tree m is a *PCNF-model* of ψ, written as $m \models \psi$, if every path P in m is a CNF-model of ϕ.

A CNF is *satisfiable* if it has a CNF-model. Two CNFs ϕ and ϕ' are *model-equivalent*, written as $\phi \equiv_m \phi'$, if and only if for all assignments m, $m \models \phi$ if and only if $m \models \phi'$. Two CNFs ϕ and ϕ' are *satisfiability-equivalent*, written as $\phi \equiv_s \phi'$, if and only if ϕ is satisfiable then ϕ' is satisfiable and vice versa. A transformation of a CNF ϕ into a CNF ϕ' is *sound* if and only if $\phi \equiv_m \phi'$. Satisfiability, model-equivalence, satisfiability-equivalence and soundness with respect to PCNFs are defined accordingly. The following properties are well known.

Proposition 1 ([21,26]). *Given PCNF* $\psi := Q_1 S_1 \ldots Q_n S_n.\ \phi$ *and CNF* ϕ' *where* $\phi \equiv_m \phi'$. *Let* $\psi' = Q_1 S_1 \ldots Q_n S_n.\phi'$. *Then* $\psi \equiv_m \psi'$.

Proposition 2 (e.g. [7]). *For CNF* ϕ *and literal* x, $\phi \wedge \{\neg x\}$ *is unsatisfiable iff* $\phi \equiv_m \phi \wedge \{x\}$.

Given PCNF ψ and $x_i \in V$. Assignment $x_i \mapsto t$, where $t \in \{false, true\}$, is *necessary* for satisfiability of ψ iff $x_i \mapsto t$ is part of every path in every PCNF-model of ψ.

Given PCNF $\psi := Q_1 S_1 \ldots Q_n S_n.\ \phi$, the assignment of literal l with $v(l) \in S_i$ yields the formula $\psi[l] := Q_1 S_1 \ldots Q_i S_i' \ldots Q_n S_n.\ \phi'$ where $S_i' := S_i \setminus \{v(l)\}$ and clauses $O(l)$ and literals $\neg l$ in $O(\neg l)$ are deleted in ϕ'.

For clause C, *universal reduction* is denoted by $UR(C) := C \setminus \{l_u \in L_\forall(C) \mid \forall l_e \in L_\exists(C), l_e < l_u\}$. A clause $C \in \phi$ where $UR(C) = \{l\}$ is *unit* and $\psi \equiv_s \psi[l]$ [6,8]. For a QBF $\psi := Q_1 S_1 \ldots Q_n S_n.\phi$, a literal l where $O(l) \neq \emptyset$ and $O(\neg l) = \emptyset$ is *pure* [8]: if $q(l) = \exists$ then $\psi \equiv_s \psi[l]$, and if $q(l) = \forall$ then $\psi \equiv_s \psi[\neg l]$.

We assume that clauses in a PCNF are fully reduced by UR. For clauses C_1, C_2 with $v \in L_\exists(C_1), \neg v \in L_\exists(C_2)$, the *Q-resolvent* C, written as $(C_1, C_2) \vdash_v C$, is defined as follows [6]: let $C' := (C_1 \cup C_2) \setminus \{v, \neg v\}$. If $\{x, \neg x\} \subseteq C'$ for $x \in V$ then no Q-resolvent exists, otherwise $C := UR(C')$. We write $\psi \vdash^* C$ if clause C can be derived from QBF ψ by Q-resolution. Adding Q-resolvents to a PCNF yields a model-equivalent formula.

Lemma 1 ([26]). *Given PCNF* ψ *and clause* C. *Then* $\psi \wedge C^2 \equiv_m \psi \wedge UR(C)$.

Proof. Our definition of PCNF-models differs from the one in [26]. There, nodes N assigning existential variables do *not* have a sibling. The proof in [26] also

---

[2] For PCNF $\psi := Q_1 S_1 \ldots Q_n S_n.\phi$, let $\psi \wedge C$ denote $Q_1 S_1 \ldots Q_n S_n.\ (\phi \wedge C)$.

applies to our semantical framework. The subtree rooted at a sibling of such node N can be deleted, and the resulting assignment tree is still a PCNF-model. □

Lemma 2. *Given PCNF ψ. If $\psi \vdash^* C$ then $\psi \equiv_m \psi \wedge C$.*

Proof. Q-resolution can be regarded as a combination of resolution for propositional logic and UR. Adding propositional resolvents yields a model-equivalent formula [7]. The claim then follows from Lemma 1 and Proposition 1. □

Given PCNF ψ and a literal x, $\psi' := QBCP(\psi, x)$ denotes a formula obtained from $\psi[x]$ by applying UR, unit clause and pure literal rule. Literal x is called an *assumption*. For clause C, $C \in QBCP(\psi, x)$ if $\psi' = QBCP(\psi, x)$ and $C \in \psi'$. We write $\emptyset \in QBCP(\psi, x)$ if the empty clause can be obtained.

3   SAT-Based FL

First we review a well-known technique for inferring unit clauses from the CNF part of a PCNF based on propositional satisfiability testing. This has already been applied to QBF [21]. We include it here as a reference for our approaches introduced in Sections 4 and 5. We show that these approaches have different effectiveness in Sections 6 and 7.

A general approach for preprocessing PCNFs can be obtained from QBF semantics and model-equivalence of CNFs. When combining Propositions 1 and 2, a SAT solver can be used to check if a unit clause is implied by the CNF part of a PCNF for QBF preprocessing [21].

Whereas propositional satisfiability testing allows to exploit Proposition 2 to full extent, a subset of all necessary assignments of a CNF can be identified by failed literal detection (FL) for SAT [1,9,18]. FL is a common approach in SAT which can be carried out in *polynomial time* based on BCP[3] as follows.

Proposition 3. *Given a CNF ϕ and literal x. If $\emptyset \in BCP(\phi, x)$ then $\phi \equiv_m \phi \wedge \{\neg x\}$. Literal x is called a* failed literal.

If x is a failed literal then every CNF-model of ϕ must set x to *false*.

Definition 1. *Given a PCNF ψ and a literal x. If $\psi \equiv_m \psi \wedge \{x\}$ then $\neg x$ is called a* failed literal. *Assigning x to* true *is a necessary assignment for ψ.*

Example 1 pointed out that Proposition 3 is not directly applicable to QBF because satisfiability might be destroyed. The reason is that assumptions are made out of quantifier ordering. This can be regarded as modifying the quantifier prefix. Carrying out $QBCP(\psi, y)$ in Example 1 is similar to shifting y to the leftmost position in the prefix, yielding $\exists y \forall x$ prior to assigning it. This might allow applications of UR impossible based on the original prefix.

In the following we introduce two approaches to detect necessary assignments by FL for QBF similar to Proposition 3, i.e. based on QBCP. The goal is to have

---

[3] "BCP" denotes QBCP without pure literal rule and universal reduction.

a polynomial-time alternative to satisfiability testing as in Propositions 1 and 2. However, as shown in Section 6, it turns out that the three approaches are actually incomparable. In the following section we apply FL with QBCP to an abstraction of the original PCNF.

4   Abstraction-Based FL

In an early approach of FL for QBF presented in [22], QBCP included neither UR nor pure literal rule. This results in limited propagation power as both UR and universal pure literals possibly trigger additional unit literals. We conjecture that such restrictions of QBCP combined with a special treatment of empty clauses like in [22] are sufficient to ensure soundness of FL.

In our approaches of FL for QBF we allow the *full* set of rules in QBCP as defined in Section 2. In the following, we analyze special cases where FL based on QBCP is sound. These results will then be used to prove soundness of FL when QBCP is applied to an abstraction of the original PCNF.

Lemma 3. *Given* $\psi := Q_1 S_1 \ldots Q_n S_n.\ \phi$ *and literal* x *where* $v(x) \in S_1$. *If* $\emptyset \in QBCP(\psi, x)$ *then* $\psi \equiv_m \psi \wedge \{\neg x\}$.

Proof. Since $x \in S_1$ we assume that x is assigned first on every path in every model of ψ. By definition of $QBCP$, $\psi[x] \equiv_s \emptyset$, i.e. ψ does not have a model where x is assigned *true* on any path. Therefore, x must be assigned *false* on all paths in all models m (if any) of ψ. Hence $\psi \equiv_m \psi \wedge \{\neg x\}$.   □

Due to Lemma 3, FL with QBCP and assumptions from the leftmost scope is sound. For universal assumptions, this can be generalized to arbitrary scopes.

Lemma 4. *Given PCNF* $\psi := Q_1 S_1 \ldots Q_n S_n.\ \phi$ *and literal* x *where* $v(x) \in S_i$ *and* $Q_i = \forall$. *If* $\emptyset \in QBCP(\psi, x)$ *then* ψ *is unsatisfiable.*

Proof. The PCNF where QBCP is carried out can be regarded as $\psi' := \forall x Q_1 S_1 \ldots Q_{i-1} S_{i-1} \forall (S_i \setminus \{x\}) \ldots Q_n S_n.\ \phi$. Note the change in the prefix by moving $\forall x$ to the front. If $\emptyset \in QBCP(\psi', x)$ then ψ' is unsatisfiable due to Lemma 3. Then ψ is unsatisfiable as well because if a PCNF with prefix pattern $\forall x \ldots \exists y$ (pattern of ψ') is unsatisfiable then also with $\exists y \ldots \forall x$ (pattern of ψ).   □

We introduce an approach where FL is carried out on an abstraction of the original PCNF. The abstraction affects the quantifier prefix.

Definition 2 (Quantifier Abstraction). *For* $\psi := Q_1 S_1 \ldots Q_{i-1} S_{i-1} Q_i S_i \ldots \ldots Q_n S_n.\ \phi$, *the quantifier abstraction of* ψ *with respect to* S_i *is* $Abs(\psi, i) := \exists (S_1 \cup \ldots \cup S_{i-1}) Q_i S_i \ldots Q_n S_n.\ \phi$.

By Definition 2 variables from scopes smaller than S_i are treated as existentially quantified. This gives an overapproximation of ψ with respect to PCNF-models, following from the definition in Section 2.

Corollary 1. *For PCNF* ψ *and PCNF-model* m: *if* $m \models \psi$ *then* $m \models Abs(\psi, i)$.

In practice, FL with assumptions from S_i and QBCP is carried out on $Abs(\psi, i)$ instead of ψ. This is called *abstraction-based FL*. Unit clauses learnt by FL on $Abs(\psi, i)$ can then be added to the original PCNF ψ (see Example 2 below). 'First we prove soundness of abstraction-based FL with respect to $Abs(\psi, i)$.

Lemma 5. *Given PCNF $\psi := Q_1 S_1 \ldots Q_n S_n. \phi$ and literal x where $v(x) \in S_i$. If $\emptyset \in QBCP(Abs(\psi, i), x)$ then $Abs(\psi, i) \equiv_m Abs(\psi, i) \wedge \{\neg x\}$.*

Proof. We distinguish cases by the quantifier type of x.

If $q(x) = Q_i = \exists$ then $Abs(\psi, i) = \exists (S_1 \cup \ldots \cup S_{i-1} \cup S_i) \ldots Q_n S_n. \phi$. Due to Lemma 3, $Abs(\psi, i) \equiv_m Abs(\psi, i) \wedge \{\neg x\}$.

If $q(x) = Q_i = \forall$ then $Abs(\psi, i) = \exists (S_1 \cup \ldots \cup S_{i-1}) \forall S_i \ldots Q_n S_n. \phi$. If $\emptyset \in QBCP(Abs(\psi, i), x)$ then $Abs(\psi, i)$ is unsatisfiable due to Lemma 4, and so is $Abs(\psi, i) \wedge \{\neg x\}$.   □

If QBCP on $Abs(\psi, i)$ with assumption x where $v(x) \in S_i$ yields the empty clause then x is a failed literal and clause $\{\neg x\}$ can be added to $Abs(\psi, i)$. Due to Corollary 1 and the second case of the proof of Lemma 5, $Abs(\psi, i)$ preserves an intuitive property of universal quantification in ψ: if one branch of a universal variable cannot lead to a solution then ψ is unsatisfiable. Relying on Corollary 1 and Lemma 5, we prove that failed literals obtained on $Abs(\psi, i)$ are also sound with respect to original PCNF ψ.

Theorem 1. *Given PCNF $\psi := Q_1 S_1 \ldots Q_n S_n. \phi$ and literal x where $v(x) \in S_i$. If $\emptyset \in QBCP(Abs(\psi, i), x)$ then $\psi \equiv_m \psi \wedge \{\neg x\}$.*

Proof. We show both directions of $\psi \equiv_m \psi \wedge \{\neg x\}$. If $m \models \psi \wedge \{\neg x\}$ then also $m \models \psi$ since ψ is less constrained than $\psi \wedge \{\neg x\}$.

For the other direction assume $\emptyset \in QBCP(Abs(\psi, i), x)$. Let m be a PCNF-model of ψ, i.e. $m \models \psi$ (if no such m exists then the claim follows immediately). By Corollary 1, also $m \models Abs(\psi, i)$. By Lemma 5, also $m \models Abs(\psi, i) \wedge \{\neg x\}$. Clause $\{\neg x\}$ is satisfied under m. Therefore, $m \models \psi \wedge \{\neg x\}$.   □

Example 2. Given PCNF[4] $\psi := \forall a_1 \exists e_2, e_3 \forall a_4 \exists e_5. \{a_1, e_2\}, \{\neg a_1, e_3\}, \{e_3, \neg e_5\}, \{a_1, e_2, \neg e_3\}, \{\neg e_2, a_4, e_5\}$. We have $\emptyset \in QBCP(Abs(\psi, 2), \neg e_3)$ since the assumption will make clauses $\{\neg a_1, e_3\}$ and $\{e_3, \neg e_5\}$ unit because a_1 is treated as existential. Clause $\{\neg e_2, a_4, e_5\}$ becomes unit due to UR. Finally clause $\{a_1, e_2\}$ is empty and unit clause $\{e_3\}$ is learnt.

Although abstraction-based FL uses all QBCP rules in contrast to [22], in general this does not result in full propagation power of QBCP on $Abs(\psi, i)$. Depending on i, i.e. the quantifier level of the current assumption, $Abs(\psi, i)$ typically has fewer universal variables than ψ. This influences detection of unit literals by UR and pure literal rule. Hence we expect more powerful QBCP for assumptions from S_1 than from S_n. For assumptions from different scopes, our approach is more dynamic than restricting the set of QBCP rules in advance as in [22].

---

[4] Unless stated otherwise, all PCNFs in examples provided in the paper are satisfiable. For brevity, pure literal detection is ignored and CNFs (clauses) are presented as sets of clauses (literals).

5   QBCP-Guided Q-Resolution

Abstraction-based FL from the previous section allows to apply QBF-specific inference rules like UR and universal pure literals to a larger extent than previous approaches where QBCP rules were restricted. However, it still lacks full propagation power as could be obtained on the original PCNF.

In this section we present an approach for FL which operates on the original PCNF, thus taking full benefits from QBF-specific QBCP rules. Because this might in general destroy satisfiability as pointed out in Example 1, we apply Q-resolution to validate failed literals detected by QBCP. This approach is inspired by CDCL in search-based QBF solvers [12,29].

An assignment A is generated by making an assumption x and carrying out QBCP with the full set of inference rules on the original PCNF. If $\emptyset \in QBCP(\psi, x)$ then candidate clauses for Q-resolution are selected from ψ entirely with respect to assignment A, i.e. its implication graph like in CDCL for SAT solving [27]. If the unit clause $\{\neg x\}$ corresponding to the negated assumption can be deduced in that way then x is a valid failed literal. This is called *QBCP-guided Q-resolution*. Soundness follows right from Lemma 2. The following example shows a nontrivial application.

Example 3. Given PCNF $\psi := \exists e_1, e_2 \forall a_3 \exists e_4, e_5. \{a_3, e_5\}, \{\neg e_2, e_4\}, \{\neg e_1, e_4\}, \{e_1, e_2, \neg e_5\}$. With assumption $\neg e_4$ we get $\emptyset \in QBCP(\psi, \neg e_4)$ since $\{\neg e_1\}, \{\neg e_2\}$ and $\{\neg e_5\}$ become unit. Finally $\{a_3, e_5\}$ is empty by UR. The negated assumption $\{e_4\}$ is then derived by resolving clauses in reverse-chronological order as they were affected by assignments: $(\{a_3, e_5\}, \{e_1, e_2, \neg e_5\}) \vdash \{e_1, e_2\}$, $(\{e_1, e_2\}, \{\neg e_2, e_4\}) \vdash \{e_1, e_4\}, (\{e_1, e_4\}, \{\neg e_1, e_4\}) \vdash \{e_4\}$.

Note that selecting Q-resolution candidates based on the current assignment generated by QBCP is only a heuristic. That is, even when it fails the negated assumption can possibly be deduced by selecting arbitrary clauses for Q-resolution.

Example 4. Given PCNF $\psi := \forall a_1 \exists e_2 \forall a_3 \exists e_4. \{a_1, e_2\}, \{e_2, a_3, e_4\}, \{e_2, a_3, \neg e_4\}$. We have $\emptyset \in QBCP(\psi, \neg e_2)$ immediately by UR in the first clause, but e_2 obviously cannot be produced by Q-resolution from that single clause affected by the assignment. However, we have $(\{e_2, a_3, e_4\}, \{e_2, a_3, \neg e_4\}) \vdash^* \{e_2\}$.

Due to Lemma 4 unsatisfiability can be concluded *without* Q-resolution if $\emptyset \in QBCP(\psi, x)$ for $x \in S_i$ where $Q_i = \forall$. This property is similar to abstraction-based FL but we expect more QBF-specific inferences in QBCP with this approach. Further the empty clause might be derived when validating a failed literal by Q-resolution. In this case, unsatisfiability follows immediately.

Example 5. Given PCNF $\psi := \forall a_1 \exists e_2, e_3, e_4, e_5. \{e_2, \neg e_5\}, \{\neg e_2, e_5\}, \{a_1, e_2\}, \{\neg a_1, e_3\}, \{\neg e_3, e_4\}, \{\neg e_3, \neg e_4\}$. We have $\emptyset \in QBCP(\psi, e_2)$ because $\{a_1, e_2\}$ is satisfied which makes $\{a_1\}$ pure and $\{e_3\}$ unit in $\{\neg a_1, e_3\}$. Finally $\{e_4\}$ is unit and $\{\neg e_3, \neg e_4\}$ becomes empty. Q-resolution as in Example 3 produces the empty clause: $(\{\neg e_3, \neg e_4\}, \{\neg e_3, e_4\}) \vdash \{\neg e_3\}, (\{\neg e_3\}, \{\neg a_1, e_3\}) \vdash \emptyset$. Note that $\emptyset \notin QBCP(Abs(\psi, 2), e_2)$ because $\{e_3\}$ does not become unit since the universal pure literal rule is not applicable as before.

6   Comparing FL Approaches

We presented one approach for FL based on SAT testing and two based on QBCP to find necessary assignments in PCNFs: SAT-based FL according to Proposition 2 (Section 3), abstraction-based FL (Section 4) and QBCP-guided Q-resolution (Section 5). As argued above, the last two benefit from QBF-specific inferences in QBCP increasingly in that order. This is due to larger numbers of universal variables in the formulae where FL is applied to.

In the following we compare the three approaches according to their effectiveness. We name examples which demonstrate that they are incomparable: one approach is able to detect necessary assignment the other one cannot.

Proposition 4. *Abstraction-based FL and QBCP-guided Q-resolution are incomparable with respect to detecting necessary assignments.*

Example 6. For the PCNF from Example 2 unit clause $\{e_3\}$ *cannot* be derived by Q-resolution, i.e. neither by QBCP-guided Q-resolution nor when allowing arbitrary candidate clauses. This is in contrast to abstraction-based FL. Note that assigning e_3 to *true* is necessary as can be seen from a semantical evaluation. Every path in every PCNF-model has to assign e_3 to *true*.

Example 7. For the PCNF from Example 3 we have $\emptyset \notin QBCP(Abs(\psi, 3), \neg e_4)$ which is in contrast to QBCP-guided Q-resolution. Due to abstraction UR is not applicable to make $\{a_3, e_5\}$ empty. Similarly, clause $\{e_4\}$ cannot be inferred when UR is excluded from QBCP rules as in [22].

Note that Proposition 4 severely affects QBF solvers relying on Q-resolution for conflict-driven clause learning (CDCL). For certain PCNFs *no* such solver will ever be able to deduce all necessary assignments.

The following result was also obtained in the more general context of clause learning when SAT solving was combined with search-based QBF solving [24].

Proposition 5. *SAT-based FL and QBCP-guided Q-resolution are incomparable with respect to detecting necessary assignments.*

Example 8. Given PCNF $\psi := \exists e_1 \forall a_2 \exists e_3. \{e_1, a_2, e_3\}, \{e_1, a_2, \neg e_3\}, \{e_1, \neg a_2, e_3\}, \{e_1, \neg a_2, \neg e_3\}$. A SAT solver will find out that the CNF with assumption $\neg e_1$ is unsatisfiable, hence $\{e_1\}$ can be learnt. This is possible neither by QBCP-guided Q-resolution nor by abstraction-based FL.

Example 9. For the PCNF from Example 3, SAT-based FL cannot learn $\{e_4\}$ because the CNF has a model with assumption $\neg e_4$.

Proposition 6. *SAT-based FL and abstraction-based FL are incomparable with respect to detecting necessary assignments.*

Example 10. For the PCNF from Example 2, SAT-based FL cannot detect $\{e_3\}$ because the CNF has a model with assumption $\neg e_3$.

Example 11. See Example 8.

7   Experiments

We implemented the three approaches of FL for QBF presented in Sections 3 to 5 in our novel preprocessor QxBF[5]. The idea is to profit from all approaches based on the observations from the previous section. The tool operates in rounds with three phases. First, QBCP with the full set of QBF-specific rules is carried out on the *original* formula, including any unit clauses learnt in earlier rounds. The second phase consists of either abstraction-based FL or QBCP-guided Q-resolution. Finally, SAT-based FL is applied in the third phase because it turned out to be most effective in practice (see below). Rounds are run in cyclic fashion until completion unless a time limit is reached.

The SAT solver PicoSAT [3] is used for SAT-based FL by Proposition 2. This allows for incremental SAT solving and optimizations based on CNF-models to reduce the number of SAT solver calls like in [21]. Additionally, unit clauses learnt by PicoSAT are propagated using QBF-specific QBCP rules within QxBF.

Table 1 compares the impact of different FL approaches on the performance of QBF solvers based on search (DepQBF [17] and QuBE7.1 [10]) and variable elimination (Quantor [2], squolem [15] and Nenofex [16]) using all benchmarks from QBFEVAL'10 [20]. For QuBE7.1 internal preprocessing was disabled (QuBE7.1-np). We used latest publicly available versions of solvers except internal versions of Nenofex and DepQBF, *all* without proprietary preprocessing[6]. Results using sQueezeBF [11], a state-of-the-art QBF preprocessor, are reported for reference. We cannot expect FL to be competitive with sQueezeBF as the latter applies a larger pool of inference rules (details are given below).

We combined abstraction-based FL and QBCP-guided Q-resolution with SAT-based FL (lines "ABS+SAT" and "QRES+SAT") in rounds and phases as described above. At most 40 seconds were assigned to each approach, totalling a maximum of 80 seconds for entire preprocessing. Additionally, heuristic limits were imposed on numbers of propagations in QBCP and decisions in PicoSAT. Performance of elimination-based solvers increases considerably both in terms of solved formulae and run time (lower part of table). Results are less impressive for search-based solvers. Further, they only differ slightly with respect to individual FL approaches when applied to DepQBF (middle part of table), with a limit of 80 seconds each. We combined only "ABS+SAT" with elimination-based solvers as the performance with "QRES+SAT" is likely similar. We did not apply "SAT" alone due to incomparability observed in Section 6. DepQBF performs best with SAT-based FL but also preprocessing times are larger. We argue that tuning the run time of abstraction-based FL and QBCP-guided Q-resolution while maintaining effectiveness is easier in practice. Run time of QBCP is close to linear with respect to formula size in contrast to SAT solving time in SAT-based FL.

Our FL approaches are not competitive compared to sQueezeBF. In the first line of Table 1, we allowed 900 seconds altogether for the combination

---

[5] Project web page: http://fmv.jku.at/qxbf/

[6] Setup: Ubuntu Linux 9.04, Intel®Q9550 2.83 GHz with 900 seconds / 3GB total time and memory limit. Exceeding the memory limit is counted as a time out.

Table 1. Solver performance with(out) time-limited failed literal preprocessing. Times are average total run times *including* preprocessing and time outs, with average preprocessing times in parentheses. The leftmost column indicates FL approaches: no preprocessing ("None"), SAT-based FL ("SAT"), abstraction-based FL ("ABS") and QBCP-guided Q-resolution ("QRES").

QBFEVAL'10: 568 formulae					
Preprocessing	Solver	Solved	Time (Preproc.)	SAT	UNSAT
sQueezeBF	DepQBF	435	233.28 (36.94)	201	234
sQueezeBF+(ABS+SAT)		434	239.84 (42.79)	201	233
SAT	DepQBF	379	322.31 (7.17)	167	212
QRES+SAT		378	322.83 (6.22)	167	211
ABS+SAT		378	323.19 (7.21)	167	211
ABS		375	327.64 (3.33)	168	207
QRES		374	327.63 (1.83)	167	207
None		372	334.60 (—)	166	206
ABS+SAT	QuBE7.1-np	320	432.22 (7.21)	143	177
None		318	434.69 (—)	143	175
ABS+SAT	Quantor	229	553.65 (7.21)	112	117
	Nenofex	224	553.37 (7.21)	104	120
None		211	573.65 (—)	103	108
	Quantor	203	590.15 (—)	99	104
ABS+SAT	squolem	154	658.28 (7.21)	63	91
None		124	708.80 (—)	53	71

of sQueezeBF and DepQBF because the former does not support setting resource limits. In this experiment, sQueezeBF alone solved 39 formulae and timed out on 15. Overall performance decreases slightly (second line) if "ABS+SAT" with 80 seconds time limit as before is applied additionally to formulae which were simplified but not solved by sQueezeBF. However, now 64 formulae were solved solely by preprocessing. This indicates that "ABS+SAT" is incomparable to sQueezeBF at least from a practical perspective. From the 514 formulae simplified but not solved by sQueezeBF, 489 were not solved by "ABS+SAT" alone. Among them, still 147 assignments were fixed on average (median 20) by "ABS+SAT". Further, the total number of remaining (i.e. neither eliminated nor assigned) variables in all 489 formulae was reduced by 2% due to "ABS+SAT".

Table 2 compares the effectiveness of the three FL approaches in more detail. We considered formulae from QBFEVAL'10 where *all* approaches ran *until completion* within 900 seconds but, differently from Table 1, *neither* with propagation *nor* decision limits. In general FL fixes substantially more assignments than QBCP on the original formula (column "None"). Abstraction-based and SAT-based FL are best according to average and median numbers of fixed assignments, but the latter is more costly in terms of run time. The large difference between average and median values is due to few benchmarks from (blackbox) model checking (instances "biu*" and "*BMC*", see also below) where SAT-based FL fixed fewer assignments. Further, QBCP-guided Q-resolution performs fewer propagations per assumptions in QBCP than abstraction-based FL. The reason is that the former typically detects spurious empty clauses earlier due to universal reduction. This also results in smaller run times.

Motivated from incomparability observed in Section 6, we compared the sets of assignments that were fixed by different FL approaches in Table 3.

Table 2. Average and median run times, fixed assignments, and propagations per assumption for FL approaches. "None" is QBCP on original formula only.

QBFEVAL'10: 524 formulae completed by all				
Preprocessing	None	ABS	QRES	SAT
Avg. Fixed	607.17	730.31	724.10	715.77
Med. Fixed	103.5	137.00	135.00	181.50
Avg. Time	0.02	3.19	0.76	10.80
Med. Time	0.00	0.16	0.02	0.20
Avg. Props/As	—	118.80	51.08	—
Med. Props/As	—	40.01	6.68	—

We considered all three pairwise combinations. Like in Table 2, we focused on formulae where *both* FL approaches of a pair ran *until completion* within 900 seconds. For each pairwise combination, only formulae where sets of fixed assignments (FA) were different were taken into account (first line). We then separated formulae by larger numbers of unique FAs (second line), i.e. FAs detected by one approach but not by the other. For example in section "ABS vs. QRES", on 121 formulae abstraction-based FL found more unique FAs than QBCP-guided Q-resolution. Equal numbers of unique FAs do not show up in that statistics. The third, fourth and fifth line report total, average and median numbers of unique FAs over formulae with different FAs. For example in section "ABS vs. QRES", abstraction-based FL detected 3752 (average 28.86, median 1) unique FAs compared to 58 (average 0.44, median 0) by QBCP-guided Q-resolution. The last two lines show average and median differences between unique FAs detected by left and right approaches in each section. Larger values indicate that the left approach is better than the right one. For example in section "ABS vs. QRES", for each of the 130 considered formulae we subtracted the number of unique FAs detected by abstraction-based FL from the one of QBCP-guided Q-resolution. On average abstraction-based FL detected 28.41 more unique FAs than QBCP-guided Q-resolution whereas the median is 1.

In general average values suggest that abstraction-based FL is better than QBCP-guided Q-resolution and SAT-based FL (sections "ABS vs. QRES" and "ABS vs. SAT") and that QBCP-guided Q-resolution is better than SAT-based FL (section "QRES vs. SAT"). But median values indicate the opposite tendency. As in Table 2, few benchmarks account for skew statistics in Table 3. For example in section "ABS vs. QRES", abstraction-based FL found 1603 unique

Table 3. Pairwise comparison of FL approaches (complete runs as in Table 2)

QBFEVAL'10: formulae with different fixed assignments (FAs)					
	ABS vs. QRES		ABS vs. SAT		QRES vs. SAT	
Formulae with Diff. FAs	130		183		220	
Formulae wrt. Unique FAs	121	9	57	126	36	180
Total Unique FAs	3752	58	24268	16648	24237	19874
Avg. Unique FAs	28.86	0.44	132.61	90.97	110.16	90.33
Med. Unique FAs	1	0	0	13	0	5
Avg. Diff. in Unique FAs	28.41		41.63		19.83	
Med. Diff. in Unique FAs	1		-14		-4.5	

FAs on instance `lognBWLARGEB1-shuffled` compared to 0 by QBCP-guided Q-resolution. In section "ABS vs. SAT", abstraction-based FL found between 1000 and 7000 unique FAs on some "`biu*`" instances from bounded model checking compared to 0 by SAT-based FL. In contrast to this, SAT-based FL found 2668 on instance `c3_BMC_p1_k256-shuffled` compared to 0 by abstraction-based FL. Similar observations can be made for section "QRES vs. SAT", hence Table 3 confirms incomparability results from Section 6.

8 Conclusion

We studied failed literal detection (FL) for QBF to infer necessary assignments. Whereas a common approach based on SAT solving turned out to be effective, it suffers from exponential run time and requires careful tuning in practice. We presented two alternatives based on abstraction and Q-resolution which rely on QBCP. The three approaches are incomparable: there are QBFs where a necessary assignment can be detected by one approach but not by the other. Experiments with our implementation in QxBF confirmed that observations. Moreover, abstraction-based FL is a polynomial-time alternative to SAT-based FL. This enables efficient dynamic applications in search-based QBF solvers. Incomparability suggests that FL could benefit from combinations of all three approaches in portfolio-style preprocessors. Combinations of FL with other preprocessing techniques for QBF are future work. Further, it is unclear if clause learning in QBF solvers could be improved. The common implementations based on Q-resolution are not optimal due to incomparability and results from [24].

We want to thank Aina Niemetz and Mathias Preiner for implementing parts of QxBF, Martina Seidl for discussions, and the reviewers for valuable comments.

References

1. Le Berre, D.: Exploiting the Real Power of Unit Propagation Lookahead. Electronic Notes in Discrete Mathematics 9, 59–80 (2001)
2. Biere, A.: Resolve and expand. In: Hoos, H.H., Mitchell, D.G. (eds.) SAT 2004. LNCS, vol. 3542, pp. 59–70. Springer, Heidelberg (2005)
3. Biere, A.: PicoSAT Essentials. JSAT 4(2-4), 75–97 (2008)
4. Biere, A., Heule, M., van Maaren, H., Walsh, T. (eds.): Handbook of Satisfiability. Frontiers in AI and Applications, vol. 185. IOS Press, Amsterdam (2009)
5. Bubeck, U., Kleine Büning, H.: Bounded Universal Expansion for Preprocessing QBF. In: Marques-Silva, J., Sakallah, K.A. (eds.) [19], pp. 244–257
6. Kleine Büning, H., Karpinski, M., Flögel, A.: Resolution for Quantified Boolean Formulas. Inf. Comput. 117(1), 12–18 (1995)
7. Kleine Büning, H., Lettmann, T.: Propositional Logic: Deduction and Algorithms. Cambridge University Press, New York (1999)
8. Cadoli, M., Schaerf, M., Giovanardi, A., Giovanardi, M.: An Algorithm to Evaluate Quantified Boolean Formulae and Its Experimental Evaluation. J. Autom. Reasoning 28(2), 101–142 (2002)
9. Freeman, J.W.: Improvements To Propositional Satisfiability Search Algorithms. PhD thesis, University of Pennsylvania (1995)

10. Giunchiglia, E., Marin, P., Narizzano, M.: QuBE7.0 (System Description). JSAT 7, 83–88 (2010)
11. Giunchiglia, E., Marin, P., Narizzano, M.: sQueezeBF: An Effective Preprocessor for QBFs Based on Equivalence Reasoning. In: Strichman, O., Szeider, S. (eds.) [28]
12. Giunchiglia, E., Narizzano, M., Tacchella, A.: Clause/Term Resolution and Learning in the Evaluation of Quantified Boolean Formulas. J. Artif. Intell. Res (JAIR) 26, 371–416 (2006)
13. Groote, J.F., Warners, J.P.: The Propositional Formula Checker HeerHugo. J. Autom. Reasoning 24(1/2), 101–125 (2000)
14. Heule, M., van Maaren, H.: Look-Ahead Based SAT Solvers. In: Biere, et al (eds.) [4]
15. Jussila, T., Biere, A., Sinz, C., Kröning, D., Wintersteiger, C.M.: A First Step Towards a Unified Proof Checker for QBF. In: Marques-Silva, J., Sakallah, K.A. (eds.) [19]
16. Lonsing, F., Biere, A.: Nenofex: Expanding NNF for QBF Solving. In: Kleine Büning, H., Zhao, X. (eds.) SAT 2008. LNCS, vol. 4996, pp. 196–210. Springer, Heidelberg (2008)
17. Lonsing, F., Biere, A.: DepQBF: A Dependency-Aware QBF Solver (System Description). JSAT 7, 71–76 (2010)
18. Lynce, I., Marques Silva, J.P.: Probing-Based Preprocessing Techniques for Propositional Satisfiability. In: ICTAI, p. 105. IEEE Computer Society, Los Alamitos (2003)
19. Marques-Silva, J., Sakallah, K.A. (eds.): SAT 2007. LNCS, vol. 4501. Springer, Heidelberg (2007)
20. Peschiera, C., Pulina, L., Tacchella, A., Bubeck, U., Kullmann, O., Lynce, I.: The Seventh QBF Solvers Evaluation (QBFEVAL 2010). In: Strichman, O., Szeider, S. (eds.) [28]
21. Pigorsch, F., Scholl, C.: An AIG-Based QBF-Solver Using SAT for Preprocessing. In: Sapatnekar, S.S. (ed.) DAC, pp. 170–175. ACM, New York (2010)
22. Rintanen, J.: Improvements to the Evaluation of Quantified Boolean Formulae. In: Dean, T. (ed.) IJCAI, pp. 1192–1197. Morgan Kaufmann, San Francisco (1999)
23. Samer, M.: Variable Dependencies of Quantified CSPs. In: Cervesato, I., Veith, H., Voronkov, A. (eds.) LPAR 2008. LNCS (LNAI), vol. 5330, pp. 512–527. Springer, Heidelberg (2008)
24. Samulowitz, H., Bacchus, F.: Using SAT in QBF. In: van Beek, P. (ed.) CP 2005. LNCS, vol. 3709, pp. 578–592. Springer, Heidelberg (2005)
25. Samulowitz, H., Bacchus, F.: Binary Clause Reasoning in QBF. In: Biere, A., Gomes, C.P. (eds.) SAT 2006. LNCS, vol. 4121, pp. 353–367. Springer, Heidelberg (2006)
26. Samulowitz, H., Davies, J., Bacchus, F.: Preprocessing QBF. In: Benhamou, F. (ed.) CP 2006. LNCS, vol. 4204, pp. 514–529. Springer, Heidelberg (2006)
27. Marques Silva, J.P., Lynce, I., Malik, S.: Conflict-Driven Clause Learning SAT Solvers. In: Biere, et al (eds.) [4], pp. 131–153
28. Strichman, O., Szeider, S. (eds.): SAT 2010. LNCS, vol. 6175. Springer, Heidelberg (2010)
29. Zhang, L., Malik, S.: Towards a Symmetric Treatment of Satisfaction and Conflicts in Quantified Boolean Formula Evaluation. In: Van Hentenryck, P. (ed.) CP 2002. LNCS, vol. 2470, pp. 200–215. Springer, Heidelberg (2002)

Reducing Chaos in SAT-Like Search:
Finding Solutions Close to a Given One

Ignasi Abío[1], Morgan Deters[2], Robert Nieuwenhuis[1], and Peter J. Stuckey[3]

[1] Technical University of Catalonia (UPC), Barcelona
[2] New York University
[3] National ICT Australia, University of Melbourne

Abstract. Motivated by our own industrial users, we attack the following challenge that is crucial in many practical planning, scheduling or timetabling applications. Assume that a solver has found a solution for a given hard problem and, due to unforeseen circumstances (e.g., rescheduling), or after an analysis by a committee, a few more constraints have to be added and the solver has to be re-run. Then it is almost always important that the new solution is "close" to the original one.

The *activity-based* variable selection heuristics used by SAT solvers make search *chaotic, i.e.*, extremely sensitive to the initial conditions. Therefore, re-running with just one additional clause added at the end of the input usually gives a completely different solution. We show that naive approaches for finding close solutions do not work at all, and that solving the Boolean optimization problem is far too inefficient: to find a reasonably close solution, state-of-the-art tools typically require much more time than was needed to solve the original problem.

Here we propose the first (to our knowledge) approach that obtains close solutions quickly. In fact, it typically finds the optimal (i.e., closest) solution in only 25% of the time the solver took in solving the original problem. Our approach requires no deep theoretical or conceptual innovations. Still, it is non-trivial to come up with and will certainly be valuable for researchers and practitioners facing the same problem.

1   Introduction

For many practical problems, good encodings into propositional logic exist that make them amenable to be solved with SAT. Due to techniques such as conflict-driven backjumping, lemma learning and restarts, state-of-the-art SAT solvers can in many cases efficiently solve large and hard real-world instances. For problems that have no good or compact direct encodings into propositional logic, several extensions of SAT are emerging. One of these extensions is *SAT Modulo Theories* (SMT), where atoms need not be propositional symbols, but may belong to *theories*, like, for example, linear arithmetic, as in the formula $x \leq 2 \ \wedge \ (x{+}y \geq 10 \ \vee \ 2x{+}3y \geq 30) \ \wedge \ y \leq 4$. In SMT, a SAT solver cooperates with *theory solvers* that can handle *conjunctions* of theory atoms (see, *e.g.*, [NOT06] for details). Another extension of SAT is the Lazy Clause Generation

K.A. Sakallah and L. Simon (Eds.): SAT 2011, LNCS 6695, pp. 273–286, 2011.
© Springer-Verlag Berlin Heidelberg 2011

approach of [OSC09], where new propositional clauses are generated on demand each time a given constraint propagates, thus frequently reducing the number of clauses needed in comparison with a direct *a priori* SAT encoding.

SAT and SAT-like solving approaches almost universally make use of *activity-based* search heuristics, which roughly speaking, select the variables that have been involved in many recent conflicts. A drawback of activity-based heuristics is that they make the search behave *chaotically* (explaining why is out of the scope of this paper), i.e., extremely sensitive to the initial conditions, the so-called *butterfly effect*.

But in practice it is almost always important that the new solution is "close" to the original one. For example, analyzing a solution may take time and effort and include discussions with other people. If someone, inspired by the solution, suggests adding a few new constraints, it is undesirable that a new solution for the extended problem has nothing in common with what was analyzed previously. Something similar happens in the context of *rescheduling*, where a solution that was intended to be used for a period of time has to be adapted due to unforeseen circumstances: changes should be minimal since many resources (people, vehicles, machines) are already allocated according to the original solution.

In this paper, Section 2 gives a short introduction to state-of-the-art SAT solving. In Section 3 we accurately define the problem and we discuss the distance metrics, e.g., what it means for a solution to be *close*. Section 4 presents the experimental setting and the large set of real-world benchmarks used along the paper. In particular, in Section 5 we use them to experimentally demonstrate the extremely chaotic behavior of SAT Solvers, and in Section 6 to evaluate a naive attempt for finding close solutions inspired by local search methods.

Since this method does not solve the problem, in Section 7 we introduce a new approach. It combines a polarity heuristic, incremental SAT and branch-and-bound. In Section 8 we compare our method with (i) SAT-based optimization and Max-SAT solvers; (ii) modeling the problem as a 0-1 integer optimization problem and using CPLEX on it. As we shall see, approaches (i) and (ii) behave very poorly,[1] but our new approach obtains close solutions very quickly. In fact, it typically finds the optimal (i.e., closest) solution in only 25% of the time the solver takes in solving the original problem.

Finally, Section 9 gives a factor analysis of our approach: experiments reveal that all ingredients contribute. Related work is discussed and conclusions are given in Section 10.

2   State-of-the-Art SAT Solvers

Let P be a fixed finite set of propositional symbols. If $p \in P$, then p and $\neg p$ are *literals* of P. The *negation* of a literal l, written $\neg l$, denotes $\neg p$ if l is p, and p if l is $\neg p$. A *clause* is a disjunction of literals $l_1 \vee \ldots \vee l_n$. A *unit clause* is a

---

[1] An earlier (rejected) submission about this work failed to explain this adequately and to show this experimentally for approach (i). In addition, here we also consider approach (ii) and compare with more related work.

clause consisting of a single literal. A (CNF) *formula* is a conjunction of one or more clauses $C_1 \wedge \ldots \wedge C_n$. A (partial truth) *assignment* M is a set of literals such that $\{p, \neg p\} \subseteq M$ for no p. A literal l is *true* in M if $l \in M$, is *false* in M if $\neg l \in M$, and is *undefined* in M otherwise. A literal is *defined* in M if it is either true or false in M. A clause C is true in M if at least one of its literals is true in M. It is false in M if all its literals are false in M, and it is *undefined* in M otherwise. A formula F is true in M, denoted $M \models F$, if all its clauses are, and then M is a *model* of F. If F has no models then it is *unsatisfiable*. If F and F' are formulas, we write $F \models F'$ if F' is true in all models of F. If C is a clause $l_1 \vee \ldots \vee l_n$, we write $\neg C$ to denote the formula $\neg l_1 \wedge \ldots \wedge \neg l_n$.

Following [NOT06], here we say that a *state* in a SAT solver is a pair of the form $M \parallel F$, where F is a finite set of clauses, and M is a (partial) assignment, where a literal l may be annotated as a *decision literal* (see below), writing it as l^d. A clause C is a *conflict* in a state $M \parallel F, C$ if $M \models \neg C$. A SAT solving procedure can be modeled by a set of rules over such states:

UnitPropagate :

$$M \parallel F, C \vee l \implies M\ l \parallel F, C \vee l \quad \text{if} \quad \begin{cases} M \models \neg C \\ l \text{ is undefined in } M \end{cases}$$

Decide :

$$M \parallel F \qquad\qquad \implies M\ l^d \parallel F \qquad \text{if} \quad \begin{cases} l \text{ or } \neg l \text{ occurs in a clause of } F \\ l \text{ is undefined in } M \end{cases}$$

Fail :

$$M \parallel F, C \qquad \implies Fail \qquad \text{if} \quad \begin{cases} M \models \neg C \\ M \text{ contains no decision literals} \end{cases}$$

Backjump :

$$M\ l^d\ N \parallel F, C \implies M\ l' \parallel F, C \quad \text{if} \quad \begin{cases} M\ l^d\ N \models \neg C, \text{ and there is} \\ \text{some clause } C' \vee l' \text{ such that:} \\ \quad F, C \models C' \vee l' \text{ and } M \models \neg C', \\ \quad l' \text{ is undefined in } M, \text{ and} \\ \quad l' \text{ or } \neg l' \text{ occurs in } F \text{ or in } M\ l^d\ N \end{cases}$$

Learn :

$$M \parallel F \qquad\qquad \implies M \parallel F, C \quad \text{if} \quad \begin{cases} \text{each atom of } C \text{ occurs in } F \text{ or in } M \\ F \models C \end{cases}$$

Forget :

$$M \parallel F, C \qquad \implies M \parallel F \qquad \text{if} \quad \{ F \models C$$

For deciding the satisfiability of an input formula F, one can generate an arbitrary derivation $\emptyset \parallel F \implies \ldots \implies S_n$, where S_n is a final state (no rule applies). Under simple conditions, this always terminates. Moreover, for every derivation ending in a final state S_n, *(i)* F is unsatisfiable iff S_n is *Fail*, and *(ii)* if S_n is of the form $M \parallel F$ then M is a model of F (cf.[NOT06] for details).

In the current state-of-the-art SAT solvers such as MiniSAT [ES04], the *variable selection heuristics* are *activity-based*: roughly, Decide is done on variables with *many* occurrences in *recent* conflicts. In this paper we also consider the choice of *polarity* for Decide, *i.e.*, whether the variable is set to true or to false.

We say that a state M is at *decision level* n if in M there are n decision literals. In Backjump, $C' \lor l'$ is called the *backjump clause*. This clause is a logical consequence to which UnitPropagate would have applied at a lower decision level, and Backjump does precisely this, after reverting to that decision level. In practice, the backjump clause is computed in a *conflict analysis* process, which is beyond the scope of this paper.

The Learn rule corresponds to adding *lemmas* (clauses that are logical consequences) such as the backjump clause. Since a lemma is aimed at preventing future similar conflicts, when these conflicts are not very likely to be found again the lemma can be removed by the Forget rule. In practice, a lemma is removed when its *activity* drops below a certain threshold; the activity can be, *e.g.*, the number of times it becomes a unit or a conflicting clause [GN02].

3   Problem Definition

Assume we have found a solution *Sol* to a problem defined by a formula (a set of clauses) F and we are given a small set of additional clauses δ. We wish to find a solution *Sol'* that is *close* to *Sol* for the clause set $F \cup \delta$.

One way for defining solutions' proximity is by considering their Hamming distance (the number of variables which take a different value). As many problems have some *hidden* auxiliary variables in their SAT encoding F, it is frequently useful to consider only the *visible* (i. e. non hidden) variables for the distance definition.

Certain applications can require slightly more involved cost functions instead of just Hamming distance. For example, a single property of the solution, seen by the user, may depend on *combinations* of visible variables. For example, in the sports scheduling problems we will use later, a property like a match may depend on a variable m_{ijr} saying that these two teams i and j meet on round r, and another two h_{ir} and h_{jr} saying whether team i and j plays at home on round r. A more accurate cost function to capture "nearness to the existing solution" in this case would count a distance of 1 if either of m_{ijr} or h_{ir} differ from their previous values, but not count 2 if both differ.

However, in this paper we have only considered Hamming distance cost functions for simplicity in the computations. In the majority of the practical cases, a close solution for some distance is also a close solution for the Hamming distance (see the previous example).

4   Benchmarks

We have considered 40 instances of real-world benchmarks coming from five different families. Each instance consists of a different SAT formula F, the first solution *Sol*, and a number of required additional constraints δ. The first four families are for scheduling a double round-robin tournament among N (16, 20 or 24) teams:

r16: 10 instances with about 3000 variables and around 55000 clauses each;

r20: 10 instances with around 5000 variables and 180000 clauses each;

R20: 10 instances with around 5000 variables, 140000 clauses each;

r24: 4 instances with around 9000 variables, 270000 clauses each.

All teams meet each other once in the first $N-1$ weeks and again in the second $N-1$ weeks, with exactly one match per team each week. A given pair of teams must play at the home of one team in one half, and at the home of the other in the other half, and such matches must be spaced at least a certain minimal number of weeks apart. Additional constraints include, e.g., that no team ever plays at home (or away) three times in a row, other (public order, sportive, TV revenues) constraints, blocking given matches on given days, etc. Instances are rather different among each other, but most of them have around 10% hidden variables. The R20 instances are also different in that their δs contain more constraints and hence the closest solution is usually not as close (see below).

The fifth family of benchmarks has six problems tt0 - tt5 coming from real-world hard curriculum-based course timetabling problems, from the International Timetabling Competition, see the Barcelogic results on formulation 2 at http://tabu.diegm.uniud.it/ctt. These problems are very different from the r ones. Their numbers of (visible) variables and clauses are:

instances	variables	visible variables	clauses
tt0	12537	1500	71919
tt1	137688	6314	667470
tt2	60968	3150	305601
tt3	556569	9810	3372803
tt4	125029	4494	1001737
tt5	124330	3381	612475

For each instance, we consider Hamming distance on the visible variables as the cost function. All experiments were performed on a 2.66MHz Xeon.

5   Chaotic Behavior of SAT

In this section we analyze what happens when simply re-executing the Solver with the new input $F \cup \delta$. Table 1 contains results on all 40 instances.

Here *Time original* denotes the time (in seconds) spent to compute the original solution *Sol*, *Time re-execution* denotes the time spent in the computation of *Sol'*. d_{opt} denotes the minimal Hamming distance from the original solution *Sol* to any solution of $F \cup \delta$. *Time ratio* is defined by the ratio between the re-execution time and the original time.

The *quality* of a solution *Sol'* at distance d of *Sol* is a real number between 0 and 1 defined by d_{opt}/d. For example, if d_{opt} is 10, then a solution at distance 50 has quality 0.2.

Table 1. Results of re-execution

Instance	Time original	d_{opt}	Quality	Time re-execution	Time ratio
r16-0	0.88	12	0.03	0.93	1.06
r16-1	1.58	14	0.04	1.27	0.80
r16-2	1.66	8	0.02	0.74	0.45
r16-3	0.97	8	0.02	1.63	1.68
r16-4	3.56	64	0.14	7.09	1.99
r16-5	0.03	12	0.22	0.04	1.33
r16-6	0.02	14	0.03	0.05	2.50
r16-7	0.4	18	0.04	0.69	1.72
r16-8	3.55	8	0.02	1.27	0.36
r16-9	1.39	12	0.03	0.61	0.44
r20-0	12.23	24	0.04	12.37	1.01
r20-1	59.6	8	0.01	20.00	0.34
r20-2	9.47	12	0.02	9.65	1.02
r20-3	12.82	14	0.03	2.83	0.22
r20-4	20.15	18	0.19	20.03	0.99
r20-5	20.48	16	0.02	8.82	0.43
r20-6	8.81	18	0.04	2.09	0.24
r20-7	10.88	20	0.03	13.46	1.24
r20-8	13.52	16	0.04	8.95	0.66
r20-9	7.04	12	0.03	12.39	1.76
R20-0	1.77	8	0.02	3.56	2.01
R20-1	2.37	88	0.17	6.30	2.66
R20-2	6.69	96	0.19	9.53	1.42
R20-3	9.46	8	0.01	5.30	0.56
R20-4	5.4	136	0.25	1.14	0.21
R20-5	1.14	1	0.00	7.04	6.18
R20-6	7.71	104	0.19	4.95	0.64
R20-7	5.45	26	0.05	0.62	0.11
R20-8	0.61	82	0.16	7.03	11.52
R20-9	7.49	94	0.16	1.78	0.24
r24-0	227.97	42	0.04	143.51	0.63
r24-1	124.28	58	0.05	315.14	2.54
r24-2	277.49	14	0.01	226.80	0.82
r24-3	200.53	8	0.01	416.14	2.08
tt-0	1.62	10	0.03	0.36	0.22
tt-1	0.96	10	0.07	0.93	0.97
tt-2	0.38	6	0.04	0.28	0.74
tt-3	16.3	8	0.01	14.20	0.87
tt-4	27.42	26	0.04	16.17	0.59
tt-5	1.75	8	0.02	1.73	0.99

These experiments show the chaotic behavior of SAT Solvers: re-running the same solver with the same set of clauses except one or two added at the end of the input file causes the solver to perform a completely different search, giving a very different execution in terms of distance of the solutions and also in computation time. In particular, qualities are typically below 0.1, that is, ten times more distant than the optimal solution.

6   Trying a Local Search-Like Solution

In local search techniques, to find close solutions one usually resumes the search at the point where the original solution was found with the hope that another solution is found in the nearby neighborhood. Therefore, at first sight, mimicking local search might seem a good option for overcoming the chaotic behavior of SAT.

More specifically, we want to re-execute the solver in the region of the search tree where the original solution was found. A simple way of implementing this idea is by changing the *variable selection heuristics* as follows. We remember the ordered sequence of decision literals of the original solution, and when the solver is re-launched with the new constraints, it always decides on the first undefined literal of the sequence, with the same polarity, until the first conflict occurs. After that, we fall back to the standard decision heuristic. Note that this will always find the same solution Sol if Sol is also a solution of $F \cup \delta$.

Unfortunately, the results do not improve significantly upon re-running from scratch as described in the previous section. Table 2 contains the results of this method. We have obtained similar results with some variations of this method (keeping the lemmas of the original execution as in the next section, keeping this heuristic, or a combination of both).

7   Our Barcelogic Approach

As we have seen in the previous sections, the naive approaches are not effective for solving this problem in practice. The good news is that an adequate combination of three quite well-known ingredients does obtain close solutions very quickly.

The first ingredient is a polarity selection heuristic: the SAT solver uses its standard heuristic for picking the next variable to decide upon, but for visible variables it sets this variable's polarity as in the original solution Sol (other optimization tools do this too: first try those values that minimize the cost function; it is also related to, but different from, *phase saving* [PD10]).

Second, a branch-and-bound wrapper is placed around the standard SAT loop. Each time the cost of the best solution discovered so far is exceeded by the current partial assignment, due to literals $l_1 \ldots l_n$ (on visible variables) that disagree with Sol, a backjump is forced from a conflict analysis on an "explanation" $\neg l_1 \vee \ldots \vee \neg l_n$ of why the cost is currently too high. In particular, this is done

Table 2. Results of a local-search-like approach

Instance	Time original	d_{opt}	Quality	Time re-execution	Time ratio
r16-0	0.88	12	0.03	0.80	0.91
r16-1	1.58	14	0.03	1.87	1.18
r16-2	1.66	8	0.02	0.66	0.40
r16-3	0.97	8	0.02	2.81	2.90
r16-4	3.56	64	0.13	3.82	1.07
r16-5	0.03	12	0.03	0.08	2.67
r16-6	0.02	14	0.03	0.00	0.00
r16-7	0.4	18	0.04	0.18	0.45
r16-8	3.55	8	0.02	0.85	0.24
r16-9	1.39	12	0.03	1.27	0.91
r20-0	12.23	24	0.04	9.50	0.78
r20-1	59.6	8	0.31	0.03	0.00
r20-2	9.47	12	0.55	0.03	0.00
r20-3	12.82	14	0.03	6.64	0.52
r20-4	20.15	18	0.33	0.03	0.00
r20-5	20.48	16	0.03	21.12	1.03
r20-6	8.81	18	0.03	17.50	1.99
r20-7	10.88	20	0.03	6.69	0.61
r20-8	13.52	16	0.03	2.15	0.16
r20-9	7.04	12	0.02	6.96	0.99
R20-0	1.77	8	0.02	2.43	1.37
R20-1	2.37	88	0.16	5.20	2.19
R20-2	6.69	96	0.15	2.26	0.34
R20-3	9.46	8	0.02	4.77	0.50
R20-4	5.4	136	0.26	6.34	1.17
R20-5	1.14	1	0.00	5.84	5.12
R20-6	7.71	104	0.20	6.18	0.80
R20-7	5.45	26	0.05	11.27	2.07
R20-8	0.61	82	0.16	4.94	8.10
R20-9	7.49	94	0.17	2.82	0.38
r24-0	227.97	42	0.04	134.14	0.59
r24-1	124.28	58	0.05	3574.00	28.76
r24-2	277.49	14	0.01	157.08	0.57
r24-3	200.53	8	0.01	296.43	1.48
tt-0	1.62	10	0.03	0.21	0.13
tt-1	0.96	10	0.29	0.29	0.30
tt-2	0.38	6	0.60	0.13	0.34
tt-3	16.3	8	0.01	18.13	1.11
tt-4	27.42	26	0.04	11.88	0.43
tt-5	1.75	8	0.01	2.36	1.35

each time a better model is found, in order to find, from then on, only lower-cost models. Here this explanation clause need not be learned.

The backjump clause itself is learned as usual. Eventually this process terminates by discovering unsatisfiability—that there is no "better" solution to the best already found. As is well-known, it may require far more time to prove optimality than it does to find an optimal solution.[2] However, good solutions can often be found in a short time. See, e.g., [MMS04, LNORC09, LNORC11] and references of these for many more details and an abstract framework for Boolean optimization.

Third, the lemmas the SAT Solver generated when finding the original solution are added; this is sound since there are only *additional* constraints, no removed ones; this latter idea is also used in the context of incremental SAT solving for, e.g., verification applications.

8   Experimental Comparison with Cplex and Other Tools

In this section we compare experimentally our approach with other tools. We first encoded $F \cup \delta$ together with the cost function as a pseudo-Boolean (0-1 Integer Programming) optimization problem and tried the state-of-the-art pseudo-Boolean solver *Bsolo* [MMS04] and the well-known commercial CPLEX solver.

We also tried several state-of-the-art Max-SAT solvers. MiniMaxSAT [HLO08] found close solutions only in a few cases. The unsatisfiable-core-based MaxSAT solvers *msuncore* [MSP09] and *PM2* [ABL09] were not competitive either, among other reasons because unsat-core-based solvers find no solution before the optimal one. We do not report here on these MaxSAT solvers' results: they were always much worse than the listed ones.

We also tried Barcelogic omitting its ingredients one by one, i.e., without keeping the lemmas from the first run or without the modified polarity heuristic. The results are described in the next section. Bsolo and CPLEX results are without the lemmas: the number of lemmas was much bigger than the number of original constraints and these solvers perform much worse if we add them.

The results are given in Table 3.

Solution quality: As before, the table lists *solution qualities* as real numbers between 0 and 1: d_{opt} denotes the minimal Hamming distance from the original solution *Sol* to any solution of $F \cup \delta$ and again we say that a solution *Sol'* at distance d of *Sol* has *quality* d_{opt}/d.

Entries in the table: The table gives results on all 40 instances for Barcelogic, Bsolo and CPLEX. For each instance, column 2 lists the time T the (Barcelogic) SAT solver took to compute the initial solution *Sol*. The third column indicates the cost of the optimal solution, d_{opt}. For each approach, the table lists the quality of the solution found after 25% of T, after 50% of T, etc., up to 800% of T. Moreover, the two average rows show the average of, respectively, the first 20 problems and the 20 other (harder) ones. The two plots of figure 1

---

[2] In fact, for some of the benchmarks in this paper *proving* optimality took days of CPU time.

Table 3. Comparative results of the three most competitive approaches: Barcelogic, Bsolo and CPLEX

	Time	d_{opt}	Barcelogic						Bsolo						Cplex					
			25	50	100	200	400	800	25	50	100	200	400	800	25	50	100	200	400	800
r16-0	0.88	12	1	1	1	1	1	1	0	0	.67	.67	1	1	0	0	0	1	1	1
r16-1	1.58	14	1	1	1	1	1	1	0	0	1	1	1	1	0	0	0	0	0	1
r16-2	1.66	8	1	1	1	1	1	1	0	1	1	1	1	1	0	1	1	1	1	1
r16-3	0.97	8	1	1	1	1	1	1	0	0	.67	1	1	1	0	0	1	1	1	1
r16-4	3.56	64	.86	.86	.94	1	1	1	.67	.67	.67	.67	.67	.67	0	0	0	0	0	0
r16-5	0.03	12	0	0	.50	.60	1	1	0	0	0	0	0	0	0	0	0	0	0	0
r16-6	0.02	14	0	0	0	0	.12	.64	0	0	0	0	0	0	0	0	0	0	0	0
r16-7	0.4	18	.82	.82	.82	.82	1	1	0	0	0	.36	.36	.36	0	0	0	0	0	0
r16-8	3.55	8	1	1	1	1	1	1	1	1	1	1	1	1	1	1	1	1	1	1
r16-9	1.39	12	1	1	1	1	1	1	0	.60	.60	.60	.60	1	0	0	1	1	1	1
r20-0	12.23	24	1	1	1	1	1	1	.34	.34	.34	.50	1	1	0	0	0	0	0	1
r20-1	59.6	8	1	1	1	1	1	1	.67	1	1	1	1	1	1	1	1	1	1	1
r20-2	9.47	12	1	1	1	1	1	1	.27	.38	.38	.60	.60	.75	0	1	1	1	1	1
r20-3	12.82	14	1	1	1	1	1	1	1	1	1	1	1	1	0	1	1	1	1	1
r20-4	20.15	18	1	1	1	1	1	1	.41	.41	.41	.43	.43	.64	0	0	.38	.38	1	1
r20-5	20.48	16	1	1	1	1	1	1	.57	.57	.57	.89	.89	1	0	0	0	0	.24	1
r20-6	8.81	18	1	1	1	1	1	1	.30	.82	.82	.82	.82	1	0	0	0	.90	.90	1
r20-7	10.88	20	1	1	1	1	1	1	.83	.83	.83	.91	1	1	0	0	.32	.32	.32	1
r20-8	13.52	16	1	1	1	1	1	1	.35	.35	.35	.35	.35	1	0	0	0	0	1	1
r20-9	7.04	12	1	1	1	1	1	1	0	.22	.25	.55	.60	.86	0	1	1	1	1	1
Av.	-	-	**.88**	**.88**	**.91**	**.92**	**.96**	**.98**	**.32**	**.46**	**.58**	**.67**	**.72**	**.81**	**.10**	**.30**	**.43**	**.53**	**.62**	**.80**
R20-0	1.77	8	1	1	1	1	1	1	0	0	1	1	1	1	0	0	1	1	1	1
R20-1	2.37	88	.57	.57	.57	.57	.72	.75	0	0	0	0	.66	.66	0	0	0	0	0	0
R20-2	6.69	96	.74	.74	.74	.80	.84	.89	0	.53	.53	.55	.55	.70	0	0	0	0	0	0
R20-3	9.46	8	1	1	1	1	1	1	1	1	1	1	1	1	1	1	1	1	1	1
R20-4	5.4	136	.65	.65	.86	.86	.91	.97	0	0	0	0	0	0	0	0	0	0	0	0
R20-5	1.14	1	1	1	1	1	1	1	0	0	0	1	1	1	0	0	0	1	1	1
R20-6	7.71	104	.80	.80	.88	.88	.88	.88	0	0	0	.42	.57	.57	0	0	0	0	0	0
R20-7	5.45	26	.93	.93	1	1	1	1	.68	.68	.68	.68	.68	.68	0	1	1	1	1	1
R20-8	0.61	82	.84	.84	.84	.85	.98	.98	0	0	0	0	.60	.60	0	0	0	0	0	0
R20-9	7.49	94	.64	.77	.77	.84	.90	.90	0	.43	.59	.59	.59	.59	0	0	0	0	0	0
r24-0	227.97	42	1	1	1	1	1	1	0	0	0	0	.57	.57	0	0	0	0	0	0
r24-1	124.28	58	.58	.58	.58	.74	.74	.74	0	.42	.42	.42	.42	.42	0	0	0	0	0	0
r24-2	277.49	14	1	1	1	1	1	1	.37	.37	.37	.37	.37	.37	0	1	1	1	1	1
r24-3	200.53	8	1	1	1	1	1	1	1	1	1	1	1	1	1	1	1	1	1	1
tt-0	1.62	10	1	1	1	1	1	1	0	0	0	0	0	0	0	0	0	0	0	1
tt-1	0.96	10	0	.36	.36	.36	.36	.36	0	0	0	0	0	0	0	0	0	0	0	0
tt-2	0.38	6	0	.60	.60	.60	.75	.75	0	0	0	0	0	0	0	0	0	0	0	0
tt-3	16.3	8	.57	.57	.57	.57	.67	.67	0	0	0	0	0	0	0	0	0	0	0	0
tt-4	27.42	26	.10	.10	.50	.50	.50	.50	0	0	0	0	0	0	0	0	0	0	0	0
tt-5	1.75	8	0	0	0	0	.13	.14	0	0	0	0	0	0	0	0	0	0	0	0
Av.	-	-	**.67**	**.73**	**.76**	**.78**	**.82**	**.83**	**.15**	**.22**	**.28**	**.35**	**.45**	**.46**	**.10**	**.20**	**.25**	**.30**	**.30**	**.35**

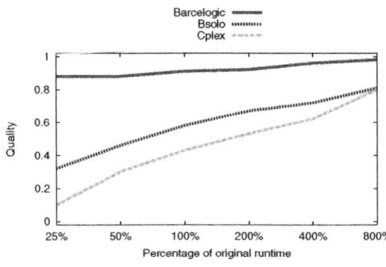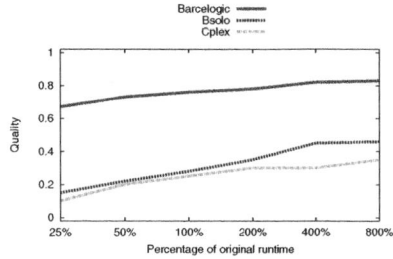

Fig. 1. Average quality of the different approaches on the first 20 problems (left) and the second 20 harder ones (right).

represent graphically these averages. They also give some intuition about how the approaches scale.

9   Factor Analysis of the Barcelogic Approach

In this section we evaluate separately the different ingredients used in our approach. More specifically, we show the experimental results of our solver with just a Branch and Bound ("B&B" in the table; first column), adding the lemmas ("B&B + lemmas"; second column), with the modified polarity heuristic ("B&B + polarity"; third column) and finally "B&B + All" (fourth column). The results are given in Table 4. As in the previous section, the table shows the quality of the solution found after 25%, 50%, etc. of the time spent in solving the original problem.

Clearly, the polarity decision heuristic hugely improves the method. On the other hand, keeping the lemmas helps significantly for the hard problems, while on the easier ones the overhead of reading the additional clauses frequently does not pay off.

Again, the two plots of figure 2 represent graphically the results of the table for average solution qualities of, respectively, the first 20 instances, and the other much harder 20 ones.

10   Related Work and Conclusions

We have studied, from a practical point of view, the problem of, given a SAT formula F with a model Sol, and a small set of additional clauses δ, finding a model of $F \cup \delta$ that is *close* to Sol.

Similar problems were studied before in a more theoretical (complexity) setting. [HHOW05] examine the problem of finding a set of diverse or similar solutions for a single problem using constraint programming. Their MOSTCLOSE

Table 4. Results of the factor analysis

	Basic B&B						B&B + lemmas						B&B + polarity						B&B + All					
	25	50	100	200	400	800	25	50	100	200	400	800	25	50	100	200	400	800	25	50	100	200	400	800
r16-0	0	0	0	.04	.04	.04	.03	.03	.03	.04	.04	.04	1	1	1	1	1	1	1	1	1	1	1	1
r16-1	0	0	.04	.04	.04	.04	0	0	.04	.04	.04	.04	1	1	1	1	1	1	1	1	1	1	1	1
r16-2	0	.02	.02	.02	.02	.02	.02	.02	.02	.02	.02	.02	1	1	1	1	1	1	1	1	1	1	1	1
r16-3	0	0	0	.03	.03	.03	0	0	.02	.02	.02	.02	1	1	1	1	1	1	1	1	1	1	1	1
r16-4	0	0	0	.14	.16	.16	0	0	.14	.14	.15	.16	.80	.80	.82	.82	.86	.91	.86	.86	.94	1	1	1
r16-5	0	0	.38	.38	.38	.38	0	0	0	.03	.03	.03	0	0	.24	1	1	1	0	0	.50	.60	1	1
r16-6	0	0	0	.03	.03	.04	0	0	0	0	0	.03	0	.12	1	1	1	1	0	0	0	0	.12	.64
r16-7	0	0	0	.04	.05	.05	0	0	0	.05	.05	.05	.69	.82	.90	.90	1	1	.82	.82	.82	.82	1	1
r16-8	0	.02	.02	.02	.02	.02	0	.02	.02	.02	.02	.02	1	1	1	1	1	1	1	1	1	1	1	1
r16-9	0	.04	.04	.04	.04	.04	0	0	0	.03	.03	.03	1	1	1	1	1	1	1	1	1	1	1	1
r20-0	0	0	0	.05	.05	.06	.04	.04	.05	.05	.05	.05	1	1	1	1	1	1	1	1	1	1	1	1
r20-1	0	.01	.01	.01	.01	.02	.01	.01	.01	.01	.01	.01	1	1	1	1	1	1	1	1	1	1	1	1
r20-2	0	0	0	.02	.02	.02	0	0	.02	.02	.02	.02	1	1	1	1	1	1	1	1	1	1	1	1
r20-3	.03	.03	.03	.03	.03	.03	0	0	.02	.02	.03	.03	1	1	1	1	1	1	1	1	1	1	1	1
r20-4	0	0	0	.20	.20	.20	0	0	.04	.04	.04	.04	1	1	1	1	1	1	1	1	1	1	1	1
r20-5	0	.02	.02	.03	.03	.03	0	.03	.03	.03	.03	.03	1	1	1	1	1	1	1	1	1	1	1	1
r20-6	.04	.04	.04	.04	.04	.04	.04	.04	.04	.04	.04	.04	1	1	1	1	1	1	1	1	1	1	1	1
r20-7	0	0	0	.03	.03	.05	.06	.06	.06	.06	.06	.06	.91	.91	1	1	1	1	1	1	1	1	1	1
r20-8	0	0	.04	.04	.04	.04	0	.03	.03	.03	.03	.03	1	1	1	1	1	1	1	1	1	1	1	1
r20-9	0	0	0	.03	.03	.03	0	0	.02	.02	.02	.02	1	1	1	1	1	1	1	1	1	1	1	1
Av.	**0**	**.01**	**.03**	**.06**	**.06**	**.07**	**.01**	**.01**	**.03**	**.04**	**.04**	**.04**	**.87**	**.88**	**.95**	**.99**	**.99**	**1**	**.88**	**.88**	**.91**	**.92**	**.96**	**.98**
R20-0	0	0	0	0	.02	.02	0	0	0	.02	.02	.02	1	1	1	1	1	1	1	1	1	1	1	1
R20-1	0	0	0	0	.18	.18	0	0	0	.15	.15	.15	0	0	.64	.64	.64	.71	.57	.57	.57	.57	.72	.75
R20-2	0	0	0	.19	.19	.20	0	0	.22	.25	.25	.25	.52	.69	.73	.75	.86	.91	.74	.74	.74	.80	.84	.89
R20-3	0	0	.01	.02	.02	.02	0	.02	.02	.02	.02	.02	1	1	1	1	1	1	1	1	1	1	1	1
R20-4	.25	.25	.25	.25	.26	.26	.23	.25	.25	.25	.25	.27	0	.44	.77	.77	.79	.85	.65	.65	.86	.86	.91	.97
R20-5	0	0	0	0	0	0	0	0	0	0	0	0	1	1	1	1	1	1	1	1	1	1	1	1
R20-6	0	0	.20	.20	.26	.26	0	.16	.16	.23	.23	.23	.81	.81	.87	.87	.90	.91	.80	.80	.88	.88	.88	.88
R20-7	.05	.05	.06	.06	.06	.06	0	.05	.05	.05	.05	.06	.93	.93	1	1	1	1	.93	.93	1	1	1	1
R20-8	0	0	0	0	0	0	0	0	0	0	.13	.13	.50	.50	.50	.59	.72	.93	.84	.84	.84	.85	.98	.98
R20-9	0	.16	.20	.20	.20	.20	.18	.21	.21	.21	.24	.24	.71	.81	.85	.85	.89	.89	.64	.77	.77	.84	.90	.90
r24-0	0	0	.04	.04	.04	.04	0	0	0	.04	.04	.04	1	1	1	1	1	1	1	1	1	1	1	1
r24-1	0	0	0	0	.05	.05	0	.05	.05	.06	.06	.06	.67	.67	.76	.76	.76	.76	.58	.58	.58	.74	.74	.74
r24-2	0	0	.01	.01	.01	.01	0	.01	.01	.01	.01	.01	1	1	1	1	1	1	1	1	1	1	1	1
r24-3	0	0	0	.01	.01	.01	.01	.01	.01	.01	.01	.01	1	1	1	1	1	1	1	1	1	1	1	1
tt-0	.03	.03	.03	.03	.04	.04	0	0	0	0	.03	.03	.05	.14	.14	.19	.25	.50	1	1	1	1	1	1
tt-1	0	0	.07	.07	.07	.07	0	0	.04	.04	.04	.04	0	.36	.36	.36	.36	.36	0	.36	.36	.36	.36	.36
tt-2	0	0	.04	.04	.04	.04	0	0	0	.02	.02	.02	0	.60	.60	.60	.60	1	0	.60	.60	.60	.75	.75
tt-3	0	0	.01	.01	.01	.01	0	0	0	.01	.01	.01	0	0	.01	.04	.11	.29	.57	.57	.57	.57	.67	.67
tt-4	0	0	.04	.04	.04	.04	0	.04	.04	.04	.04	.04	0	0	.06	.41	.41	.41	.10	.10	.50	.50	.50	.50
tt-5	0	0	0	.02	.02	.02	0	0	.02	.02	.02	.02	0	0	0	0	.14	.14	0	0	0	0	.13	.14
Av.	**.02**	**.02**	**.05**	**.06**	**.08**	**.08**	**.02**	**.04**	**.05**	**.07**	**.08**	**.08**	**.51**	**.60**	**.66**	**.69**	**.72**	**.78**	**.67**	**.73**	**.76**	**.78**	**.82**	**.83**

question is very similar to the problem we examine looking for the closest solution to an existing solution, but both solutions are for the same problem.
They outline two approaches: a reformulation approach that at least doubles
the size of the problem, and a more efficient heuristic approach which is simply
a branch and bound search. Our results show that this by itself is not enough in
the SAT context. Distance-SAT [BM06] explores the decision problem, given a
formula G and an *arbitrary* partial interpretation I, is there a model of G that
disagrees with I on at most k variables? [BM06] tries on random and handcrafted

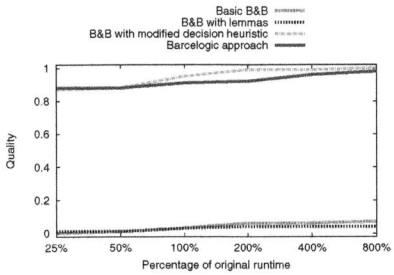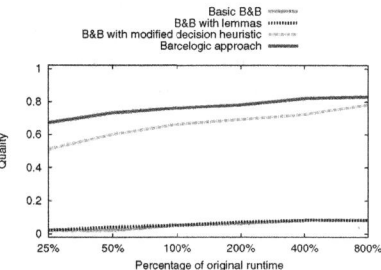

Fig. 2. Average quality of the factor analysis on the first 20 problems (left) and the second 20 harder ones (right)

problems two algorithms based on the classical Davis/Logemann/Loveland (DLL) procedure [DLL62], but a translation into CNF is reported to work better. For our case, where deciding SAT for G is already hard, such a translation is rather hopeless. One clearly needs to exploit that in our problem I is a model of a *known* subformula of G that is almost the same as G.

Indeed, our experiments reveal that, while state-of-the-art Boolean optimization solvers behave poorly, our Barcelogic approach behaves very well, frequently finding the optimal (i.e., closest) solution in only 25% of the time the SAT solver took in solving the original problem.

Acknowledgements

NICTA is funded by the Australian Government as represented by the Department of Broadband, Communications and the Digital Economy and the Australian Research Council.

Abío and Nieuwenhuis are partially supported by Spanish Min. of Educ. and Science through the LogicTools-2 project (TIN2007-68093-C02-01). Abío is also partially supported by FPU grant.

References

[ABL09]   Ansótegui, C., Bonet, M.L., Levy, J.: Solving (weighted) partial maxsat through satisfiability testing. In: Marques-Silva, J., Sakallah, K.A. (eds.) SAT 2007. LNCS, vol. 4501, pp. 427–440. Springer, Heidelberg (2007)

[BM06]   Bailleux, O., Marquis, P.: Some computational aspects of distance-sat. J. Autom. Reasoning 37(4), 231–260 (2006)

[DLL62]   Davis, M., Logemann, G., Loveland, D.: A Machine Program for Theorem-Proving. Comm. of the ACM, CACM 5(7), 394–397 (1962)

[ES04]   Eén, N., Sörensson, N.: An Extensible SAT-solver. In: Giunchiglia, E., Tacchella, A. (eds.) SAT 2003. LNCS, vol. 2919, pp. 502–518. Springer, Heidelberg (2004)

[GN02]   Goldberg, E., Novikov, Y.: BerkMin: A Fast and Robust SAT-Solver. In: 2002 Conference on Design, Automation, and Test in Europe, DATE 2002, pp. 142–149. IEEE Computer Society, Los Alamitos (2002)

[HHOW05]   Hebrard, E., Hnich, B., O'Sullivan, B., Walsh, T.: Finding diverse and similar solutions in constraint programming. In: 20th National Conf. on Artificial Intelligence (AAAI), pp. 372–377 (2005)

[HLO08]   Heras, F., Larrosa, J., Oliveras, A.: MiniMaxSAT: An efficient Weighted Max-SAT Solver. J. Artificial Intell. Research 31, 1–32 (2008)

[LNORC09]   Larrosa, J., Nieuwenhuis, R., Oliveras, A., Rodríguez-Carbonell, E.: Branch and bound for boolean optimization and the generation of optimality certificates. In: Kullmann, O. (ed.) SAT 2009. LNCS, vol. 5584, pp. 453–466. Springer, Heidelberg (2009)

[LNORC11]   Larrosa, J., Nieuwenhuis, R., Oliveras, A., Rodríguez-Carbonell, E.: A framework for certified boolean branch-and-bound optimization. J. Autom. Reasoning 46(1), 81–102 (2011)

[MMS04]   Manquinho, V., Marques-Silva, J.: Satisfiability-based algorithms for boolean optimization. Ann. Math. Artif. Intell. 40(3-4), 353–372 (2004)

[MSP09]   Manquinho, V., Silva, J.M., Planes, J.: Algorithms for weighted boolean optimization. In: Marques-Silva, J., Sakallah, K.A. (eds.) SAT 2007. LNCS, vol. 4501, pp. 495–508. Springer, Heidelberg (2007)

[NOT06]   Nieuwenhuis, R., Oliveras, A., Tinelli, C.: Solving SAT and SAT Modulo Theories: From an abstract Davis–Putnam–Logemann–Loveland procedure to DPLL(T). Journal of the ACM, JACM 53(6), 937–977 (2006)

[OSC09]   Ohrimenko, O., Stuckey, P.J., Codish, M.: Propagation via lazy clause generation. Constraints 14(3), 357–391 (2009)

[PD10]   Pipatsrisawat, K., Darwiche, A.: On modern clause-learning satisfiability solvers. J. Autom. Reason. 44(3), 277–301 (2010)

Generating Diverse Solutions in SAT

Alexander Nadel

Intel Corporation, P.O. Box 1659, Haifa 31015 Israel
alexander.nadel@intel.com

Abstract. This paper considers the DIVERSEkSET problem in SAT, that is, the problem of efficiently generating a number of diverse solutions (satisfying assignments) given a propositional formula. We provide an extensive analysis of existing algorithms for this problem in a newly developed framework and propose new algorithms. While existing algorithms adapt modern SAT solvers to solve DIVERSEkSET by changing their polarity selection heuristic, our new algorithms adapt the variable ordering strategy as well. Our experimental results demonstrate that the proposed algorithms improve the diversification quality of the solutions on large industrial instances of DIVERSEkSET arising in SAT-based semi-formal verification of hardware.

1 Introduction

SAT solving is a core reasoning engine in a variety of applications [1]. The basic functionality of a SAT solver consists of solving the following decision problem: given a propositional formula in Conjunctive Normal Form (CNF), decide whether it has a satisfying assignment (also called a *model* or *solution*). However, major industrial applications require additional abilities from the solver. This paper considers the DIVERSEkSET problem in SAT: given a satisfiable propositional formula in CNF, return a user-given number of solutions that are as diverse as possible.

In [2] we proposed a number of algorithms for solving DIVERSEkSET in SAT in the context of the SAT-based semiformal hardware verification flow, where the DIVERSEkSET solver is the core reasoning engine. The flow has practical importance, since it is able to find bugs in complex industrial designs that are missed by both Bounded Model Checking (BMC) and simulation [2]. The main idea of [2] is that, given a complex property that cannot be verified by BMC, since BMC cannot reach a sufficient bound, one can advance towards the property along multiple paths via user-given waypoints. The paths must be as diverse as possible in order not to miss bugs. A DIVERSEkSET solver is used to extract such paths. Diversification quality is defined in [2] as the normalized sum of the Hamming distances between each pair of solutions. The DIVERSEkSET algorithms proposed in [2] adapt a modern conflict-driven clause-learning (CDCL) SAT solver, invoking it only once to generate all the models. The algorithms are *polarity-based* in the sense that diversification is achieved solely by changing the polarity selection heuristic, where the polarity is the value assigned to each

K.A. Sakallah and L. Simon (Eds.): SAT 2011, LNCS 6695, pp. 287–301, 2011.

new decision variable. We proposed *randomized* and *guided* polarity-based approaches. Randomized approaches select the polarity randomly on all or some of the occasions, while guided approaches select the polarity so as to explicitly guide the solver away from previous solutions.

The first contribution of this paper is the development of a convenient framework for analyzing DIVERSE*k*SET algorithms and the analysis of existing algorithms using this framework. (The analysis in [2] is very brief, since [2] is mostly dedicated to a particular application of DIVERSE*k*SET.) In particular, our framework allows one to measure diversification quality online (i.e. while the solver is running, as opposed to offline, after the solver has finished) and to estimate the contribution of each variable to diversification quality.

We analyze the empirical behavior of DIVERSE*k*SET algorithms on 66 CNF instances generated by Intel's semiformal verification flow for generating 2 to 100 models. This is in contrast to [2], which reports about experiments with 10 models only, since this number was used by the semiformal application. Our experimental setup provides us the valuable ability to analyze the behavior of the algorithms as a function of the number of models. Our analysis can also be empirically helpful for semiformal verification, since the number of required models is expected to grow as more computational resources become available. In order to improve readability, we present and discuss relevant experiments immediately after describing a certain family of algorithms instead of concentrating all the experimental results in one section. The instances we used have 213,047 variables and 738,862 clauses on average, while the largest instance has 910,868 variables and 3,251,382 clauses. All the instances are available from the author. All the algorithms were implemented in Intel's CDCL SAT solver Eureka and run on Intel® Xeon® machines with 4Ghz CPU frequency and 32Gb of memory.

The second contribution of this paper is the introduction and analysis of new algorithms for DIVERSE*k*SET. Our new algorithms are *variable-based*; that is, they change the variable ordering in addition to the polarity selection heuristic. We propose guided and randomized variable-based methods, which can be local or global with respect to the default decision heuristic. Our algorithms improve diversification quality. We observe a trade-off between diversification quality and run-time. Moderate improvement of diversification quality can be achieved with negligible run-time cost, while more significant improvement in quality requires additional run-time. We show how one can control the trade-off between quality and run-time.

The rest of the paper is organized as follows. Section 2 reviews some related work and provides the necessary background. Definitions are provided in Section 3. Existing algorithms are analyzed in Section 4. Section 5 is dedicated to the new variable-based algorithms. The conclusion and directions for future work appear in Section 6.

2   Related Work and Background

As far as we know, the only work that considers the DIVERSE*k*SET problem in SAT is our previous work [2]. However, the problem of finding a user-given number of diverse solutions has been studied in the Constraint Satisfaction Problem (CSP) domain (e.g., [3, 4]). A guided value-based method for solving DIVERSEkSET for CSP is proposed in [3]. A randomized value-based method is also known in the CSP community [4] (we did not find a paper introducing it). A number of efficient value-based methods are proposed in [4] in the context of automatic generation of architectural tests.

A number of works (e.g., [5–7]) are dedicated to the related problem of generating a (nearly) uniformly distributed sampling of the solution space in various domains, including SAT [7]. We denote by kSAMPLING the problem of generating k out of N solutions, where each solution should be selected with the probability as close as possible to $1/N$. It is important to realize the difference between DIVERSEkSET and kSAMPLING (explained in the context of CSP in [4]). Consider the problem of finding two diverse/uniformly distributed solutions given a tautological formula. Consider an algorithm which returns the following two models: (1) All the variables are assigned 1; (2) All the variables are assigned 0. This algorithm returns the optimal solution for DIVERSEkSET for a tautological formula. However, it is unsatisfactory for the sake of kSAMPLING, since the solutions are predefined. Still, since a set of solutions for kSAMPLING can be used as an approximation for a set of solutions for DIVERSEkSET, one can evaluate the performance of existing algorithms for kSAMPLING on DIVERSEkSET instances.

The DIVERSEkSET algorithms presented in this paper are built on top of a modern CDCL SAT solver. Modern SAT solvers are extremely efficient on huge industrial instances. Among the key features that enable the solvers to be so efficient, despite the apparent difficulty of solving huge instances of NP-complete problems, are *dynamic behavior* and *search locality*, that is, the ability to maintain the set of assigned variables and recorded clauses relevant to the currently explored space. This effect is achieved through various techniques, such as 1UIP conflict clause learning [8], non-chronological backtracking [9], rapid restarts [10], variable decision heuristics (also known as variable ordering heuristics) and polarity selection heuristics.

The variable decision heuristics of a modern SAT solver are dynamic [8]. Their goal is to improve the locality of the search by picking variables that participated in recent conflict analysis. One can distinguish between variable-based decision heuristics and clause-based decision heuristics (mixed variable- and clause-based heuristics are also in use). Variable-based heuristics are based on VSIDS [8]. VSIDS maintains a score for each literal. The score is increased for a variable that participates in conflict analysis. Once in a while the scores are decreased. Consider now the clause-based heuristic CBH [11]. CBH maintains a clause list containing both the initial and the conflict clauses. Whenever a new conflict clause is derived, CBH moves the clauses that participated in conflict analysis, along with the new conflict clause, to the top of the list. The next decision variable is picked from the topmost non-satisfied clause using the variable with the greatest VSIDS score. CBH tends to pick interrelated variables, a fact which makes the search more local.

Most modern SAT solvers (including Eureka, which we use for our experiments) employ the phase-saving heuristic [12–14] as their polarity selection heuristic. The phase-saving heuristic for variable v always chooses the last value v was assigned. This strategy tries to refocus the search on subspaces that the solver has knowledge about.

3   Definitions

We start with defining some auxiliary notions. Given two boolean values σ_1 and σ_2, a pair $\{\sigma_1, \sigma_2\}$ is *different* if $\sigma_1 \neq \sigma_2$. Assume we are given a propositional formula in CNF F with q variables V and r satisfying assignments $M = \{\mu_1, \ldots, \mu_r\}$ (also known as *models* or *solutions*) for F. We define $\mu_m^u \in \{0, 1\}$, where $u \in V$ and $1 \leq m \leq r$, to be a value assigned to the variable u in μ_m.

The Hamming distance between two models μ_i and μ_j is defined to be the number of different pairs amongst $\{\mu_i^u, \mu_j^u\}$ for $u \in V$. Diversification quality is defined in [2] as the sum of the Hamming distances between each pair of models, normalized to the range $[0 \ldots 1]$ by dividing by $q\binom{r}{2}$.

We use the same measure for diversification quality but calculate it differently, keeping in mind two goals. First, we want to be able to estimate the contribution of each variable to quality. Second, we want to be able to measure quality online as well as offline. An offline version of our definitions is presented next. Afterwards we show how to generalize our definitions so that they can be used online as well.

Let the *variable (diversification) quality* S_m^u, given a variable u and m models, be the number of different pairs amongst the pairs of values assigned to u (namely, $\{\mu_i^u, \mu_j^u\}$, where $1 \leq i, j \leq r$ and $i < j$). Note that the variable quality S_{m+1}^u for $m \geq 1$ models is the variable quality for m models plus the number of different pairs amongst $\{\mu_{m+1}^u, \mu_i^u\}$ for $1 \leq i \leq m$. Let p_m^u and n_m^u be the number of times u was assigned 1 and 0, respectively, in m models. We have the following recursive definition for variable quality for $m \geq 1$:

$$S_1^u = 0; \quad S_{m+1}^u = \begin{cases} S_m^u + n_m^u & \text{if } \mu_{m+1}^u = 1; \\ S_m^u + p_m^u & \text{if } \mu_{m+1}^u = 0. \end{cases}$$

We provide an alternative definition for variable quality, which is sometimes more useful. It is not hard to see that $p_m^u \times n_m^u$ is exactly the number of different pairs amongst $\{\mu_i^u, \mu_j^u\}$. Hence, we have:

$$S_m^u = p_m^u \times n_m^u$$

Now we can define the *(diversification) quality* Q_m for $1 \leq m \leq r$ models as the sum of all the variable qualities, normalized to the range $[0 \ldots 1]$:

$$Q_m = \frac{\sum_{u \in V} S_m^u}{\binom{m}{2} q}$$

We provide another useful notion of a potential of a variable. Given a variable $u \in V$ and $m \geq 1$ models, $\Pi_m^u = p_m^u - n_m^u$ is the *potential* of u. The potential of a variable is the difference between the number of 1's and 0's assigned to u in all the models. The *absolute potential* of u is the absolute value of the potential $|\Pi_m^u|$. We will see later that the potential and the quality of a variable are closely connected. Fig. 1 provides a simple example of applying our definitions.

$$\begin{array}{cccc} & \mu_1 & \mu_2 & \mu_3 \\ v & 0 & 0 & 0 \\ u & 1 & 1 & 0 \end{array}$$

Fig. 1. An example of applying our definitions, given two variables and three models. We have: $p_3^v = 0$; $n_3^v = 3$; $p_3^u = 2$; $n_3^u = 1$. The variable qualities are: $S_3^v = p_3^v \times n_3^v = 0$; $S_3^u = p_3^u \times n_3^u = 2$. The quality is: $Q_3 = (0 + 2)/(3 \times 2) = 1/3$. The potentials are: $\Pi_3^u = 1$; $\Pi_3^v = -3$.

Now we show how to modify our definitions so as to allow using them both online and offline. The algorithms presented in this paper invoke a CDCL SAT solver once to generate all the models and restart the search immediately after a new model is discovered. We call the algorithms/solvers which follow the above-mentioned scheme *compact*. Suppose that a compact DIVERSE*k*SET solver has found $m > 0$ models and is searching for a new model. Such a solver maintains the current partial assignment. We modify the notions of p_m^u/n_m^u, the variable quality S_m^u, and the variable potential Π_m^u simply by considering the current partial assignment as another model when counting the number of 1's and 0's assigned to a variable (the modification is required for assigned variables only). To generalize the notion of overall quality, one needs to make an adjustment, since the number of satisfying assignments is now different for assigned and unassigned variables. Let q_1 be the number of unassigned variables and q_2 be the number of assigned variables. Then we have: $Q_m = (\sum_{u \in V} S_m^u)/((\binom{m}{2})q_1 + (\binom{m+1}{2})q_2)$. Note that the online versions of our definitions are a strict generalization of our notions (namely, those of p_m^u, n_m^u, S_m^u, Π_m^u, and Q_m) in the sense that they converge with the offline versions after the solver has completed its run. When mentioning these notions in the paper, we are referring to their online version when a DIVERSE*k*SET solver invocation is analyzed online and either to their online or offline version if the solver has finished.

4   Analizing Existing Algorithms

First, we refer to two non-compact algorithms, DPLL-BASED SAMPLING (mentioned in [6] without a reference) and XOR-SAMPLE [7], that were designed for the *k*SAMPLING problem, but which can also be applied to DIVERSE*k*SET. Both DPLL-BASED SAMPLING and XOR-SAMPLE invoke a SAT solver once to generate each model. Diversification is achieved by randomizing the first value assigned to each variable for DPLL-BASED SAMPLING and by adding random XOR constraints for XOR-SAMPLE. DPLL-BASED SAMPLING and XOR-SAMPLE are shown in [2, 15] to be inferior to compact DIVERSEkSET algorithms in terms of both quality and run-time. In addition, [15] analyzes a compact DIVERSEkSET algorithm AllSAT-Sampling [2], which yields a low quality but is very fast. The present work concentrates on compact algorithms that yield a relatively high quality.

PRAND [2] (Rand-k-SAT in [2]) is a compact randomized polarity-based algorithm. It operates by overriding the traditional polarity selection heuristic to select the polarity randomly on all occasions. PRAND can be thought of as a generalization of DPLL-BASED SAMPLING.

PGUIDE [2] (Guide-k-SAT in [2]) is a compact guided polarity-based algorithm. It is designed to greedily improve quality. PGUIDE does not change the default behavior of the SAT solver before the first model is encountered. Assume PGUIDE is about to decide on the polarity of a newly assigned variable u when $m > 0$ models have already been found. If $\Pi_m^u > 0$, u is assigned 0; if $\Pi_m^u < 0$, u is assigned 1; if $\Pi_m^u = 0$, u is assigned a random value. Prop. 1 shows that this simple algorithm improves quality whenever the variable decision heuristic picks an unassigned variable with a non-zero potential. Note that this useful property does not hold for PRAND, hence PGUIDE is expected to result in better quality.

Proposition 1. *Assume that a compact* DIVERSEkSET *solver employing* PGUIDE *is running and that it has encountered $m > 0$ models. Let u be an unassigned variable picked by the variable decision heuristic. Let Q_m^1 and Q_m^0 be the qualities if the current partial assignment is extended by assigning a value 1 and 0, respectively, to u. Then, if $\Pi_m^u > 0$ then $Q_m^0 > Q_m^1$; if $\Pi_m^u < 0$ then $Q_m^1 > Q_m^0$; if $\Pi_m^u = 0$ then $Q_m^1 = Q_m^0$.*

Proof. Assume $\Pi_m^u > 0$. The recursive definition of variable quality implies that assigning u the value 1 or 0 will change its variable quality by n_m^u or p_m^u, respectively. The assumption $\Pi_m^u > 0$ implies that $p_m^u > n_m^u$. Hence, the change in the variable quality of u will be greater if u is assigned 0 than if it is assigned 1. Note that the change in the overall quality is proportional to the change in the variable quality of u, since the variable quality of u is the only addendum that changes in the dividend of the definition of quality, while the change in the divisor is independent of the value assigned to u. Hence we have $Q_m^0 > Q_m^1$. The proof for the other two cases is similar.   □

PGUIDE is designed to correct the potential of one variable at a time, where by correcting the potential we mean bringing the absolute potential closer to 0. This operation improves the quality of one specific variable. Such a strategy yields the optimal overall quality given a tautological formula. However, it does not take into account dependencies between variables, which appear in real-world well-structured formulas.

PBCPGUIDE [2] (BCP-aware Guide-k-SAT in [2]) is a refinement of PGUIDE which takes into consideration dependencies between variables by taking into account the impact of Boolean Constraint Propagation (BCP) on quality. It performs

BCP for both polarities of a new decision variable u alternatively and measures the new quality for each polarity. It then continues with the polarity that yielded the better quality. PBCPGUIDE's flow is detailed in [2]. Note that unlike PGUIDE, PBCPGUIDE is designed to take into account dependencies between variables. Applying PBCPGUIDE might have a significant negative impact on performance, since it has to perform BCP two or three times per decision. To be able to control the trade-off between quality and run-time, one can limit PBCPGUIDE usage as follows. PBCPGUIDE will be used until a user-given number of conflicts T is encountered by the solver. Afterwards, the algorithm will switch to plain PGUIDE until the next model is encountered. Then PBCPGUIDE is reinvoked until T conflicts are encountered again.

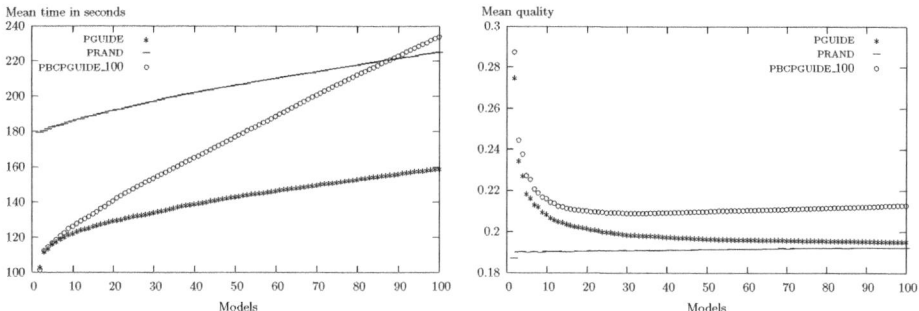

Fig. 2. Comparing polarity-based algorithms

Compare the empirical behavior of PGUIDE and PRAND in Fig. 2 and Table 1. Both the mean run-time and the mean quality of PGUIDE is consistently better than that of PRAND for any number of models. The difference in quality is especially significant for the first models, but it goes down quickly as the number of models increases and seems to approach an asymptote.

Compare the behavior of PBCPGUIDE to that of PGUIDE in Table 1. Four versions of PBCPGUIDE with different threshold values $T \in \{10, 100, 1000, 10000\}$ (called PBCPGUIDE_T for each T) were tested. Predictably, PBCPGUIDE has a positive impact on quality, but deteriorates performance, where the effect is more significant for larger threshold values. Interestingly, while PBCPGUIDE_1000 and PBCPGUIDE_10000 yield almost the same quality, the run-time of PBCPGUIDE_10000 is significantly inferior. Hence it is not worth using a threshold greater than 1000. The balance between quality and run-time achieved by PBCPGUIDE_100 is attractive. For example, for generating 100 models, PBCPGUIDE_100 yields a better quality than PGUIDE by 9.1% and is slower by only 47.2%. Compare the behavior of PBCPGUIDE_100 as a function of the number of models with that of PGUIDE in Fig. 2. The run-time function of PBCPGUIDE_100 goes up much more quickly than that of PGUIDE; however, the gap for 100 models is still reasonable. The quality function of PBCPGUIDE_100 is always higher than that of PGUIDE. Interestingly, the gap increases with the number of models. Moreover, while PGUIDE's quality function is monotonically decreasing, PBCPGUIDE_100's quality function's tail is increasing. This is related to the fact that, unlike PGUIDE, PBCPGUIDE takes into account dependencies between variables. Further explanation regarding the behavior of the quality functions of PGUIDE and PBCPGUIDE_100 and the difference between them is provided in [15].

Table 1. Mean quality and mean run-time for DIVERSE*k*SET algorithms, given 100, 50, and 10 models on 66 benchmarks from the semiformal verification of hardware. All the numbers except the last row are relative to the behavior of PGUIDE. For example, PBCPGUIDE_100 yields better quality than PGUIDE by 9.1%, but is slower by 47.2% for generating 100 models. The algorithms are sorted by the quality obtained when generating 100 models. The absolute mean quality and mean run-time in seconds of PGUIDE are provided in the last row.

| | 100 | | 50 | | 10 | |
Algorithm	Quality	Time	Quality	Time	Quality	Time
PBCPGUIDE_100-VRANDGLOB_30	1.123	4.649	1.119	4.536	1.087	4.435
PBCPGUIDE_100-VRANDGLOB_20	1.121	3.829	1.114	3.653	1.081	3.469
PBCPGUIDE_100-VRANDGLOB_10	1.117	2.925	1.109	2.694	1.074	2.474
PBCPGUIDE_10000	1.111	7.628	1.111	6.731	1.1	4.559
PBCPGUIDE_1000	1.11	3.465	1.11	2.677	1.082	1.505
PBCPGUIDE_100-VGUIDEGLOB_200	1.107	1.707	1.09	1.509	1.063	1.259
PBCPGUIDE_100-VGUIDEGLOB_100	1.107	1.665	1.09	1.472	1.058	1.224
PBCPGUIDE_100-VRANDGLOB_2	1.106	2.007	1.093	1.766	1.05	1.52
PBCPGUIDE_100-VGUIDEGLOB_10	1.105	1.656	1.087	1.44	1.055	1.171
PBCPGUIDE_100-VRANDLOC	1.102	1.742	1.083	1.537	1.036	1.348
PGUIDE-VRANDGLOB_30	1.099	4.058	1.095	4.181	1.062	4.243
PBCPGUIDE_100-VGUIDELOC	1.097	1.537	1.078	1.317	1.047	1.121
PGUIDE-VRANDGLOB_20	1.091	3.331	1.086	3.407	1.055	3.422
PBCPGUIDE_100	1.091	1.472	1.068	1.239	1.036	1.033
PBCPGUIDE_100-VRANDLOC-NAÏVE	1.085	1.733	1.07	1.53	1.041	1.375
PGUIDE-VRANDGLOB_10	1.077	2.4	1.072	2.426	1.041	2.42
PBCPGUIDE_100-VGUIDELOC-NAÏVE	1.076	1.632	1.06	1.367	1.045	1.118
PGUIDE-VGUIDEGLOB_200	1.054	1.288	1.057	1.282	1.04	1.203
PGUIDE-VGUIDEGLOB_100	1.053	1.262	1.051	1.253	1.033	1.199
PBCPGUIDE_10	1.042	1.071	1.039	1.03	1.03	1.016
PGUIDE-VRANDGLOB_2	1.042	1.484	1.042	1.483	1.021	1.485
PGUIDE-VGUIDEGLOB_10	1.041	1.202	1.039	1.202	1.031	1.132
PGUIDE-VRANDLOC	1.036	1.263	1.033	1.274	1.01	1.296
PGUIDE-VGUIDELOC	1.014	1.004	1.016	1.008	1.021	1.013
PGUIDE	1	1	1	1	1	1
PRAND	0.9839	1.413	0.9731	1.439	0.9123	1.519
The absolute numbers for PGUIDE	0.1954	159	0.1968	143	0.2086	122

5   Variable-Based Methods

This section introduces a number of *variable-based* compact algorithms for DI-VERSE*k*SET. We show how diversification quality can be improved by changing the variable ordering. We propose local and global variable-based methods. Local methods select the next decision variable from a subset of variables considered as relevant by the variable decision heuristic, while global methods consider a wider set of variables. Local variable-based methods are expected to result in a moderate quality improvement, but be run-time-efficient. Global variable-based methods are expected to be more costly in terms of performance, but yield better diversification quality. We propose both guided and randomized variable-based algorithms. Guided variable-based methods select variables with the largest absolute potential. Randomized variable-based methods add a certain degree of randomness to the variable ordering. We take PGUIDE and PBCPGUIDE_100 as the baseline algorithms in the sense that we integrate our variable-based methods into a solver that already uses PGUIDE or PBCPGUIDE_100.

Section 5.1 proposes local variable-based algorithms intended to be integrated with PGUIDE. Section 5.2 shows how to modify these algorithms of Section 5.1 in order to combine them with PBCPGUIDE_100. Section 5.3 presents the global variable-based algorithms. Our algorithms are integrated within the CBH decision heuristic.

5.1 Local Variable-Based Methods for PGUIDE

VGUIDELOC is a *guided local variable-based* algorithm. VGUIDELOC changes the variable decision heuristic after $m > 0$ models have already been found as follows. It picks an unassigned variable with the maximal absolute potential from the topmost non-satisfied clause in the clause list. If more than one variable have the same absolute potential, a variable with the greatest VSIDS score is picked as in the original CBH. VGUIDELOC is designed to increase PGUIDE's positive impact on quality, since, according to Prop. 2, picking a variable with a larger absolute potential improves quality by a larger margin. PGUIDE-VGUIDELOC (that is, the combination of PGUIDE and VGUIDELOC) is not expected to significantly deteriorate the performance of the solver, since this strategy makes only minimal changes to the default CBH heuristic. Prop. 3 yields that VGUIDELOC picks variables with the worst variable quality in the clause and corrects its potential. However, since PGUIDE is unaware of dependencies between variables, PGUIDE-VGUIDELOC might deteriorate the quality of other variables assigned by BCP. This fact might hurt the ability of PGUIDE-VGUIDELOC to improve overall quality.

VRANDLOC is a *randomized local variable-based* algorithm. VRANDLOC picks a random variable from the topmost non-satisfied clause. PGUIDE-VRANDLOC (that is, the combination of PGUIDE and VRANDLOC) is fairer than both plain PGUIDE and PGUIDE-VGUIDELOC with respect to variable ordering, since PGUIDE-VRANDLOC may choose variables that would rarely or never be chosen by the other two methods due to their low VSIDS score or low absolute potential. Randomized variable-based method will work better than the guided method when there are many hidden dependencies between variables. PGUIDE-VRANDLOC is expected to have a negative impact on the performance of the solver, since it violates the locality principle, yet this impact should not be too significant, because the variables are still picked from the same clause.

Below, after proving a useful lemma, we provide two propositions that are essential for understanding the ideas behind VGUIDELOC.

Lemma 1. *Assume that a compact* DIVERSEkSET *solver employing* PGUIDE *is running and that it has encountered* $m > 0$ *models. Let* v *and* u *be two unassigned variables, such that* $|\Pi_m^u| > |\Pi_m^v|$. *Then* $max(p_m^u, n_m^u) > max(p_m^v, n_m^v)$.

Proof. The definition of the potential implies that for every variable t it holds that $|\Pi_m^t| = max(p_m^t, n_m^t) - min(p_m^t, n_m^t)$. Since $p_m^t + n_m^t = m$, we have $|\Pi_m^t| = max(p_m^t, n_m^t) - (m - max(p_m^t, n_m^t)) = 2 \times max(p_m^t, n_m^t) - m$. The latter equation implies that if $|\Pi_m^u| > |\Pi_m^v|$ then $max(p_m^u, n_m^u) > max(p_m^v, n_m^v)$. □

Proposition 2. *Assume that a compact* DIVERSEkSET *solver employing* PGUIDE *is running and that it has encountered* $m > 0$ *models. Let* v *and* u *be two unassigned variables, such that* $|\Pi_m^u| > |\Pi_m^v|$. *Assume that the solver is about to assign either* u *or* v. *Let the quality after* u *or* v *is assigned be* Q_m^u *or* Q_m^v, *respectively. Then,* $Q_m^u > Q_m^v$.

Proof. Recall from the proof of Prop.1 that the change in overall quality is proportional to the change in the quality of the variable picked by the variable decision heuristic. PGUIDE's flow implies that if it holds that $p_m^t > n_m^t$ or $n_m^t > p_m^t$ for an unassigned variable t picked by the decision heuristic, then PGUIDE will pick the value 0 or 1, respectively, for t. This latter fact and the recursive definition of variable quality imply that the change in variable quality following an assignment of t by PGUIDE is $max(p_m^t, n_m^t)$. By Lemma 1, $max(p_m^u, n_m^u) > max(p_m^v, n_m^v)$. □

Proposition 3. *Assume that a compact* DIVERSEkSET *solver employing* PGUIDE *is running and that it has encountered* $m > 0$ *models. Let* u *and* v *be two unassigned variables. If* $|\Pi_m^u| > |\Pi_m^v|$, *then* $S_m^u < S_m^v$.

Proof. Assume to the contrary that $S_m^u \geq S_m^v$. We denote $max(p_m^u, n_m^u)$ and $max(p_m^v, n_m^v)$ by x and y, respectively. The definition $S_m^t = p_m^t \times n_m^t$ and our assumption imply that $x(m - x) \geq y(m - y)$. By Lemma 1, if $|\Pi_m^u| > |\Pi_m^v|$ then $x > y$, hence one can express x as $x = y + \delta$, where $\delta > 0$. Substituting the latter equality into $x(m-x) \geq y(m-y)$ gives us: $x(m-x) \geq (x-\delta)(m-x+\delta)$. Opening parenthesis and simplifying gives us the following inequality: $\delta(2x - m - \delta) \leq 0$. Since $\delta > 0$, we have $2x - m - \delta \leq 0$. Substituting $\delta = x - y$ into the latter inequality gives us the following one: $m \geq x + y$. Definitions imply that $max(p_m^t, n_m^t) \geq m/2$ for any unassigned t, hence $x, y \geq m/2$. Since $x > y$, it must hold that $x > m/2$. Since $y \geq m/2$ and $x > m/2$, it cannot hold that $m \geq x + y$. Contradiction. □

Compare the empirical behavior of PGUIDE-VGUIDELOC and PGUIDE-VRANDLOC to that of plain PGUIDE in Table 1 and Fig. 3. Both PGUIDE-VGUIDELOC and PGUIDE-VRANDLOC yield a consistent improvement in quality over PGUIDE.

PGUIDE-VGUIDELOC's run-time penalty over PGUIDE is negligible, while PGUIDE-VRANDLOC is 26% slower than PGUIDE for 100 models. PGUIDE-VRANDLOC yields better quality than PGUIDE-VGUIDELOC when the number of models exceeds 14. As the number of models increases, PGUIDE-VRANDLOC's advantage in quality becomes more significant. Hence, in the long run, being fairer with respect to variable ordering contributes to PGUIDE's impact on quality more than correcting the potential of one variable at a time, even though the selected variable has the largest absolute potential in the topmost clause. The guided variable-based approach is more efficient in terms of run-time than the randomized approach, since the guided method is closer to the efficient default variable decision heuristic: while the randomized method chooses variables, independently of their VSIDS score, the guided method prefers variables with the highest VSIDS score out of all the variables with the same potential. Interestingly, both PGUIDE-VGUIDELOC and PGUIDE-VRANDLOC are inferior to PBCPGUIDE_10 in terms of quality. In the next section we show how one can combine the variable-based algorithms with PBCPGUIDE.

5.2   Local Variable-Based Methods for PBCPGUIDE

We considered a number of ways of combining local variable-based methods with PBCPGUIDE. First, one could integrate VGUIDELOC and VRANDLOC as is into PBCPGUIDE. However, it is unclear whether the resulting algorithms (named PBCPGUIDE_100-VGUIDELOC-NAÏVE and PBCPGUIDE_100-VRANDLOC-NAÏVE, respectively) would yield better quality than plain PBCPGUIDE, since, unlike PGUIDE, PBCPGUIDE tries to improve quality by taking into account dependencies between variables per se. It might turn out that additional variable-based considerations are not required. Second, one could apply VGUIDELOC and VRANDLOC only after the threshold on the number of conflicts for PBCPGUIDE is reached. We dub the resulting strategies PBCPGUIDE_100-VGUIDELOC and PBCPGUIDE_100-VRANDLOC, respectively. They utilize the power of both PBCPGUIDE and the combination of VGUIDELOC and VRANDLOC with PGUIDE to improve quality by taking into account dependencies between variables, yet they should not be as costly as applying PBCPGUIDE with a larger threshold.

Consider the behavior of PBCPGUIDE_100-VGUIDELOC and PBCPGUIDE_100-VRANDLOC in Table 1 and Fig. 3. These methods improve the quality of plain PBCPGUIDE_100 at the expense of run-time. Note that while the guided method is faster than the randomized method, the randomized method is preferable to the guided method in terms of quality. Recall that we observed (and explained) a similar behavior when we combined the variable-based methods with PGUIDE. Compare the empirical behavior of PBCPGUIDE_100-VGUIDELOC-NAÏVE and PBCPGUIDE_100-VRANDLOC-NAÏVE with plain PBCPGUIDE_100 in Table 1. Not only do these strategies not improve the quality of PBCPGUIDE_100, but they deteriorate both the quality and the run-time for 100 models as well. Hence the variable-based methods that try to take into account dependencies between variables interfere with PBCPGUIDE.

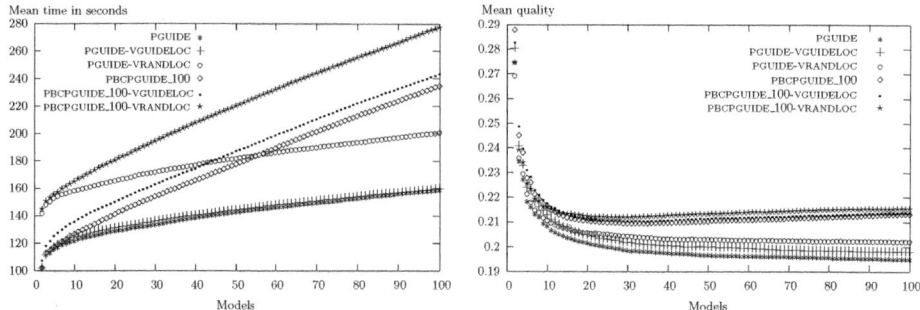

Fig. 3. Combining local variable-based algorithms with PGUIDE and PBCPGUIDE_100

It would be interesting to try a BCP-aware guided variable-based method, which would apply BCP for both polarities for more than one variable and pick the variable and polarity that yield the best quality. We did not implement such an algorithm, since a straightforward implementation would be extremely costly in terms of run-time. Applying BCP in parallel and borrowing techniques from look-ahead SAT solvers [16], designed to consider a wider set of variables at each decision point, are appealing directions for future work to improve quality. The present work proposes another way to improve quality: global variable-based methods which apply the same principles as the local methods but consider a wider set of variables at each decision point. These global variable-based methods are discussed in the next section.

5.3   Global Variable-Based Methods

VGUIDEGLOB is a *global guided variable-based* algorithm. It picks a variable v with the largest absolute potential out of a wider selection of clauses than its local counterpart. VGUIDEGLOB picks a variable with the greatest potential from the N topmost clauses, including satisfied clauses, unless the N topmost clauses are satisfied, in which case the algorithm considers all the clauses in the list up to and including the topmost non-satisfied clause. The primary criteria for breaking ties is to prefer a variable from a clause that is as close as possible in the list to the topmost non-satisfied clause. The secondary criteria for breaking ties is to prefer variables with better VSIDS scores. The idea behind the tie-breaking strategies is to make the heuristic as efficient as possible by making it as close as possible to the original CBH. For our experiments, we used $N \in \{10, 100, 200\}$. We refer to the combination of PGUIDE and PBCPGUIDE_100 with VGUIDEGLOB with a parameter value N as PGUIDE-VGUIDEGLOB_N and PBCPGUIDE_100-VGUIDEGLOB_N, respectively.

VRANDGLOB is a *global randomized variable-based* algorithm. It picks an unassigned variable at random in $T\%$ of the cases out of all the decisions, where T is a parameter. The default decision heuristic is used in the rest of the cases. Note that this strategy is independent of the decision heuristic. We tried $T \in \{2, 10, 20, 30\}$ for our experiments. We refer to the combination of PGUIDE and PBCPGUIDE_100 with VRANDGLOB with a parameter value T as PGUIDE-VRANDGLOB_T and PBCPGUIDE_100-VRANDGLOB_T, respectively.

We combined global variable-based methods with PBCPGUIDE by applying them only after the threshold on the number of conflicts for PBCPGUIDE was reached, since this was found in Section 5.2 to be the optimal strategy for integrating local variable-based methods with PBCPGUIDE.

Consider the empirical behavior of the combination of the global variable-based methods with PGUIDE and PBCPGUIDE_100 in Table 1. The following conclusions equally hold for combining global variable-based methods with either PGUIDE or PBCPGUIDE_100. As expected, the global methods yield better quality than the local methods (combined with the respective polarity-based algorithm), but do so at the expense of run-time. This observation holds for both randomized and guided methods. The parameter N or T for randomized or guided methods, respectively, can be used to control the trade-off between quality and run-time. The larger the corresponding parameter, the better the quality and the worse the run-time. However, increasing the parameter too much does not yield added benefit, since the improvement in quality becomes marginal, while the run-time continues to increase. Compare now the global guided methods vs. the global randomized methods. The randomized methods are costly in terms of run-time, but yield better quality (when the threshold N is higher than 2). Recall that we observed the same behavior while analyzing local methods.

Interestingly, PBCPGUIDE_100-VRANDGLOB_T, when T is sufficiently large, outperforms plain PBCPGUIDE with threshold 10000 in terms of both quality and run-time for 100 and 50 models. This result shows that, in the long run, it pays first to use PBCPGUIDE_100 and then to switch to PGUIDE-VRANDGLOB, rather than to use plain PBCPGUIDE with a larger threshold. Understanding the reasons for this phenomenon is left for future research.

6   Conclusions and Future Work

This work is the first full-blown paper dedicated to the DIVERSEkSET problem in SAT, that is, the problem of efficiently generating a number of diverse solutions (satisfying assignments) given a propositional formula. We proposed a framework for analyzing DIVERSEkSET algorithms in SAT and carried out an extensive empirical evaluation of existing and new algorithms on large industrial instances of DIVERSEkSET arising in SAT-based semiformal verification of hardware [2].

Our analysis showed that adapting the SAT solver's polarity selection heuristic in a guided way, that is, explicitly avoiding previous solutions, is preferable to randomizing the polarity. Considering the dependencies between variables by taking into account the effect of BCP improves diversification quality, but deteriorates run-time.

We introduced a number of variable-based algorithms that improve diversification quality by adapting the variable decision heuristic in addition to the polarity selection heuristic. We distinguished between randomized and guided algorithms as well as between local and global algorithms. Randomized and global algorithms are more costly in terms of run-time but yield better quality than guided and local algorithms. Overall, while a moderate improvement in quality over purely polarity-based methods can be achieved at a negligible run-time cost, obtaining a more significant improvement in quality requires additional run-time. We showed how one can control the trade-off between quality and run-time to achieve an attractive balance. The eventual choice of algorithms should depend on the needs of each specific application.

The following directions for future research seem attractive. Parallelizing DIVERSEkSET algorithms should be helpful in improving both quality and run-time. It would be interesting to investigate ways to adapt various components of the SAT solver (such as conflict analysis schemes or restart strategies) to DIVERSEkSET. Borrowing techniques from look-ahead SAT solvers and developing DIVERSEkSET algorithms on top of such solvers is another interesting direction.

Acknowledgments

The author would like to thank Amit Palti for supporting this work, Paul Inbar for editing the paper, and Jim Grundy, Vadim Ryvchin, and Yevgeny Schreiber for providing useful suggestions that helped to improve this work.

References

1. Biere, A., Heule, M., van Maaren, H., Walsh, T. (eds.): Handbook of Satisfiability. Frontiers in Artificial Intelligence and Applications, vol. 185. IOS Press, Amsterdam (2009)
2. Agbaria, S., Carmi, D., Cohen, O., Korchemny, D., Lifshits, M., Nadel, A.: SAT-based semiformal verification of hardware. In: Bloem, R., Sharygina, N. (eds.) Proceedings of the 10th International Conference on Formal Methods in Computer-Aided Design (FMCAD 2010), pp. 25–32 (October 2010)
3. Hebrard, E., Hnich, B., O'Sullivan, B., Walsh, T.: Finding diverse and similar solutions in constraint programming. In: Veloso, M.M., Kambhampati, S. (eds.) AAAI, pp. 372–377. AAAI Press / The MIT Press (2005)
4. Schreiber, Y.: Value-ordering heuristics: Search performance vs. Solution diversity. In: Cohen, D. (ed.) CP 2010. LNCS, vol. 6308, pp. 429–444. Springer, Heidelberg (2010)
5. Dechter, R., Kask, K., Bin, E., Emek, R.: Generating random solutions for constraint satisfaction problems. In: AAAI/IAAI, pp. 15–21 (2002)
6. Kitchen, N., Kuehlmann, A.: Stimulus generation for constrained random simulation. In: ICCAD, pp. 258–265 (2007)
7. Gomes, C.P., Sabharwal, A., Selman, B.: Near-uniform sampling of combinatorial spaces using XOR constraints. In: Schölkopf, B., Platt, J.C., Hoffman, T. (eds.) NIPS, pp. 481–488 (2006)
8. Moskewicz, M.W., Madigan, C.F., Zhao, Y., Zhang, L., Malik, S.: Chaff: Engineering an efficient SAT solver. In: DAC, pp. 530–535. ACM, New York (2001)
9. Silva, J.P.M., Sakallah, K.A.: GRASP: A search algorithm for propositional satisfiability. IEEE Transactions on Computers 48, 506–521 (1999)

10. Huang, J.: The effect of restarts on the efficiency of clause learning. In: Proceedings of the 20th International Joint Conference on Artificial Intelligence, pp. 2318–2323 (2007)

11. Dershowitz, N., Hanna, Z., Nadel, A.: A clause-based heuristic for SAT solvers. In: Bacchus, F., Walsh, T. (eds.) SAT 2005. LNCS, vol. 3569, pp. 46–60. Springer, Heidelberg (2005)

12. Frost, D., Dechter, R.: In search of the best constraint satisfaction search. In: AAAI, pp. 301–306 (1994)

13. Strichman, O.: Tuning SAT checkers for bounded model checking. In: Emerson, E.A., Sistla, A.P. (eds.) CAV 2000. LNCS, vol. 1855, pp. 480–494. Springer, Heidelberg (2000)

14. Pipatsrisawat, K., Darwiche, A.: A lightweight component caching scheme for satisfiability solvers. In: Marques-Silva, J., Sakallah, K.A. (eds.) SAT 2007. LNCS, vol. 4501, pp. 294–299. Springer, Heidelberg (2007)

15. Nadel, A.: Generating diverse solutions in SAT: Paper addendum (2011), http://www.cs.tau.ac.il/research/alexander.nadel/multiple_cex_addendum.pdf

16. Heule, M., van Maaren, H.: Look-ahead based SAT solvers [1], pp. 155–184

Captain Jack: New Variable Selection Heuristics in Local Search for SAT

Dave A.D. Tompkins[1], Adrian Balint[2], and Holger H. Hoos[1]

[1] Department of Computer Science
University of British Columbia, Canada
{davet,hoos}@cs.ubc.ca
[2] Institute of Theoretical Computer Science,
Ulm University, Germany
adrian.balint@uni-ulm.de

Abstract. Stochastic local search (SLS) methods are well known for their ability to find models of randomly generated instances of the propositional satisfiability problem (SAT) very effectively. Two well-known SLS-based SAT solvers are SPARROW, one of the best-performing solvers for random 3-SAT instances, and VE-SAMPLER, which achieved significant performance improvements over previous SLS solvers on SAT-encoded software verification problems. Here, we introduce a new highly parametric algorithm, CAPTAIN JACK, which extends the parameter space of SPARROW to incorporate elements from VE-SAMPLER and introduces new variable selection heuristics. CAPTAIN JACK has a rich design space and can be configured automatically to perform well on various types of SAT instances. We demonstrate that the design space of CAPTAIN JACK is easy to interpret and thus facilitates valuable insight into the configurations automatically optimized for different instance sets. We provide evidence that CAPTAIN JACK can outperform well-known SLS-based SAT solvers on uniform random k-SAT and 'industrial-like' random instances.

1 Introduction

The propositional satisfiability problem (SAT) is one of the most prominent problems in computer science, not only because it is a prototypical $\mathcal{NP}$-complete problem, but also because of its simplicity, expressiveness and practical relevance. Problem instances from domains such as software verification can be easily encoded into SAT, and there is much interest in developing SAT solvers that can solve these practical problems effectively. There is also much interest in random instances; they have been frequently studied and are underlying one of the three categories in the SAT competition.

Two popular approaches for solving SAT are conflict driven clause learning (CDCL) and stochastic local search (SLS), and in this work we focus on the latter. SLS solvers are usually incomplete, *i.e.*, they cannot determine with certainty that a given propositional formula is unsatisfiable. SLS algorithms for SAT typically start by randomly assigning to every variable appearing in a given formula a value of either true or false. Then, in each subsequent search step a variable is selected to have its truth assignment *flipped* from true to false or vice versa.

K.A. Sakallah and L. Simon (Eds.): SAT 2011, LNCS 6695, pp. 302–316, 2011.

Because SLS is the most effective approach for solving random satisfiable instances, there has been much interest in studying the performance of SLS-based solvers on random instances from the so-called uniform random k-SAT distribution [6], especially 3-SAT instances at, or near, the solubility phase transition, where there is an equal probability of generating a satisfiable or unsatisfiable instance [12]. One of the best known SLS solvers for large random 3-SAT instances near the phase transition is SPARROW [3], which replaced the variable selection heuristic in GNOVELTY$^+$ [13] with a probabilistic distribution-based mechanism. We will describe SPARROW in Section 2.1.

While SLS is currently not competitive with CDCL on large satisfiable industrial instances, there has been some recent success in closing the gap. The VE-SAMPLER algorithm [18] was able to achieve a significant improvement in this area by solving the CBMC software verification benchmark instances over ten times faster than the previous best known SLS algorithm SATENSTEIN-LS [10]. In Section 2.2, we will describe some of the specifics of VE-SAMPLER.

SATENSTEIN-LS and VE-SAMPLER are examples of *highly parametric* algorithms that are designed to be *automatically* configured, using an automated algorithm configurator that takes as inputs an algorithm, its configuration space and an instance set and then attempts to find the best-performing configuration of the algorithm on the given instance set. Following this approach, which is a prominent special case of computer-aided algorithm design [7], the traditional role of an algorithm designer is redefined to be more focused on constructing rich and interesting *spaces* of algorithms.

It is well established that SAT instances drawn from different sources and distributions have different characteristics, and the efficacy of a solver on an instance often depends on those characteristics. Portfolio-based approaches, such as SATZILLA [19], exploit this phenomenon by selecting one or more solvers to be used for solving a given instance based on characteristics of that instance. One of our goals in this work was to create a highly parametric algorithm that can help explore the algorithmic differences (or similarities) between configurations that achieve good results on different types of instances (*e.g.*, random *vs* application or random 3-SAT *vs* 7-SAT). To achieve this goal, our algorithm should have a parameter space that is not only easy to understand, but also contains configurations that achieve good performance on a wide variety of instance types; without good performance, the resulting configurations are not very meaningful, and without intuitively understandable configurations, it is difficult to draw conclusions from automatically optimized configurations. In Section 4, we introduce CAPTAIN JACK, a new, highly parametric algorithm that attempts to strike a balance between these two objectives. It was named for the fictional pirate Captain Jack Sparrow, because it incorporates elements from SPARROW, as well as because it can achieve good performance on a wide variety of instances, and is hence a *jack-of-all-trades*.

In Section 6, we present evidence that CAPTAIN JACK does achieve good performance on nine different instance sets, and is now the best known SLS-based SAT algorithm for large random 3-SAT instances and a class of recently proposed 'industrial-like' instances [1]. Moreover, we show how the resulting configurations found by CAPTAIN JACK provide interesting insight into the configurations found for each of the instance sets; for example, we found evidence that the CAPTAIN JACK configuration optimized for the previously mentioned industrial-like instances exhibits characteristics consistent with those obtained for some practical software verification instances.

Additional information and experimental data, including source code and instance sets, can be found at www.cs.ubc.ca/research/captain-jack.

2   Background

Throughout this work, we use the approach for modeling and representing SLS algorithms introduced by Tompkins and Hoos [18], according to which each search step involves three heuristic stages. First, the variables are *filtered* so that only a subset of variables are considered as flip candidates. Next, the candidates are evaluated according to one or more *variable expressions (VEs)*, where each VE is a mathematical expression that can include *properties* of the variables. Finally, once the VEs have been evaluated, a *variable selection mechanism (VSM)* is employed to select the variable to be flipped. A *controller* determines for each search step which filter, VEs and VSM are used.

Some variable properties are defined via a VE that contains other properties, such as *score*, which is equivalent to the VE $\langle make - break \rangle$, where the properties *make* and *break* measure the number of clauses that would become satisfied and unsatisfied, respectively, if the variable were to be flipped. The value of a property can depend on the specific context in which the variable is selected and additional state information of the algorithm. For example, algorithms with dynamic clause penalties, such as PAWS [15] and GNOVELTY$^+$ [13], use a *penalized* property *penScore*, whose value depends on the full variable assignment and on the clause penalties (weights).

Variable properties can be loosely classified as either *greedy* properties, which tend to increase the number of satisfied clauses during search, such as *make* and *break*, or *diversification* properties, which tend to better explore the search space and avoid local minima, such as *age* and *flips*. The *age* property is defined as the number of search steps that have occurred since the given variable was last flipped. The *flips* property (*a.k.a. flipcount*) measures how many times a variable has been flipped. In Section 4, we will describe several new greedy and diversification properties.

2.1   SPARROW

The SPARROW algorithm [3], named after the city of Ulm's mascot, is based on the GNOVELTY$^+$ algorithm [13]. GNOVELTY$^+$ combines a clause penalty-based scheme similar to PAWS [15] with the promising variable scheme of G^{2}WSAT [11] (see [13] and the GNOVELTY$^+$ source code, version 1.2, from the 2009 SAT Competition for more details). The behaviour of SPARROW differs from GNOVELTY$^+$ only when there are no (penalized) promising variables, in which case a novel VE and a probabilistic VSM is used instead of the NOVELTY-based component in GNOVELTY$^+$. The VE used by SPARROW is $\langle sparrowScore \cdot sparrowAge \rangle$, where *sparrowScore* and *sparrowAge* are defined as follows[1]:

$$sparrowScore = \begin{cases} c_b^{penScore} & \text{if } penScore < 0 \\ 1 & \text{otherwise} \end{cases}, \qquad sparrowAge = 1 + \left(\frac{age}{c_d} \right)^{c_e}.$$

---

[1] The definition and notation we use differs slightly from the published version of SPARROW [3], but accurately reflects the source code implementation.

SPARROW uses a *distribution-based VSM*, where each variable is selected proportionally to the VE $\langle sparrowScore \cdot sparrowAge \rangle$. When we use a similar approach to select an element from a set where the elements are assigned fixed weights, we will refer to it as *weighted selection*.

The full parameter space for SPARROW includes the three constants mentioned above and the smoothing probability (ps) inherited from GNOVELTY$^+$. Balint and Fröhlich proposed $(c_b, c_d, c_e, ps) := (2, 4, 10^5, 0.347)$ as a good configuration for large 3-SAT instances, which was found by manual tuning on selected 3-SAT competition benchmark instances [3].

2.2   VE-SAMPLER and PARAMILS

VE-SAMPLER [18] was developed to demonstrate the power of using new and innovative variable properties and VEs in SLS algorithms. It was inspired by the VW2 algorithm [14], which was observed to be very effective on the CBMC software verification instances with the VE $\langle break + c \cdot flips \rangle$. Based on the WALKSAT architecture, VE-SAMPLER uses the selection of an unsatisfied clause as a filter. In each search step, a controller selects one of six VEs using weighted selection; one of these VEs corresponds to a (freebie) random walk step, and the remaining five are all of the general form:

$$\|greedy\|^{a_1} + \text{clw}(s, m, l) \cdot \|diversification\|^{a_2} ,$$

where $\|p\|$ indicates a property p that is normalized to values between zero and one, and $\text{clw}(s, m, l)$ is a simple mechanism that selects between three coefficients (s, m, l) depending on whether the clause length is respectively less than, equal to, or greater than three. Each VE uses one greedy property (chosen from a set of five) and one diversification property (chosen from a set of thirteen), except for one VE that uses two greedy properties (see [18] for details). VE-SAMPLER uses *maximum VSM*, where the variable with the maximum evaluated VE is selected. At the time of this writing, VE-SAMPLER is the fastest SLS-based SAT solver on the CBMC and SWV software verification instances [18]; it has subsequently been shown to have good performance on random 3-SAT instances [16], although on these, it does not reach the performance of SPARROW or SATENSTEIN-LS.

The configuration space of VE-SAMPLER is enormous, with over 10^{50} unique configurations. The configurator used on VE-SAMPLER was PARAMILS [9,8], an SLS-based procedure that searches the configuration space of a given algorithm. The primary inputs to PARAMILS are a target solver (binary), a set of target (training) instances, a solver cutoff time, an evaluation function and a configuration file that specifies the configuration space (each solver parameter along with a set of possible values). The evaluation function used was the penalized average runtime (PAR-10), where instances not solved within the cutoff are counted as ten times the given cutoff time. The primary output of PARAMILS is the best configuration of the target algorithm that PARAMILS found by the search process, *i.e.*, the configuration that achieves the best PAR-10 performance on the instances in the given set.

To ensure that the results from PARAMILS generalize to instances other than those used during the optimization process, we use a set of test instances to report final results that is disjoint from the training set used when running PARAMILS.

3   Design Considerations underlying CAPTAIN JACK

Many SLS for SAT algorithms switch between greedy (intensification) steps and diversification steps, or use diversification properties as tie-breakers in greedy steps. SPARROW and VE-SAMPLER have *mixed steps* that combine a greedy and a diversification property in a VE; the likelihood of a variable being selected can be dominated by the greedy property, the diversification property or neither of them. CAPTAIN JACK allows for all three types of steps (greedy, diversification and mixed), and we introduce parameters to control the balance between the three. Mixed steps are rarely used in SLS algorithms for SAT, and we were interested in observing the proportion of mixed steps PARAMILS would select, and how that proportion would depend on the target instance set. We were also curious whether PARAMILS would select a distribution-based VSM as in SPARROW or a maximum VSM as in VE-SAMPLER; therefore, CAPTAIN JACK supports both types of VSMs.

One of the objectives of our earlier work had been to encourage the use of more diverse variable properties [18]; this was achieved in VE-SAMPLER by means of categorical parameters that select properties from a given set. One problem with this approach is that a configuration only has a few selected properties, which may include duplicates, and it is hard to assess the viability of each individual property. To help avoid this problem in CAPTAIN JACK, each property has a parameterized weight that controls how frequently it is selected. This makes it easier to assess which properties are important for a given instance set. It also renders CAPTAIN JACK an excellent framework for introducing and testing new variable properties; after these are added to the configuration space, the automated configurator can gradually introduce them into configurations by means of modifying their weights relative to other properties.

One significant departure in CAPTAIN JACK from SPARROW is the absence of penalized clause weights. In preliminary experiments, we observed that with penalized clauses the proportion of (greedy) promising steps was significantly higher, reducing the impact of the core CAPTAIN JACK components we were interested in exploring. In addition, we found the penalized promising variable mechanism as implemented in GNOVELTY$^+$ (and, hence, SPARROW) problematic and memory intensive for large instances.

Finally, when designing a highly parametric algorithm with the aim of configuring it automatically, it is important to understand the limitations of the automatic algorithm configurators currently available. The state-of-the-art configurator PARAMILS, which we used in this work, tends to have difficulties with configuration spaces like that of VE-SAMPLER, characterized by many categorical parameters (*e.g.*, property selection) and complex interaction between parameters (*e.g.*, the normalization and non-linear transformation for each property). The same holds for all other configurators we are currently aware of. To render CAPTAIN JACK more easily configurable, we decided to use very few categorical parameters, no conditional parameter dependencies and smoother interactions between parameters (as introduced by the previously mentioned property weights).

4   Captain Jack

In each search step of CAPTAIN JACK, the controller makes four core algorithmic decisions that determine the behaviour of the solver:

1. if promising steps are enabled and promising variables exist, select one; otherwise,
2. determine if a greedy, diversification or mixed step will occur;
3. select the greedy and/or diversification properties to use; and
4. determine the VSM (maximum or distribution-based) to be used and select the variable accordingly.

First, if promising variables exist, a straightforward G^2WSAT greedy search step [11] is taken, in which the promising variable with the best *score* is selected, breaking ties by the *age* property; this step is skipped if promising variables are turned off, which is one of the many configurable parameters. If no promising variables exist, an unsatisfied clause is selected at random (*i.e.*, the same filter as in WALKSAT is used).

The second decision in CAPTAIN JACK is which *type* of step to take: a greedy step, a diversification step or a mixed step. Each type of step has a parameterized weight, and the type of step is determined by weighted selection. For instances with variable clause lengths, the weights also depend on the length of the selected clause. CAPTAIN JACK uses a clause length range classification similar to VE-SAMPLER, where each clause falls into one of the following four ranges: $\{\leq 2, 3, 4 \ldots 9, \geq 10\}$; thus, for instances with variable clause lengths, there are 12 weights that determine the step type.

The next decision is to select the greedy and/or diversification properties. There are 9 greedy and 17 diversification properties, as described in Table 1. Each property is assigned a parameterized weight and is selected by weighted selection. For greedy and diversification steps, only one property is selected, while for mixed steps one greedy and one diversification property is selected, and the product of those two is computed.

The final decision in CAPTAIN JACK is whether to use a distribution-based VSM or a maximum VSM. The probability of using a maximum VSM is a parameterized value that is determined by the clause size and the type of the search step, resulting in 12 total probabilities for instances with variable clause lengths.

Table 1 gives an overview of the variable properties used in CAPTAIN JACK. As in VE-SAMPLER, when using interchangeable properties, special care must be taken to adjust the values of the *score* properties that can have negative values, and the *break* and *flips* properties, where large property values indicate undesirable choices. In CAPTAIN JACK, we opted for simplicity over potentially more effective normalizations and used very straightforward adjustments. For example, a constant is added to the *score* property values so the minimal candidate variable *score* is one (see website for more details).

In addition to the five greedy properties used in VE-SAMPLER, we introduced four new greedy properties (see upper part of Table 1). The *sparrowScore* property assigns a constant value of one to non-promising variables with a positive score; *sparrowScore*$_2$ replaces this constant with a new parameter (c_s). The *scoreRatio* has two obvious forms, and we added a separate Boolean parameter (b_s) to determine which of the two ratios should be used. *relMake* and *relBreak* were already used in VE-SAMPLER, as the *relative* number of clauses affected, normalized by the number of occurrences of the variable (*numOcc* and *numOcc'* are the number of clauses the variable currently appears in

Table 1. Variable properties in CAPTAIN JACK *Top:* greedy, *bottom:* diversification

make	number of clauses that become satisfied if the variable is flipped
break	number of clauses that become unsatisfied if the variable is flipped
score	increase in the number of satisfied clauses if the variable is flipped ($make - break$)
sparrowScore$_2$	from the SPARROW algorithm: if $score \leq 0$, $c_b{}^{score}$, otherwise c_s
scoreRatio	if b_s, $(make/(make + break))$, otherwise $(make/(break + \epsilon))$
relMake	($make/numOcc$): *make* adjusted by the number of occurrences of the variable (see text)
relBreak	$(break/(numOcc' + \epsilon))$
relScore	($relMake - relBreak$)
relScoreRatio	same as *scoreRatio*, with *relMake* and *relBreak*
rand	a random number between zero and one
flat	no property value, *i.e.*, one
fair	1 for the 'next' variable in the clause, otherwise 0 (see text)
last	0 if the variable was flipped the last time the clause was selected, otherwise 1
age	number of search steps since the variable was last flipped
age$_1$	number of search steps since the flip prior to the most recent flip
age$_5$	number of search steps since the fifth most recent flip
ageRange	the *age* property with less sensitivity $\lfloor age/c_r \rfloor$
sparrowAge	from the SPARROW algorithm: $1 + (age/c_d)^{c_e}$
tabu	0 if the variable is tabu ($age < c_t$), otherwise 1
flips	number of times the variable has been flipped
flops	number of times the variable appeared in a selected clause, but was not flipped
normFlops	similar to *flops*, but is increased by $1/clauseLen$ each time it is not flipped
resetFlops	same as *flops*, but reset to zero whenever the variable is flipped (excl. promising steps)
relFlips	($flips/numOcc$)
relFlops	($relFlops/numOcc$)
relNormFlops	($normFlops/numOcc$)

as false and true, respectively). The properties *relScore* and *relScoreRatio* are defined similarly.

Furthermore, we introduced ten new diversification properties (see lower part of Table 1). Whenever a clause is selected and the *fair* property is selected, the 'next' variable in the clause is assigned a property value of one, and all others are zero; this is implemented by maintaining a counter for each clause and simply selecting the next variable in sequence. The *last* property is zero for the variable that was flipped at the most recent time the same clause was selected, regardless of the property based on which that previous selection was made. In VE-SAMPLER, the *age'* property was used to keep track of the number of steps since the flip prior to the most recent flip. Here, we call this property *age$_1$* and generalize it to the *age$_k$* family of properties. In CAPTAIN JACK we wanted to explore larger values of k and added the *age$_5$* property. The *ageRange* property uses a divisor parameter (c_r) and a floor function to achieve a coarser evaluation of *age*. In VE-SAMPLER, the *filtCount* property was used to maintain how often a variable appears in the filtered variables (*i.e.*, the selected clause). The *flops* property is very similar, but is instead only incremented when the variable appears in the selected clause and is not flipped. The *normFlops* property is related to *flops*, but is incremented by $(1/clauseLen)$ when the variable is not flipped. The *resetFlops* property is the same as *flops*, but is reset to zero whenever the variable is selected (excluding when it is flipped as a promising variable). Finally, the *relFlips*, *relFlops* and *relNormFlops* properties are all normalized analogously to *relMake*.

5 Experimental Setup

For our experiments we used nine benchmark sets: six random uniform k-SAT sets, one random industrial-like set and two sets of SAT-encoded software verification problems.

For the random instances, we generated two sets each for $k = 3, 5, 7$; one set for smaller instances at the solubility phase transition and one set of larger, slightly underconstrained instances. For the phase transition sets, we used clauses/variables ratios of 4.26, 21.11 and 87.79 for 3-, 5-, and 7-SAT, respectively, as specified by Mertens *et al.* [12]. For the underconstrained sets, we chose ratios of 4.2, 20 and 85, as previously used in the SAT competition. To select the instance size n for these sets, we took into consideration both the sizes used in the competition and the time required to solve instances. We selected $n = 1\,000$ (1k) and $10\,000$ (10k) for 3-SAT, 100 and 500 for 5-SAT, and 60 and 90 for 7-SAT. For each set, we generated instances with the 2009 SAT competition generator and removed instances that were not solved by TNM within $10\,000$ seconds; we randomly selected 250 instances for training and 250 for testing.

The CBMC and SWV software verification instance sets have been previously studied in the literature [10,8,18]. While PICOSAT [4] can solve any of the instances in these sets in less than two seconds, they are known to be challenging for SLS-based solvers; in fact, about 50% of the instances in SWV cannot be solved consistently by any SLS-based SAT solver we are aware of.

Finally, we used the double power-law generator provided by Ansótegui *et al.* to generate a set of random industrial-like instances we dub IL50k; we chose this generator since it produces variable length clauses that have properties similar to industrial problems [1]. Our set contains satisfiable instances with the same characteristics as the instances used by Ansótegui *et al.* $(\frac{m}{n}, \beta, \overline{k}) := (2.650, 0.75, 5)$, but with $5 \cdot 10^4$ (50k) variables instead of 500k; we randomly chose 50 training and 50 test instances.

We compared the performance of CAPTAIN JACK against six state-of-the-art SLS solvers. These include the three top SLS solvers from the 2009 SAT Competition random category, TNM, GNOVELTY$^+$2 and ADAPTG^{2}WSAT2009^{++} (henceforth, as AG^{2}2009^{++}), for which we used the parameterless competition versions (see the competition booklet for details). We also selected the UBCSAT [17] implementation of SPARROW [3]; it behaves exactly like the original implementation, but is more efficient and exposes additional parameters. The final two solvers are the highly parametric VE-SAMPLER [18] and SATENSTEIN-LS [10] solvers. For VE-SAMPLER, we used the three native (*i.e.*, non-interpreted) implementations configured for CBMC, SWV and random 3-SAT instances at the phase transition (R3SAT in [10]). We chose not to reconfigure the interpreted version of VE-SAMPLER on our new sets, which would require developing native implementations for a fair comparison.

We used the FOCUSEDILS 2.3.5 variant of the PARAMILS framework to configure SPARROW, SATENSTEIN-LS and CAPTAIN JACK on each of the nine sets. In each of the configuration experiments for SPARROW and SATENSTEIN-LS, we performed 24 independent PARAMILS runs of at least 7 CPU days each, from which we selected the one with the best performance on the respective training set. For CAPTAIN JACK, we used a training protocol comprising three sequential stages, designed to deal with successively harder instances and larger fractions of the given training set. These three

Table 2. Solver Evaluation on Test Sets. Each cell summarizes the test-set performance of the solver for 10 runs on each instance in the set with a cutoff of 600 seconds. The top row shows the penalized average runtime (PAR-10): the mean of all runs over all instances with timeouts replaced with a penalized value of $6\,000 (= 10 \cdot 600)$ seconds. The second row shows the mean of the median runtimes for each instance in the set, where if any instances has a median at the cutoff time, the median is included but marked ($^+$). The third row indicates the percentage of all runs completed within the timeout. The algorithms indicated ([*]) have been optimized by PARAMILS on each target set. Unfortunately, GNOVELTY$^+$2 crashed on the CBMC and SWV instances and could therefore not be evaluated on those.

Algorithm	3-SAT		5-SAT		7-SAT		IL50k	CBMC	SWV
	1k	10k	100	500	60	90			
	PAR-10 m.m. % sol.	PAR-10 m.m. % sol.	PAR-10 m.m. % sol.	PAR-10 m.m. % sol.	PAR-10 m.m. % sol.	PAR-10 m.m. % sol.	PAR-10 m.m. % sol.	PAR-10 m.m. % sol.	PAR-10 m.m. % sol.
CAPTAIN JACK[*]	**69.0** 15.4$^+$ 99.0%	**43.3** **24.4** **99.8%**	0.31 0.23 100%	12.1 9.2 100%	1.5 1.1 100%	68.9 22.3$^+$ 99.4%	**0.83** **0.78** 100%	0.35 0.31 100%	4533 464$^+$ 24.5%
SPARROW[*]	63.0 11.9$^+$ 99.1%	220 58.3$^+$ 97.1%	**0.18** **0.14** 100%	12.9 9.1 100%	**0.81** **0.59** 100%	32.4 12.9 99.7%	1.4 1.3 100%	69.4 3.9 99.0%	4413 446$^+$ 26.5%
VE-SAMPLER[*]	82.4 15.0$^+$ 98.8%	n/a	n/a	n/a	n/a	n/a	n/a	**0.08** **0.07** 100%	2856 295$^+$ **52.5%**
SATENSTEIN[*]	33.5 8.2$^+$ **99.6%**	72.1 30.0 99.3%	0.21 0.15 100%	**3.3** **2.6** 100%	0.92 0.68 100%	34.1 13.6 99.7%	1.2 1.1 100%	0.62 0.54 100%	4640 464$^+$ 22.7%
TNM	75.4 14.2$^+$ 98.9%	691 154$^+$ 90.3%	0.22 0.17 100%	27.2 17.5 99.9%	1.9 1.6 100%	30.6 13.3 99.8%	350 44.4$^+$ 94.4%	525 73.5$^+$ 91.8%	4640 464$^+$ 22.7%
GNOVELTY$^+$2	78.6 16.3$^+$ 98.9%	2604 382$^+$ 58.8%	0.27 0.20 100%	11.2 8.3 100%	1.1 0.78 100%	22.1 **10.3** 99.9%	1901 295$^+$ 70.4%	n/a	n/a
AG^{2}2009^{++}	76.3 12.4$^+$ 98.8%	2373 331$^+$ 62.1%	0.21 0.17 100%	24.4 15.2 99.9%	1.5 1.4 100%	**15.6** 11.9 **100%**	6.9 6.5 100%	3267 383$^+$ 46.7%	4217 440$^+$ 30.0%

stages consisted of 24 independent PARAMILS runs for one, three and three CPU days, respectively (see website for further details).

The PARAMILS training was conducted on Westgrid clusters (see website for details), but otherwise all solver evaluations and times reported were performed using the EDACC framework [2] running on bwGRiD [5] (Intel two socket 4-Core Xeon E5540 CPUs 2.83 GHz, with 16GB RAM running Linux).

6   Results and Discussion

In Table 2, we present the results from our evaluation of CAPTAIN JACK and several state-of-the-art algorithms on the instance sets described in Section 5. As previously mentioned, CAPTAIN JACK was designed to perform well on a wide variety of instances, and this is reflected in the results. We note that the configurations evaluated here are, for the most part, not the best that exist in the CAPTAIN JACK design space. For example, in results not reported here, but available online, we ran PARAMILS for an additional three days on each set and were able to obtain configurations that achieved modest improvements in PAR-10 (6%-9%) on 3sat10k, 5sat500 and IL50k and more significant improvement (20%) on CBMC and SWV. Furthermore, as we will discuss later in this section, we found evidence that in some cases, the best CAPTAIN JACK configuration on a given type of instances was not the one optimized for that type, which clearly indicates that PARAMILS at least in some cases produced sub-optimal configurations.

The relative performance of CAPTAIN JACK is the best on 3sat10k and IL50k, and CAPTAIN JACK is now the best known SLS solver for those sets. On the sets 5sat100, 7sat60 and 7sat90 CAPTAIN JACK performs significantly worse than SPARROW. We conjecture that this is connected to the relatively small number of variables in these instances, but further investigation is needed to explain this phenomenon.

Finally, we were quite surprised by the strong performance of SPARROW on IL50k, as results obtained for the default configuration of SPARROW were much worse. Interestingly, the configuration of SPARROW found by PARAMILS on IL50k uses no smoothing, a rare situation for high-performance SLS algorithms using clause penalties – a phenomenon that might warrant further investigation.

Next, we investigated the CAPTAIN JACK configurations found by PARAMILS for each of our instance sets (see Table 3). As mentioned earlier, we do not believe that these configurations are optimal, and longer runs of PARAMILS could produce rather different configurations. Nevertheless, we believe that qualitative differences between these configurations, which are based on multiple, long runs of PARAMILS, are likely meaningful.

In CAPTAIN JACK we introduced several new variable properties, and we were curious to see which role these would play in the configurations found by PARAMILS. Clearly, the age_5 property is very effective, which was surprising to us, especially since the value of 5 was selected somewhat arbitrarily; this suggests that the age_k family of properties merits further study. The slight modification we introduced in $sparrowScore_2$, which allows for positive non-promising variables to have a parameterized value (c_s), also appeared to be quite effective, and most configurations had a value of c_s slightly greater than one. Considering that the original $sparrowScore$ property had been developed for solving random 3-SAT, it is perhaps not surprising that $sparrowScore_2$ is prominently used only in the random k-SAT configurations. Conversely, our new properties $scoreRatio$ and $relScoreRatio$ appeared to be useful only on the software verification benchmarks, both of which had b_s set to true (see Table 1). Our intuition was that the *fair* and *flops*-based properties would be good for diversification, but that they should be used sparingly; our results are consistent with this intuition, but further study is warranted. Overall, the diversification properties most often selected are the *age* variants, which is also the most prominent diversification property in the literature. It is also very clear that the *flips* properties are very important for instances with variable-length clauses and non-uniform variable distributions.

We were wondering whether PARAMILS would prefer a 'traditional' approach for SLS-based SAT solving with mostly greedy steps and a few diversification steps, as it did for 7sat90; however, most of the optimized configurations turned out to favour mixed steps. This supports previous evidence that exploring new and innovative

Table 3. Parameter Settings in CAPTAIN JACK. *(Top:)* The weight for each search step type (greedy / mixed / diversification), and for each type, the percent of steps where the maximum VSM was selected (as opposed to a distribution-based VSM). For instances with variable length clauses, the values depend on the selected clause length $\{\leq 2, 3, 4 \ldots 9, \geq 10\}$. *(Bottom:)* The weight for each variable property (greedy and diversification properties are selected independently). All weights have been normalized to appear as percentages. Non-applicable values are shown as a dash (-), and weights with value zero are blank. The most significant properties (with a combined weight of $\geq 75\%$) are in bold. As an example, for 3sat10k, 94% of the (non-promising) search steps are mixed steps, and in 90% of those steps the maximum VSM is used. *make* and age_5 are each independently selected as the greedy and diversification property for 40% of the mixed steps. This means that for 13.5% $(0.94 \cdot 0.9 \cdot 0.4 \cdot 0.4)$ of the (non-promising) search steps the variable with the maximum value of $\langle make \cdot age_5 \rangle$ is selected.

step type / VSM	3-SAT 1k	10k	5-SAT 100	500	7-SAT 60	90	IL50k ≤2	3	4-9	≥10	CBMC ≤2	3	4-9	≥10	SWV ≤2	3	4-9	≥10
	%	%	%	%	%	%	≤2	3	4-9	≥10	≤2	3	4-9	≥10	≤2	3	4-9	≥10
greedy step		3	6	67		93	10			8	8	67	89	33	31		32	94
mixed step	100	94	94	33	100	1	80	100	67	31	62	17		66	62	100	64	3
div. step		3				6	10		33	62	31	17	11	1	8		4	3
greedy: % max	-	80	30	50	-	50	80	-	-	100	80	0	40	0	100	-	0	90
mixed: % max	10	90	90	10	90	0	90	10	0	70	60	70	-	10	60	100	0	30
div.: % max	-	0	-	-	-	20	10	-	20	0	30	10	80	0	10	-	50	0

property	3-SAT 1k	10k	5-SAT 100	500	7-SAT 60	90	IL50k	CBMC	SWV
	%	%	%	%	%	%	%	%	%
make		**40**	2.8			0.7	0.5	13	
break	1.2	10	2.8		**50**	**47**		6.3	6.2
score	9.9	10	2.8	20		3	**70**	25	0.4
sparrowScore$_2$	**79**	**40**	**90**	**78**	**50**	**47**	0.5	0.8	3.1
scoreRatio	9.9		1.4	2.4		1.5		0.8	**25**
relMake	-	-	-	-	-	-	18	1.6	0.4
relBreak	-	-	-	-	-	-		8.8	3.1
relScore	-	-	-	-	-	-	1.1	1.6	12
relScoreRatio	-	-	-	-	-	-	1.1	**51**	**50**
rand	15	0.6	0.5	0.3	0.6	**52**	1.3		1.4
flat		**9.9**	1	**44**	0.3			0.3	0.3
fair	0.2		0.5	1.4	0.3		0.7	5.2	0.7
last	15	4.9	4.1	0.7	10	**13**	0.3		0.3
age	3.6	**9.9**	**33**	1.4	2.5	**13**		1.3	2.8
age$_1$	0.5	**20**	16	1.4			**43**		
age$_5$	15	**40**	4.1	0.7	**20**			**42**	**44**
ageRange	7.3	0.6	1	2.7	0.6		0.3	1.3	
sparrowAge	**29**		0.3	0.7	**20**	1.6	2.7	0.3	
tabu	3.6	4.9	**33**	**44**	**41**	3.2	0.7	0.7	
flips	3.6		2	0.7	1.3		**43**	0.3	**44**
flops	7.3		0.5		2.5	1.6	0.3	0.3	0.3
normFlops	0.5		0.3	2.7	0.3	**13**			
resetFlops	0.5	**9.9**	4.1	0.3	0.3	3.2		0.3	0.7
relFlips	-	-	-	-	-	-	5.4	**42**	1.4
relFlops	-	-	-	-	-	-	2.7	5.2	0.3
relNormFlops	-	-	-	-	-	-		0.7	2.8

methods for mixing properties in VEs is a promising area of research. We introduced clause-length-dependent behaviour in CAPTAIN JACK to see if we could observe interesting trends, *e.g.*, we hypothesized that mixed or diversification steps could be more important for larger clause lengths, but the results are inconclusive. However, it appears that IL50k and CBMC benefit from more diversification steps than the random k-SAT sets, and – as we will observe in Table 4 – it appears that SWV does as well. For random k-SAT, the configurations for the underconstrained sets allow for more greedy steps, which is consistent with the understanding that these instances are relatively easier to search than those at the phase transition. The proportion of solely greedy steps is actually higher (12%) for 3sat10k if we consider that *flat* is selected as the diversification property for 10% of the mixed steps; this is an example of the kind of inter-parameter dependency that we were attempting to minimize in CAPTAIN JACK, but is impossible to eliminate completely.

There is no clear preference between maximum *vs* distribution-based VSMs. It would be interesting re-run PARAMILS on CAPTAIN JACK with two different restricted configuration spaces that force all steps to be either a maximum or distribution-based VSM; we could then observe which VSM would achieve better performance, and study the differences between the resulting configurations. We note that CAPTAIN JACK is well suited for this kind of analysis, which we are confident will lead to an improved understanding of the performance potential inherent in various algorithm components, and of their interaction when solving various types of SAT instances.

There are a few additional parameter settings, such as the parameters c_s, c_r and b_s in Table 1, that are not shown in Table 3 and can be found online. One of these is the Boolean parameter to control if G^2WSAT promising steps are taken; only the CBMC and SWV configurations do not take promising steps. This may suggest that promising steps may be more suited for randomly generated instances (including IL50k), which is consistent with the observation that AG^22009^{++}, which relies heavily on promising steps, is the worst performer on CBMC in Table 2.

Another method for evaluating the differences between our configurations is to cross-test each of the configurations on each of the sets, and we present the results of these experiments in Table 4. These results also indicate how 'specialized' each of our optimized configurations are. The most surprising result is that in several cases, the best configuration for an instance set is actually *not* the one optimized for the respective training set. As previously stated, this clearly indicates that PARAMILS does not always find optimal configurations within the design space of CAPTAIN JACK. The most interesting such configuration is the one obtained for IL50k, which performs very well on the SWV set. This suggests that the industrial-like instances could indeed be very useful for optimizing performance on harder industrial instances. This similarity is reinforced further as the SWV configuration is the second-best configuration (by a large margin) on the IL50k set. Finally, because the IL50k set has an *average* clause length of 5, we hypothesized that the 5-SAT configurations might perform well on IL50k or vice-versa, but this appears not to be the case. This further highlights the fact that the structural aspects of real verification instances captured by the industrial-like instances, albeit simplistic, are at least to some degree informative.

One final experiment we performed was to test if algorithms trained on the IL50k instances would be able to solve larger instances from the same distribution. Ansótegui *et al.* [1] generated larger (500k) instances, and demonstrated that GNOVELTY$^+$ was unable to solve these. We generated a set of ten such instances (IL500k) that, aside from the number of variables, have the same characteristics as the IL50k set. Because

Table 4. Cross-Testing of CAPTAIN JACK configurations. Each configuration of CAPTAIN JACK was run once on each instance in each test set with a cutoff of 600 seconds. We report the ratio of the resulting PAR-10 to the PAR-10 for the targeted configuration. Configurations that outperform the targeted configuration for the set are in bold.

Configuration	3-SAT		5-SAT		7-SAT		IL50k	CBMC	SWV
	1k	10k	100	500	60	90			
CJ [3sat1k]	1	61.5	1.38	95.7	1.08	1.03	157	5876	1.02
CJ [3sat10k]	2.65	1	1.41	545	1.99	3.99	167	1890	1.02
CJ [5sat100]	2.56	135	1	93.2	1.18	**0.72**	170	7108	1.03
CJ [5sat500]	24.3	200	1.35	1	1.00	**0.97**	1271	10014	1.00
CJ [7sat60]	99.1	200	**0.82**	539	1	2.33	786	9989	1.02
CJ [7sat90]	105	200	1.82	12.1	1.44	1	1929	3088	**0.98**
CJ [IL50k]	16.6	200	4.50	567	2.20	15.8	1	1106	**0.83**
CJ [CBMC]	19.9	200	6.71	483	2.97	7.70	1236	1	1.02
CJ [SWV]	148	200	17.6	567	9.47	79.2	2.29	2.43	1

the instances are so large, we observed that many SLS solvers (*e.g.*, from the SAT competition, but also SATENSTEIN-LS) encounter technical problems when trying to solve them. We ran the IL50k configurations of CAPTAIN JACK and SPARROW and the UBCSAT implementation of ADAPTG^{2}WSAT$^+$ on each of the 10 instances with a cutoff of 12 hours per instance. ADAPTG^{2}WSAT$^+$ solved only 5 instances and SPARROW was able to solve 9. However, CAPTAIN JACK was able to solve all 10 instances in a combined time of 77 minutes. For perspective, PICOSAT [4] solved all 10 instances in a combined time of 2 minutes and showed little variation in runtime per instance compared to the SLS solvers.

7   Conclusions and Future Work

In this work, we have introduced CAPTAIN JACK, a highly parametric SLS algorithm that can be automatically configured to perform well on various types of SAT instances and is currently the best known SLS algorithm for solving large random 3-SAT and 'industrial-like' instances. We designed CAPTAIN JACK in a way that would aid us in exploring which components and heuristic mechanisms give rise to strong performance on different types of SAT instances and made several interesting observations in this respect. We also introduced several new variable properties and provided evidence these can be very effective; in particular, our results suggest that the family of age_k properties merits further investigation. Finally, we provided preliminary evidence that training on smaller industrial-like instances may be a viable approach to improving SLS algorithm performance on larger industrial problems.

Our results reported here provided further evidence that mixed VEs can be very effective; while CAPTAIN JACK, SPARROW and VE-SAMPLER combine only two variable properties, we believe that it may be interesting to investigate more complex combinations. We also see potential in developing an *adaptive* CAPTAIN JACK that adjusts its balance between diversification and intensification throughout the search and incorporates a mixed VE that combines property values accordingly. Ultimately, we hope that CAPTAIN JACK will provide further insight into SLS algorithm development, and that algorithm developers will be able to gain insight and inspiration from examining

CAPTAIN JACK configurations that are effective for solving particular instance sets. We believe that such work will lead to new and specialized lightweight algorithms, similar to SPARROW, that improve the state-of-the-art for solving SAT.

Acknowledgments

We would like to thank the anonymous reviewers for their valuable feedback. This research has been enabled by the use of computing resources provided by the bwGRiD project [5], WestGrid and Compute/Calcul Canada. Furthermore, HH acknowledges funding received through the NSERC Discovery Grants Program.

References

1. Ansótegui, C., Bonet, M.L., Levy, J.: Towards industrial-like random SAT instances. In: IJ-CAI 2009, pp. 387–392 (2009)
2. Balint, A., Diepold, D., Gall, D., Gerber, S., Kapler, G., Retz, R.: EDACC - an advanced platform for the experiment design, administration and analysis of empirical algorithms. In: LION-2011 (to appear)
3. Balint, A., Fröhlich, A.: Improving stochastic local search for SAT with a new probability distribution. In: Strichman, O., Szeider, S. (eds.) SAT 2010. LNCS, vol. 6175, pp. 10–15. Springer, Heidelberg (2010)
4. Biere, A.: PicoSAT essentials. JSAT 4, 75–97 (2008)
5. bwGRiD: Member of the German D-Grid initiative, funded by the Ministry of Education and Research and the Ministry for Science, Research and Arts Baden-Württemberg
6. Chvátal, V., Szemerédi, E.: Many hard examples for resolution. Journal of the ACM 35(4), 759–768 (1988)
7. Hoos, H.H.: Computer-aided design of high-performance algorithms. Tech. Rep. TR-2008-16, University of British Columbia (2008)
8. Hutter, F., Hoos, H.H., Leyton-Brown, K., Stützle, T.: ParamILS: An automatic algorithm configuration framework. Journal of Artificial Intelligence Research 36, 267–306 (2009)
9. Hutter, F., Hoos, H.H., Stützle, T.: Automatic algorithm configuration based on local search. In: AAAI 2007, pp. 1152–1157 (2007)
10. KhudaBukhsh, A.R., Xu, L., Hoos, H.H., Leyton-Brown, K.: SATenstein: Automatically building local search SAT solvers from components. In: IJCAI 2009, pp. 517–524 (2009)
11. Li, C.M., Huang, W.Q.: Diversification and determinism in local search for satisfiability. In: Bacchus, F., Walsh, T. (eds.) SAT 2005. LNCS, vol. 3569, pp. 158–172. Springer, Heidelberg (2005)
12. Mertens, S., Mézard, M., Zecchina, R.: Threshold values of random k-SAT from the cavity method. Random Structures & Algorithms 28, 340–373 (2006)
13. Pham, D.N., Thornton, J., Gretton, C., Sattar, A.: Combining adaptive and dynamic local search for satisfiability. JSAT 4, 149–172 (2008)
14. Prestwich, S.: Random walk with continuously smoothed variable weights. In: Bacchus, F., Walsh, T. (eds.) SAT 2005. LNCS, vol. 3569, pp. 203–215. Springer, Heidelberg (2005)
15. Thornton, J., Pham, D.N., Bain, S., Ferreira Jr., V.: Using cost distributions to guide weight decay in local search for SAT. In: Ho, T.-B., Zhou, Z.-H. (eds.) PRICAI 2008. LNCS (LNAI), vol. 5351, pp. 405–416. Springer, Heidelberg (2008)
16. Tompkins, D.A.D.: Dynamic Local Search for SAT: Design, Insights and Analysis. Ph.D. thesis, University of British Columbia (2010)

17. Tompkins, D.A.D., Hoos, H.H.: UBCSAT: An implementation and experimentation environment for SLS algorithms for SAT and MAX-SAT. In: H. Hoos, H., Mitchell, D.G. (eds.) SAT 2004. LNCS, vol. 3542, pp. 306–320. Springer, Heidelberg (2005)
18. Tompkins, D.A.D., Hoos, H.H.: Dynamic scoring functions with variable expressions: New SLS methods for solving SAT. In: Strichman, O., Szeider, S. (eds.) SAT 2010. LNCS, vol. 6175, pp. 278–292. Springer, Heidelberg (2010)
19. Xu, L., Hutter, F., Hoos, H.H., Leyton-Brown, K.: SATzilla: Portfolio-based algorithm selection for SAT. Journal of Artificial Intelligence Research 32, 565–606 (2008)

Careful Ranking of Multiple Solvers with Timeouts and Ties

Allen Van Gelder

University of California, Santa Cruz, CA 95064
http://www.cse.ucsc.edu/~avg

Abstract. In several fields, Satisfiability being one, there are regular competitions to compare multiple solvers in a common setting. Due to the fact some benchmarks of interest are too difficult for all solvers to complete within available time, time-outs occur and must be considered.

Through some strange evolution, time-outs became the only factor that was considered in evaluation. Previous work in SAT 2010 observed that this evaluation method is unreliable and lacks a way to attach statistical significance to its conclusions. However, the proposed alternative was quite complicated and is unlikely to see general use.

This paper describes a simpler system, called *careful ranking*, that permits a measure of statistical significance, and still meets many of the practical requirements of an evaluation system. It incorporates one of the main ideas of the previous work: that outcomes had to be freed of assumptions about timing distributions, so that non-parametric methods were necessary. Unlike the previous work, it incorporates ties.

The *careful ranking* system has several important non-mathematical properties that are desired in an evaluation system: (1) the relative ranking of two solvers cannot be influenced by a third solver; (2) after the competition results are published, a researcher can run a new solver on the same benchmarks and determine where the new solver would have ranked; (3) small timing differences can be ignored; (4) the computations should be easy to understand and reproduce. Voting systems proposed in the literature lack some or all of these properties.

A property of *careful ranking* is that the pairwise ranking might contain cycles. Whether this is a bug or a feature is a matter of opinion. Whether it occurs among leaders in practice is a matter of experience.

The system is implemented and has been applied to the SAT 2009 Competition. No cycles occurred among the leaders, but there was a cycle among some low-ranking solvers. To measure robustness, the new and current systems were computed with a range of simulated time-outs, to see how often the top rankings changed. That is, times above the simulated time-out are reclassified as time-outs and the rankings are computed with this data. *Careful ranking* exhibited many fewer changes.

1 Introduction and Overview

Empirical comparison of computational performance is an important technique for advancing the state of the art in software. In Propositional Satisfiability

K.A. Sakallah and L. Simon (Eds.): SAT 2011, LNCS 6695, pp. 317–328, 2011.
© Springer-Verlag Berlin Heidelberg 2011

and several related fields testing programs on benchmarks is complicated by the fact that time limits must be set, because some benchmarks of interest are too difficult for all programs to complete within available time. Programs can fail to complete a test due to exhausting time or some other resource, often memory. There is no clearly correct way to integrate the results of failed tests and completed tests, to compute a single "figure of merit."

To focus the discussion, let us assume that the property we wish to measure is speed of solution, and we are evaluating the results of a SAT competition. If every program could complete every test, we would simply add up the times for each program and rank them according to this total, with smallest being best. From this point of view, time-outs and other failures are defects in the experiment.

In actuality, there is not time for every program to complete every test, and failures do occur. This leads to what is called *censored data* in the literature: there "really" is a value for the time the program would have taken on a benchmark, we just did not find out what that value is. The question is, what is a good way to rank the programs, based on the data that *is* available. Logically, we would want this ranking method to produce the same results as the ideal experiment, to the extent possible.

The ranking method that has been used in recent SAT competitions, which we shall call *solution-count ranking*,[1] is to set some time limit *ad hoc*, and simply count how many tests are successfully completed. Through some strange evolution, time-outs, which are the manifestations of defects in the experiment, became the only factor that was considered in evaluation. Total CPU time is used as a tie-breaker only if solution counts are equal.

Previous work by Nikolić observed that **solution-count ranking** is unreliable and lacks a way to attach statistical significance to its conclusions [Nik10]. However, the proposed alternative was quite complicated and had some practical drawbacks. The purpose of this paper is to describe and propose a simpler system that meets the practical requirements for ranking solvers in a SAT competition (endorsed by a survey of solver developers and users),[1] and also gives information about the statistical significance of the results, or lack thereof.

Definition 1. *Practical requirements*

1. The relative ranking of two solvers cannot be influenced by a third solver.
2. After the competition results are published, a researcher can run a new solver on the same benchmarks and determine where the new solver would have ranked.
3. Small timing differences can be ignored.
4. The computations should be easy to understand and reproduce.

One earlier method, called the *purse* method,[2] lacked properties (1) and (2) and fell into disfavor after a few trials.   □

---

[1] See http://www.satcompetition.org/2009/spec2009.html, where it is called "Lexicographical NBSOLVED, sum ti."

[2] See http://www.satcompetition.org/2007/rules07.html

The methodology we propose, called *careful ranking*, incorporates one of the main ideas of Nikolić: that outcomes must be free of assumptions about timing distributions, because we have no information about these distributions. Non-parametric methods are necessary. Unlike the previous work, our proposal incorporates ties to account for timing differences that are considered inconsequential for ranking purposes.

The **careful ranking** system has the important non-mathematical properties given in Definition 1. The main ingredient of **careful ranking** is that all pairs of competitors are compared in isolation, leading to a pair of "scores" that sum to zero. A large positive score indicates a significantly faster solver. The null hypothesis is that both solvers are equally fast "overall," or "in the long run." The *expected value* of the score is zero, under this hypothesis. The difference between zero and the observed score may be converted into a standard measure of statistical significance.

For a k-way competition, there are $k(k-1)/2$ pairwise matches. The results are expressed with a dominance matrix, as described in Section 5. The final ranking is extracted from this matrix.

There is a meta-ranking question to be addressed. How can we compare various ranking methods, since we do not know the "true answers?" The method we propose, and use, is to measure sensitivity to changes in the time limit. We do not know what would have happened if we used a larger time limit. But what would have happened under all *shorter* time limits can be determined from the available data.

The **careful ranking** system is implemented[3] and has been applied to the SAT 2009 Competition. The implementation is csh scripts, sed, and awk, which should be portable. No cycles occurred among the leaders, but there was a cycle among some low-ranking solvers. To measure robustness, the new and current systems were computed with a range of simulated time-outs, to see how often the top rankings changed. That is, times above the simulated time-out are reclassified as time-outs and the rankings are computed with this data.

2 Related Work

There is a large body of work on various aspects of experimental comparisons. We restrict ourselves to immediately related work on ranking solvers. Non-mathematical considerations for a scoring method are discussed in general terms by Le Berre and Simon [LBS04], and influenced several aspects of the method proposed here. One such aspect is our provision for many timing differences to be treated as a tie, because it appears that many people consider calling one solver the winner in these cases is a distortion. The reaction to this perceived distortion has been to reduce the importance of speed to nearly nothing, as long as the solver stays within the time limit. We hope that treating "minor" timing differences as ties will make the technique more acceptable than prior techniques that used time as the major consideration.

---

[3] Code is at http://www.cse.ucsc.edu/~avg/CarefulRanking/

Brglez and co-authors [BLS05, BO07] replicate instances into classes to gather statistics. Their goals are quite different from ranking a competition. Nikolić [Nik10] extends these ideas to compare more than two programs. The non-mathematical, practical issues mentioned in Definition 1 are not considered in these papers.

Pulina conducted an extensive empirical evaluation of several scoring methods [Pul06]. One criterion he used is similar to the one we use, decreasing the time limit and measuring stability. Our proposed method is significantly different from those he analyzed. Most or all of the comparison methods he studied lacked the independence from a third solver. Thus a later researcher could not see where new work fit into a previous competition.

Pulina introduced the idea of viewing the ranking problem as a *voting situation*: each benchmark "votes" for the solvers (the "candidates") by a preference ballot that ranks them by solution time. This is a very attractive idea, but unfortunately, none of the well-known proposed voting methods satisfy the criteria of Definition 1 and elaborated further in the URL given there. There is a vast literature on this subject, as surveyed by Levin and Nalebuff [LN95], and more recently treated by Pomerol and Barba-Romero [PBR00] and Tideman [Tid06].

A detailed comparison with all proposals would take us far afield, so we restrict attention to the Schulze method, which has enjoyed recent popularity [Sch03]. That popularity is not surprising, because the Schulze method, unlike many other proposals (such as Borda), permits voters to vote equal preferences among subsets of the candidates (e.g., $D=1$, $(A,C)=2$, $B=7$ is a valid ballot in a field of 10).

Suppose a competition is being run with six solvers and 63 benchmarks with Schulze ranking (the example may use many combinations of numbers), and the following events transpire. After 60 benchmarks have been run on all solvers, the Schulze ranking is computed and solver A is uniquely winning. On each of the last three benchmarks, solver A has the best performance of any solver and solver D has the *worst* performance of any solver. (For example, solver D might time out on the last three benchmarks). However, when the Schulze ranking is computed using all 63 benchmarks, D wins. No, this is not a typo. See Appendix B of [Sch03] for complete details[4].

It is impossible to imagine that any organizers of a competition would adopt the Schulze method, if they know about this possibility. Moreover, this is not a quirk in the Schulze method. It is known to be present in a large class of methods that satisfy the *Condorcet principle* [LN95, Tid06]. The phenomenon is known as the *no-show paradox* [Mou88], because solver A would have been better off without the "support" of the last three benchmarks, on all of which A was the clear winner.

The above example is possible under Schulze ranking and many other voting systems because it does not satisfy criterion (1) in Definition 1, that other solvers should not be able to affect the *relative ranking* of solvers A and D.

---

[4] The example cited has some pairwise ties among candidates, for simplicity of presentation, but these ties can be removed by "fuzzing" without changing the outcomes.

3   What Is a Tie?

Say we are comparing solvers (R, S) on a set of benchmarks, $\{B_i \mid i = 1, \ldots, n\}$, with time limit τ. The data is two lists of numbers, $t_i(R)$ and $t_i(S)$, the solution times of the two solvers on B_i. Numbers are floating point and include Inf to denote a failure of any kind. (We choose not to distinguish among failure reasons, except that a *wrong answer* means the solver is disqualified and its matches are not scored.) All data other than Inf is between 0 and τ, and we call these *finite* times.

We interpret the lists $t_i(R)$ and $t_i(S)$ as a series of n mini-matches, each with a stake of one point. A tie awards 0 to each solver. A win for R gives R a score of 1 and gives S a score of -1, and the reverse if S wins.

Clearly if $t_i(R) = t_i(S)$, the result is a tie. The question is what other outcomes should be considered a tie. The current method treats any pair of finite times as a tie; the only win is a finite time vs. Inf. The opposite extreme is that any $t_i(R) < t_i(S)$ is a win for R. Nikolić performed a theoretical analysis that depended on a complete absence of ties, so he "discarded" benchmarks where all times were under 5 seconds, which got rid of all the exact equalities with finite times, and then treated any finite time difference as a win [Nik10]. This is essentially the same as saying any pair of times under 5 seconds is a tie.

Our thesis is that some time differences should be considered "inconsequential" in the sense that someone trying to select the better solver between R and S for use in an application would not be influenced these time differences. We hypothesize that on longer runs, larger time differences would be considered inconsequential, so we want to define a **tie zone** whose width grows as run times get longer. We also believe that most people agree that below some threshold, *all* time differences are inconsequential. The user decides where to set this threshold, which we call **noise**, and which is the only user-specified parameter needed to specify the tie zone.

The growth rate we choose is founded in recurrent-event theory. We model the solver's computation as a long series of search events with independent outcomes. The probability that a search event has a successful outcome is very small, and the solver terminates upon the first successful search event. This is the well-known Poisson process. The *standard deviation* of the time to termination is proportional to the *square root* of the *average* time to termination. If two solvers have the same (theoretical) average time to termination on benchmark B_i, then their time difference is a random variable with mean zero and standard deviation proportional to the square root of their common average. We propose that observed time differences less than some number of standard deviations should be considered as ties, because they do not provide compelling evidence that one solver's average is really shorter than the other's. This is purely an heuristic model, of course.

Once we accept the idea that it is sensible for the tie zone to grow proportionally to $\sqrt{(t_i(R) + t_i(S))/2}$, the square root of the average of the two observed solving times, all that remains is to choose a constant of proportionality. The user makes this choice indirectly by specifying a scalar parameter called **noise**.

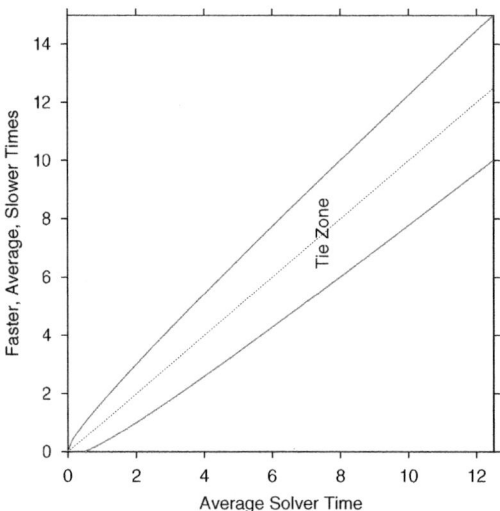

Fig. 1. Tie Zone for **noise** = 1 minute; times in minutes. Lower curve: faster solver time; upper curve: slower solver time; middle line: average of upper and lower curves.

As stated above, the intuition is that all solution times at or below the **noise** level should be treated as indistinguishable. Figure 1 shows the tie zone for **noise** = one minute. Any pair of times, both under one minute, fall into the tie zone.

To generate the desired tie zone, we define:

$$\alpha = \sqrt{\textbf{noise}/2}$$
$$\Delta = \alpha \sqrt{(t_i(R) + t_i(S))/2}.$$

Then the "tie zone" extends from $(t_i(R)+t_i(S))/2 - \Delta$ to $(t_i(R)+t_i(S))/2 + \Delta$. For R to win it is necessary that

$$t_i(R) < (t_i(R) + t_i(S))/2 - \Delta$$

Since $t_i(R) \geq 0$, S is assured of (at least) a tie whenever $t_i(S) \leq \textbf{noise}$, as was desired by the user.

4   Pairwise Matches

Say we are comparing solvers (R, S) on a set of benchmarks, $\{B_i \mid i = 1, \ldots, n\}$, with time limit τ. The data is two lists of numbers, $t_i(R)$ and $t_i(S)$. As described in Section 3, we interpret the lists $t_i(R)$ and $t_i(S)$ as a series of n mini-matches, with each outcome for R being -1, 0, or 1. The (algebraic) total is the raw score for the match, denoted $raw(R, S)$. For simplicity, all pairs are processed, so S is compared with R at some point to get $raw(S, R)$, which of course equals $-raw(R, S)$.

The value of $raw(R, S)$ can be used to test the null hypothesis, which is that R and S have an equal probability of winning a random mini-match. We also need the number of decisive (non-tie) mini-matches, denoted $decisive(R, S)$. Then the Student t parameter is given by

$$\text{Student } t = \frac{raw(R, S)}{\sqrt{decisive(R, S)}}, \tag{1}$$

which expresses the raw score in standard deviations. Statistically, the match is modeled as $decisive(R, S)$ fair coin flips. If $decisive(R, S)$ is large, the distribution is close to Gaussian. We are certainly justified in rejecting the null hypothesis when $|t| \geq 2$, without figuring out the exact value of p, which is the probability of observing a t-value this large or larger. In this case $p < 0.03$.

To summarize, if Student $t \geq 2$ in (1), we may conclude that R is faster than S with high confidence, on a space of benchmarks for which the actual benchmarks used are representative. If Student $t = 1$ we may have "medium" confidence, because a value this large or larger would occur with probability about 0.16 if the solvers were really equally fast on average.

5   Competition Ranking

We propose to create a k-way competition ranking of solvers S_1, ..., S_k by forming a $k \times k$ matrix M in which

$$M_{i,j} = \begin{cases} 1 & \text{if } raw(S_i, S_j) > 0 \\ 0.5 & \text{if } raw(S_i, S_j) = 0 \\ 0 & \text{if } raw(S_i, S_j) < 0 \end{cases} \tag{2}$$

This matrix can be interpreted as specifying a directed graph (also called M), where solvers are vertices and an edge from S_i to S_j exists wherever $M_{i,j} \neq 0$. If $M_{i,j} = M_{j,i} = 0.5$, there are edges in both directions. If this graph is acyclic, it defines a total order among the solvers, which we call the *dominance* order. In practice, we usually are not concerned about establishing a total order among *all* participants; it is sufficient if there is a total order among the leaders, perhaps the 5–6 top ranks.

First, let us focus on the case that the leaders do not have any tied matches, not even with a non-leader. Things are slightly more complicated otherwise. In the case of no leader ties, adding up the rows of the leaders provides a "definitive" ranking. That is, if M restricted to the leaders defines an acyclic graph, each row sum is unique, and ranking the leaders by row sums is unambiguous.

However, it is possible, even in the case of no leader ties, that the graph has a cycle [Nik10]. This possibility is present because ties are not transitive in mini-matches. In other words, on a specific benchmark B_i, it is possible that S_1 ties with S_2 and S_2 ties with S_3, but S_1 wins or loses against S_3. We can easily create a set of timings on three benchmarks so that $raw(S_1, S_2) = 1$, $raw(S_2, S_3) = 1$, and $raw(S_3, S_1) = 1$. Longer and more complex cyclic structures can be constructed, as well. If a cyclic structure is present, then at least two row sums must be equal (still in the case of no leader ties in matches).

The conclusion from the preceding discussion is that row sums can provide quick hints, are easily interpreted, but may be inconclusive. If used carelessly in the presence of ties, they can be misleading. On the positive side, we expect them to be adequate for most situations. But "most" is not good enough, so we need a procedure that always gives an unambiguous result.

Example 2. This small example shows some complications that can arise, involving pairwise ties and cycles. Let us assume that the time-out is 15 and that a difference of 3 is a winning margin in the range of times shown below, while a difference of 2 is a tie. The left side shows times for three solvers on three benchmarks. The middle shows the raw scores. The right side shows the dominance matrix.

	S_1	S_2	S_3
B_1	10	13	14
B_1	14	12	10
B_3	12	11	14

	S_1	S_2	S_3
S_1	0	1	0
S_2	-1	0	1
S_3	0	-1	0

	S_1	S_2	S_3
S_1	0	1	0.5
S_2	0	0	1
S_3	0.5	0	0

Although S_1 beats S_2 and S_2 beats S_3, still S_3 ties S_1, so all three are cyclically related. However, no row-sums are equal. □

Treating M simply as a connected, directed graph, its vertices (the solvers) can be partitioned into strongly connected components. (For small graphs, the famous linear-time procedure is unnecessary; matrix multiplications and additions suffice.) The component graph, obtained by collapsing every strongly connected component to a single node, defines a total order.

We propose that all solvers living in the same **strongly connected component (SCC)** of the graph M described in (2) shall be equally ranked; otherwise the relative ranking is determined by the component graph. This policy provides an unambiguous specification for all situations.

If a tie-break is necessary (e.g., an indivisible trophy is awarded), we recommend that all solvers in a single SCC shall be ranked among themselves by the sums of their raw scores within the SCC. That is, if $S_1, \ldots, S_k$ comprise an SCC, then

$$TieBreak(S_i) = \sum_{j=1}^{k} raw(S_i, S_j)$$

This amounts to treating each mini-match among $S_1, \ldots, S_k$ as a single-point contest between two solvers in a round-robin event similar to teams in a league playing a season, so we call this the **round-robin tie-break** method. It is also known as Copeland's method in the voting-system literature [PBR00]. The advantage of this method is that it is easily understood and familiar. Its disadvantage is that the comparative ranking of S_1 and S_2 depends on mini-matches involving other solvers in the SCC.

Table 1. The dominance matrix for 16 solvers in the final phase, based on **careful ranking**

		1	2	3	4	5	6	7	8	9	10	11	12	13	14	15	16
1	CircUs	0	0	0	0	0	0	0	1	1	1	1	1	0	0	1	0
2	LySAT_i	1	0	1	1	1	0	1	1	1	1	1	1	0	1	1	0
3	MXC	1	0	0	0	0	0	1	1	1	1	1	1	0	1	1	0
4	ManySAT_1.1	1	0	1	0	0	0	1	1	1	1	1	1	0	1	1	0
5	MiniSAT_09z	1	0	1	1	0	0	1	1	1	1	1	1	0	1	1	0
6	MiniSat_2.1	1	1	1	1	1	0	1	1	1	1	1	1	0	1	1	0
7	Rsat	1	0	0	0	0	0	0	1	1	1	1	1	0	1	1	0
8	SAT07_Rsat	0	0	0	0	0	0	0	0	1	0	1	1	0	0	0	0
9	SAT07_picosat	0	0	0	0	0	0	0	0	0	0	0	0	0	0	0	0
10	SATzilla	0	0	0	0	0	0	0	1	1	0	1	1	0	0	1	0
11	SApperloT	0	0	0	0	0	0	0	0	1	0	0	0	0	0	0	0
12	clasp	0	0	0	0	0	0	0	0	1	0	1	0	0	0	0	0
13	glucose	1	1	1	1	1	1	1	1	1	1	1	1	0	1	1	0
14	kw	1	0	0	0	0	0	0	1	1	1	1	1	0	0	0	0
15	minisat_cumr	0	0	0	0	0	0	0	1	1	0	1	1	0	1	0	0
16	precosat	1	1	1	1	1	1	1	1	1	1	1	1	1	1	1	0

In practice, we expect SCCs to be about 2–4 solvers. Outcomes perceived as being "unfair" seem unlikely, because all the solvers involved are peers. In Example 2, the round-robin tie-break makes $S_1 > S_2 > S_3$. Notice that tweaking the times by ± 0.1 does not change the result, using this method. However, with the solution-count method, S_1 and S_2 are tied with the times as shown, but tweaking can make either one the winner.

6 Results on the SAT 2009 Competition

The final round of the Application section in the SAT 2009 Competition[5] was conducted with a time limit of 10000 seconds, used 292 benchmarks, and involved 16 solvers. The organizers were Daniel Le Berre, Laurent Simon, and Olivier Roussel. The discussion uses abbreviated solver names; please see the web page for complete names. The solvers were ranked for the competition using the **solution-count ranking** method described in Section 1. We computed the rankings that would have resulted using **careful ranking**. The dominance matrix discussed in Section 5 is shown in Table 1.

Examination of this matrix shows that solvers 1, 10, 14, and 15 are in one strongly connected component, so they share ranks 9–12, according to Section 5. All other solvers are not in any cycles, so have unique ranks.

6.1 Robustness of Ranking

We analyzed the robustness of the ranking methods by counting how many times there was some change in the top three ranks as the time limit was varied

---

[5] See http://www.satcompetition.org/2009/

Table 2. Numbers of changes in top three ranks for two ranking methods and various time limits (seconds)

time range	solution-count	careful rank
1600–2000	8	4
2000–4000	10	0
4000–6000	4	0
6000–8000	0	0
8000–10000	1	0

continuously. We note that precosat stayed in first place for all time limits 4000 seconds and above, in both ranking methods. Table 2 summarizes the numbers of changes in various ranges. (Returning to an earlier permutation is considered a change, too.) It seems clear that **careful ranking** is more robust by this criterion.

We offer this intuitive explanation for why **careful ranking** gives less variations as the time limit changes. With **solution-count ranking**, a mini-match victory is only temporary, as the time limit increases: S_1 wins the mini-match against S_2 only if S_1 succeeds and S_2 times out. But for a high enough time limit S_2 also succeeds (in theory), and the victory is wiped out. However, with **careful ranking**, once the time limit is sufficiently above the solving time of S_1 and S_2 still has not succeeded, the victory is permanent for this mini-match.

6.2   Differences in Ranking

The two ranking methods, **careful ranking** and **solution-count ranking**, disagreed on the third place solver with the final time limit of 10,000 seconds. MiniSat_2.1 held third place behind glucose for all time limits above 2000 under **careful ranking**.

Under the **solution-count ranking**, MiniSat_2.1 and LySAT_i exchanged places two and three several times, with LySAT_i finally taking the lead after about 8100. By the 10,000 mark MiniSat_2.1 was in sixth place.

Under the **solution-count ranking**, precosat and glucose were apparently "neck and neck," as they each solved 204 instances. The tie-break was on cumulative CPU time, and precosat won. Other solvers were in the 190's well separated from the two leaders.

Quite a different picture emerges under **careful ranking**. We show three matches with their statistics. "Std. Devs." refers to the Student t from (1).

Winner	Loser	Raw Score	Std. Devs.	Prob. Faster
precosat	glucose	16	1.65	0.97
glucose	MiniSat_2.1	8	0.83	0.79
MiniSat_2.1	LySAT_i	8	0.86	0.80

In this ranking, precosat has a more convincing win than any of the others.

6.3 Tie-Break Illustration

Recall that Table 1 shows that solvers 1, 10, 14, and 15 are in one strongly connected component, sharing ranks 9–12, For purposes of illustration, we apply the round-robin tie-break procedure described in Section 5 to these four solvers, although in practice it is probably not important to break this tie.

The left side just below shows the raw scores for solvers involved. The right side shows the dominance subgraph.

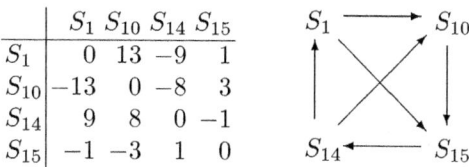

	S_1	S_{10}	S_{14}	S_{15}
S_1	0	13	−9	1
S_{10}	−13	0	−8	3
S_{14}	9	8	0	−1
S_{15}	−1	−3	1	0

The round-robin ranking, based on the row-sums, gives $S_{14} > S_{10} > S_{15} > S_1$.

7 Conclusion

This paper described a new ranking system that provides a measure of statistical significance, allows for small timing differences to be treated as ties, and ensures that a pairwise comparison between two solvers is not influenced by a third solver. The latter property also allows later researchers to replicate the competition conditions and find out where their solver would have ranked. Another application of this technique is to evaluate whether software changes from one version to another caused a performance difference that is statistically significant, or whether the difference is in a range that might well just be random.

Acknowledgment. We thank Daniel Le Berre, Laurent Simon, and Olivier Roussel for making the 2009 competition data available in text form.

References

[BLS05] Brglez, F., Li, X.Y., Stallmann, M.F.: On SAT instance classes and a method for reliable performance experiments with sat solvers. Annals of Mathematics and Artificial Intelligence 43, 1–34 (2005)

[BO07] Brglez, F., Osborne, J.A.: Performance testing of combinatorial solvers with isomorph class instances. In: Workshop on Experimental Computer Science, San Diego. ACM, New York (2007) (co-located with FCRC 2007)

[LBS04] Le Berre, D., Simon, L.: The essentials of the sat 2003 competition. In: Proc. SAT (2004)

[LN95] Levin, J., Nalebuff, B.: An introduction to vote-counting schemes. The Journal of Economic Perspectives 9, 3–26 (1995)

[Mou88] Moulin, H.: Condorcet's principle implies the no-show paradox. The Journal of Economic Theory 45, 53–64 (1988)

[Nik10] Nikolić, M.: Statistical methodology for comparison of SAT solvers. In: Strichman, O., Szeider, S. (eds.) SAT 2010. LNCS, vol. 6175, pp. 209–222. Springer, Heidelberg (2010)

[PBR00]   Pomerol, J.-C., Barba-Romero, S.: Multicriterion Decision in Management: Principles and Practice. Springer, Heidelberg (2000)

[Pul06]   Pulina, L.: Empirical evaluation of scoring methods. In: Third European Starting AI Researcher Symposium (2006)

[Sch03]   Schulze, M.: A new monotonic and clone-independent single-winner election method. In: Tideman, N. (ed.) Voting Matters, vol. (17), pp. 9–19 (October 2003), `http://www.votingmatters.org.uk`

[Tid06]   Tideman, N.: Collective Decisions and Voting: the Potential for Public Choice. Ashgate (2006)

Generalized Conflict-Clause Strengthening
for Satisfiability Solvers

Allen Van Gelder

University of California, Santa Cruz, CA 95064
http://www.cse.ucsc.edu/~avg

Abstract. The dominant propositional satisfiability solvers of the past decade use a technique often called conflict-driven clause learning (CDCL), although nomenclature varies. The first half of the decade concentrated on deriving the best clause from the conflict graph that the technique constructs, also with much emphasis on speed. In the second half of the decade efforts have emerged to exploit other information that is derived by the technique as a by-product of generating the conflict graph and learning a conflict clause. The main thrust has been to strengthen the conflict clause by eliminating some of its literals, a process often called conflict-clause minimization, but more accurately described as conflict-clause width reduction, or strengthening.

This paper first introduces implication sequences as a general framework to represent all the information derived by the CDCL technique, some of which is not represented in the conflict graph. Then the paper analyzes the structure of this information. The first main result is that any conflict clause that is a logical consequence of an implication sequence may be derived by a particularly simple form of resolution, known as linear input regular. A key observation needed for this result is that the set of clauses in any implication sequence is Horn-renamable. The second main result is that, given an implication sequence, and a clause C derived (learned) from it, it is *NP*-hard to find a minimum-cardinality subset of C that is also derivable. This is in sharp contrast to the known fact that such a minimum subset can be found quickly if the derivation is restricted to using only clauses in the conflict graph.

1 Introduction

More and more, propositional satisfiability solvers (SAT solvers, for short) are making their way into other applications as tools. The leading methodology, often called *conflict-driven clause learning* (CDCL), is well established, yet continues to evolve. The underlying idea is to derive *conflict clauses* as a by-product of failed search lines; these clauses are added to the set of clauses representing the formula to be solved. Recall that in SAT testing a *formula* is a conjunctively joined set of *clauses*, each of which is a set of disjunctively joined *literals*, abbreviated as CNF format. The number of literals in a clause is its *width*. The field has become extremely technical, but we shall try to present the main ideas of our new findings informally, to be accessible to non-specialists.

K.A. Sakallah and L. Simon (Eds.): SAT 2011, LNCS 6695, pp. 329–342, 2011.
© Springer-Verlag Berlin Heidelberg 2011

One active line of research is how to "strengthen", or reduce the width of, such conflict clauses. We put "strengthen" in quotes because one finds several different terms in the literature, including "improve," "subsume," "reduce," and "minimize." In most cases, what is meant is to find another soundly derived clause whose literals comprise a proper subset of the literals in the conflict clause (sometimes the reference clause to be reduced is an original clause). The clause is a constraint that can be satisfied by making any one of its literals true, so reducing the number of literals creates a stronger constraint. For the strengthening to be logically sound, no solutions to the overall formula may be eliminated. The stronger constraint simply replaces the original conflict clause. Han and Somenzi provide a good review of this emerging subfield [9].

In recent papers two rather surprising observations have emerged concerning solution of typical industrial instances, with thousands of variables and over a million clauses, in some cases: (1) conflict clauses can be reduced 32% in width, on average, permitting a substantial savings in memory [2]; (2) these reductions can be discovered and applied to achieve substantial net savings in time, as well [16,17]. the method sketched in a poster by Sörensson and Eén, The reduction method has come to be known as *recursive conflict-clause minimization*. It uses only the same clauses as were used to *derive* the conflict clause.

The subject of this paper, and some other recent papers, is how to bring additional clauses from the formula into the strengthening process, clauses which were *not* used in the derivation of the conflict clause. We call this *generalized* conflict-clause strengthening. Audemard *et al.* describe one method, which they call *inverse arcs*, to use other clauses beneficially. In their method, the newly derived conflict clause is not necessarily a subset of the original, but it removes certain literals from the original to enable a longer back-jump during the back-track, which ensues immediately after the conflict clause is recorded [1]. Han and Somenzi go somewhat in the other direction, by using (possibly intermediate) clauses derived during conflict analysis to strengthen clauses that existed before the analysis began [9].

One motivation for considering subsets of the originally derived conflict clause (rather than including new literals, as in [1]) is that this clause is known to have a property called *1-empowering* [15]. Derivable subsets of a 1-empowering clause are also 1-empowering.

Implication sequences are introduced in Section 2 as a new formalization of part of the operation of CDCL SAT solvers. Implication sequences are supersets of the previous formalization, which we call *antecedent sequences* in this paper. The additional clauses that are included are called *volunteers*. Section 3 shows that implication sequences are Horn Renamable, after reviewing Horn clauses and Horn renamability. The main technical results are given in Section 4. There it is shown that every clause that is a logical consequence of an implication sequence has a certain simple and short form of resolution derivation. Then it is shown that finding a minimum-cardinality conflict clause that satisfies additional natural conditions is *NP*-hard.

1.1   Terminology and Notation

Let $\mathcal{V}$ be a set of propositional variables. Propositional variables may take on the truth values *true* (or 1) and *false* (or 0). A *literal* is either v or its negation, $\overline{v}$, where v is a variable in $\mathcal{V}$ or $v = \perp$, which denotes *false*, but is treated as a positive literal to make the notation more uniform. Instead of $\overline{\perp}$ we write $\top$, for readability. We consider $\overline{\overline{v}}$ to be synonymous with v. (To distinguish propositional variables from literals, usually letters near the middle of the alphabet (p, q, r, etc.) denote literals, and letters near the end of the alphabet denote propositional variables.)

A *clause* is a disjunctively connected set of literals, and is non-tautological unless specified otherwise. The literals comprising a clause may be shown between square brackets. The *width* of a clause is the number of literals in it. A *CNF formula* (*formula* for short) is a conjunctively connected set of clauses.

An *assignment* is a partial function from variables to truth values. It is often represented by a set of literals that are assigned *true* by the assignment. A *total assignment* assigns values to all variables. An assignment $\mathcal{A}$ is said to *satisfy* a literal p if $\mathcal{A}$ assigns *true* to p, and it is said to *falsify* p if it assigns *true* to $\overline{p}$. The terminology extends to logical expressions in the natural way. A formula is said to *satisfiable* if it is satisfied by some assignment; otherwise the formula is *unsatisfiable*.

Resolution is denoted as follows. For two clauses, $C_1 = [r, p_1, \ldots, p_k]$ and $C_2 = [\overline{r}, q_1, \ldots, q_j]$, r is called the *clashing literal* and *resolution on r* yields the *resolvent*: $\mathbf{res}_r(C_1, C_2) = [p_1, \ldots, p_k, q_1, \ldots, q_j]$, which must not be tautological, unless stated otherwise. A *resolution proof* is a sequence of resolutions whose operand clauses are in the formula under consideration or derived earlier in the proof. A *resolution refutation* (*refutation* for short) is a resolution proof that derives an empty clause.

Unit-clause propagation consists of doing all possible resolutions in which at least one operand is a unit clause. The effect is to reduce the width of the second operand by one, which may result in a new unit clause, whose effects are similarly propagated. If the second operand is also a unit clause, the empty clause is derived.

In the course of unit-clause propagation, the first clause that shrinks to width one or zero is called the *antecedent* of the associated unit literal in many papers. (Some papers use the term "reason" instead of "antecedent".)

2   Implication Sequences

Although CDCL solvers have many technical details, the part we are concerned with can be described in terms of implication sequences, which are composed of propagation sequences. We define propagation sequences and implication sequences abstractly, but the action of a CDCL solver actually creates such sequences.

Definition 2.1. An *assumption clause* is a special clause that serves only a notation purpose, of the form $[q, \top]$. This records that literal q is assumed to be *true*, and is assigned *true* at this point in whatever sequence contains the clause. For uniformity, q is called the *satisfied literal* of such a clause. In addition, $[\top]$ is a placeholder assumption clause that assumes nothing.

A *unit clause* is a clause in which all literals except one have been assigned *false* (falsified). The remaining literal is called the *implied literal*, as well as the *satisfied literal* of this clause. The complements of the falsified literals in the clause are called *reason literals* for this clause.

A *falsified clause* is a clause in which all literals have been assigned *false*. However, for uniformity, we add $\bot$ as an extra literal, and call it the *implied literal* and *satisfied literal*, so that a falsified clause can be processed as though it were a unit clause. $\qquad\Box$

Definition 2.2. A *propagation sequence* is a sequence of clauses C_i, $i = 1$, ..., m, that begins with an assumption clause and continues with zero or more standard clauses that have become unit clauses or falsified clauses. The unit-clause propagation begins with the assumption, as well as variable assignments that were made prior to the propagation sequence, as its unit clauses. The clauses C_i, for $i > 1$, appear in the propagation sequence in the order they were found to be unit or falsified. To some extent, this order is solver dependent. A propagation sequence ends when no further unit clauses or falsified clauses can be derived by unit-clause propagation.

The sequence may not be unique, but once C_1 is chosen, the set of clauses in the propagation sequence is unique. If no falsified clause is derived, then the final assignment, as a set of literals, is unique. Assignments made in one propagation sequence carry over into subsequent propagation sequences that are part of the same implication sequence, which is defined next. $\qquad\Box$

Definition 2.3. An *implication sequence* is a sequence of one or more propagation sequences in which the last propagation sequence contains at least one falsified clause, and no earlier propagation sequence contains a falsified clause. Each propagation sequence is usually called a *level* (or *decision level*) in the implication sequence, with level numbers beginning at one for the first assumption (and zero before any assumption). An implication sequence may also be viewed as the *concatenation* of its propagation sequences; which view is taken should be clear from the context. Within an implication sequence, clauses (other than assumption clauses) are named as follows: (1) The clause that is earliest in the implication sequence among those that contain q as their satisfied literal is called the *antecedent* of q, and is said to *satisfy* q. If the antecedent is not an assumption clause it also is said to *imply* q (the word *force* is sometimes seen). (2) Other clauses that contain q as their (only) satisfied literal are called *volunteers*.[1] These clauses are said to *re-imply* q.

---

[1] In gardening lexicon, a *volunteer* is a plant that was not intentionally planted but is not objectionable, whereas a *weed* is objectionable.

Notice that $[q]$ is a unit clause that implies or re-implies q, while $[q, \top]$ denotes an "assumption" (decision or guess) to make q *true* in the computation, but has no logical effect on whether the sequence is satisfiable. □

Note that many solvers stop processing before an implication sequence is complete, if a falsified clause is discovered, and many do not record volunteers. However, the assignments that *were* recorded, and their order, determine, at least implicitly, which clauses are in the implication sequence, as defined.

Example 2.1. This example illustrates the definition of implication sequence, using these clauses, which are part of a formula.

$$C_1 = [\overline{v}, \overline{x}, \overline{y}, z] \quad C_2 = [y, \overline{u}, \overline{w}, \overline{x}] \quad C_3 = [z, \overline{x}, \overline{v}]$$

$$C_4 = [w, \overline{v}, \overline{y}] \quad C_5 = [u, \overline{v}, \overline{w}] \quad C_6 = [x, \overline{t}, \overline{u}] \qquad (1)$$

The following is a possible implication sequence, with one level per line. The implied or re-implied literal is shown in parentheses for each clause.

$$1 \ [v, \top]$$
$$2 \ [w, \top], \quad C_5(u)$$
$$3 \ [t, \top], \quad C_6(x) \quad C_2(y) \quad C_3(z) \quad C_4(w) \quad C_1(\bot) \qquad (2)$$

C_5 becomes a unit clause at level 2 with u as the implied literal, and v and w as reason literals. C_4 is a volunteer because it re-implies w. It appears in the sequence at a point where all of its literals are assigned. The order in which C_4 and C_3 appear depends on the solver, as they are both eligible as soon as y is assigned *true*. This example is continued in Example 2.2. □

2.1   DPLL and Implication Sequences

Before the modern era of SAT solving the predominant solver methodology was a backtracking search that came to be called DPLL, or a variant of that procedure. "DPLL"" stands for Davis, Putnam, Logemann, and Loveland, who originated the procedure in two classical papers [6,5]. We briefly review this for unsatisfiable formulas in terms of implication sequences.

DPLL builds an implication sequence as just described, and in addition keeps track of whether each assumption is a *left branch* or a *right branch* in the search tree of assignments that it is exploring. When an implication sequence is concluded with a falsified clause on a left branch with assumption p, the procedure retracts the entire propagation sequence including p, and starts a new propagation sequence with the right-branch assumption $\overline{p}$. Every left-branch assumption is followed up with the complementary right-branch assumption. In Example 2.1, the level-3 propagation sequence would be retracted and an alternative level-3 propagation sequence would be initiated with the assumption $\overline{t}$.

DPLL is naturally expressed with a recursive procedure. Early attempts to enhance DPLL used essentially the same backtracking method, and attempted to prune the search by deriving various clauses.

2.2   CDCL and Implication Sequences

CDCL began with GRASP [14], was soon improved by Chaff [13], and quickly became the dominant SAT solving methodology of the modern era. Many papers mistakenly describe this method as DPLL enhanced with clause learning. Although DPLL can be "annotated" to derive the same clauses as GRASP, it might be forced also to derive exponentially many *additional* clauses. Therefore, as claimed by the original GRASP authors, the way CDCL derives and uses conflict clauses makes it an essentially different method. Stepping through the process with an appropriate example quickly illustrates the difference.

Example 2.2. We continue with Example 2.1 at the conclusion of its implication sequence, (1). The immediate goal is to derive a *conflict clause* that has exactly one literal that was falsified during the latest propagation sequence, which is level 3 in the example (see (2)). We illustrate the *1-UIP scheme*, which is most popular. A sequence of resolutions begins with the falsified clause, C_1, and works backwards through *antecedent* clauses that are also at level 3.

$$D_1 = \mathbf{res}_y(C_2, C_1) = [\bot, \overline{v}, \overline{x}, \overline{u}, \overline{w}, \overline{z}]$$
$$D_2 = \mathbf{res}_z(C_3, D_1) = [\bot, \overline{v}, \overline{x}, \overline{u}, \overline{w}]$$

The literal x is called the *first unique implication point* (1-UIP) and D_2 is called the *1-UIP conflict clause* because D_2 has $\overline{x}$ as its only literal that was assigned on level 3. D_2 is called an *asserting clause* because, after all level-3 assignments are retracted, D_2 becomes a unit clause. The CDCL solver now "learns" D_2, that is, D_2 is now considered part of the formula.

So far, this could fit into the framework of DPLL, but now the CDCL difference emerges. All assignments made on level 3 are retracted. D_2 is now a unit clause, as one literal became unassigned. Instead of starting another propagation sequence with some assumption, the level-2 propagation sequence is continued with the new unit clause D_2 and implied literal $\overline{x}$.

$$1 \; [v, \top]$$
$$2 \; [w, \top], \quad C_5(u) \quad D_2(\overline{x}) \quad \ldots \tag{3}$$

Notice that $\overline{x}$ is not the complement of any previous assumption. If $\overline{x}$ causes further unit (or empty) clauses to be derived, they append to the level-2 propagation sequence. If unit-clause propagation dies out without falsifying a clause, then a new propagation sequence, with a new assumption literal, is initiated.

In standard CDCL, volunteers are ignored. Thus the position of C_4 on level 3 does not matter. The continuations in Example 2.3 and Example 2.4 illustrate issues that must be considered if volunteers are to be incorporated into the clause-learning process. The *inverse arcs* technique [1] was a first step in this direction.   □

Example 2.3. The formula and implication sequence are the same as in Example 2.2. This example shows that volunteers can create a cyclic structure that complicates correct reasoning.

The conflict clause derived from the above implication sequence is

$$D_2 = [\overline{v}, \overline{w}, \overline{u}, \overline{x}].$$

As things stand now, backtracking will go to level 2, where D_2 has one unassigned literal, but cannot go further due to the presence of $\overline{u}$ and $\overline{v}$. Can D_2 be strengthened to permit backtracking to level 1?

Clause C_4 meets all the criteria of Audemard *et al.* [1] for a usable inverse arc: the reason literal y appears as an implied literal at level 3, the level of the conflict, and its antecedent, C_2, participated in the derivation of the conflict clause; the reason literal v appears at level 1, which *precedes* the level in which w became satisfied; finally, w was satisfied at level 2, the current backtrack level.

The motivation is that resolving w out of the conflict clause makes progress toward permitting a longer back jump, while introducing $\overline{y}$ might not be a problem because $\overline{y}$ was able to be resolved out during the derivation of the conflict clause.

However, care must be taken to actually perform the steps, and not simply delete w, assuming the steps will succeed as hoped. (In the `minisat2` conflict-clause reduction, literals are simply deleted, and this is sound because only antecedents are used.) The derivation may continue:

$$
\begin{aligned}
D_3 &= \mathbf{res}_u(C_5, D_2) &=& \quad [\bot, \overline{v}, \overline{x}, \overline{w}] \\
D_4 &= \mathbf{res}_w(C_4, D_3) &=& \quad [\bot, \overline{v}, \overline{x}, \overline{y}] \\
D_5 &= \mathbf{res}_y(C_2, D_4) &=& \quad [\bot, \overline{v}, \overline{x}, \overline{u}, \overline{w}]
\end{aligned}
$$

D_4 re-introduced $\overline{y}$ at level 3, so it is not an asserting clause, like D_3 is. The extra level-3 literal had to be resolved out using C_2. But the resolvent D_5 is just the same clause as D_2, so the procedure is in a cycle. Indeed, $[\overline{v}, \overline{x}]$ would be an unsound derivation. A more favorable case is shown in Example 2.4.   □

Example 2.4. A slight change to the clauses in Example 2.3 illustrates how a volunteer *can* be useful. Clause C_7 replaces clause C_4.

$$
\begin{aligned}
C_1 &= [\overline{v}, \overline{x}, \overline{y}, \overline{z}] \quad & C_2 &= [y, \overline{u}, \overline{w}, \overline{x}] \quad & C_3 &= [z, \overline{x}, \overline{v}] \\
C_7 &= [w, \overline{v}, \overline{z}] \quad & C_5 &= [u, \overline{v}, \overline{w}] \quad & C_6 &= [x, \overline{t}, \overline{u}]
\end{aligned}
$$

We assume the same implication sequence as earlier examples, but with C_7 in the place of C_4. The conflict clause D_2 is the same, since its derivation ignores volunteers. C_7 also meets all the criteria of Audemard *et al.* [1] for a usable inverse arc (z plays the former role of y). The derivation may continue:

$$
\begin{aligned}
D_3 &= \mathbf{res}_u(C_5, D_2) &=& \quad [\bot, \overline{v}, \overline{x}, \overline{w}] \\
D_6 &= \mathbf{res}_w(C_7, D_3) &=& \quad [\bot, \overline{v}, \overline{x}, \overline{z}] \\
D_7 &= \mathbf{res}_z(C_3, D_6) &=& \quad [\bot, \overline{v}, \overline{x}]
\end{aligned}
$$

This time, $\overline{z}$ at level 3 has been re-introduced in D_6, making it non-asserting, so C_3 must be used to resolve out the extra level-3 literal, producing D_7. D_7 is asserting and is stronger than the previously derived asserting clauses, D_2 and D_3. The end result is that D_7 is soundly derived as the conflict clause, and a back-jump to level 1 is possible. That is, after retracting all assignments made at the current level 3, the procedure determines that none of the assignments at level 2 influence D_7, so these are all retracted, as well, and D_7 is added as an additional unit clause at level 1, with $\overline{x}$ as the satisfied literal.

$$1 \, [v, \top], \quad D_7(\overline{x}) \quad \ldots \tag{4}$$

Notice that $\overline{x}$ is not the complement of any previous assumption. If $\overline{x}$ causes further unit (or empty) clauses to be derived, they append to the level-1 prop-agation sequence. Thus the CDCL procedure has departed decisively from the DPLL framework. In fact, it would be perfectly proper for the next assumption literal to be w or t again. ☐

Examples 2.3 and 2.4 demonstrate the importance of discovering cycles, if vol-unteers are to be included in conflict-clause reduction.

2.3   Traditional Use of Implication Sequences

In the standard methodology originated in GRASP [14], and continued in Chaff [13], Minisat [7], and other solvers, a conflict graph is constructed using only antecedents, besides one chosen falsified clause. Several papers formalize this technique [18,3]. For any literal that has been assigned false, there is precisely one antecedent in which its complement is the (true) implied literal (the an-tecedent might be an assumption clause). The antecedent necessarily precedes all occurrences of this false literal.

Definition 2.4. Let $C = \{C_i\}$ be an implication sequence of clauses. Let C_A be the subsequence of decisions and antecedents, and let C_V be the subsequence of volunteer clauses. We call C_A an *antecedent sequence* to distinguish it from the implication sequence. We suppose that the final decision in C led to one or more falsified clauses, the earliest being in C_A. Any additional falsified clauses are in C_V. ☐

It is an easy matter to define an acyclic graph in which satisfied literals are vertices, with $\bot$ being the satisfied literal of the chosen falsified clause. If q is a vertex and its antecedent is $[q, \overline{p_1}, \ldots, \overline{p_k}]$, there are directed edges from q to the vertices for $p_1, \ldots, p_k$; if q is an assumption, there are no outgoing edges.[2] Vertices are included in the conflict graph only if they are reachable from the $\bot$ vertex.

We have defined an antecedent sequence to be an implication sequence in which all volunteers have been discarded. Since we have a one-to-one correspon-dence between vertices and antecedents, we might regard the antecedents as *being* the vertices, instead of the satisfied literals being the vertices. Then the vertices of the conflict graph comprise a subset of the antecedent sequence, which in turn is a subset of the full implication sequence.

---

[2] This edge orientation is opposite that seen in several *papers*, but is consistent with the *solvers'* actual data structure.

3   Implication Sequences Are Horn Renamable

The key insight for this paper is that the set of clauses in any implication sequence is Horn renamable. Thus the rich body of theory for Horn-clause reasoning can be brought to bear. Recall that a *Horn clause* has one or zero positive literals. A *Horn set* is a set of Horn clauses. A set of clauses is called *Horn renamable* if flipping the polarities of all occurrences of certain variables turns it into a Horn set. It is known from the early days of theorem proving [10,4,12] that:

Theorem 3.1. *Positive unit resolution* is complete for Horn sets; that is, the empty clause is derivable from a Horn set if and only if it is derivable by a resolution proof in which one operand is always a positive unit clause.   □

Corollary 3.2. *Unit resolution* is complete for renamable Horn sets; that is, the empty clause is derivable from a renamable Horn set if and only if it is derivable by a resolution proof in which one operand is always a unit clause.   □

The following simple lemma may be known to some researchers, at least for antecedent sequences.[3] We state it here for self-containment and because it appears not to be widely known and is so far unpublished.

Lemma 3.3. An implication sequence is Horn renamable.

Proof: Flip every negative satisfied literal and flip every negative assumption literal. Now every clause is a Horn clause whose positive literal is its satisfied literal.   ■

It is unnecessary to do this flipping in the actual computation, but for convenience of presentation, we assume without loss of generality that satisfied literals are always positive. (The attentive reader may have noticed this in the examples; Lemma 3.3 justifies the practice.)

4   Conflict-Clause Strengthening Problem

Let $C = C_1, C_2, \ldots, C_m$ be an implication sequence of clauses. As in Definition 2.4, let C_A be the antecedent sequence of C; that is, the subsequence of decisions and antecedents. Let C_V be the subsequence of volunteer clauses. We suppose that the final decision in C led to one or more falsified clauses, the earliest being in C_A. Any additional falsified clauses are in C_V.

Let γ_0 be the conflict clause derived by the CDCL solver using the 1-UIP scheme [14,18,3], or any scheme that derives asserting clauses (recall Section 2.2). That is, γ_0 is derived from the conflict graph based on clauses in C_A reachable from the falsified clause; we call these clauses C_A^*. We know that adding $\neg(\gamma_0)$ as

---

[3] Previous papers use the term "implication graph" for the graph associated with antecedents, but we avoid this term because our "implication sequence" includes volunteers.

unit clauses to C_A^* makes an inconsistent set. The *conflict-clause strengthening problem* is to find another, stronger, conflict clause, $\gamma \subset \gamma_0$, where subset is strict. There are several versions, depending on what is allowed.

Suppose the problem is cast as finding a minimum-width $\gamma \subseteq \gamma_0$ that is logically implied by C_A^*, or equivalently, such that adding $\neg(\gamma)$ as unit clauses to C_A^* causes inconsistency. Then it is a "folklore theorem" that this problem can be solved in P-time and that the minimum-width clause is unique [16,17]. The procedure implemented in MiniSat 2.0 [8] is believed to achieve this. This procedure is now called *recursive conflict-clause reduction*.

A more ambitious goal is to require that $\gamma \subseteq \gamma_0$ be the minimum-width clause that is logically implied by *all of C*. That is, by including the volunteer clauses in C_V, a smaller subset of γ_0 may be logically implied, or equivalently, a smaller subset of $\neg(\gamma_0)$ may be sufficient to produce inconsistency, as illustrated in Example 2.4. We now define this problem formally in the *NP* framework as a decision problem.

Definition 4.1. The decision form of the *general minimum conflict clause problem* is defined as follows.

Input: An implication sequence C (Definition 2.3), a conflict clause γ_0 as described above, and a positive integer K.

Question: Is there a clause $\gamma \subset \gamma_0$ with at most K literals such that $\neg(\gamma) \cup C$ is inconsistent?

Note that $\neg(\gamma)$ is treated as a set of unit clauses in this notation. ☐

Before addressing the complexity of the strengthening problem, we show in Section 4.1 that *any* clause that is a logical consequence of an implication sequence C has a simple, short derivation of a particular kind.

4.1   Implication Sequences and Linear Input Regular Derivations

The property stated in the next theorem is known for *antecedent* sequences (i.e., C_A^* in the above discussion), due to Beame *et al.* [3]. The next theorem shows that it holds for entire implication sequences. The proof idea reduces the problem to one covered by Beame *et al.*.

Definition 4.2. A *linear input regular* (LIR) resolution derivation is a sequence in which each derived clause after the first uses an "input" clause as the first operand and the previous derived clause as the second operand, and does not resolve on any literal more than once. (The terminology follows Biere [2], but such derivations were less descriptively called "trivial resolutions" by Beame *et al.* [3].) An "input" clause is one that was in the original formula or was derived before the present derivation began. ☐

Theorem 4.3. Let C be an implication sequence and let the clause γ be a logical consequence of the clauses in C. (Note that assumption clauses do not play any role in determining logical consequences.) Then γ (or a subset of γ) can be derived by a LIR resolution from C.

Proof: Assume W.L.O.G. (in view of Lemma 3.3) that C and γ are Horn. Add $\neg(\gamma)$ to the antecedents and volunteers of C, and find a refutation by *positive* unit resolution, which is known to be complete for Horn clauses. (Theorem 3.1). Derive each positive unit clause only once. The result is a conflict graph in which every implied literal has a unique antecedent. For purposes of forming the conflict graph, every positive literal of $\neg(\gamma)$ is treated as a decision, i.e., it has no antecedent. If γ has a positive literal x, then $[\overline{x}] \in \neg(\gamma)$ is treated as a unit clause in the input clause set. Following the terminology and results of Beame *et al.* [3], the conflict graph has a cut in which the literals of $\neg(\gamma)$ comprise the "reason" side of the cut and the remaining literals comprise the "conflict" side of the cut. Therefore, γ can be derived by LIR. Note that if γ has a positive literal x it becomes a *negative* unit clause $[\overline{x}]$. Although it cannot play the role of the required positive unit clause for resolution, eventually the positive unit clause $[x]$ gets implied, and then the two can resolve. This completes the proof. ∎

The implication of this theorem is that conflict clauses that are derivable from implication sequences that include volunteers have short, non-redundant, derivations. For example, the procedures described by Audemard *et al.* [1] for using "inverse arcs" apparently involve redundant derivations, as illustrated in Example 2.4. The above theorem tells us that resolving on the same literal more than once is unnecessary if a proper order is used.

Example 4.1. Again consider the clauses and the same implication sequence as in Example 2.4, where the use of the volunteer C_7 was successful, but required resolving on some literals more than once.

$$C_1 = [\overline{v}, \overline{x}, \overline{y}, \overline{z}] \quad C_2 = [y, \overline{u}, \overline{w}, \overline{x}] \quad C_3 = [z, \overline{x}, \overline{v}]$$
$$C_7 = [w, \overline{v}, \overline{z}] \quad C_5 = [u, \overline{v}, \overline{w}] \quad C_6 = [x, \overline{t}, \overline{u}]$$

Here is a linear input regular derivation from C_1, the falsified clause:

$$D_1 = \mathbf{res}_y(C_2, C_1) = [\bot, \overline{v}, \overline{x}, \overline{u}, \overline{w}, \overline{z}]$$
$$D_8 = \mathbf{res}_u(C_5, D_1) = [\bot, \overline{v}, \overline{x}, \overline{w}, \overline{z}]$$
$$D_9 = \mathbf{res}_w(C_7, D_8) = [\bot, \overline{v}, \overline{x}, \overline{z}]$$
$$D_{10} = \mathbf{res}_z(C_3, D_9) = [\bot, \overline{v}, \overline{x}]$$

The key difference from Example 2.4 is that resolution on $\overline{z}$ at level 3 was delayed, so that it did not need to be re-introduced. □

After the initial conflict clause has been derived, there are several published methods for reducing it. The method used in MiniSat 2.0 amounts to doing additional resolutions (possibly redundantly) on literals that were implied in earlier propagation sequences, yielding a subset of the original conflict clause [8,16]. It is now known that the redundancy is efficiently avoidable [17]. Volunteer clauses are not used. (Although Theorem 4.3 guarantees that a LIR proof exists, even when volunteer clauses are included, it does not tell how to find it.)

Definition 4.4. Given a set of Horn clauses H, define directed edges between variables by $v \rightarrow w$ whenever H contains some clause in which v occurs positively and w occurs negatively. If the resulting graph is acyclic, then H is said to be *acyclic*. A Horn renamable set of clauses is acyclic if it is acyclic Horn after some renaming. □

Antecedent sequences are always acyclic. Although implication sequences often are not acyclic, Theorem 4.3 guarantees that any logical consequence can be derived from a subset of clauses that *is* acyclic.

Audemard *et al.* [1] described a method for using certain volunteers to resolve away literals that were assigned in the propagation sequence at the *backtrack level*, to enable longer back jumping. Their method might resolve on literals more than once, and might produce a a conflict clause that is not a subset of the original.

The general conflict-clause strengthening problem addressed in this paper has the goal of reducing the final conflict clause to be a small subset of the original, using volunteers in some cases, to achieve greater reductions than are possible with antecedents alone.

4.2 The General Minimum Conflict Clause Problem Is *NP*-Complete

Our next result is that the decision form of the general minimum conflict clause problem, stated in Definition 4.1, is *NP*-complete. That is, finding a minimum-cardinality subset $\gamma \subset \gamma_0$, where γ_0 is a conflict clause derived from a general implication sequence, is *NP*-hard. This finding stands in sharp contrast with the fact that the problem can be solved a low-degree polynomial time for implication sequences *without* volunteers. Kleine Büning and Lettmann give a theorem with somewhat the same flavor [11, Problem MI, p. 245], but Theorem 4.6 below is not a corollary, because it requires that (A) the input clauses comprise an implication sequence C that could be generated by a CDCL solver and (B) the clause to be minimized must be a subset of a specified conflict clause γ_0, that could be derived by the same CDCL solver, rather than being any subset of variables. CDCL-derivable conflict clauses are not arbitrary; it is known that they have a property called 1-*empowering* [15]. Thus Theorem 4.6 has several additional restrictions not found in "Problem MI." The proof uses reduction from the well known Hitting Set problem, whose formal definition follows.

Definition 4.5. The decision form of the *Hitting Set problem* is:

Input: A collection of sets S_i, $i = 1, \ldots, m$ whose union is $U = \{x_j \mid j = 1, \ldots, n\}$ and an integer M such that $0 < M < n$.

Question: Is there a subset $H \subset U$ with at most M elements such that H intersects each S_i? □

Theorem 4.6. The general minimum conflict clause (GMCC) problem is *NP*-complete. The problem remains *NP*-complete if the implication sequence C is restricted to be an acyclic clause set, as defined in Definition 4.4.

Proof: The problem is in *NP* because, if a clause γ is presented as a certificate, then the set of clauses $\neg(\gamma) \cup C$ is renamable-Horn, so can be checked for inconsistency with unit-clause propagation in P-time. To show *NP*-hardness, reduce from Hitting Set (Definition 4.5).

Using the notation in the definition, the transformation arranges that each $x_j \in U$ is an assumption and γ_0 contains each $\overline{x_j}$, as well as some "control" literals. Clauses of the form $[s_i, \overline{x_j}]$ are generated to specify set membership, i.e., $x_j \in S_i$ in the Hitting Set instance. Control variables y_1, y_2, y_3, and z ensure that the desired conflict clause γ_0 is derived by a CDCL solver. Additional "control" clauses, including one volunteer clause, ensure that a sufficiently small-width γ is logically implied if and only if the x_j that occur in $\neg(\gamma)$ provide a sufficiently small H. The formal details follow.

Transform a Hitting Set instance $(\{S_i\}, M)$ into a GMCC instance (C, γ_0, K) with the following steps:

(1) Output the following propositional clause sequence over the variables x_j, $j = 1, \ldots, n$; s_i, $i = 1, \ldots, m$; y_k, $k = 1, 2, 3$; and z.

S-clauses:	For each x_j in order, $j = 1, \ldots, n$: output the decision clause $[x_j, \top]$, then, for each S_i such that $x_j \in S_i$, output $[s_i, \overline{x_j}]$.
decision y-clause:	output $[y_1, \top]$.
first z-clause:	output $[z, \overline{y_1}, \overline{x_1}, \overline{x_2}, \ldots, \overline{x_n}]$.
second y-clause:	output $[y_2, \overline{y_1}]$.
volunteer z-clause:	output $[z, \overline{y_2}, \overline{s_1}, \overline{s_2}, \ldots, \overline{s_m}]$.
third y-clause:	output $[y_3, \overline{y_2}]$.
all-negative clause:	output $[\overline{y_1}, \overline{y_3}, \overline{z}]$.

The above clauses comprise the sequence C, which is easily seen to be an implication sequence.

(2) Output the 1-UIP conflict clause $\gamma_0 = [\overline{y_1}, \overline{x_1}, \ldots, \overline{x_n}]$.

(2) Output $K = M + 1$.

The output C, is clearly an acyclic Horn clause set, and can clearly be computed in time quadratic in the length of the Hitting set instance. It is straightforward to show that (C, γ_0, K) is a yes instance of GMCC if and only if $(\{S_i\}, M)$ has a hitting set of size at most $M = K - 1$. ∎

Keep in mind that the sequence C is not the whole formula presented to the CDCL solver, just one "run" to a conflict clause. In general, given any specific deterministic solver of this class, the transformation can be tweaked and the rest of the formula can be specified to force the solver into the desired sequence of decisions, implications and re-implications.

5   Conclusion

We considered the structure of the set of *all* fully assigned clauses at the time that a conflict-driven clause-learning (CDCL) solver derives (learns) a conflict clause. These clauses can be organized into an implication sequence that faithfully represents the actions of a CDCL solver, such as GRASP, Chaff, Minisat, and others. However, these solvers ignore the information available in many of these clauses, which we name "volunteers." We showed that the set of clauses in an implication sequence is always Horn renamable. It followed from this that *any* clause that is logically implied by the clauses of the implication sequence has a linear input regular derivation (Definition 4.2). We also showed that in this environment trying to squeeze a derived clause, such as a conflict clause, down to its absolutely minimum width is *NP*-hard.

Acknowledgment. We thank the anonymous referees for helpful comments.

References

1. Audemard, G., Bordeaux, L., Hamadi, Y., Jabbour, S., Sais, L.: A generalized framework for conflict analysis. In: Kleine Büning, H., Zhao, X. (eds.) SAT 2008. LNCS, vol. 4996, pp. 21–27. Springer, Heidelberg (2008)
2. Biere, A.: Picosat essentials. J. Satisfiability, Boolean Modeling and Comp. 4, 75–97 (2008)
3. Beame, P., Kautz, H., Sabharwal, A.: Towards understanding and harnessing the potential of clause learning. J. Artificial Intelligence Research 22 (2004)
4. Chang, C.-L., Lee, R.C.-T.: Symbolic Logic and Mechanical Theorem Proving (1973)
5. Davis, M., Logemann, G., Loveland, D.: A machine program for theorem-proving. Communications of the ACM 5, 394–397 (1962)
6. Davis, M., Putnam, H.: A computing procedure for quantification theory. Journal of the Association for Computing Machinery 7, 201–215 (1960)
7. Eén, N., Sörensson, N.: An extensible SAT-solver. In: Giunchiglia, E., Tacchella, A. (eds.) SAT 2003. LNCS, vol. 2919, pp. 502–518. Springer, Heidelberg (2004)
8. Eén, N., Sörensson, N.: MiniSat v.1.13 – a SAT solver with conflict-clause minimization. Poster at SAT (2005)
9. Han, H., Somenzi, F.: On-the-fly clause improvement. In: Kullmann, O. (ed.) SAT 2009. LNCS, vol. 5584, pp. 209–222. Springer, Heidelberg (2009)
10. Henschen, L.J., Wos, L.: Unit refutations and Horn sets. JACM 21 (1974)
11. Kleine Büning, H., Lettmann, T.: Propositional Logic: Deduction and Algorithms (1999)
12. Loveland, D.W.: Automated Theorem Proving: a Logical Basis. North-Holland, Amsterdam (1978)
13. Moskewicz, M., Madigan, C., Zhao, Y., Zhang, L., Malik, S.: Chaff: Engineering an efficient SAT solver. In: 39th Design Automation Conference (June 2001)
14. Marques-Silva, J.P., Sakallah, K.A.: GRASP–a search algorithm for propositional satisfiability. IEEE Transactions on Computers 48, 506–521 (1999)
15. Pipatsrisawat, K., Darwiche, A.: A new clause learning scheme for efficient unsatisfiability proofs. In: AAAI (2008)
16. Sörensson, N., Biere, A.: Minimizing learned clauses. In: Kullmann, O. (ed.) SAT 2009. LNCS, vol. 5584, pp. 237–243. Springer, Heidelberg (2009)
17. Van Gelder, A.: Improved conflict-clause minimization leads to improved propositional proof traces. In: Kullmann, O. (ed.) SAT 2009. LNCS, vol. 5584, pp. 141–146. Springer, Heidelberg (2009)
18. Zhang, L., Madigan, C., Moskewicz, M., Malik, S.: Efficient conflict driven learning in a boolean satisfiability solver. In: ICCAD (November 2001)

Empirical Study of the Anatomy of Modern Sat Solvers

Hadi Katebi[1], Karem A. Sakallah[1], and João P. Marques-Silva[2]

[1] EECS Department, University of Michigan
{hadik,karem}@umich.edu
[2] CSI/CASL, University College Dublin
jpms@ucd.ie

Abstract. Boolean Satisfiability (SAT) solving has dramatically evolved in the past decade and a half. The outcome, today, is manifested in dozens of high performance and relatively scalable SAT solvers. The significant success of SAT solving technology, specially on *practical* problem instances, is credited to the aggregation of different SAT enhancements. In this paper, we revisit the organization of modern conflict-driven clause learning (CDCL) solvers, focusing on the principal techniques that have contributed to their impressive performance. We also examine the interaction between input instances and SAT algorithms to better understand the factors that contribute to the difficulty of SAT benchmarks. At the end, the paper empirically evaluates different SAT techniques on a comprehensive suite of benchmarks taken from a range of representative applications. The diversity of our benchmarks enables us to make fair conclusions on the relation between SAT algorithms and SAT instances.

1 Introduction

SAT solving, today, plays a significant role in modeling and solving real world applications. Although first to be proved NP-complete, SAT gained significant attention due to its practical importance, and managed to achieve major advancements in its algorithms and data structures, specially over the past 15 years. There are currently a number of highly scalable SAT solvers, all based on the classic DPLL search framework. These solvers, known as *conflict-driven clause learning (CDCL)* solvers, can generally handle problem instances with several million variables and clauses.

Modern CDCL solvers differ in many aspects, but they all share four major features. These features, proposed at different stages of SAT development, are:

- Conflict-driven clause learning [23,24]
- Random search restarts [17]
- Boolean constraint propagation using lazy data structures [27]
- Conflict-based adaptive branching [27]

Centered around the above four features, and spurred in large part by SAT competitions and races, a number of performance techniques have also been incorporated in different solvers including:

K.A. Sakallah and L. Simon (Eds.): SAT 2011, LNCS 6695, pp. 343–356, 2011.
© Springer-Verlag Berlin Heidelberg 2011

- Random branching combined with adaptive branching [14]
- Random initial scoring for conflict-based adaptive branching [14]
- Conflict clause minimization [36]
- Literal phase saving [31]
- Random restart strategies [1,6,34]

With the above enhancements, SAT solving has seen dramatic progress. However, modern solvers still fail, unpredictably, on many practical problem instances. Furthermore, even for cases where a solver manages to process an instance, it is generally not obvious what features of the solver contributed most to the instance's tractability. And while most researchers in the field would acknowledge that the above enhancements are generally helpful, there is still some debate about their relative importance. Attempts at "dissecting" modern SAT solvers to isolate the relative contribution to overall performance of the various components of their intricate algorithms have been quite rare. An early attempt is reported in [20], but to our knowledge very little has been reported in the open literature since. In this paper, we review all the aforementioned features of modern CDCL solvers, and experimentally characterize their contribution in solving a suite of 1000 benchmarks chosen from 12 diverse application areas. The diversity of our benchmarks allows us to better understand the behavior of modern solvers and their interaction with input instances. The immediate aim of this article is to experimentally verify the validity of some of the widely-accepted "facts" in the SAT community, and to report possible anomalies. As a larger goal, we hope to raise enough incentive for the theoretical computer science community to develop appropriate theoretical/analytical models that can better explain the remarkable success and the unexpected failures of modern SAT solvers.

The remainder of this paper is organized as follows. Section 2 briefly recounts the major developments in SAT technology, and discusses various performance techniques. Section 3 presents the methodology of our study. Section 4 describes our benchmark suite and articulates the rationale behind our choice. The results of the experiments, obtained using a configurable version of **MiniSAT**, are presented and analyzed in Section 5. Finally, the paper ends with conclusions in Section 6.

2   Major Features of CDCL Solvers

The pioneering techniques to solve the SAT problem, referred to as the DPLL algorithm, go back to the early 1960s [12,11]. DPLL is composed of three main features: *branching, unit propagation* (or *Boolean constraint propagation* (BCP)), and *backtracking*. Branching is essential to move forward in the search space, and backtracking is used to return from futile portions of the space. Unit propagation speeds up the search by deducing appropriate consequences, i.e. *implications*, of branching choices. This basic framework was subsequently extended with several algorithmic enhancements that greatly increased its performance and scalability. In the remainder of this section, we review four of the major enhancements, and highlight several of their extensions. The features discussed in this section have been shown, through extensive empirical evidence, to be critical for scalability and performance. These features are presented in chronological order of their appearance.

2.1 Conflict-Driven Clause Learning

The first major enhancement to DPLL came in 1996 with the debut of the **GRASP** solver [23,24]. **GRASP** introduced a new *learning* mechanism from *conflicting* assignments. The learning procedure in **GRASP** consists of the following steps:

- Analyzing the conflict and deriving an effective learned clause
- Attaching the newly derived learned clause to the original formula clauses
- Performing non-chronological backtracking

Instead of simply negating all the literals of a conflicting assignment, **GRASP** identifies a small set of assignments that are sufficient to expose the conflict by building an *implication graph*. When this so-called *effective* learning is complete, **GRASP** attaches the new learned clause to the original formula clauses, and backtracks non-chronologically to the decision level where the conflict is resolved.

Recent solvers, such as **MiniSAT** 2.2.0 [13,14], perform learning by following the exact same steps as proposed in **GRASP**, but also employ additional enhancements in conflict analysis. One such enhancement is *conflict clause minimization* [36] which aims at eliminating redundant literals from a conflict clause. There are two types of conflict minimization implemented in **MiniSAT**: *local* and *recursive*. In local, *self-subsuming resolution* is applied in reverse assignment order, using antecedents marked in the implication graph. In recursive, the conflict clause is recursively minimized by deleting the literals whose antecedents are dominated by other literals of the clause in the implication graph.

2.2 Random Restarts

In 1998, an experimental study [16], conducted by Gomes et al., revealed that the running times of complete search algorithms, such as SAT, often show a non-negligible amount of unpredictability; there always exists a probability of encountering a problem that takes exponentially more time to solve than any other problems encountered before. They explained this behavior by a phenomenon called *heavy-tailed cost distribution*. To avoid heavy tails (mitigate against exponential run times), Gomes et al. suggested the use of a controlled amount of *randomization* in search algorithms [17]. This allows search procedures to escape from regions of the space that contain no solutions. In SAT solving, randomization takes place in the form of restarts. When a SAT solver encounters a certain number of conflicts, it restarts the search by backtracking to the root level of the search tree. The limit on the number of conflicts varies in different solvers, but one common policy, also adopted in **MiniSAT**, is to use the Luby [1] sequence. Other restarting strategies, such as adaptive [6] and problem-specific [34], are also addressed in more recent publications.

2.3   Boolean Constraint Propagation Using Lazy Data Structures

Triggered by the observation that the run time of constraint solvers was mostly dominated by Boolean constraint propagation, a new efficient and highly scalable data structure and related algorithms were introduced by the **Chaff** solver [27] in 2001. The new scheme, referred to as *two-literal watching*, asserts that the status of a clause, required for the propagation process, can be maintained by watching just two of the literals of the clause that are not assigned to 0. The status is updated only when one of the watched literals is assigned to 0. Using this scheme, the clause becomes unit when no non-0-assigned literal other than the other currently watched literal is found. This scheme was in contrast to earlier mechanisms which determined the status of a clause by monitoring a counter that kept track of assignments to the clause's literals. The two-literal watching scheme enabled the status of clauses to be updated lazily and led to a significant reduction in the overhead of BCP.

2.4   Conflict-Based Adaptive Branching

Branching heuristics can have a significant effect on the performance of SAT solvers. Ranging from random decision strategies to complicated cost optimization functions, branching heuristics aim to minimize the number of decision steps, while imposing a minimal computational overhead. One effective heuristic, introduced in **GRASP**, is *dynamic largest individual sum (DLIS)* [22]. DLIS maintains counts of literals in unresolved clauses, and selects the literal with the highest count as its next branching decision. A more recent and more effective decision strategy, however, is *Variable State Independent Decaying Sum (VSIDS)*, introduced in **Chaff** [27]. Unlike previous strategies, VSIDS is highly coupled with the clause learning procedure. It attempts to satisfy conflict clauses (particularly, more recent ones) by keeping a counter for each literal, incrementing the counters at the time of a conflict for the literals that appear in the conflict, and choosing the literal with the highest counter at each round of decision. Since VSIDS updates counters only when a conflict is encountered, it has the advantage of incurring very low overhead.

The original VSIDS, as introduced in **Chaff**, kept a counter for each literal. In **MiniSAT**, counters, called *activities*, are associated with variables. Furthermore, **MiniSAT** takes advantage of *literal phase saving* [31] to avoid solving independent subproblems multiple times, when non-chronological backtracking occurs. First introduced by **RSat** [30], phase saving caches the literals that are erased from the list of assignments during backtracking, and uses them to decide on the phase of the variable that the branching heuristic suggests next. Using this strategy, SAT solvers maintain the information of the variables that are not related to the current conflict, but forced to be erased from the list of assignments by backtracking.

3   MiniSAT Configurations

For the experiments in our study, we chose **MiniSAT** 2.2.0 as the constraint solver. By default, **MiniSAT** performs conflict-driven clause learning and provides the following user-specified options:

- `rnd-freq`: This option applies a controlled amount of random decisions (0% to 100%) to VSIDS. 0 is default.
- `rnd-init`: When enabled, the activities of variables are initialized randomly. By default, all activities are initialized to 0.
- `ccmin-mode`: This is used to set the level of conflict minimization, (0) none, (1) basic (local) and (2) deep (recursive). Deep minimization is default.
- `phase-saving`: This option controls the level of phase saving, (0) none, (1) limited, and (2) full. In full, all the literals erased from the list of assignments during backtracking are cached. In limited, only the literals assigned in the latest decision level are saved. Full phase saving is default.
- `luby`: If deactivated, a power of 2 function (i.e., 2^x) with a base interval of 100 is applied as the restarting sequence. Luby is default.

We will refer to the default configuration of **MiniSAT** as CDCL. To assess the contribution of the four major enhancements to DPLL described in Section 2, we instrumented **MiniSAT** with the following additional options:

- **Disable clause learning** (`dis-learn`): When activated, **MiniSAT** reverts to DPLL-style search, i.e, it no longer performs clause learning, or non-chronological backtracking. In our implementation, we still account for conflict analysis, since VSIDS requires this procedure to correctly update variable counts. Note that, since learning is disabled, we discard the result of conflict analysis (i.e., the derived learned clause).
- **Disable restarts** (`dis-restart`): **MiniSAT** applies a Luby restart mechanism with a base interval of 100. In other words, it restarts the search whenever the number of conflicts reaches 100, 100, 200, 100, 100, 200, 400, ... By using this option, restarting is disabled during search.
- **Disable two-watched-literals** (`dis-2WL`): Enabling this option forces **MiniSAT** to perform counter-based BCP.
- **Disable VSIDS** (`DLIS`): When activated, **MiniSAT** applies the DLIS branching heuristic; otherwise it defaults to the VSIDS heuristic.

In our study, we conducted two sets of experiments. In the first set, we measured the relative contribution of each of the four major CDCL features by disabling them one at a time to determine the impact of a feature's absence on performance. These configurations of **MiniSAT** are denoted by ¬CL (no clause learning), ¬RST (no restarts), ¬2WL (counter-based BCP), and ¬VSIDS) (DLIS branching). Our reference for comparison was the default CDCL configuration which enables all of these features. In the second set of experiments, we started with CDCL under default settings for all options and explored the effect of a) adding randomness to VSIDS branching, b) adding randomness to the initial variable activities, c) adjusting the amount of conflict clause minimization, d) changing the level of phase saving, and e) modifying the restart policy.

Table 1. Benchmark families

Family	Instances	SAT	UNS	UNK	Description
atpg	100	28	72	0	Circuit testing
bioinf	30	8	12	10	Bioinformatics
config	50	15	35	0	Product configuration
crypto	30	26	3	1	Cryptanalysis
equiv	30	5	25	0	Equivalence checking
fpga	50	25	22	3	FPGA routing
hbmc	250	88	146	16	Hardware bounded model checking
hverif	200	125	75	0	Hardware verification
netcfg	10	7	2	1	Network configuration
plan	80	51	24	5	Planning
sverif	120	57	52	11	Software verification
termrw	50	26	22	2	Term rewriting
Total:	1000	461	490	49	

4   Benchmarks

We assembled a suite of 1000 CNF instances from 12 diverse application areas. The list of benchmark families, along with the total number of instances (column "Instances"), and the number of satisfiable, unsatisfiable and unknown instances (columns "SAT", "UNS" and "UNK", respectively) are shown in Table 1[1]. These benchmarks were chosen based on a number of factors including:

- Representation of real-world problem domains where SAT had been successfully applied over the last decade and a half.
- Representation of benchmark archives that are used to rank solvers in SAT Competitions (http://www.satcompetition.org/) and SAT Races (http://baldur.iti.uka.de/sat-race-2010/).
- Inclusion of a reasonable number of *easy* problem instances to enable all solver configurations to finish on at least some instances.
- Weighting the participation of each family (in terms of the number of instances representing it) by the relative success of applying SAT solving technology to that family in the recent past.

Our suite consists of benchmarks dated from the early 1990s to today. The oldest benchmarks are from the atpg, plan, equiv and fpga families [19,33,28]. Of these, atpg has seen the most progress in the processing time of its instances. Other families, such as config [35], hbmc [7], hverif [37,21] and sverif [4], represent application areas where SAT was extensively applied over the years. The remaining benchmarks, netcfg [29], termrw [15], crypto [25], and bioinf [8,10],

---

[1] The status of each instance was determined by consulting publicly-available data at various benchmark archives. We were unable to determine the status of 28 instances and tagged them with UNK even though they may be known to be SAT or UNS.

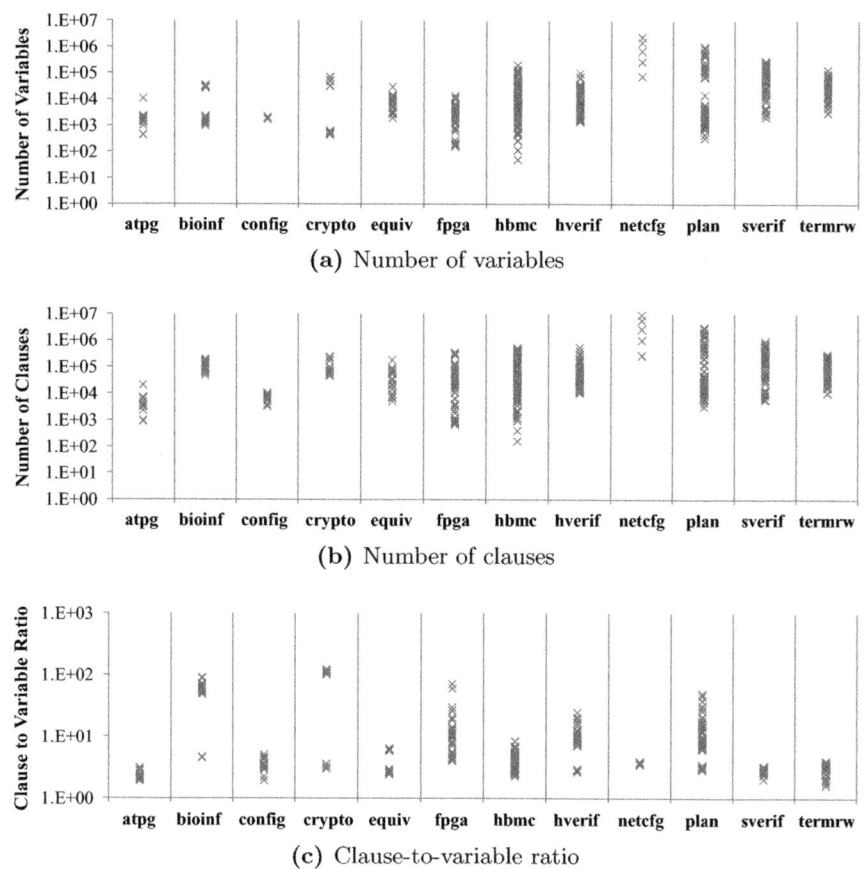

(a) Number of variables

(b) Number of clauses

(c) Clause-to-variable ratio

Fig. 1. Benchmark Statistics

correspond to more recent application domains. The majority of the instances in our suite have also appeared in SAT competitions. Note that we did not include random benchmarks since a) such benchmarks, especially random 3-SAT, have been studied extensively [26], and b) real-world applications are rarely random.

Figures 1 and 2 provide a variety of statistics for the benchmark families. The benchmarks cover a wide range with the smallest instance (50 variables and 159 clauses) coming from hbmc and the largest (2,270,930 variables and 8,901,845) from netcfg. For the clause size distributions in Figure 2, we did not include the percentage of 1-literal clauses, since they are eliminated prior to the search.

5   Experimental Evaluation

Our experiments were conducted on a cluster of servers at University College Dublin (UCD) consisting of 3GHz CPUs with 32GB memory and running the

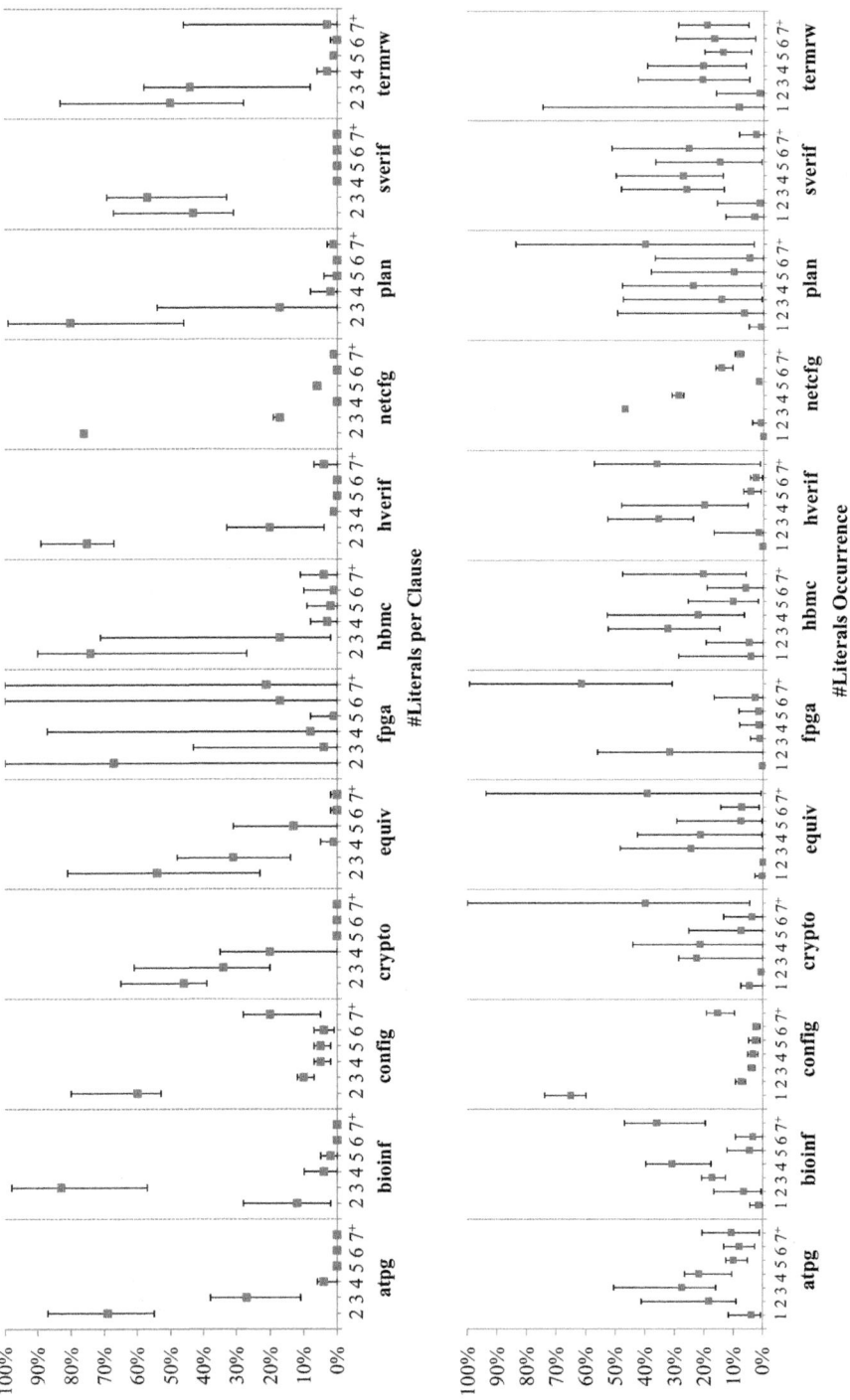

Fig. 2. The distribution of the average #clauses of a given size and #literals of a given occurrence for benchmark families

Table 2. Number of instances solved by disabling major CDCL features

Family	Runs	¬CL	¬VSIDS	¬2WL	¬RST	CDCL
atpg	1000	965	1000	1000	1000	1000
bioinf	300	19	34	88	141	150
config	500	472	500	500	500	500
crypto	300	52	22	113	235	237
equiv	300	50	92	187	224	231
fpga	500	325	403	444	441	470
hbmc	2500	762	1872	2241	2307	2333
hverif	2000	1413	1700	1934	1967	1984
netcfg	100	0	20	60	74	87
plan	800	327	449	559	564	650
sverif	1200	336	592	937	754	1006
termrw	500	116	248	346	446	420
Total:	10000	4837	6932	8409	8653	9068

64-bit Linux operating system. To obtain meaningful statistical data, we used a script that re-orders the variables and clauses in a CNF instance using a random seed[2] to create ten different versions of each benchmark. We then applied fifteen different configurations of **MiniSAT** to each benchmark version for a total of 150,000 separate runs. Each run was allowed a maximum of 1000 CPU seconds.

5.1   Relative Contribution of Major CDCL Features

Table 2 and Figure 3 summarize the results of the first set of experiments. The goal here was to determine the relative contribution to overall performance, measured by the number of solved instances within the 1000-second time-out, of each of the four CDCL features. This goal was achieved indirectly by disabling the features one at a time as described earlier. Examination of these results leads to the following conclusions:

– The number of instances solved by disabling each of the features suggests the following ordering of their relative importance to solver performance: CL > VSIDS > 2WL > RST. Specifically, disabling clause learning yields the worst performance (finishing on only 4837 instances) followed by disabling VSIDS (6932 instances solved), two-watched-literals (8409 instances solved) and restarts (8653 instances solved). Another way of stating this is to note that the solver configurations that include clause learning (namely, ¬VSIDS, ¬2WL, and ¬RST) dominate the configuration that excludes it. This is not true of the other configurations, i.e., including a feature does not always yield improved performance over excluding that feature. A more direct measure of the relative importance of these features is to compare the configurations

---

[2] We obtained the reorder.c script and a seed generator from Laurent Simon. The script was originally written by Edward Hirsh and later modified by Simon to handle large benchmarks.

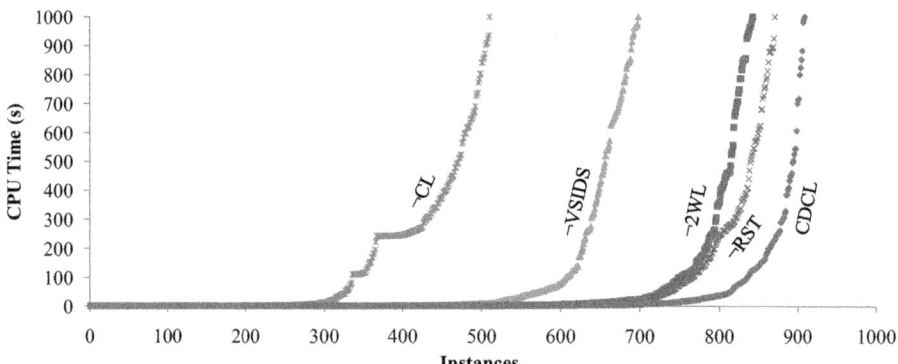

Fig. 3. The run time distribution of the four major CDCL features (data points for timed-out runs are not shown to reduce clutter). These run times are averages over 10 runs per benchmark, and account for time-outs using *maximum likelihood estimation* (MLE) [32]. With a 90% confidence level, 71% of those averages are accurate to within 25%. Higher accuracy can always be obtained by increasing the number of runs.

in which they are disabled against the CDCL configuration in which they are all enabled. Using this measure, we see that enabling CL, VSIDS, 2WL, and RST leads, respectively, to the solution of 4231, 2136, 659, and 415 additional instances.

- Configurations ¬VSIDS and CDCL differ only in the branching heuristic and allow a direct comparison between DLIS and VSIDS. The number of instances solved with VSIDS (9068 in configuration CDCL) is significantly higher than the number solved with DLIS (6932 in configuration ¬VSIDS). Two factors contribute to this performance advantage: a) the much lower overhead of VSIDS compared to DLIS since it only updates activities whenever conflicts arise whereas DLIS updates literal counters every time a literal is assigned/unassigned, b) the selection of literals occurring in the most recent conflicts as opposed to literals occurring the most in unresolved clauses.
- Configurations ¬2WL and CDCL differ only in the implementation of BCP and allow a direct comparison between counter-based and two-watched-literal unit propagation. The number of instances solved with 2WL (9068 in configuration CDCL) is higher than the number solved with the counter-based approach (8409 in configuration ¬2WL). This performance improvement is also due to two factors: a) unlike the counter-based approach which requires updating clause status during branching and backtracking, 2WL propagation needs to update clause status only during branching, and b) 2WL propagation only needs to perform status updates when watched literals are assigned to 0.
- Configurations ¬RST and CDCL differ only in whether restarts are disabled or enabled (using the Luby strategy) and show that the impact of restarts, compared with the other major features, is rather modest. Enabling Luby restarts allows 9068 instances to be solved compared to 8653 instances solved

Table 3. Number of instances solved under different **MiniSAT** options

Family	CDCL	rnd-freq				rnd-init	ccmin-mode		phase-saving		no-luby
		25	50	75	100		none	basic	none	limited	
atpg	1000	1000	1000	1000	1000	1000	1000	1000	1000	1000	1000
bioinf	150	133	107	72	46	150	139	149	150	150	148
config	500	500	500	500	50	500	500	500	500	500	500
crypto	237	67	63	49	35	228	214	223	219	234	**243**
equiv	231	221	216	181	162	231	220	222	224	**235**	224
fpga	470	456	453	444	421	470	**471**	468	454	463	462
hbmc	2333	2328	2322	2225	2057	2328	2328	2333	2318	2326	2315
hverif	1984	**1989**	**1993**	**1997**	1949	1984	**1993**	1991	1971	**1997**	1960
netcfg	87	76	75	60	72	80	76	77	74	74	67
plan	650	619	593	526	490	647	637	640	606	636	586
sverif	1006	915	858	762	302	1004	1003	996	976	967	944
termrw	420	416	407	378	291	420	416	417	**426**	**424**	**444**
Total:	9068	8720	8587	8194	7325	9042	8997	9016	8918	9006	8893

when restarts are disabled. To better understand the behavior of random restarts, we examined their effect separately on the SAT and UNS instances. Of the 10000 instances, Luby restarts (configuration CDCL) solved 4533 SAT instances and 4535 UNS instances and timed out on the remaining 932. When restarts were disabled, 4230 SAT and 4423 UNS instances were solved and 1347 instances timed out. These results suggest that, surprisingly, restarts do help for both SAT and UNS instances, but that they are more helpful for SAT instances. However, additional analysis shows that the effect of restarts is not always predictable. For instances, only 420 instances (250 SAT and 170 UNS) of the termrw family were solved with restarts whereas 446 (252 SAT and 194 UNS) were solved when restarts were disabled.

– Of the four features, CL and 2WL showed consistent improvement across all instances when they were enabled. In contrast, the performance of VSIDS and RST was more variable. On reflection, this is to be expected as VSIDS and RST are heuristics whereas CL and 2WL are algorithmic optimizations.

As expected, enabling these four features (the CDCL configuration) yields the best performance and explains why most competitive SAT solvers include them in their implementations.

5.2 The Impact of Additional Options in CDCL Solvers

Table 3 reports the number of instances solved by **MiniSAT** (configuration CDCL) when several of its options deviate from their default settings. Bolded entries in the table indicate option settings that led to better performance than the default. These results show that, overall, **MiniSAT** performs best under the default settings. In some cases, however, changing a default setting yields slightly improved performance. For example, adding some randomness to VSIDS helped solve up to 13 more instances of the hverif family. Similarly, relaxing conflict clause minimization helped solve up to 9 more instances of the same family. Relaxing phase saving was modestly helpful for the equiv, hverif and termrw families. Finally, applying a power of 2 rather than the Luby restart strategy helps solve more instances in the crypto and termrw families. Still, Luby is generally more effective, confirming the earlier results reported by Huang [18].

One surprising anomaly in these experiments is the observation that a completely random branching strategy (option rnd-freq=100) solved more instances (7325) than the DLIS heuristic (6932). However, DLIS branching solved 477 instances that random branching failed to process! Such mixed results are hard to explain without further detailed analysis of the specific instances involved and any particular attributes they may have.

Finally, unlike the first set of experiments, it is not possible to draw general conclusions from these results as it seems that the optimal values of such settings need to be determined by trial and error. The options analyzed here are best viewed as refinements added on top of the four major features of CDCL. This is partly justified by noting that, unlike CL, VSIDS, 2WL and RST, the inclusion or exclusion of these refinements has, at best, a modest impact on performance.

6   Conclusions

Much effort has been devoted over the past fifteen years to improve the capacity and performance of SAT solvers that are architected around the CDCL framework. On the other hand, few researchers have explored the interactions among the various algorithmic and heuristic components of a modern CDCL solver to determine their relative importance. And while such solvers are successful in processing many practical instances, they still fail, unpredictably, on many others. The question of why CDCL works well on certain instances and not so well on others is rarely addressed in the literature. One of the few attempts to provide a theoretical explanation for the success of clause learning is due to Beame et al. [5] who show that, as a proof system, clause learning is more powerful than regular and therefore DP resolution.

This paper should be viewed as a preliminary attempt to understand the impact on performance of the primary and secondary features of a modern CDCL solver. The ultimate goal should be the development of analytical/theoretical models that relate the performance of a CDCL solver to key attributes of its input SAT instances. Such attributes include the symmetries of CNF formulas [2], the cut width of graph representations of CNF instances [9], and the *scale-free* graph structure of industrial instances [3]. This will help spur further algorithmic improvements as well as the development of customized SAT solvers that can take advantage of such structural attributes.

Acknowledgement

This work was partially supported by SFI grant BEACON (09/PI/12618) and by the United States National Science Foundation under Grant No. 0705103.

This paper is partly based on, and further extends, the article "Anatomy and Empirical Evaluation of Modern SAT Solvers," in Bull. of Euro. Assoc. for Theor. Computer Science, vol. 103, pp. 96-121, February 2011.

References

1. Alistair, M.L., Sinclair, A., Zuckerman, D.: Optimal speedup of las vegas algorithms. Information Processing Letters 47, 173–180 (1993)
2. Aloul, F., Sakallah, K., Markov, I.: Efficient symmetry breaking for boolean satisfiability. IEEE Transactions on Computers 55(5), 549–558 (2006)
3. Ansótegui, C., Bonet, M.L., Levy, J.: On the structure of industrial SAT instances. In: Gent, I.P. (ed.) CP 2009. LNCS, vol. 5732, pp. 127–141. Springer, Heidelberg (2009)
4. Babic, D., Hu, A.J.: Calysto: scalable and precise extended static checking. In: International Conference on Software Engineering, pp. 211–220 (2008)
5. Beame, P., Kautz, H.A., Sabharwal, A.: Towards understanding and harnessing the potential of clause learning. Journal of Artificial Intelligence Research 22, 319–351 (2004)
6. Biere, A.: Adaptive restart strategies for conflict driven SAT solvers. In: Kleine Büning, H., Zhao, X. (eds.) SAT 2008. LNCS, vol. 4996, pp. 28–33. Springer, Heidelberg (2008)
7. Biere, A., Cimatti, A., Clarke, E., Strichman, O., Zhu, Y.: Bounded Model Checking. In: Advances in Computers. Academic Press, London (2003)
8. Bonet, M.L., John, K.S.: Efficiently calculating evolutionary tree measures using SAT. In: Kullmann, O. (ed.) SAT 2009. LNCS, vol. 5584, pp. 4–17. Springer, Heidelberg (2009)
9. Broering, E., Lokam, S.V.: Width-based algorithms for SAT and CIRCUIT-SAT. In: Giunchiglia, E., Tacchella, A. (eds.) SAT 2003. LNCS, vol. 2919, pp. 162–171. Springer, Heidelberg (2004)
10. Corblin, F., Bordeaux, L., Fanchon, E., Hamadi, Y., Trilling, L.: Connections and integration with SAT solvers: A survey and a case study in computational biology. In: Hybrid Optimization: the 10 years of CPAIOR. Springer, Heidelberg (2010)
11. Davis, M., Logemann, G., Loveland, D.: A machine program for theorem-proving. Communications of the ACM 5, 394–397 (1962)
12. Davis, M., Putnam, H.: A computing procedure for quantification theory. Journal of the ACM 7, 201–215 (1960)
13. Eén, N., Sörensson, N.: An extensible SAT-solver. In: Giunchiglia, E., Tacchella, A. (eds.) SAT 2003. LNCS, vol. 2919, pp. 502–518. Springer, Heidelberg (2004)
14. Eén, N., Sörensson, N.: MiniSAT, version 2.2.0 (2010), http://minisat.se/downloads/minisat-2.2.0.tar.gz
15. Fuhs, C., Giesl, J., Middeldorp, A., Schneider-Kamp, P., Thiemann, R., Zankl, H.: SAT solving for termination analysis with polynomial interpretations. In: Marques-Silva, J., Sakallah, K.A. (eds.) SAT 2007. LNCS, vol. 4501, pp. 340–354. Springer, Heidelberg (2007)
16. Gomes, C.P., Selman, B., Crato, N.: Heavy-tailed distributions in combinatorial search. In: Smolka, G. (ed.) CP 1997. LNCS, vol. 1330, pp. 121–135. Springer, Heidelberg (1997)
17. Gomes, C.P., Selman, B., Kautz, H.: Boosting combinatorial search through randomization. In: National Conference on Artificial Intelligence, pp. 431–437 (July 1998)

18. Huang, J.: The effect of restarts on the efficiency of clause learning. In: Proceedings of the 20th International Joint Conference on Artifical Intelligence, pp. 2318–2323. Morgan Kaufmann Publishers Inc., San Francisco (2007)

19. Larrabee, T.: Test pattern generation using Boolean satisfiability. IEEE Transactions on Computer-Aided Design 11(1), 4–15 (1992)

20. Lynce, I., Marques-Silva, J.: Building state-of-the-art sat solvers. In: Proceedings of the 15th Eureopean Conference on Artificial Intelligence (ECAI 2002), pp. 166–170. IOS Press, Amsterdam (2002)

21. Manolios, P., Srinivasan, S.K.: A parameterized benchmark suite of hard pipelined-machine-verification problems. In: Borrione, D., Paul, W. (eds.) CHARME 2005. LNCS, vol. 3725, pp. 363–366. Springer, Heidelberg (2005)

22. Marques-Silva, J.: The impact of branching heuristics in propositional satisfiability algorithms. In: Barahona, P., Alferes, J.J. (eds.) EPIA 1999. LNCS (LNAI), vol. 1695, pp. 62–74. Springer, Heidelberg (1999)

23. Marques-Silva, J., Sakallah, K.A.: GRASP: A new search algorithm for satisfiability. In: Srivas, M., Camilleri, A. (eds.) FMCAD 1996. LNCS, vol. 1166, pp. 220–227. Springer, Heidelberg (1996)

24. Marques-Silva, J., Sakallah, K.A.: GRASP-A search algorithm for propositional satisfiability. IEEE Transactions on Computers 48(5), 506–521 (1999)

25. Mironov, I., Zhang, L.: Applications of SAT solvers to cryptanalysis of hash functions. In: Biere, A., Gomes, C.P. (eds.) SAT 2006. LNCS, vol. 4121, pp. 102–115. Springer, Heidelberg (2006)

26. Mitchell, D., Selman, B., Levesque, H.: Hard and easy distributions of sat problems. In: National Conference on Artificial Intelligence, pp. 459–465 (1992)

27. Moskewicz, M., Madigan, C., Zhao, Y., Zhang, L., Malik, S.: Engineering an efficient SAT solver. In: Design Automation Conference, pp. 530–535 (June 2001)

28. Nam, G.-J., Sakallah, K.A., Rutenbar, R.A.: Satisfiability-based layout revisited: Detailed routing of complex FPGA s via search-based boolean SAT. In: International Symposium on Field-Programmable Gate Arrays (February 1999)

29. Narain, S.: Network configuration management via model finding. In: Conference on Systems Administration, pp. 155–168 (2005)

30. Pipatsrisawat, K., Darwiche, A.: Rsat 1.03: Sat solver description. Technical Report D–152, Automated Reasoning Group, Computer Science Department, UCLA (2006)

31. Pipatsrisawat, K., Darwiche, A.: A lightweight component caching scheme for satisfiability solvers. In: Marques-Silva, J., Sakallah, K.A. (eds.) SAT 2007. LNCS, vol. 4501, pp. 294–299. Springer, Heidelberg (2007)

32. Rice, J.A.: Mathematical Statistics and Data Analysis. Duxbury Press, Boston (2006)

33. Selman, B., Kautz, H.: Planning as satisfiability. In: European Conference on Artificial Intelligence, pp. 359–363 (1992)

34. Sinz, C., Iser, M.: Problem-sensitive restart heuristics for the DPLL procedure. In: Kullmann, O. (ed.) SAT 2009. LNCS, vol. 5584, pp. 356–362. Springer, Heidelberg (2009)

35. Sinz, C., Kaiser, A., Küchlin, W.: Formal methods for the validation of automotive product configuration data. AI EDAM 17(1), 75–97 (2003)

36. Sörensson, N., Biere, A.: Minimizing learned clauses. In: Kullmann, O. (ed.) SAT 2009. LNCS, vol. 5584, pp. 237–243. Springer, Heidelberg (2009)

37. Velev, M.N., Bryant, R.E.: Effective use of boolean satisfiability procedures in the formal verification of superscalar and vliw microprocessors. J. Symb. Comput. 35(2), 73–106 (2003)

Translating Pseudo-Boolean Constraints into CNF

Amir Aavani

Simon Fraser University
aaa78@sfu.ca

1   Introduction

A Pseudo-Boolean constraint (PB-constraint) is a generalization of a clause. A PB-constraint is an inequality (equality) on a linear combination of Boolean literals ($\sum_{i=1}^{n} a_i l_i \; OP \; b$) where $a_1, \cdots a_n$ and b are constant integers, $l_1, \cdots, l_n$ are literals and OP is a comparison operator. The left-hand side of a PB-constraint under assignment $\mathcal{A}$ is equal to the sum of the coefficients whose corresponding literals are mapped to true by $\mathcal{A}$. This kind of constraints has been widely used in expressing NP-complete problems. Several approaches have been proposed to translate a PB-constraint to CNF, [3], [2].

In this paper, we propose a new encoding for translating PB-constraints whose comparison operator is "=" to CNF. The CNF produced by the proposed encoding has small size, and also the constraints for which one can expect the SAT solvers to perform well on the produced CNF can be characterized. We show that there are many constraints for which the proposed encoding has a good performance. It worths mentioning that an arbitrary PB-constraint can be rewritten as a single equivalent PB-constraint whose comparison operator is "=" and all its constant integers are positive.

Definition 1. Given constraint Q on set of variables X, we call the pair $\langle v, C \rangle$, where v is a Boolean variable, C is a set of clauses on $X \cup Y \cup \{v\}$ and Y is a set of propositional variables, a *valid translation* if for every satisfying total assignment $\mathcal{A}$ to $X \cup Y \cup \{v\}$ for C, $\mathcal{A}$ satisfies Q iff it maps v to *true*, i.e., $C \models v \Leftrightarrow Q$.

2   Proposed Method

Let a PBMod-constraint be an equation in the following form:

$$\sum_{i=1}^{n} a_i' l_i = b' \; (\text{mod M}). \tag{1}$$

where $0 \le a_i' < M$ for all $1 \le i \le n$ and $0 \le b' < M$. Total Assignment $\mathcal{A}$ is a solution to (1) iff the value of left-hand side summation under $\mathcal{A}$ minus the value of right-hand side of the equation, b', is a multiple of M.

K.A. Sakallah and L. Simon (Eds.): SAT 2011, LNCS 6695, pp. 357–359, 2011.
© Springer-Verlag Berlin Heidelberg 2011

Definition 2. The PBMod-constraint $Q(M) : \sum a_i' l_i = b' (\text{mod } M)$ is called to be the *conversion* of the PB-constraint $Q : \sum a_i l_i = b$, modulo M iff

1. $a_i' = a_i \bmod M$
2. $b' = b \bmod M$

Proposition 1. Let $M = \{M_1, \cdots, M_m\}$ be a set of m relatively prime integers. The set of assignments satisfying $Q : \sum a_i l_i = b$ is exactly the same as the set of assignments satisfying all the m PBMod-constraints $Q(M_k)$ if $\prod_{k=1}^m M_k > S = \sum a_i$.

One candidate for the set M is a subset of prime numbers. One can enumerate the prime numbers and add them to the set of modulos, $M^P = \{2, 3, ..., P_m\}$, until their multiplication exceeds S. The next proposition gives us an estimation for the size of set M^P as well as the maximum value in M^P.

Proposition 2. Let $M^P = \{2, \cdots, P_m\}$ be the set of primes s.t. $\prod_{p \in M^P} p \geq S$. Then:

1. $m = |M^P| \leq \log S$.
2. $P_m < (\log S)^2$.

Theorem 1. Let $Q : \sum a_i l_i = b$ be a PB-constraint. Also let $M^P = \{P_1, \cdots, P_m\}$ be as above, and the pair $\langle v_k, C_k \rangle$ be a *valid translation* for PBMod-constraint $Q(P_k)$. Then, the pair $\langle v, C \rangle$ is a valid translation for PB-constraint Q where $C = \cup_k C_k \cup C'$ and C' is the set of clauses describing $v \Leftrightarrow (v_1 \wedge v_2 \cdots \wedge v_m)$.

Translation of PBMod-constraint Through DP The translation presented here is similar to translation through BDD, described in [3]. Tseitin variable, D_m^l, is defined inductively as follows: +

$$
D_m^l = \begin{cases} \top & \text{if } l \text{ and } m \text{ are both zero;} \\ \bot & l = 0 \text{ and } m > 0; \\ (D_{(m-a_l) \bmod M}^{l-1} \wedge x_l) \vee (D_m^{l-1} \wedge \neg x_l) & Otherwise \end{cases}
$$

3 Performance of Unit Propagation

There are three situations in which UP is able to infer the input variables values of a PB-constraint Q:

1. Unit Propagation Detects Inconsistency: If Q is unsatisfiable, UP may be able to infer that there is no assignment satisfying Q.
2. Unit Propagation Solves Constraint: UP may be able to infer the whole solution for Q if there is just a single satisfying solution to Q.
3. Unit Propagation Infers the Value for an Input Variable: UP may be able to infer that the value of input variable x_k is *true/false* if x_k takes the same value in all the solutions to Q. This is a generalization of previous case.

It can be shown that for each of the above cases, there are at least $(\frac{\sum a_i}{\log \sum a_i})^{n+1} = \frac{2^n Poly(n)}{Poly(n)^{n+1}}$ different PB-constraints in the form $\sum a_1 l_i = b$ such that CNFs, produced using the proposed approach, allow UP to infer input variables.

4   Conclusion

Our translation produces a polynomial size CNF w.r.t. the input size. We also argued that for exponentially many instances, produced CNFs are arc-consistent. This number is much bigger for our encoding comparing to the existing encodings. Interested readers are invited to read the complete version of this paper [1].

References

1. Aavani, A.: Translating Pseudo-Boolean Constraints into CNF,
   http://arxiv.org/abs/1104.1479
2. Bailleux, O., Boufkhad, Y., Roussel, O.: New encodings of pseudo-boolean constraints into CNF. In: Kullmann, O. (ed.) SAT 2009. LNCS, vol. 5584, pp. 181–194. Springer, Heidelberg (2009)
3. Eén, N., Sörensson, N.: Translating pseudo-boolean constraints into SAT. Journal on Satisfiability, Boolean Modeling and Computation 2(3-4), 1–25 (2006)

Analyzing the Instances of the MaxSAT Evaluation[*]

Josep Argelich[1], Chu Min Li[2], Felip Manyà[3], and Jordi Planes[1]

[1] Universitat de Lleida, Lleida, Spain
[2] MIS, Université de Picardie Jules Verne, Amiens, France
[3] Artificial Intelligence Research Institute (IIIA-CSIC), Bellaterra, Spain

The MaxSAT Evaluation [1] is an affiliated event of the SAT Conference that is held every year since 2006, and is devoted to empirically evaluate exact MaxSAT algorithms solving any of the following problems: MaxSAT, Weighted MaxSAT (WMaxSAT), Partial MaxSAT (PMaxSAT), and Weighted Partial MaxSAT (WPMaxSAT).

The objective of this paper is to analyze the instances of the 2010 MaxSAT Evaluation in order to gain new insights into their computational hardness, answer some questions that have been asked to us as organizers, and evaluate how appropriate are the current settings of parameters such as timeout and available RAM memory. To this end, we conducted a number of experiments, which were performed on a cluster with 160 2 GHz AMD Opteron 248 Processors with 1 GB of RAM memory.

In the experiments, we considered the 2,675 instances of the 2010 MaxSAT Evaluation: 544 MaxSAT instances, 349 WMaxSAT instances, 1,122 PMaxSAT instances, and 660 WPMaxSAT instances. Instances were assigned to one of the following three categories: random, crafted and industrial. We used the 17 solvers that participated in MaxSAT-2010. They can be classified into three main types: branch and bound (B&B) solvers, satisfiability-based (sat-based) and unsatisfiability-based (unsat-based) solvers. In the first type, we find 10 solvers: akmaxsat, akmaxsat_ls, IncMaxSatz, IncWMaxSatz, Maxsat_Power, LS_Power, WMaxsat_Power, LSW_Power, WMaxSatz-2009, and WMaxSatz+. In the second type, we find 2 solvers: SAT4J-Maxsat, and QMaxSAT. In the third type, we find 5 solvers: WPM1, PM2, WPM2, wbo 1.4a, and wbo 1.4b.

In what follows, we summarize the experiments, point out the lessons we have learned, and suggest to introduce some modifications in forthcoming evaluations:

Experiment 1: Historical evolution. We compared how fast are the best solvers of the last evaluation compared with the best solvers that participated in previous evaluations but have not been submitted to MaxSAT-2010 on the PMaxSAT instances of the random, crafted and industrial categories. The results provide evidence that some older solvers are yet highly competitive in some categories, and suggest that we should consider the best previous solver for each problem and category until it is beaten by new solvers. On the other hand, taking into account the number of unsolved instances not yet solved by any participating solver, we should report the number of instances that have been solved for the first time in the results of the evaluation.

[*] Research supported by Generalitat de Catalunya (2009-SGR-1434), *Ministerio de Ciencia e Innovación* (CONSOLIDER CSD2007-0022, INGENIO 2010, Acción Integrada HA2008-0017, TIN2009-14704-C03-01, and TIN2010-20967-C04-01/03), and the *Secretaría General de Universidades del Ministerio de Educación: Programa Nacional de Movilidad de Recursos Humanos.*

K.A. Sakallah and L. Simon (Eds.): SAT 2011, LNCS 6695, pp. 360–361, 2011.

Experiment 2: Analysis of the timeout. We evaluated the impact of setting a timeout of 7,200 seconds instead of the timeout of previous evaluations (1,800 seconds). The idea is to find out if it is necessary to change the current timeout because it introduces a bias in favor of some solvers. The results indicate that the current timeout is adequate for the evaluation. The introduction of a higher timeout could complicate the development of the evaluation without introducing significant differences in the results.

Experiment 3: Analysis of RAM memory. The amount of available RAM memory may produce quite different performance profiles. The cluster used in the evaluation has 2 processors per node, and they share 1 GB of RAM memory. So, we decided to evaluate the impact of setting 1GB of RAM memory instead of 512MB. The results indicate that the fact of doubling the available RAM memory does not lead to remarkable differences in performance. However, it is interesting to double the memory from time to time in order to detect anomalous situations: Maxsat_Power and LSW_Power showed a much better performance profile with 1GB, due to the way these solvers manage dynamic memory but not to the solving techniques they implement. On the other hand, it would be interesting to perform the evaluation with a cluster allowing 4GB or more of RAM memory to every solver, but this is beyond the reach of the organizers for the time being.

Experiment 4: Size of instance sets. The submitted instance sets have different size, and we rank solvers by the total number of solved instances. This may bias the results in that there may be sets of instances with a large number of instances and sets with just a few instances. Therefore, for ranking solvers by their ability to solve instances from different sets, we normalized the results taking into account the number of instances in each set. We observed that in some cases the resulting rankings are different, and propose, for future evaluations, to set a maximum of 100 instances per set, and present the results using both the ranking based on total number of solved instances and the ranking based on percentage of solved instances.

Experiment 5: Parameters of instances. We have analyzed several parameters of the instance sets: the median number of variables and clauses, the mean size of the first core found, the mean value of the solutions, and the core size multiplied by the solution. The results indicate that the problem size, the first size of the unsatisfiable core, and the number of unsatisfiability cores can give very useful indications when selecting a MaxSAT solver: when the instance has fewer than, e.g. 5,000 clauses, use a B&B solver; otherwise, search for a unsatisfiable core of the instance, if the core contains more than, e.g. 10 clauses, again use a B&B solver, if the hard clauses of the instances are of very simple form (e.g. binary clauses with negative literals), always use a B&B solver. In all other cases, use a sat-based or unsat-based solver. Regarding sat-based solvers, which are good on large size instances, it seems to be decisive the quality of the first upper bound.

Reference

1. Argelich, J., Li, C.M., Manyà, F., Planes, J.: The first and second Max-SAT evaluations. Journal on Satisfiability, Boolean Modeling and Computation 4, 251–278 (2008)

Model Counting Using the Inclusion-Exclusion Principle

Huxley Bennett and Sriram Sankaranarayanan[*]

University of Colorado, Boulder, CO.
`first.lastname@colorado.edu`

The inclusion-exclusion principle is a well-known mathematical principle used to count the number of elements in the union of a collection of sets in terms of intersections of sub-collections. We present an algorithm for counting the number of solutions of a given k-SAT formula using the inclusion-exclusion principle. The key contribution of our work consists of a novel subsumption pruning technique. Subsumption pruning exploits the alternating structure of the terms involved in the inclusion-exclusion principle to discover term cancellations that can account for the individual contributions of a large number of terms in a single step.

The Inclusion-Exclusion Principle and #SAT

Given sets $A_1, \ldots, A_m, m > 0$, the inclusion-exclusion principle states that $|\bigcup_{i=1}^{m} A_i| = \sum_{i=1}^{m} |A_i| - \sum_{1 \leq i < j \leq m} |A_i \cap A_j| + \cdots (-1)^{m+1} |A_1 \cap A_2 \cap \cdots \cap A_m|$. It is well-known that this principle can be applied to count the number of solutions of a given k-CNF formula [4,3].

Let φ be a k-CNF formula consisting of variables $x_1, \ldots, x_n$ and clauses $C_1, \ldots, C_m$. Each clause $\mathrm{lits}(C_i) : \{\ell_{(1,k)}, \ldots, \ell_{(i,k)}\}$ is a set of k literals, each literal of the form $\ell_i : x_j$ or $\ell_i : \neg x_j$. We will count the number $N_U : \#\mathrm{UNSAT}(\varphi)$ of solutions that *do not satisfy* φ using the inclusion-exclusion principle. Let $A_1, \ldots, A_m$ denote the sets of variable assignments which *dis-satisfy* the clauses $C_1, \ldots, C_m$, respectively, in φ. Therefore, $N_U = |\bigcup_{i=1}^{m} A_i|$ can be calculated using the inclusion-exclusion principle, as a summation ranging over all subsets of clauses $S \subseteq \{C_1, \ldots, C_m\}$:

$$N_U = \sum_{S \subseteq \{C_1, \ldots, C_m\}} t(S) \quad \text{where} \quad t(S) = \begin{cases} 0 & \text{if } \exists\, j, \{x_j, \neg x_j\} \subseteq \mathrm{lits}(S) \\ ((-1)^{|S|+1} \cdot 2^{n-|\mathrm{lits}(S)|}) & \text{otherwise} \end{cases},$$

where $\mathrm{lits}(S)$ represents all the literals appearing in the clauses of S. Given N_U, we may obtain the number of satisfying solutions as $2^n - N_U$. Note that the number of terms involved in the summation is exponential in the formula size.

One solution to improving the complexity of this procedure is to prune away terms involving subsets S where $N(S) = 0$ in the summation above. This is achieved by avoiding subsets S which include *interfering clauses* C_j, C_k that contain a variable x_i and its negation $\neg x_i$. Such an optimization has been proposed elsewhere [3,4]. In this work, we present yet another optimization through subsumption pruning.

[*] This work was partially supported by the National Science Foundation (NSF) award CNS-1016994.

K.A. Sakallah and L. Simon (Eds.): SAT 2011, LNCS 6695, pp. 362–363, 2011.

Tree Exploration and Subsumption Pruning

We now present a brief sketch of our subsumption pruning technique. More details are available from an extended version of this paper [1]. Our technique arranges the terms in the inclusion-exclusion formula as a tree and performs a recursive depth-first tree exploration to consider *non-interfering* clause sequences of the form $[C_{i1}; \ldots; C_{id}]$. Each node v in the search tree is defined by its current clause sequence $S : [C_{i1}; C_{i2}; \ldots; C_{id}]$, where $d \geq 1$ is the depth of the node. The node is associated with the term $t(S) = (-1)^{d+1} 2^{n - |\text{lits}(S)|}$. Through the search, we maintain the invariant that S is interference free and that $1 \leq i_1 < \cdots < i_d \leq m$.

Consider a node $S : [C_{i1}; \ldots; C_{ij}]$. Let $T : [S; C_l]$ be a child of S extended by adding the clause C_l. We say that S *subsumes* T iff $\text{lits}(S) = \text{lits}(T)$. In other words, every literal in the clause C_l is already contained in some clause in S. The main theorem in this paper takes advantage of subsumptions to make a drastic improvement on the basic scheme given previously:

Theorem 1. *Let T_j be a subsumed child of S in the search tree. Considering any child $T_l : [S; C_l]$ of S, where $l > j$ and the corresponding child $T_l' : [T_j : C_l]$, then $t(T_l') = -t(T_l)$ and $t(\text{subtree}(T_l')) = -t(\text{subtree}(T_l))$. We conclude that $t(\text{subtree}(S)) = \sum_{i=1}^{j-1} t(\text{subtree}(T_i))$.*

This theorem, whose proof is in the extended version of the paper, concludes that if S subsumes one of its children T_j, then due to the alternating sum involved in the inclusion-exclusion principle that the children of S and T_j cancel each other out. In practice this means that we need only explore the children $T_1, \ldots, T_{j-1}$ of S, a significant improvement over evaluating all of the children of S, especially when j is small.

Preliminary experimental evaluations of our technique is reported in our extended report [1]. We present a summary of these results obtained over randomly generated k-SAT instances. (A) The application of our subsumption pruning technique provides a significant speedup (2-3x) on most of the larger instances. Nevertheless, the technique itself is limited in the size of formulae that can be handled, especially when compared to other approaches to counting using DPLL [2]. (B) An integration of our technique inside DPLL-based model counters compares favorably to existing DPLL-based model counters CDP and Relsat.

Currently, we are in the process of evaluating our approach over structured benchmarks and analyzing the expected running times for our technique over randomly generated formulae.

References

1. Bennett, H., Sankaranarayanan, S.: Model counting using the inclusion-exclusion principle, Draft (available upon request) (2011)
2. Gomes, C.P., Sabharwal, A., Selman, B.: Model counting. In: Handbook of Satisfiability, ch. 20. IOS Press, Amsterdam (2008)
3. Iwama, K.: CNF-satisfiability test by counting and polynomial average time. SIAM Journal on Computing 18(2), 385–391 (1989)
4. Lozinskii, E.L.: Counting propositional models. Information Processing Letters 41, 327–332 (1992)

Phase Transitions in Knowledge Compilation: An Experimental Study[*]

Jian Gao[1], Minghao Yin[1], and Ke Xu[2]

[1] College of Computer Science
Northeast Normal University, Changchun, 130024, China
jiangao.cn@hotmail.com, ymh@nenu.edu.cn
[2] State Key Lab. of Software Development Environment
Beihang University, Beijing, 100191, China
kexu@nlsde.buaa.edu.cn

Introduction and Background

Phase transitions, as a kind of well-known phenomena in artificial intelligence, have attracted a great amount of attention in recent years [1,2]. Many NP-complete problems, such as random SAT and random Constraint Satisfaction Problems (CSPs), have a critical point that separates overconstrained and underconstrained regions, and soluble-to-insoluble phase transition occurs at this critical point, which is always accompanied with the transitions of CPU runtimes. Both systematic search algorithms and local search algorithms suffer an easy-hard-easy pattern when solving those problems. In fact, the easy-hard-easy patterns are not only expressed in terms of the time, but also in terms of the space. That is phase transition in knowledge compilation.

Knowledge compilation [3] is used to compile solutions of a problem into a tractable language. Many target languages of knowledge compilation have been proposed for compiling SAT instances and CSPs. Easy-hard-easy patterns in those languages have been shown in the early studies [4,5]. Schrag and Crawford [4] studied phase transitions in compiling 3-SAT instances to prime implicates (PIs) and showed the critical point occurs when the ratio (r) of #clauses (m) to #variables (n) is around 2.0. While recent studies have proposed many more succinct languages [3], such as Ordered Binary Decision Diagram (OBDD) [6], deterministic, Decomposable Negation Normal Form (d-DNNF) [7] and Deterministic Finite-state Automaton (DFA). Differ from PIs, these languages covert solutions into more compact forms using the property of solution symmetry. In this paper, we investigate easy-hard-easy patterns in empirical results of compiling random SAT and CSP into OBDD, d-DNNF and DFA.

Main Results

First, we show experimental results concerning random 3-SAT instances. Fig. 1 depicts the easy-hard-easy pattern when compiling instances with n=30 into the three

---

[*] A full version of this paper is available at http://arxiv.org/abs/1104.0843. This work is supported by the National Science Foudation of China under grant No. 60803102 and 60973033. Correspondence to: Minghao Yin (Northeast Normal University) or Ke Xu (Beihang University).

K.A. Sakallah and L. Simon (Eds.): SAT 2011, LNCS 6695, pp. 364–366, 2011.
© Springer-Verlag Berlin Heidelberg 2011

target languages, where the number of nodes in the compilation results are used to measure the sizes. The peak points of those curves are with same value of the ratio r, which is 1.8. Additionally, random k-SAT instances are also considered. We can observe that those target languages share the same critical point of the easy-hard-easy pattern, where $r \approx 2.6$ for $k=4$ and $r \approx 4.5$ for $k=5$. Based on those results, we conjecture that all target languages belonging to subsets of Decomposable Negation Normal Form (DNNF) suffer the compilation phase transition with the same critical point. An explanation of this phenomenon is the inherent changes on interchangeable structure of solutions. We observe that the number of interchangeable solutions with respect to 2 variables has a great impact on the sizes of compilation results.

Next, we show the sizes of compilation results increase exponentially as n grows linearly. We convert 3-SAT instances into d-DNNFs, and take 6 values of r uniformly. For each r, we vary n from 10 to 60 at increments of 5. Fig. 2 shows the results with the logarithmic vertical axis. Curves are all nearly linear, so the size grows exponentially in the general cases. As r is close to 1.8, slopes of lines grow larger, and the sizes around phase transition regions grow fastest. Besides, we surmise there also exists a phase transition separates polynomial and exponential sizes. For 3-SAT, the critical point of the polynomial-to-exponential phase transition is around $r=0.3$.

Furthermore, we show that the easy-hard-easy pattern also exists in compiling random CSPs. We employ RB model [7] to generate random CSP instances. The RB model is described by constraint arity k, the number of variables n, domain size $d=n^{\alpha}$, constraint number $m=rn\ln n$, and the constraint tightness p. We fix k, α, p, and compile CSPs into DFAs. The peak point of DFA sizes is fixed as the number of variables increases. For instance, when $k=2, \alpha=1.2, p=0.5$, the peak point occurs at $r=0.52$.

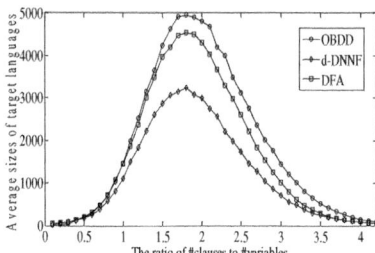

Fig. 1. The easy hard easy pattern   **Fig. 2.** Exponential increments of sizes

References

1. Cheeseman, P., Kanefsky, B., Taylor, W.M.: Where the Really Hard Problems Are. In: Proc. IJCAI 1991, pp. 331–340. Morgan Kaufmann Publishers, Inc., San Francisco (1991)
2. Achlioptas, D., Naor, A., Peres, Y.: Rigorous location of phase transitions in hard optimization problems. Nature 435, 759–764 (2005)
3. Darwiche, A., Marquis, P.: A Knowledge Compilation Map. J. Artif. Intell. Res. 17, 229–264 (2002)
4. Schrag, R., Crawford, J.M.: Implicates and prime implicates in Random 3-SAT. Artificial Intelligence 81, 199–222 (1996)

5. Darwiche, A.: A Compiler for Deterministic, Decomposable Negation Normal Form. In: Proc. AAAI/IAAI 2002, pp. 627–634. AAAI Press, Menlo Park (2002)
6. Narodytska, N., Walsh, T.: Constraint and Variable Ordering Heuristics for Compiling Configuration Problems. In: Proc. IJCAI 2007, pp. 149–154. Morgan Kaufmann Publishers, Inc., San Francisco (2007)
7. Xu, K., Li, W.: Exact Phase Transitions in Random Constraint Satisfaction Problems. J. Artif. Intell. Res. 12, 93–103 (2000)

EagleUP: Solving Random 3-SAT Using SLS with Unit Propagation

Oliver Gableske[1] and Marijn J.H. Heule[2]

[1] Institute of Theoretical Computer Science, Ulm University, Germany
[2] Algorithmics Group, Delft University of Technology, The Netherlands

While the application of unit propagation (UP) is of vital importance in systematic search solvers to solve structured problems of the SAT competitions [8], its application in stochastic local search (SLS) solvers is rare. Examples for combining UP with SLS solvers are UnitWalk [4] and QingTing [6] and both solvers show strong performance on structured instances. Despite the success on structured formulas, the application of UP in SLS only seemed to weaken the performance on random k-CNF formulas. The approach described in this abstract briefly presents how UP can be embedded in SLS solvers to boost their performance on random 3-CNF formulas as well.

In summary, several questions must be answered in order to embed UP into a given SLS solver: When to call for UP in the ongoing SLS search? What variable assignments is UP supposed to propagate in what order? And finally, how to use the result of a UP call? For the following explanations, we assume the ongoing search on a CNF formula F of a G2WSAT solver [5] with current assignment α.

When to call for UP? It is reasonable to interrupt the SLS search as soon as it cannot improve its current assignment α anymore. This is usually the case as soon as it has no further promising variables to flip (we call this a *dead end*).

What variable assignments is UP supposed to propagate in what order? The idea is to use the dead end assignment α and propagate all the assignments made herein, because such an assignment usually satisfies a large portion of the clauses of F. Propagation stops as soon as UP runs into a conflict. The result is a (partial) assignment β, that hosts all variable assignments that could be propagated by UP without running into a conflict.

The variable ordering we use is computed according to a recursive weight heuristic. The general idea of recursive weight heuristics is to help systematic search SAT solvers identify variables with strong impact on the formula. Such solvers usually pick a single variable in every node of their search tree and assign it to a not yet explored value. Picking variables with strong impact will then give a large reduction of the remaining formula, and therefore, a strong reduction in the size of the remaining search space.

To create a variable ordering, we use the recursive weight heuristic RW [3,1]. We create a variable ordering θ_{RW_5} such that with $x_i, x_j \in \mathrm{VAR}(F)$: $\theta_{\mathrm{RW}_5}(x_i) < \theta_{\mathrm{RW}_5}(x_j) \Leftrightarrow \mathrm{RW}_5(x_i) > \mathrm{RW}_5(x_j)$. The ordering θ_{RW_5} has to be computed exactly once (i.e., it is static) and prioritizes variables with high impact on F.

The reason why we prefer variables with high impact is that assigning these variables first creates new unit clauses sooner. Creating new unit clauses sooner then means that UP relies less often on the dead end assignment α in future

K.A. Sakallah and L. Simon (Eds.): SAT 2011, LNCS 6695, pp. 367–368, 2011.

iterations. Given a satisfiable formula, this is helpful since this dead end, currently not satisfying all clauses of the formula, must contain erroneous variable assignments. The less often UP relies on it, the smaller the chance of propagating one of the contained errors.

How to use the result of a UP call? Comparing α and β on all variables that are assigned in β can provide a set of variables that has changed its assignment during the unit propagation. Let us call this set $\mathcal{M}$. In short $\mathcal{M} = \{x \in \mathcal{V} \mid \beta(x) \text{ is defined and } \alpha(x) \neq \beta(x)\}$. The idea is to multi-flip all variables in $\mathcal{M}$ and continue the SLS search from this new position in the search space. In general, applying UP helps the SLS solver to escape from the current dead end to an assignment that has increased consistency regarding the formula.

There is, however, an additional problem that arises from the application of the static variable ordering θ_{RW_5}. Empirical tests show that a G2WSAT solver will encounter a dead end in about every third flip on a 3-CNF formula. Calling for UP in every third flip will most likely give similar results, and is therefore considered to be a waste of computational time. In order to overcome that problem, it is sufficient to increase the number of flips between two calls to UP (we call this a cool-down period and denote it by $\mathfrak{c}$).

Our approach to compute these cool-down periods uses the Cauchy probability distribution. Let $\gamma \in \mathbb{R}, \gamma > 0$ and $\omega \in \mathbb{R}$. The *cumulative distribution function* of the Cauchy distribution is $C : \mathbb{R} \mapsto \mathbb{R}$, $C(z) = P(Z < z) = 0.5 + 1/\pi \cdot \arctan((z - \omega)/\gamma)$. In summary, whenever UP was performed, a random number $a \in [0, 1)$ is picked and the next cool-down period is then computed as $\mathfrak{c} = \min\{z \mid C(z) \geq a\}$.

We have performed an empirical study to test the feasibility of our approach. The results show an average speedup of 18% when using a UP enhanced G2WSAT solver in comparison to its pure SLS variant for large size random 3-CNF formulas. Preliminary tests on crafted and application formulas suggest that our approach to combine UP with SLS can be of advantage here as well. The results are available at [7]. This abstract is available as a full paper at [2].

References

1. Athanasiou, D., Fernandez, M.A.: Recursive Weight Heuristic for Random k-SAT. Technical report from Delft University (2010),
   http://www.st.ewi.tudelft.nl/sat/reports/RecursiveWeightHeurKSAT.pdf
2. Full paper, http://www.uni-ulm.de/in/theo/mitarbeiter/olivergableske.html
3. Mijnders, S., De Wilde, B., Heule, M.J.H.: Symbiosis of search and heuristics for random 3-SAT. In: LaSh 2010 (2010)
4. Hirsch, E.A., Kojevnikov, A.: UnitWalk: A New SAT Solver that Uses Local Search Guided by Unit Clause Elimination. AMAI 43(1-4), 91–111 (2005)
5. Li, C.-M., Huang, W.Q.: Diversification and determinism in local search for satisfiability. In: Bacchus, F., Walsh, T. (eds.) SAT 2005. LNCS, vol. 3569, pp. 158–172. Springer, Heidelberg (2005)
6. Li, X.Y., Stallmann, M.F., Brglez, F.: A local search SAT solver using an effective switching strategy and an efficient unit propagation. In: Giunchiglia, E., Tacchella, A. (eds.) SAT 2003. LNCS, vol. 2919, pp. 53–68. Springer, Heidelberg (2004)
7. The results of our study, http://edacc.informatik.uni-ulm.de/EDACC3/index
8. The SAT competition homepage, http://www.satcompetition.org

Non-Model-Based Algorithm Portfolios for SAT

Yuri Malitsky[1], Ashish Sabharwal[2], Horst Samulowitz[2], and Meinolf Sellmann[2]

[1] Brown University, Dept. of Computer Science, Providence, RI 02912, USA
`ynm@cs.brown.edu`
[2] IBM Watson Research Center, Yorktown Heights, NY 10598, USA
{`ashish.sabharwal,samulowitz,meinolf`}`@us.ibm.com`

When tackling a computationally challenging combinatorial problem, one often observes that some solution approaches work well on some instances, while other approaches work better on other instances. This observation has given rise to the idea of building algorithm portfolios [5]. Leyton-Brown et al. [1], for instance, proposed to select one of the algorithms in the portfolio based on some features of the instance to be solved. This approach has been blessed with tremendous success in the past. Especially in SAT, the SATzilla portfolios [7] have performed extremely well in past SAT Competitions [6].

We investigate alternate ways of building algorithm portfolios that differ substantially from the way SATzilla assembles a portfolio. The key idea behind SATzilla is to train a runtime prediction model for each constituent solver, based on a number of well-engineered features of SAT instances. Given a new instance, SATzilla predicts the runtime of each candidate solver based on instance features and the trained models, and chooses the solver that is predicted to perform the best. In contrast, we consider *non-model-based machine learning techniques* such as simple k-nearest-neighbor (k-NN) classification to determine which solver to use to tackle a given instance.

Our motivation stems from two observations: (a) accurately predicting the runtime of sophisticated SAT solvers is a very challenging task; indeed, the runtime predictor underlying SATzilla can be even orders of magnitude off from the true runtime; and (b) while fast and accurate runtime prediction is certainly sufficient for building a solver portfolio, it is by no means necessary. In fact, it would suffice entirely if we could predict the fastest solver without having any knowledge of how long it will actually take to solve the given instance. This idea has found success in fields adjoining SAT, for example in portfolios for the quantified Boolean formula (QBF) problem [4], for general constraint satisfaction problems (CSPs) [3], and to some extent even for SAT itself [2].

Our portfolio works as follows. In the learning phase, we are given a pool $\mathcal{T}$ of training instances, a function that provides features for any given problem instance (we use the 48 core SATzilla features here), a set $\mathcal{S}$ of constituent solvers forming the portfolio, and a timeout t. We compute the runtime (with cutoff t) for all solvers on all instances as well as normalization parameters so that all features for all instances in the training set populate the interval $[0, 1]$.

At runtime, given a new instance I, we compute its features, normalize them, and compute the set $T_I \subset \mathcal{T}$ consisting of k training instances closest to I in terms of Euclidean distance. Then, for each solver $S \in \mathcal{S}$, we compute the

K.A. Sakallah and L. Simon (Eds.): SAT 2011, LNCS 6695, pp. 369–370, 2011.
© Springer-Verlag Berlin Heidelberg 2011

Table 1. Performance comparison of pure solvers, portfolios, and virtual best solver

	Pure Solvers							Portfolios		VBS
	agw-sat0	agw-sat+	gnov-elty+	**kcnfs**	march	pico-sat	SAT-enstein	SAT-zilla	**12-NN**	
PAR10	6400	6667	6362	**5813**	6524	7384	7089	4399	**3940**	3454
Avg Time	678	698	677	**659**	688	752	722	534	**529**	480
# Solved	268	255	270	**298**	262	220	234	366	**390**	413
% Solved	47.0	44.7	47.4	**52.3**	46.0	38.6	41.1	64.2	**68.4**	72.5

penalized runtime (PAR10 score) of S on T_I, and select the solver that has the lowest PAR10 score as our recommended solver to use on I. The choice of k can have an impact on the performance of the portfolio. We therefore learn a "good" value of k for the training set T by performing cross-validation with 100 random sub-samples of base-validation splits in a 70-30 ratio.

Extensive empirical results are omitted due to lack of space. Table 1 shows one representative sample of our results, comparing against SATzilla2009_R, the Gold Medal winning solver in the random category of SAT Competition 2009 [6]. We base our portfolio on the same set of solvers as SATzilla2009_R, use the 2,247 random category instances from SAT Competitions 2002-2007 as our training set and the 570 random category instances from SAT Competition 2009 as the test set, and a 1,200 second timeout. Experiments were run on Intel dual-core, dual processor, Dell Poweredge 1855 blade servers with 8GB of memory each.

As Table 1 shows, SATzilla outperforms individual solvers dramatically, solving 68 more instances (366) than the best performing individual solver, kcnfs. Our k-NN approach pushes the performance level substantially further, solving 390 instances within 1200 seconds whereby the VBS can solve only 23 more. In other words, SATzilla closes 55% of the gap between the best individual solver and the best possible portfolio. Simple k-NN closes 80% of this gap. We conclude that this easy non-model-based approach marks a significant improvement over a portfolio approach that has dominated SAT Competitions for half a decade.

References

1. Leyton-Brown, K., Nudelman, E., Andrew, G., McFadden, J., Shoham, Y.: A Portfolio Approach to Algorithm Selection. In: IJCAI, pp. 1542–1543 (2003)
2. Nikolić, M., Marić, F., Janičić, P.: Instance-Based Selection of Policies for SAT Solvers. In: Kullmann, O. (ed.) SAT 2009. LNCS, vol. 5584, pp. 326–340. Springer, Heidelberg (2009)
3. O'Mahony, E., Hebrard, E., Holland, A., Nugent, C., O'Sullivan, B.: Using Case-based Reasoning in an Algorithm Portfolio for Constraint Solving. In: Irish Conference on AI and Cognitive Science (2008)
4. Pulina, L., Tacchella, A.: A Multi-Engine Solver for Quantified Boolean Formulas. In: Bessière, C. (ed.) CP 2007. LNCS, vol. 4741, pp. 574–589. Springer, Heidelberg (2007)
5. Rice The, J.R.: algorithm selection problem. Advances in Computers, 65–118 (1976)
6. SAT Competition, http://www.satcomptition.org
7. Xu, L., Hutter, F., Hoos, H.H., Leyton-Brown, K.: SATzilla: Portfolio-based Algorithm Selection for SAT. JAIR 32(1), 565–606 (2008)

The Order Encoding:
From Tractable CSP to Tractable SAT

Justyna Petke* and Peter Jeavons

Oxford University Computing Laboratory, Parks Road, Oxford, UK
justyna.petke@comlab.ox.ac.uk

Many mathematical and practical problems can be expressed as *constraint satisfaction problems* (CSPs). The general CSP is known to be NP-complete, but many different conditions have been identified which are sufficient to ensure that classes of instances satisfying those conditions are tractable, that is, solvable in polynomial time [1,2,3,4,7]. The increasing efficiency of SAT-solvers has led to the development of SAT-based constraint solvers and various SAT encodings for CSPs [6]. However, most previous comparison between such encodings has been purely empirical. In a recent paper we showed that current SAT-solvers will decide the satisfiability of the *direct encoding* of any CSP instance with bounded width in expected polynomial time [5]. In this paper we give a theory-based argument to prefer the *order encoding* instead for certain other families of tractable constraint satisfaction problems. We consider problems of the form $\mathrm{CSP}(C)$, consisting of all CSP instances whose constraint relations belong to some fixed set of relations C, known as a *constraint language*. Schaefer's well-known dichotomy theorem [7] identifies all the tractable constraint languages over a Boolean domain, that is, all the tractable language classes for SAT.

A *sparse encoding* of a CSP instance introduces a new Boolean variable, $x^=_{v,a}$, for each possible variable assignment, $v = a$. The *log encoding* introduces a Boolean variable for each *bit* in the value of a CSP variable. It turns out that under such encodings tractable CSPs cannot be translated into tractable language classes of SAT. In particular, we have shown that:

Proposition 1. *No sparse encoding of a CSP instance with domain size > 2 belongs to a tractable language class of SAT. Moreover, the log encoding of any CSP instance with domain size > 7 containing certain unary constraints does not belong to any tractable language class of SAT.*

In the order encoding [8] each Boolean variable, $x^\leq_{v,c}$, represents a comparison, $v \leq c$. Under that encoding we have shown that certain tractable CSP classes are translated to tractable language classes of SAT, and hence efficiently solvable.

For example, a CSP instance is called *constant-closed* if every constraint in it allows some fixed constant value d to be assigned to all variables in its scope.

Theorem 1. *If all the constraints in a CSP instance are constant-closed for the lowest domain value, then its order encoding will be constant-closed for the value* True.

Hence, we have shown that using the order encoding to translate a CSP instance that is constant-closed for the lowest domain value gives a set of clauses satisfying

---

* The provision of an EPSRC Doctoral Training Award to Justyna Petke is gratefully acknowledged.

K.A. Sakallah and L. Simon (Eds.): SAT 2011, LNCS 6695, pp. 371–372, 2011.
© Springer-Verlag Berlin Heidelberg 2011

the first condition of Schaefer's Dichotomy Theorem. Similarly, constraints that are constant-closed under the highest domain value translate under the order encoding to clauses that satisfy the second condition of that theorem.

A rather more interesting family of tractable constraint satisfaction problems is the class of CSPs whose constraints are all *max-closed*.

Lemma 1 ([4]). *If the domain of the variables is $\{True, False\}$, with $False <$ $True$, then a constraint is min-closed if and only if it is logically equivalent to a conjunction of Horn clauses over literals representing comparisons.*

Theorem 2. *If a CSP instance P contains max-closed constraints only, then its order encoding will be min-closed.*

Hence, max-closed constraints translate using the order encoding to clauses satisfying the third condition of Schaefer's Dichotomy Theorem. By symmetry between min-closed and max-closed constraints, min-closed constraints translate to clauses satisfying the fourth condition of Schaefer's Dichotomy Theorem.

Connected-row-convex constraints were first defined in [3] using a standard matrix representation of binary relations. Here is an alternative characterisation:

Lemma 2 ([2]). *A constraint is connected-row-convex if and only if it is logically equivalent to a conjunction of 2-CNF clauses over literals representing comparisons.*

Connected-row-convex constraints translate to clauses satisfying the fifth condition of Schaefer's Dichotomy Theorem due to the following result:

Theorem 3. *If a CSP instance P contains only connected-row-convex constraints, then its order encoding will be connected-row-convex.*

The final, sixth, condition in Schaefer's Dichotomy Theorem can never be satisfied using the order encoding, since (for all domains with 3 or more elements) it is already broken by the consistency clauses, $\neg(x^{\leq}_{v,c-1}) \vee (x^{\leq}_{v,c})$. Hence we have given a complete list of all constraint languages which are encoded to tractable language classes for SAT using the order encoding.

References

1. Cohen, D., Jeavons, P.: The complexity of constraint languages. In: Handbook of Constraint Programming, ch. 8, pp. 245–280. Elsevier, Amsterdam (2006)
2. Cohen, D., et al.: Building tractable disjunctive constraints. Journal of the ACM 47, 826–853 (2000)
3. Deville, Y., et al.: Constraint satisfaction over connected row convex constraints. In: Proceedings of IJCAI 1997, pp. 405–411 (1997)
4. Jeavons, P., Cooper, M.C.: Tractable constraints on ordered domains. Artificial Intelligence Journal, 327–339 (1995)
5. Petke, J., Jeavons, P.: Local consistency and SAT-solvers. In: Cohen, D. (ed.) CP 2010. LNCS, vol. 6308, pp. 398–413. Springer, Heidelberg (2010)
6. Prestwich, S.D.: CNF encodings. In: Handbook of Satisfiability, ch. 2, pp. 75–97. IOS Press, Amsterdam (2009)
7. Schaefer, T.J.: The Complexity of Satisfiability Problems. In: Proceedings of the 10th ACM Symposium on Theory of Computing - STOC 1978, pp. 216–226. ACM, New York (1978)
8. Tamura, N., et al.: Compiling finite linear CSP into SAT. Constraints Journal 14, 254–272 (2009)

Applying UCT to Boolean Satisfiability

Alessandro Previti[1], Raghuram Ramanujan[2,*], Marco Schaerf[1],
and Bart Selman[2,*]

[1] Dipartimento di Informatica e Sistemistica Antonio Ruberti,
Sapienza, Università di Roma,
Roma, Italy
elsandro84@gmail.com, marco.schaerf@uniroma1.it
[2] Department of Computer Science
Cornell University
Ithaca, New York
{raghu,selman}@cs.cornell.edu

In this paper we perform a preliminary investigation into the application of
sampling-based search algorithms to satisfiability testing of propositional formulas in Conjunctive Normal Form (CNF). In particular, we adapt the Upper
Confidence bounds applied to Trees (UCT) algorithm [5] which has been successfully used in many game playing programs including MoGo, one of the strongest
computer Go players [3].

Rather than explore the search space in a depth-first fashion, in the style of
DPLL [2], UCT repeatedly starts from the root node and incrementally builds
a tree based on estimates of node utilities and node visit frequencies computed
from previous iterations. In most implementations of UCT, the estimated utility of a new node is computed using Monte-Carlo methods, i.e., by generating
random completions of the search (termed "playouts") and averaging their outcomes. This utility is revised each time the search revisits the node using the
estimated values of the children. This technique is especially effective when no
adequate heuristic is available to perform this value estimation task.

In this paper, we introduce and study an algorithm called UCTSAT that em-
ploys the UCT search control mechanism but replaces the playouts with a heuristic to estimate the initial utility of a node. The heuristic we use is the fraction
of the total set of clauses that are satisfied by the partial assignment associated
with the node; this fraction is computed after the application of unit propagation. While we do not expect UCTSAT to outperform the highly-optimized,
state of the art SAT solvers (especially with respect to CPU time), we believe
that the development of an algorithm based on a radically different search technique is important for at least two reasons: (a) the hardness of SAT instances
is related to the algorithm used [1], and hence UCTSAT, which uses a different search strategy, can provide useful and new insights into the complexity of
SAT instances; and (b) because such an algorithm can be useful when included

* Supported by NSF Expeditions in Computing award for Computational Sustain-
ability, 0832782; NSF IIS grant 0514429; and IISI, Cornell Univ. (AFOSR grant
FA9550-04-1-0151).

in a portfolio of algorithms (see, for example, [6]) where very different solution techniques can help expand the range of applicability of the portfolio.

As such, we focus our efforts on understanding whether UCTSAT is capable of solving SAT instances using smaller search trees than DPLL. To simplify the comparisons, we contrast our algorithm against a no-frills implementation of DPLL. We set the exploration bias parameter in UCTSAT to 0 as this yielded the best performance on average. We also experimented with varying the number of atoms that UCTSAT assigned at a given node in the search tree and discovered that setting more than one atom at once hurt the performance of the algorithm.

We compared the performance of DPLL and UCTSAT on problem instances drawn from the SATLIB repository [4]. On uniform random 3-SAT and flat-graph coloring instances of various sizes, we found little difference in the sizes of the search trees constructed by the two algorithms. We believe that this is due to the unstructured nature of these instances — UCTSAT works well when each exploration of the tree yields information that can be successfully used in subsequent iterations. In instances drawn from real-world problems (namely, single-stuck-at-fault analysis problems) that exhibit structure, we discovered that UCTSAT constructs significantly smaller search trees than DPLL — this is illustrated in table 1.

Table 1. Average tree sizes (number of nodes) for SSA circuit fault analysis instances

Instance	DPLL	UCTSAT
ssa-7552-038	9183	173
ssa-7552-158	6564	134
ssa-7552-159	5513	147
ssa-7552-160	4095	164

References

1. Aguirre, A., Vardi, M.Y.: Random 3-SAT and bDDs: The plot thickens further. In: Walsh, T. (ed.) CP 2001. LNCS, vol. 2239, pp. 121–136. Springer, Heidelberg (2001)
2. Davis, M., Logemann, G., Loveland, D.: A machine program for theorem proving. Communications of the ACM 5(7), 394–397 (1962)
3. Gelly, S., Silver, D.: Achieving master level play in 9 x 9 computer go. In: Fox, D., Gomes, C.P. (eds.) AAAI, pp. 1537–1540. AAAI Press, Menlo Park (2008)
4. Hoos, H.H., Stützle, T.: SATLIB: An Online Resource for Research on SAT 2000: Highlights of Satisfiability Research in the year 2000. In: Frontiers in Artificial Intelligence and Applications, pp. 283–292. Kluwer Academic, Dordrecht (2000), http://www.cs.ubc.ca/~hoos/SATLIB/index-ubc.html
5. Kocsis, L., Szepesvári, C.: Bandit based monte-carlo planning. In: Fürnkranz, J., Scheffer, T., Spiliopoulou, M. (eds.) ECML 2006. LNCS (LNAI), vol. 4212, pp. 282–293. Springer, Heidelberg (2006)
6. Xu, L., Hutter, F., Hoos, H.H., Leyton-Brown, K.: SATzilla: portfolio-based algorithm selection for SAT. Journal of Artificial Intelligence Research 32(1), 565–606 (2008)

A Compact and Efficient SAT-Encoding of Finite Domain CSP

Tomoya Tanjo[1], Naoyuki Tamura[2], and Mutsunori Banbara[2]

[1] Graduate School of Engineering, Kobe University, Japan
[2] Information Science and Technology Center, Kobe University, Japan
tanjo@stu.kobe-u.ac.jp, tamura@kobe-u.ac.jp, banbara@kobe-u.ac.jp

Extended Abstract

A (finite) Constraint Satisfaction Problem (CSP) is a combinatorial problem to find an assignment which satisfies all given constraints over finite domains. A SAT-based CSP solver is a program which solves a CSP by encoding it to SAT and searching solutions by SAT solvers. Remarkable improvements in the efficiency of SAT solvers make SAT-based CSP solvers applicable for solving hard and practical problems. A number of SAT encoding methods have been therefore proposed: direct encoding, support encoding, log encoding, log-support encoding, and order encoding.

Among them, *order encoding* [4] has showed a good performance for a wide variety of problems, including Open-Shop Scheduling problems, two-dimensional strip packing problems, and test case generation. Its effectiveness has also been shown by the fact that a SAT-based CSP solver Sugar [1] became a winner in several categories of the 2008 and 2009 International CSP Solver Competitions.

However, in the order encoding, the size of SAT-encoded instances becomes huge when the domain size of the original CSP is large. On the other hand, the *log encoding* [3,1] uses a bit-wise representation for integer variables. The size of SAT-encoded instances is therefore compact (linear to $\log d$), but its performance is slow in general because it requires many inference steps to "ripple" carries.

In this paper, we propose a new encoding, named *compact order encoding*, aiming to be compact and efficient. The basic idea of the compact order encoding is the use of a numeric system of base $B \geq 2$. That is, each integer variable x is represented by a summation $\sum_{i=0}^{m-1} B^i x_i$ where $m = \lceil \log_B d \rceil$ and $0 \leq x_i < B$ for all x_i, and each x_i is encoded by the order encoding.

Each ternary constraints of addition and multiplication can be encoded into at most $O(B^2 \log_B d)$ and $O(B^3 \log_B d + B^2 \log_B^2 d)$ clauses respectively which are much less than $O(d^2)$ clauses of the order encoding. The compact order encoding can generate much efficient SAT instance than the log encoding in general because it requires fewer carry propagations. Please note that the compact order encoding with base $B = 2$ is equivalent to the log encoding, and the one with base $B \geq d$ is equivalent to the order encoding.

---

[1] http://bach.istc.kobe-u.ac.jp/sugar/

K.A. Sakallah and L. Simon (Eds.): SAT 2011, LNCS 6695, pp. 375–376, 2011.
© Springer-Verlag Berlin Heidelberg 2011

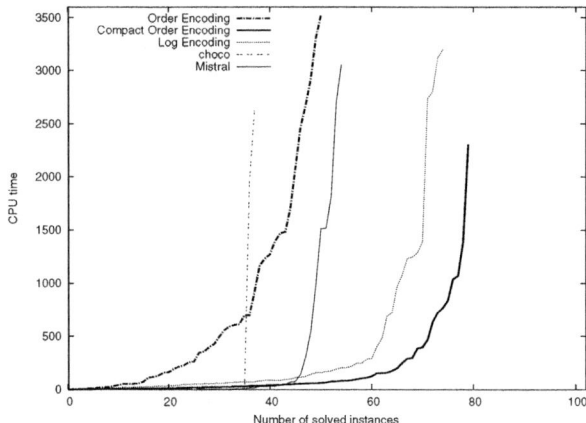

Fig. 1. Cactus plot of different encodings, choco, and Mistral for OSS instances

To evaluate the effectiveness and scalability of our encoding, we used the most difficult series of Open-Shop Scheduling (OSS) benchmark set by Brucker *et al.* We also used the instances with very large domain sizes, which are generated from OSS instances by multiplying the process times by constant factor $s \in \{1, 10, 20, 100, 200, 1000\}$. The performance of the compact order encoding with $m = 2$ (i.e. $B = \lceil d^{\frac{1}{2}} \rceil$) is compared with those of the order and log encodings in addition to the state-of-the-art CSP solvers choco 2.11 [5] and Mistral 1.550 [2].

Fig. 1 shows the cactus plot of benchmark results in which the number of solved instances is on the x-axis and the CPU time is on the y-axis. The compact order encoding solved the most instances for almost any CPU time limit and it solved large instances which could not be solved by order solvers.

As future work, we plan to investigate the choice of appropriate base B for solving a wide variety of problems.

References

1. Gelder, A.V.: Another look at graph coloring via propositional satisfiability. Discrete Applied Mathematics 156(2), 230–243 (2008)
2. Hebrard, E.: Mistral, a constraint satisfaction library. In: Proceedings of the 3rd International CSP Solver Competition. pp. 31–39 (2008)
3. Iwama, K., Miyazaki, S.: SAT-variable complexity of hard combinatorial problems. In: Proceedings of the IFIP 13th World Computer Congress, pp. 253–258 (1994)
4. Tamura, N., Taga, A., Kitagawa, S., Banbara, M.: Compiling finite linear CSP into SAT. Constraints 14(2), 254–272 (2009)
5. The choco team: choco: an open source Java constraint programming library. In: Proceedings of the 3rd International CSP Solver Competition, pp. 7–13 (2008)

Learning Polarity from Structure in SAT

Bryan Silverthorn and Risto Miikkulainen

Department of Computer Science
The University of Texas at Austin
{bsilvert,risto}@cs.utexas.edu

1   Introduction

Few instances of a computational problem are *sui generis*; most instead belong to some distribution of related instances, and information gained from solving past instances from the distribution may be leveraged to solve future instances more efficiently. Algorithm portfolio methods and algorithm synthesis systems are two examples of this idea. This paper proposes and demonstrates a third approach. It shows that, for related instances of satisfiability (SAT), variables' values in satisfying assignments can be correlated with structural features of their appearances in those instances, such as the mean polarities of their literals and the statistics of their constraint graph neighborhoods. Experiments on widely-used benchmark collections show that these features can be used by a standard classifier to generate better initial assignments and substantially improve the average performance of a modern solver.

2   Polarity Prediction as Supervised Learning

SAT variables that are semantically related may also share structurally similar positions. Consider two variables: do both appear exclusively in short clauses? Are both seen mostly in negated literals? Do both appear infrequently in their expressions? If a variable in an unsolved instance appears similar to variables in previously-solved instances, most of which take on a particular value in known satisfying assignments, then that information may inform the search process.

This paper transforms variable initialization into the supervised learning problem of predicting the satisfying value of each variable given a vector of real-valued features, including the mean and standard deviation of the lengths of the clauses in which the variable appears; the mean and standard deviation of the ratios of positive literals in such clauses; the number of such clauses; the number of times that the variable appears in a Horn clause; the number of times that the variable appears as a consequent; the ratio of positive literals involving that variable; and the degree of the corresponding node in the variable graph.

3   Classification for Assignment Initialization

To measure the impact of this learning framework, logistic regression is used to initialize the starting assignment of the TNM [4] local search solver for each

K.A. Sakallah and L. Simon (Eds.): SAT 2011, LNCS 6695, pp. 377–378, 2011.

of eight different collections of benchmarks, drawn from SATLIB [1] and from those used previously in the literature [2,3]. Each classifier is trained on assignments obtained by solving a fraction of the instances in each benchmark collection. Assignments are then generated either deterministically, by computing the maximum-likelihood assignment according to the classifier, or probabilistically, by sampling according to its class probabilities.

Table 1. The median number of search actions required to solve the instances in each benchmark collection, using deterministic or random classifier-based initialization, given as the fraction of the corresponding result obtained under the solver's default initialization scheme. Lower values are therefore better, and values less than one represent improvements in search speed. These ratios were averaged over 64 random 50% train/test splits, for which both mean (μ) and standard deviation (σ) are reported. Runs were limited to 2,000,000 steps. Both classifier-based initialization schemes reduce search cost, often substantially, for the benchmarks tested.

| | Cost Ratio | | | |
| | Deterministic | | Random | |
Collection	μ	σ	μ	σ
SAPS-newQCP	0.98	0.14	0.98	0.13
SAPS-SWGCP	0.21	0.04	0.97	0.21
satenstein-cbmc	0.01	0.00	0.23	0.11
satlib-bms	0.97	0.19	0.98	0.18
satlib-cbs-m403-b10	0.58	0.08	0.82	0.10
satlib-cbs-m449-b90	0.87	0.10	0.96	0.10
satlib-ii	0.97	0.32	0.76	0.26
satlib-rti	0.76	0.10	0.93	0.12

Table 1 reports the effects of different initialization schemes on the number of required solver steps. These empirical results strongly suggest that classifier-based initialization can reduce the net computational cost of solving large collections of related SAT instances.

References

1. Hoos, H.H., Stützle, T.: SATLIB: An online resource for research on SAT (2000), http://www.cs.ubc.ca/~hoos/SATLIB/benchm.html
2. Hutter, F., Hoos, H.H., Leyton-Brown, K., Stützle, T.: ParamILS: An automatic algorithm configuration framework. JAIR (2009)
3. KhudaBukhsh, A., Xu, L., Hoos, H., Leyton-Brown, K.: SATenstein: Automatically building local search sat solvers from components. In: IJCAI (2009)
4. Wei, W., Li, C.M.: Switching between two adaptive noise mechanisms in local search for SAT. In: SAT 2009 Competitive Events Booklet (2009)

Author Index

GPSR Compliance

The European Union's (EU) General Product Safety Regulation (GPSR) is a set of rules that requires consumer products to be safe and our obligations to ensure this.

If you have any concerns about our products, you can contact us on ProductSafety@springernature.com

In case Publisher is established outside the EU, the EU authorized representative is:

Springer Nature Customer Service Center GmbH
Europaplatz 3
69115 Heidelberg, Germany

Batch number: 09490872

Printed by Printforce, the Netherlands